Experiment and Calculation of Reinforced Concrete at Elevated Temperatures

Experiment and Calculation of Reinforced Concrete at Elevated Temperatures

Zhenhai Guo
Xudong Shi

清华大学百年校庆
TSINGHUA UNIVERSITY
CENTENARY CELEBRATION

ELSEVIER

AMSTERDAM • BOSTON • HEIDELBERG • LONDON
NEW YORK • OXFORD • PARIS • SAN DIEGO
SAN FRANCISCO • SINGAPORE • SYDNEY • TOKYO
Butterworth-Heinemann is an imprint of Elsevier

Butterworth-Heinemann is an imprint of Elsevier
225 Wyman Street, Waltham, MA 02451, USA
The Boulevard, Langford Lane, Kidlington, Oxford, OX5 1GB, UK

Library of Congress Cataloging-in-Publication Data
Application submitted

British Library Cataloguing-in-Publication Data
A catalogue record for this book is available from the British Library.

ISBN: 978-0-12-386962-3

For information on all Butterworth–Heinemann publications
visit our Web site at www.elsevierdirect.com

Printed in the United States of America

11 12 13 14 10 9 8 7 6 5 4 3 2 1

CONTENTS

PART II Temperature Field on a Cross Section of a Structural Member

PREFACE

Reinforced concrete structures are the most common component in structural engineering. Engineering experience and research achievements, improvements in manufacturing technology, and applications of new and efficient materials have led to a great deal of progress in reinforced concrete structures. The performance index is increasing continuously, structural configurations are more varied, and the scope of applications is expanding greatly. Today, reinforced concrete structures are used widely not only in various civil and public buildings, single- and multi-story industrial buildings, and high-rise and large-span buildings, but also in bridges, communication installations, and hydraulic and underground engineering. Reinforced or prestressed concrete structures are also used in special structures, e.g., TV towers, electrical transmission poles, silos, chimneys, even reactor and containment vessels in nuclear power plants, and very large hydraulic forging presses.

Generally, concrete structures work at room temperature (<60 °C) and they can be designed or their safety can be checked using the current codes.[0-1] However, some structures, e.g., metallurgical and chemical plants, chimneys, nuclear reactors and their containment structures, and hydraulic forging presses, work constantly in high temperature environments (100–500 °C). In addition, building fire accidents occasionally occur due to natural or man-made causes. These accidents cause the structure in a building bearing a high temperature attack to reach maximum temperatures of 1000 °C or higher within a short time (e.g., 1 h). When the concrete structure reaches elevated temperatures, it experiences cracking, increased deformation, and reduced strength, because of serious deterioration of material behavior and internal force redistribution of the structure. Then the structure may fail and even collapse, and this will result in significant economic losses and loss of life.

The research work related to this field is still limited in China and no corresponding design code is available for engineers. Therefore, the development requirements of construction engineering cannot be met, and research on the behavior of concrete materials and structures at elevated temperatures has become an important and urgent task.

The authors and several postgraduates have completed several research projects in this field since 1989. These projects are financially supported by the 863 High Science and Technology Plan of the National Science Committee, National Natural Science Foundation, and Doctoral Research Foundation of the Education Ministry of China. This book is a systematic collection and summary of the experimental and theoretical research results of these projects. The postgraduate students who took part in the projects are: Quiping Shi, Xudong Shi, Yütao Guo, and Jianping Yang (doctoral students) and Wei Li, Li Jiang, Huadong Li, Jianlin Nan, Tongguang Lü, Jieying Zhang, and Jinfeng Sun (masters students). In addition, many undergraduate students took part in the experimental work during their graduation projects.

The behavior of a concrete structure at high temperature is much more complicated than that at room temperature, and its theoretical analysis is quite difficult. When the environmental temperature of a structure is elevated under some conditions, a corresponding dynamic nonuniform temperature field is formed, the strength and deformation behavior of the materials (concrete and reinforcement) deteriorates significantly at high temperatures, and the internal forces undergo severe redistribution. Furthermore, temperature and load (or stress) show an obvious coupling effect, and the constitutive relationships of the materials and the mechanical behavior of the structure vary considerably under various temperature–load paths. Therefore, the mechanical behavior of concrete material, members, and structures is presented first in this book, according to the experimental results in which temperature and load act together. The working mechanisms are analyzed, experimental data are collected, general regularities are deduced, theoretical analyses are developed, and simplified calculation methods are given for engineering applications.

There are four parts in this book, which are introduced separately: the mechanical behavior of concrete and reinforcement at elevated temperatures and the coupling temperature–stress constitutive relationship of concrete; the theoretical analysis and calculation charts and tables for the temperature fields of the member section; the testing method at elevated temperature and behavior regularities of various basic structural members and statically indeterminate structures; and the theoretical analysis along the temperature–load history and practical method of calculating the ultimate strength of the members and structures at elevated temperature, and evaluation criteria and method of grading damage in structures after fire.

This book provides important concepts, experimental data, and the method and parameters of the theoretical analysis for engineers and technicians engaged in research, design, or construction, when they design and calculate the resistance of a structure at high temperature (fire) or deal with an existing structure after a fire accident. Moreover, this book may be used as a textbook for a concrete structure course by university lecturers, postgraduate, and undergraduate students of structural engineering.

The Chinese version of this book was written by two authors. Zhenhai Guo was responsible for Chapters 1, 2, 3, 4, 8, 9, 11, 13 (part), and 14, and Xudong Shi was responsible for Chapters 5, 6, 7, 10, 12, and 13 (part). Zhenhai Guo revised and completed the manuscript. This English version was translated from the Chinese version by Zhenhai Guo.

The authors sincerely thank the organizations that financially supported and assisted with the related projects and publication: 863 High Science and Technology Plan of the National Science Committee, National Natural Science Foundation, Doctoral Research Foundation of the Education Ministry of China, and Tsinghua University Press and Elsevier Inc. We also thank the postgraduates, undergraduates, technicians, and workers who worked with us and helped us with the experimental and theoretical research. In addition, I would like to give special thanks to my wife, retired professor Suying Qian, who not only patiently typed the entire manuscript, including many calculation tables and complicated equations, to form the original computer text, but also takes care thoughtfully of my life.

The experimental investigations, theoretical analyses, design, and calculation methods for concrete structures at elevated temperatures are still not completely satisfactory. Many problems need to be resolved with more extensive research. Because the authors' research results, understanding, and analysis abilities are limited, some errors may be found in this book. We are grateful to the specialists and readers for their comments and criticism.

Zhenhai Guo
Tsinghua University, Beijing

INTRODUCTION

0.1 THERMAL PROBLEMS IN STRUCTURAL ENGINEERING

Generally, a concrete structure works at room temperature within the construction period and during its long service life. The absolute value of its environmental temperature is low and does not fluctuate. A structure designed following the current codes[0-1] can satisfy the safety and service performance requirements. However, if the environmental temperature increases too much or the temperature difference varies periodically, the structure may fail as the service performance deteriorates or strength decreases. Sometimes a structure may suffer local damage or even collapse.

In structural engineering, thermal problems due to changes in temperature can be classified into three categories[0-2]:

1. The temperature changes periodically or occasionally beyond the normal value. For example, on a surface exposed to the sun in a high-rise or long-span building, the temperature increases when the sun shines and reduces when the sun sets, and the air around the building increases in the summer and decreases in the winter, which causes the interior of the structure to suffer periodic temperature differences. Because of the accumulation of hydration heat of cement in concrete during the hardening process, and the temperature variation caused by circumferential water, air, and sunshine, a nonuniform temperature field is formed in massive hydraulic structures (dams, etc.). Although the maximum temperature in the interior of these structures is not very high (generally less than 60 °C), the strain induced by the temperature change (±30 °C) is much larger than the value of the ultimate tensile strain of concrete.[0-3] This is sufficient to cause cracking of the concrete, increase deformation of the structure, induce redistribution of the internal forces, and influence the service performance of the building.
2. High temperature action is maintained for a long time within the working conditions of the building. For example, some structures in metallurgy and chemical industry workshops are subject to radiation of high temperatures throughout the year, and the temperature on the surface of the structure may reach 200 °C or even higher. When a chimney spurts smoke of high temperature, the temperature of the internal lining may reach 500–600 °C and the temperature on the external surface may reach 100–200 °C. In the reactor vessel and containment structures of a nuclear power plant, the temperature may reach 120 °C or even higher at local positions.
3. High temperature impact in a short time is caused by occasional accidents. For example, a fire in a building may last a few hours, and the maximum temperature of the fire may reach 1000 °C, even higher within only 1 h. If a chemical or nuclear explosion or an accident at a nuclear plant occurs, the temperature may reach several thousand degrees centigrade or even higher within a matter of seconds.

The structural temperature effects of the above three categories have different temperature ranges and variable rules, which cause considerable differences in the behavior of the material and structure and the level of structural damage. They can be dealt with using theoretical analyses, design methods, and structural construction, respectively. There are corresponding design codes or specifications[0-4,0-7] for practical use in many countries, but in China, there are design codes and research monographs[0-8,0-10] for the first two categories only. There is a design code[0-11] for fire prevention in a building, but it cannot deal with the analysis and design of the fire resistance of structures.

This book mainly deals with the third category of structural thermal problems. It introduces the experimental and theoretical research results on concrete and reinforcement materials and the concrete members and structures under high accidental temperatures (i.e., fire). The general regularity and mechanical mechanism of their behavior at elevated temperatures are also presented. Analysis methods and calculations are provided for the temperature fields, the resistant behavior of structural members, statically indeterminate structures at elevated temperatures, and damage evaluation after a fire. The related principles, analysis methods, and experimental data in the book can also be used as a reference for other categories of structural thermal problems.

0.2 HARMFULNESS AND RESOLUTION OF STRUCTURE AFTER FIRE

The discovery of fire accelerated the evolution of humankind, promoted civilization and social development, and led to technologic progress, economic development, and a prosperous modern society. However, loss of control of fire may result in catastrophe. In Chinese history, many cities were burnt down during wars, and cities built over several hundred years were ruined in one day. In England, the Great Fire in London (1667), caused by a fire in a bakery, brought disaster to one-third of the buildings in the city.

Different types of fire accidents cause enormous loss of human lives, natural resources, and social properties. Therefore, humankind has struggled with fire for a long time and accumulated many experiences and created effective ways to control it. In modern society, the techniques and equipment for fire prevention and extinguishment are constantly being updated and their efficiency and performance also improve constantly, as scientific techniques make progress. Nevertheless, serious fire accidents still occur for various natural, technical, and man-made reasons, and fires occur most frequently and cause the most damage to buildings in a populated city. For example, more than 10,000 building fires occur every year in China, causing the loss of more than 1000 lives and several billion Yuan in economic losses annually.

There are many causes of fire in buildings, e.g., lightning strikes, material self-combustion, dust explosion, loss of control of fire in residential and industrial settings, negligence in storage or use of fuel, mistakes in operating electrical equipment, failure of an aged insulating layer, a secondary accident after an earthquake or war, and even arson. Fire prevention should cover all aspects.

Fire is a combustion phenomenon caused after a combustible substance is inflamed and reacts intensely together with an oxidant, generally oxygen in the air. In the burning process, enormous heat is generated, the temperature of the surrounding air and various materials rises quickly, and some materials in the space may burn successively and a more serious fire may result. As the flames and smoke associated with high temperature evolve, the fire spreads to the adjacent spaces, even the whole building. When firefighters are successful, either all combustibles inside a building burn out, or oxygen is exhausted or isolated, with the effect that the fire declines gradually and is eventually extinguished. This describes one cycle of a fire accident but it is possible that burning and declining are repeated several times in one fire accident.

When a fire occurs in a building, the temperature increases, a nonuniform temperature field forms, and the structural material deteriorates in the interior of the structure. This causes damage and strength reduction at different levels in the structure. When the structure acts as the load-bearing and support system of the building, it should maintain sufficient strength for a certain time period during a fire, so that firefighters can fight the fire, rescue the injured and deceased victims safely, and save valuable property.

Therefore, it is considered that a structure fails in fire resistance, if one of three limit states[0-12,0-14] is reached:

1. The limit state of the load-bearing capacity. The load-bearing capacity of a structure is reduced at elevated temperatures, so that the service load can no longer be supported because of structural collapse, instability, or excessive deflection (e.g., 1/30th of the clear span).
2. The limit state of obstructing fire. The integrity of a structure is compromised by fire, wide cracks and holes are formed, and the spread of fire and smoke cannot be stopped.
3. The limit state of heat insulation. When the temperature on a surface unexposed to the fire of the structure increases excessively (e.g., 140 °C for an average value or 180 °C for the maximum value), it may cause a fire in the adjacent rooms and a fire spreads.

The time taken for a structural or architectural member to reach one of the three limits under fire following the standard time–temperature curve (Eqn. (5.1)) is called the endurance limit (in hours). According to the Code of Fire Prevention in Buildings,[0-11] a building can be classified into four grades depending on its importance, and the minimum endurance limit (0.5–4 h) is stipulated for different members.

Several types of materials are used in structural engineering. Timber structures are combustible, cannot prevent a fire, and can even enhance a fire after it has started. Although a steel structure is not combustible, the temperature of the steel members under fire rises quickly and causes loss of the load-bearing capacity or failure of local stability, even collapse of the whole structure, because heat conducts very quickly through steel and the structural members are composed of thin-wall shaped steel components and plates. However, the main part of reinforced concrete is concrete itself, which is a material of high heat inertia and the main structural members usually have thick sections. So, the temperature in the interior of the member elevates slowly during a fire and the temperature elevation of the reinforcement is delayed by the outer cover. Therefore, the loss of strength of the material is less significant, the load-bearing capacity of the member decreases slowly, and the fire-resistant behavior and the endurance limit of reinforced concrete structures are much better than for steel and timber structures.

If a fire continues for a long time, the damage and failure phenomena of different levels will appear successively in the concrete structure: cracking and loosening on the surface, damage to the sides and corners, explosive spalling of the cover, reinforcement exposure, member deflection, gradual separation of the surface layers from the main body, damage area penetrating into the interior of the member, and, finally, caving in, local holes, and ultimately collapse of the entire structure may result.

Many valuable lessons and experiences have been gained from previous fire accidents and effective methods to prevent and fight fire have been developed. However, preventing fire is not always possible, so one should also depend on effective methods to fight fires. After years of research, many effective measures have been created in both of these aspects:

1. Prevention of fire occurrence and spread. For example, maintaining sufficient distance between buildings, separating the longer and larger areas into several parts and building firewalls between them, selecting facilities and furniture made of incombustible materials, spraying or smearing fire-protective material on the surface of combustibles, installing hydrants and water systems, installing automatic alarms and sprinkler facilities, studying the regularity of combustion, and limiting the spread of fire.
2. Research and enhancement of the fire resistance of buildings and structural members. For example, establishing large testing furnaces for measuring the endurance limit of full-scale members, selecting reasonable materials and improving the detail in construction of the structure, setting up insulating material on the member surface, conducting

experimental and theoretical research systematically on the thermal behavior of the materials, investigating the behavior of the members and the structures at elevated temperatures and their analysis method, and developing practical methods to calculate the bearing capacity and endurance limit of the structure under fire.

In order to enhance and solve the fire resistance ability of the structure and its members, several development stages are conducted: during the initial stage, only the construction measures tested by experience are used, e.g., increasing the thickness of the concrete cover of the reinforcement and using heat-resisting concrete; at the next stage, large experimental installations are established for loading tests of full-scale members at elevated temperatures, and the endurance limit or bearing capacity of the structural member under fire is directly measured. A recent trend is to emphasize theoretical analysis based on the experimental investigation, which includes the development of thermal–mechanical constitutive models for the materials, determination of the time–temperature curve of fire, conducting nonlinear analysis of the transient temperature field, and analyzing the whole process of the structure and its members under fire. Because the behavior of concrete structures at elevated temperatures is quite complicated, the theoretical analysis is not yet satisfactory and more study and improvements are needed.

0.3 BEHAVIOR CHARACTERISTICS OF REINFORCED CONCRETE STRUCTURES AT ELEVATED TEMPERATURE

According to the existing results of experimental and theoretical research and the experience of engineering practice, the behavior of reinforced concrete structures at elevated temperatures (and fire) is considerably different from that at ambient temperature. The characteristics of reinforced concrete structures at elevated temperatures are as follows.

1. Temperature distributed nonuniformly in the interior

Since the thermal conductivity of concrete is low, the temperature on the surface of the structure rises very quickly during a fire, but the temperature in the interior increases slowly. So, a nonuniform temperature field is formed in the structure, especially a large temperature gradient in the outer layer. In addition, the temperature field varies continuously for the duration of the fire (see Part 2).

The main factors determining the temperature field of the structure are the temperature–time process, the shape and size of the members, and the thermal behavior of the concrete material. The internal forces, deformation, and small cracks in the structure have less influence on the temperature field. On the contrary, the temperature field of the structure influences considerably the internal forces, deformation, and its bearing capacity. Therefore, the analysis of the temperature field of a structure can be conducted independently and earlier than that of internal forces and deformation.

2. Serious deterioration of material behavior

The values of the strength and elastic modulus of concrete and reinforcement at elevated temperatures decrease considerably and the deformation of both materials increases correspondingly. In addition, the external damage of concrete, e.g., due to cracking, loosening, and spalling, appears successively and gradually becomes more severe as the temperature increases (see Part 1). This is the main cause of serious reduction in the bearing capacity and the endurance limit of the structure and its members at elevated temperatures.

3. Coupling effect of stress, strain, temperature, and time

When a structure at ambient temperatures is analyzed, it is necessary to study only the stress–strain relationship of the material. However, the value and the duration of high temperature strongly influence the strength and deformation of the material. Furthermore, different heating and loading histories cause various values of the strength and deformation of the material. A coupling effect is then composed of four factors: stress, strain, temperature, and time for the concrete. Therefore, in order to accurately analyze the behavior of the structure at elevated temperatures, it is necessary to develop a corresponding coupling thermal–mechanical constitutive relationship for concrete (see Part 3). This complicates the analysis of the structure and its members significantly.

4. Redistribution of the stress on the member section and the internal forces of the structure

The nonuniform temperature field of the member section inevitably results in unequal temperature strain and stress redistribution on the section. In a statically indeterminate structure, the temperature deformation of the material at high temperature is restrained by the adjacent material at a different temperature, and the joint and the support. So, the redistribution of internal forces (bending moment, shear force, axial force, and even torque) of the structure is serious. As the temperature changes with time, a continuous process of redistribution of internal forces is then set in motion. Finally, the failure mechanism and pattern of the structure is different from that at room temperature, which influences the ultimate bearing capacity of the structure at elevated temperatures (see Chapter 10).

5. Process and pattern of failure

Generally, a concrete structure under ambient temperatures fails slowly with apparent signs. The structure and its members at elevated temperatures fail suddenly because the deformation increases quickly, the failure duration is short, and fewer warning signs appear. The structural member after failure shows large residual deformation, which can be seen clearly by the naked eye (see Part 3).

REFERENCES

[0-1] National Standard of People's Republic of China, Code for Design of Concrete Structure GB 50010-2002, Construction Industry Press of China, Beijing (2002).

[0-2] Z. Guo, The Principle of Reinforced Concrete, Tsinghui University Press, Beijing (1999).

[0-3] Z. Guo, Strength and Deformation of Concrete (Experimental Basis and Constitutive Relationship), Tsinghua University Press, Beijing (1997).

[0-4] Federation International de la Précontrainte, The Design and Construction of Prestressed Concrete Reactor Vessels (March 1978), FIP/3/3.

[0-5] ASME Boiler and Pressure Vessels Code, An American National Standard (ACI Standard 359–374). Section III, Division 2 (January 1975).

[0-6] Federation International de la Précontrainte, FIP/CEB Recommendations for the Design of Reinforced and Prestressed Concrete Structural Members for Fire Resistance, FIP/1/1 First ed. (June 1975).

[0-7] Federation International de la Précontrainte, FIP/CEB Report on Methods of Assessment of the Fire Resistance of Concrete Structural Members, FIP, (1978).

[0-8] Professional Standard of Electric Industry of People's Republic of China, Design Code for Hydraulic Concrete Structures DL/T 5057–1996, Electric Power Press of China, Beijing (1997).

[0-9] B. Zhu, T. Wang, B. Ding, Z. Guo, Thermal Stress and Temperature Control of Hydraulic Concrete Structure, Hydraulic and Electric Power Press, Beijing (1976).

[0-10] Professional Standard of Metallurgical Industry of People's Republic of China, Code for Heat-Resistance Design of Reinforced Concrete Structure of Metallurgical Workshop YS12-79 (Tentative), Metallurgical Industry Press, Beijing (1979).

[0-11] National Standard of People's Republic of China, Code for Architectural Design of Fire Prevention GBJ 16-87, Construction Industry Press of China, Beijing (1988).

[0-12] S. Wang (Ed.), Fire Prevention of Building, Construction Industry Press of China, Beijing (2000).

[0-13] B.K. Bardhan-Roy, Fire resistance—design and detailing, in: F.K. Kong (Ed.), Handbook of Structural Concrete, Pitman, London (1983) 14-1–14-46.

[0-14] Y. Li, D. Ma, J. Xu, Calculation and Construction for Fire Prevention Design of Building Structure, Construction Industry Press of China, Beijing (1991).

PART I

Mechanical Behavior of Materials at Elevated Temperatures

CHAPTER 1

Strength of Concrete at Elevated Temperatures

1.1 TESTING METHOD AND DEVICE

1.1.1 General Testing Program

The typical device for testing the mechanical behavior of concrete at elevated temperatures is not yet available, and no complete device is available on the market. Generally, research organizations develop their own device or have one designed and manufactured in a factory. Because the temperature range, content, method of testing, and the shape and size of the specimen are different for each research program, the arrangement and construction of existing testing device are all different.[1-1,1-3]

A device for testing the mechanical behavior of concrete at elevated temperatures must include three systems: a system for heating and temperature control, a system to support and load the specimen, and a system of measuring and recording instruments. The general arrangement of the material behavior testing system for elevated temperatures in the Structural Engineering Laboratory of Tsinghua University is shown in Fig. 1-1. The device had been developed by the University and the construction and technical specifications are discussed in this chapter.

1. Heating and temperature control system

The system includes two sets of furnaces and the corresponding temperature controllers, which are used for preheating and accurately testing the specimen, respectively. The preheating furnace is an electrical resistance furnace of box-type SX2-12-10, which was produced by Beijing Factory of Electrical Furnace. The clear space in the furnace chamber measures 500 mm × 300 mm × 200 mm, the specified power is 12 kW, and the maximum temperature is 1000 °C. The average velocity of the temperature increase is 8–10 °C/min when

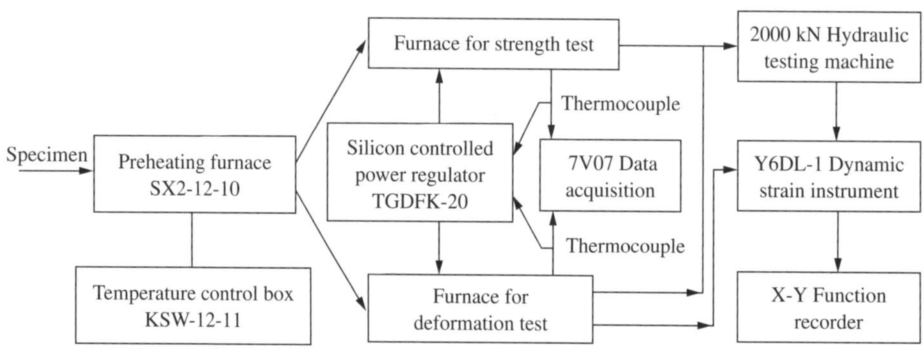

FIGURE 1-1 General scheme of the devices for testing the behavior of concrete at elevated temperatures.[1-4]

Experiment and Calculation of Reinforced Concrete at Elevated Temperatures

no specimen is placed in the chamber. The corresponding temperature control box (type KSW-12-11) is composed of a relay; the maximum temperature is 950 °C, and the fluctuation temperature range is ±8–10 °C during constant temperature testing. The temperature–time curve of the preheating furnace is shown in Fig. 1-2.

There are two furnaces for material behavior tests. One is used for testing the compressive strength and splitting tensile strength of cubic concrete specimens (edge length 100 mm), and the other is used for testing the compressive stress–strain curve of prism concrete specimens (100 mm × 100 mm × 300 mm). The furnaces were designed and produced by the authors, and their construction and technical specifications are given in Section 1.1.2. Corresponding to the furnaces, a silicon-controlled power regulator (type TGDFK-20, manufactured by the Beijing Factory of Voltage Regulator) is used to control the temperature; it has a maximum power of 20 kW. The power regulator includes a Proportional-Integral-Differential (PID) integrator, and can automatically slow down the elevating velocity of the temperature and avoid the impact of thermal inertia when the temperature in the chamber is approaching the required temperature. The power regulator can also be operated by hand at any time to adjust the elevating velocity or to reduce the temperature fluctuation range. The controlled temperature is accurate within ±2 °C.

In some other research organizations[1-3] only one furnace is used for both heating and loading the specimen during testing. However, individual preheating and testing furnaces have obvious technical advantages, although the number of facilities and investment necessary are high. The chamber volume of the preheating furnace is large enough to contain 3–6 specimens of the same set; the furnace heats them together and keeps the temperature constant. Each specimen is taken out individually and moved into the testing furnace. It is then loaded after accurately controlling and readjusting the temperature. Therefore, the temperature is uniformly distributed in the interior of the specimen. The testing temperature is controlled accurately, the experimental results for specimens of the same set show good repetition, and high testing efficiency is achieved.

2. Measuring and recording data

The data required during material behavior testing of concrete at elevated temperatures are the temperature of the furnace chamber, the temperature of the interior of the specimen, the load (or stress), and deformation of the specimen.

In order to measure the temperature in the chamber and transfer it as a feedback signal to the temperature controller, an accurate and reliable armored nickel–chromium and nickel–silicon thermocouple (type WRNK-541) is applied. The gradation of the thermocouple is EU-2, the length is 300 mm and the diameter is 6 mm, and the maximum temperature is 900 °C. The thermocouple is connected to the outer wall of the furnace through a movable cartridge flange. The

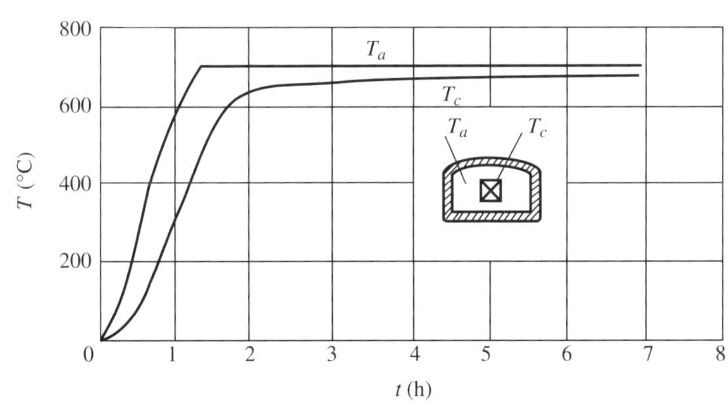

FIGURE 1-2 The temperature–time curve of the preheating furnace and the temperature in the interior of the specimen.

wire used is composed of a knitted compensatory metal, which terminates in a plug. The temperature in the interior of the specimen is measured by a transducer made of a fine and light thermocouple of nickel–chromium and nickel–silicon wires, which is buried inside the specimen in advance. The thermocouple is connected to the 7V07 data acquisition device or the isolated measurement pods (IMPs) to measure and record the temperature during testing.

The load transducer is placed on the specimen and the upper compression head. After the value of the load applied to the specimen is measured and recorded, the average compressive stress (σ) can be calculated easily.

In the deformation testing of concrete at elevated temperatures, the prism specimen is put in the middle of the furnace, which is at a uniform temperature. Because many types of deformation transducers installed inside furnaces cannot bear high temperatures, an indirect method can be used to measure the deformation (or strain) of the specimen. In method (1), a pair of stretched arms can be set on the upper and lower compression heads located at both ends of the specimen; the deformation of the specimen is transferred outside the furnace (Fig. 1-3(a)) and can be measured and recorded by the displacement transducer and the corresponding instrument. The deformation modification of the compression heads should be calibrated in advance and deducted from the measured deformation when calculating the strain of the specimen.

In method (2), two pairs of long and short bars made of stainless steel are used for measuring. The upper end of the longer bar is fixed on the upper part of the specimen and the lower end extends out of the furnace and is connected to the displacement transducer. The upper end of the shorter bar is fixed on the lower part of the specimen and the lower end also extends out of the furnace but is connected to the positioning piece (Fig. 1-3(b)[1-6]). As the long and short bars have equal temperature deformation within the same temperature area, the temperature deformation (strain) of the specimen can be calculated by deducting the temperature (expansion) deformation of the longer bar between the distance of the upper and lower parts of the specimen from the deformation value measured by the transducer. Compared with method (1), method (2) avoids errors caused by the rotation of the stretched

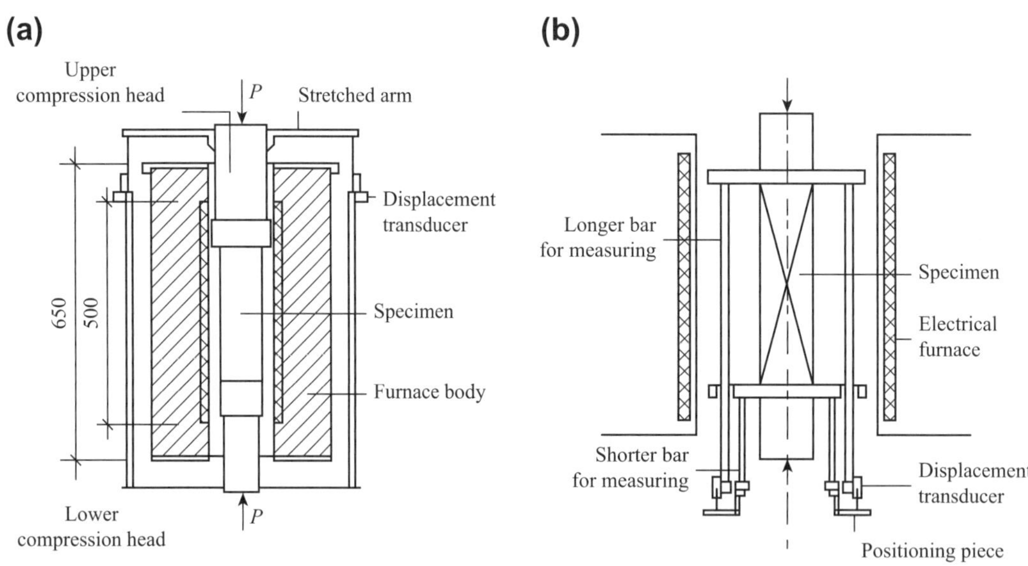

FIGURE 1-3 Furnace and measuring method of deformation testing: (a) testing furnace and measuring method (1); (b) measuring method (2).

arms and the deformation of the upper and lower compression heads that are placed in the nonuniform temperature area of the furnace. In addition, the deformation modification part of the longer bar is located in the area of the furnace at uniform temperature and the deformation of this part can be calculated accurately or calibrated in advance. Thus, method (2) has advantages over method (1), but, in order to install the measuring bars in the chamber, the section size of the concrete prism specimen used is reduced to 80 mm × 80 mm.

1.1.2 Design and Manufacture of the Furnace for Material Testing

The furnace for strength testing is a special facility used for compression and splitting tension tests on cubic concrete specimens at elevated temperatures. The size and construction of the furnace body are shown in Fig. 1-4. The main components include the case, chamber and opening bricks, heating wire, isolated layers, and thermocouple.

The cover and case constitute the outer shell of the furnace. The periphery measures 410 mm × 410 mm × 500 mm, which fits the working table of the hydraulic testing machine.

The chamber and opening bricks are made of refractory material. The chamber brick is tubular with an internal diameter of 150 mm, which is slightly longer than the diagonal of the specimen section (10 mm × 100 mm). The chamber brick is 20 mm thick and contains 28 longitudinal holes along the circumference, which permit the continuous heating wire to pass through. Its height is 250 mm and the height of the uniform temperature area in the middle exceeds the height of the specimen (100 mm). The opening bricks are placed to protect the upper and lower ends of the chamber brick.

The total power of the furnace is 3 kW. Two heating wires of 1.5 kW each pass reciprocally

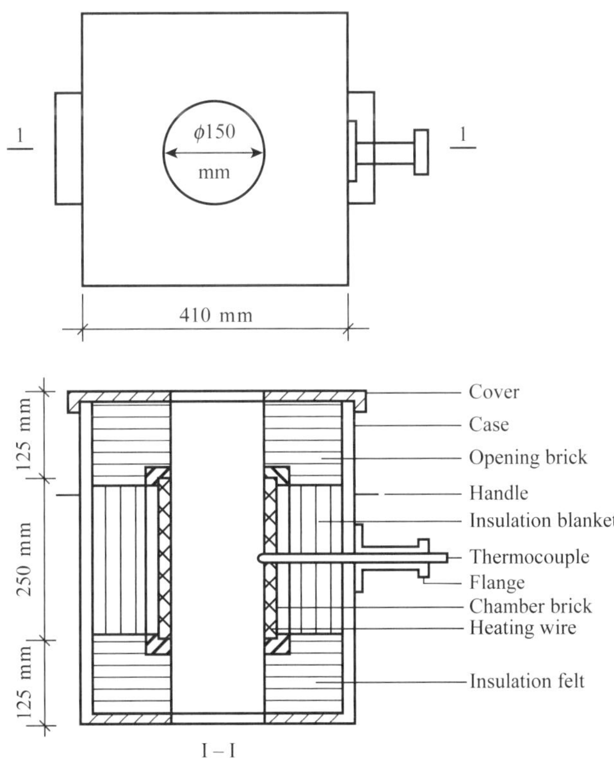

FIGURE 1-4 Construction of the furnace for strength testing.

through the longitudinal holes within the chamber brick. The ends of the wires are connected in parallel to each other and link up with the silicon-controlled power regulator, which supplies electrical energy for heating. The armored thermocouple passes through the hole and enters the chamber to measure the temperature, and is then fixed with the flange. The thermocouple is used not only for measuring the temperature but also as the feedback signal, which provides input to the power regulator and controls the temperature of the furnace chamber.

Insulation is applied to the furnace body to increase thermal efficiency and to ensure safety during use. The insulation blanket of siliceous aluminum fiber is wrapped around the chamber brick, and several layers of insulation felt of the same fiber are put on and under the chamber brick. These insulation materials are manufactured by Beijing Factory of Refractory. The insulating effect of the furnace lasts even after many uses. When the temperature in the chamber remains at 900 °C for 2 h, the temperature on the outer surface of the furnace is lower than 50 °C.

The construction and materials of the furnace for deformation testing are the same as those used for strength testing. For measuring deformation, the height of the prism concrete specimen is 300 mm, therefore the height of the chamber is 500 mm, which is created using two chamber bricks. Three heating wires of 2 kW each pass reciprocally through the longitudinal holes within the chamber bricks. The ends of the wires are connected in parallel to each other and are linked to the power regulator. The total power of the furnace is 6 kW. The size of the outer shell of the furnace is 410 mm × 410 mm × 650 mm.

Since the two test furnaces were manufactured, in cooperation with other facilities (Fig. 1-1), several hundred heating–loading tests have been completed. They have shown that the construction of the furnace body is reasonable, the insulation behavior is perfect, the accuracy of the controlling temperature is satisfactory, the testing procedure is convenient, the testing efficiency is high, the investment is low, and the results are obtained quickly.

The complete device for testing the mechanical behavior of concrete at elevated temperatures has been described above; its testing functions and technical specifications follow:

1. Test items: cubic compressive strength (f_{cu}^T), splitting tensile strength (f_t^T), complete compressive stress–strain curve, corresponding prismatic compressive strength (f_c), and peak strain (ε_p^T).
2. Specimen size: cube of 100 mm × 100 mm × 100 mm, prism of 100 mm × 100 mm × 300 mm.
3. Maximum axial compressive force: 2000 kN.
4. Maximum testing temperature: 900 °C.
5. Maximum heating velocity: 10 °C/min.
6. Precision of temperature control: ±2 °C.
7. Testing conditions: load under constant temperature, heat under constant load, constant temperature and load, cycles of elevating and reducing temperature, and variety of loading–heating paths.

1.2 COMPRESSIVE STRENGTH AT ELEVATED TEMPERATURES

1.2.1 General Phenomena During Heating

The compressive strength of concrete is the most basic and important mechanical behavior. It acts as a fundamental parameter to identify the strength grade and quality index of the concrete, and to determine the values of other types of mechanical behavior, e.g., tensile strength, elastic modulus, and peak strain. Similarly, the compressive strength and the stress–strain relationship of concrete at different temperatures are also the basis for studying the behavior of concrete structures and components at elevated temperatures.

The mechanical behavior of concrete at elevated temperatures can be obtained using certain tests using the testing device described above; they are presented individually in this chapter and in Chapters 2 and 3. For the sake of easy comparison and analysis, the raw materials and the mix of the concrete are the same for all the specimens in the various tests. The testing temperature ranges from 100 °C to 900 °C.

There are two strength grades (C20 and C40) of concrete specimens, and two types of coarse aggregates (limestone and granite) for each grade of concrete. The maximum size of the aggregates was 15–20 mm. The same types of cement and sand were used for all concretes. The mixes and the cubic strength under normal temperatures during testing of the four concretes are listed in Table 1-1.

The concrete was mixed and poured into a steel mold by hand, and was then compacted using a vibrator and the surface was daubed. The concrete specimens were left for 2 days in the laboratory and then kept in the standard curing room for 28 days after being demolded. The specimens were prepared for testing after air-drying for more than 3 days at room temperature.

In order to obtain the strength of concrete at a certain temperature (e.g., T = 700 °C), the temperature in the interior of the specimen should be distributed uniformly. However, heat conducts slowly in concrete as it is a material of thermal inertia (see Section 5.2.1). When a cubic specimen (edge length 100 mm) is heated in the preheating furnace, the temperature at the center reaches only 430 °C in 75 minutes, whereas the temperature in the chamber has already reached 700 °C (Fig. 1-2). Obviously, the temperature distributes nonuniformly in the interior of the specimen. The temperature in the chamber fluctuates within 700 ± 10 °C, but the internal temperature of the specimen increases gradually with the reduction in elevating velocity. The temperature at the center of the specimen is approximately 650 °C and 677 °C in 2 and 6 hours, respectively. Thus, it is considered that the temperature in the whole specimen distributes uniformly after 6 h. Several tests at different temperature values demonstrate that a uniform temperature in a cubic specimen (edge length 100 mm) can be achieved after heating to a certain temperature and keeping it constant for 6 h.

Therefore, the procedure for testing the compressive strength of concrete at elevated temperatures is as follows: cubic specimens are heated to the predetermined temperature and kept constant for 6 h in the preheating furnace; each specimen is taken out individually using a special clamp and put into the strength test furnace; the specimen is loaded and the compressive strength at elevated temperatures (f_{cu}^T) is measured after accurately adjusting the furnace temperature, which takes about 20 min. The loading velocity of the specimen (0.25 MPa/s) is the same as that at normal temperatures.

When the concrete specimen is heated from room temperature to 900 °C, its physical condition changes gradually in the testing process. The color and surface damage of the specimen at different temperatures are shown in Table 1-2.[1-4]

The weight of the concrete specimen decreases gradually in the heating process. The weight loss (in %) of a cubic specimen after keeping it for 6 h at a predetermined temperature is shown in Fig. 1-5.[1-6] When the testing temperature is in the range of 20–200 °C, the concrete loses weight quickly, mainly because the free water in the specimen evaporates; when the temperature is in the range of 200–500 °C, the weight is lost slowly as the chemically combined

TABLE 1-1 Mixes and Cubic Strength at Normal Temperature During Testing of Concretes

	Mix (kg/m^3)					
Type of concrete	325# Cement	Water	Sand	Limestone	Granite	f_{cu} (N/mm^2)
C20L	320	195	665	1235		30.50
C20G	320	195	665		1235	28.80
C40L	435	200	571	1200		55.00
C40G	435	200	571		1200	54.10

TABLE 1-2 Surface Features of Concrete at Different Temperatures

Temperature (°C)	Color	Cracks	Lost surface	Broken corners	Loose
100	Same as normal temperature	No	No	No	No
300	Slightly white	Fine, few	Not yet	No	No
500	Gray–white	Fine, more	Few	No	Slightly
700	Dark red	Obvious, more	Less	Few	Obvious
900	Red	Wide, more, directionless	Lose after knocking	Every corner, different level	Serious, broken by finger (after cooling)

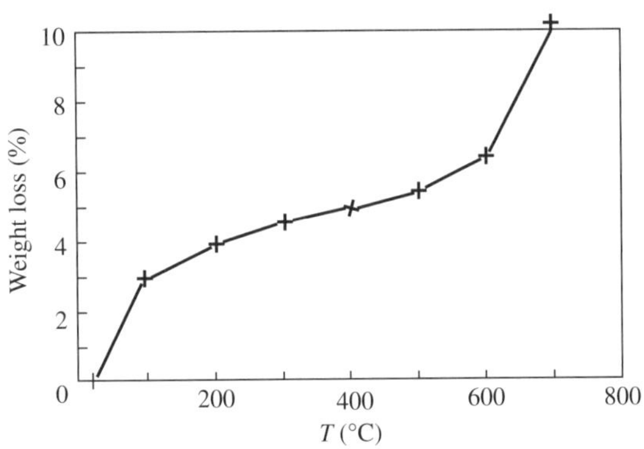

FIGURE 1-5 Weight loss of concrete (limestone aggregates) at elevated temperatures.

water in the cement mortar separates; when the temperature reaches 500 °C, the calcium hydroxide component produced from the hydration of cement decomposes and dehydrates; when the temperature exceeds 600 °C, the magnesium and calcium carbonates in the dolomite and calcite of the aggregates begin to decompose, so the aggregates become unstable and the weight loss may reach 10%; when the temperature is even higher, the outer layer of the concrete is damaged and spalls off, causing further weight loss.

The cubic specimen of concrete is loaded and failed at different temperatures, and then cooled down to room temperature. The failure pattern is shown in Fig. 1-6. The specimen tested at normal temperature presents an upright reverse pyramid and the middle section cracks and spalls off, whereas the top and bottom surfaces show no signs of failure and the corners and edges remain intact because both surfaces are confined by the steel plates that are in contact with them. The failure appearance of the specimen tested at $T < 300$ °C shows little difference from that at normal temperature. For specimens tested at higher temperature, the cracks and the broken corners and edges on both compressive surfaces appear serious and the core section in the central part reduces gradually. As the testing temperature approaches $T > 800$ °C, the specimen breaks at the end of the loading or breaks into pieces when it is taken out of the furnace.

1.2.2 Cubic Compressive Strength at Elevated Temperatures

The cubic compressive strength of concretes at different temperature f_{cu}^T are listed in Table 1-3. Two grades of concretes with two types of aggregates each are included. As the temperature of the specimen increases gradually during testing, all the internal materials of the concrete experience a series of physical and chemical reactions, and the compressive strength shows a complicated variable regulation (in contrast to Fig. 1-7, discussed later).

When the testing temperature T equals 100 °C, the ratio between the compressive strength of concrete at elevated temperatures and at normal temperature, f_{cu}^T/f_{cu} is between 0.88 and 0.94. As the free water contained in the concrete evaporates gradually, capillary cracks and

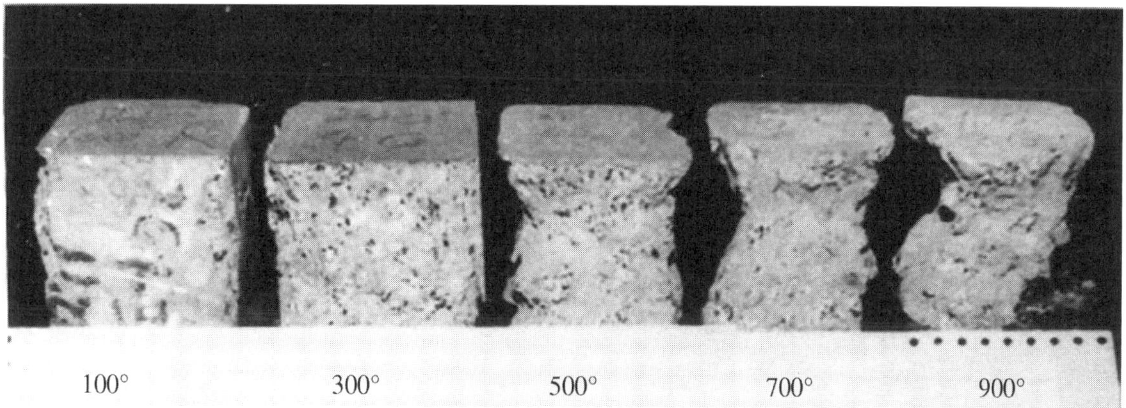

FIGURE 1-6 Failure patterns of compressive cubic specimens of concrete at elevated temperatures.

TABLE 1-3 Cubic Compressive Strength (f_{cu}^{T}, N/mm²) of Concrete at Elevated Temperatures[1-4]

Concrete	Room temperature	60 °C	100 °C	150 °C	200 °C	300 °C	400 °C	500 °C	700 °C	900 °C
C20L	30.5	28.3	28.2	29.0	31.1	32.5	30.6	24.7	10.6	3.6
C20G	28.8		26.1			30.3		22.8	8.3	2.4
C40L	55.0		50.3			56.7		43.7	21.4	5.0
C40G	54.1		48.2			54.3		40.9	13.8	2.9

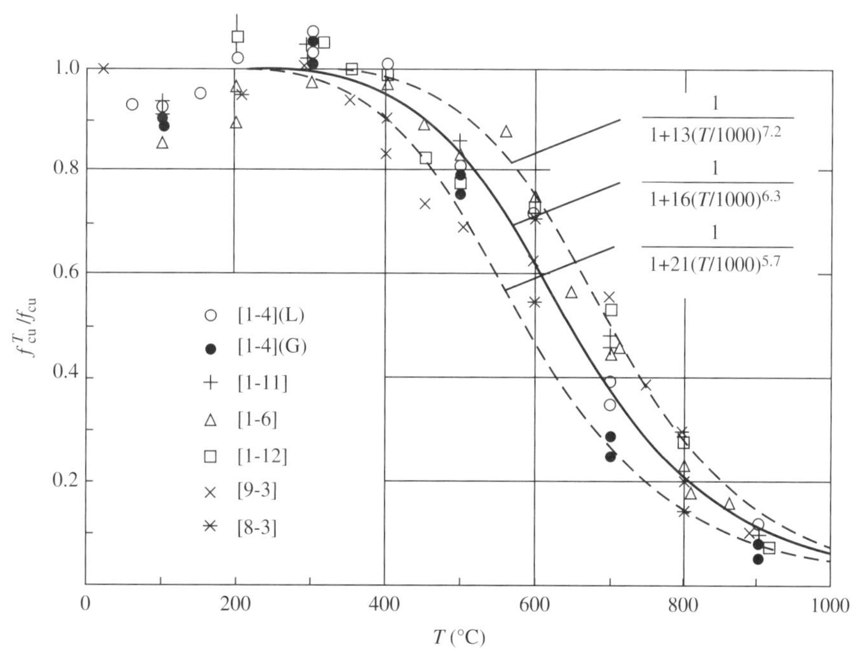

FIGURE 1-7 Cubic compressive strength of concrete at elevated temperatures.

porosity are formed in the interior of the specimen. The pressure of the water and vapor in the cracks and holes increases as the temperature rises causing tensile force in the surrounding solid materials. In addition, the stress concentration occurs at the tips of the cracks and accelerates the expansion of the cracks. Therefore, the compressive strength of the concrete reduces slightly.

When $T=200–300$°C, $f_{cu}^T/f_{cu}=0.95–1.08$, and the free water in the specimen is evaporated. Because the expansion coefficients[0-2] of the coarse aggregate and the cement mortar of the concrete are not equal, the difference in thermal deformation between them causes cracks to form on the boundary of the aggregate and reduces the compressive strength of the concrete. On the other hand, as the combined water in the cement gelatin starts to release, it strengthens the adhesive action of the cement particles and relaxes the stress concentration at the crack tip, so it helps to increase the strength of the concrete. These contradictory factors act simultaneously, so that the compressive strength of the concrete first increases slightly and later reduces within the temperature range. It shows a complicated variation. The fluctuating amplitude of the strength depends on the composite components, quality of the raw materials, and mix of concrete.

When T is greater than 400 °C, the compressive strength of the concrete decreases obviously, and $f_{cu}^T/f_{cu}=0.75–0.85$ when T = 500 °C. The difference in thermal deformation between the aggregate and the cement mortar increases continuously, and the cracks on the boundaries expand and extend. In the meantime, the water in the calcium hydroxide and other chemical compounds available after cement hydration is released with volume expansion, so the cracks expand and the compressive strength reduces more quickly.

When $T \geq 600$ °C, the quartz components in the unhydrated cement particles and the aggregates decompose and crystal is formed, accompanied by considerable expansion. Cracks also appear in the interior of some aggregates and expand as the temperature increases. The compressive strength of concrete reduces sharply:

$$T=700\ °\text{C},\ f_{cu}^T/f_{cu}=0.30–0.50;$$

$$T=800\ °\text{C},\ f_{cu}^T/f_{cu}=0.15–0.28;$$

$$T=900\ °\text{C},\ f_{cu}^T/f_{cu}=0.05–0.12.$$

The thermal action results in strength loss and deterioration of the deformation behavior of concrete. According to the analyses above, the main reasons for this can be summarized as follows:

- Crack and porosity form in the interior of concrete after water evaporates.
- The thermal behavior of the coarse aggregate and the cement mortar are different, which causes a deformation difference and internal stress between them and results in cracking at their boundary.
- Coarse aggregate expands and cracks at high temperature. This internal damage in the concrete develops and accumulates continuously, and tends to be more serious as the temperature increases.

The compressive strength of concrete varies as the temperature changes; its regularity described earlier is consistent with the existing experimental conclusions worldwide.[0-6,1-3,1-8,1-9] However, the thermal behavior of concrete depends on many factors, such as the component materials, mineral chemical composition, mix, and moisture. Furthermore, the testing device and the method, the shape and size of the specimen, the heating velocity, and the time the concrete sample was subjected to a predetermined temperature are different in each research experiment. Therefore, the experimental value (f_{cu}^T/f_{cu}) of concrete strength unavoidably shows a certain deviation.

Various factors influence the compressive strength of concrete at elevated temperatures, and they are analyzed according to the experimental results.

- Strength grade of the concrete. The relative strength (f_{cu}^T/f_{cu}) of concrete (C20–C50) at the same temperature decreases with higher grades

(Table 1-3), but there is only less than 10% difference between them. The relative strength of high-strength concrete (>C60) decreases to $f_{cu}^T/f_{cu} \approx 0.8$ when $T = 100$–300 °C, and the specimen may spall off and break suddenly when $T > 400$–500 °C.[1-8] Extra attention should be paid to this.

- Type of aggregate. Concrete containing silicon aggregate (e.g., granite) has a slightly lower strength (f_{cu}^T/f_{cu}) compared with concrete containing calcium aggregate (e.g., limestone) at the same temperature. The difference between them is approximately 6% when $T > 700$ °C, as shown in Table 1-3. Concrete containing light-weight aggregate has a much higher strength than concrete containing ordinary aggregate.[1-9,1-10]
- Other factors. The larger the water/cement ratio or the water content of the concrete, the lower the strength (f_{cu}^T/f_{cu}); the slower the heating velocity and the longer the time the specimen is held at a high temperature, the lower the strength. Cooling and heating–cooling cycles make the strength decrease continuous; the compressive stress loaded before heating the specimen increases the strength of the concrete (see Section 3.2).

All the experimental data from tests using the same experimental devices and the same conditions are collected and summarized in Fig. 1-7. The figure shows the variable regularity and the deviation in the compressive strength of concrete at different testing temperatures. A fractional expression is used as its mathematic model[1]:

$$\frac{f_{cu}^T}{f_{cu}} = \frac{1}{1 + a\left(\frac{T}{1000}\right)^b} \tag{1.1}$$

The parameters of the formula are obtained from regression analysis and the values are rounded to

$$a = 16, \; b = 6.3 \tag{1.2}$$

The corresponding parameters in the deviation curves for the upper and the lower limits are shown in Fig. 1-7.

The compressive strength of concrete at elevated temperatures calculated by this formula is nearly constant $f_{cu}^T \approx f_{cu}$ when $T < 300$ °C. Although it cannot accurately reflect the fluctuation in the practical strength within the temperature range, it fits the experimental regularity quite well when $T > 300$ °C, and the application scope is not limited ($T \rightarrow \infty$). Therefore, this formula is suitable for analysis of a fire-resistant structure (at elevated temperatures). If a more accurate value of the strength is needed for a concrete of specified grade strength, raw materials, or mix, special specimens should be manufactured and tested, and then the values of the parameters in the formula can be calibrated according to the experimental data.

The specimen experiences successively heating, maintaining constant temperature, and loading in the strength testing of concrete at elevated temperatures. The whole process takes 7–10 h and the heating time is very brief. When the concrete is heated for a long time (see Section 0.1), the compressive strength decreases gradually as the heating time continues because of the accumulation of internal damage within the concrete. For higher temperatures, the reducing amplitude of the strength is larger. It is demonstrated in certain tests[1-13] that most of the strength loss appears within the first 2 days in the heating period, and the strength gradually tends to stabilize afterward.

1.2.3 Compressive Strength After Cooling

When a structure experiences an accidental high temperature and the fire eventually cools down to normal temperature, the residual strength of the concrete at this moment is the main basis to evaluate the damage level and the safety of the structure and has considerable influence in working out a strengthening scheme.

The procedure for testing the residual strength of concrete after it is cooled is:

1. Put a cubic specimen into the preheating furnace and heat it to the predetermined temperature.
2. Maintain a constant temperature for 6 h.

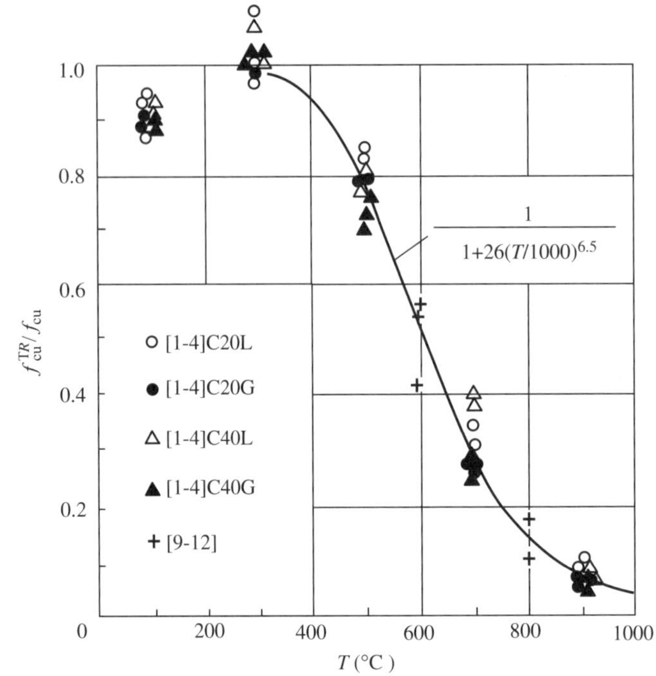

FIGURE 1-8 Compressive strength of a cubic concrete specimen after it is cooled.

3. Open the door of the furnace. The temperature in the chamber cools down naturally to room temperature.
4. Take the specimen out after 24 h and then load it at room temperature.
5. Measure the compressive strength (f^{TR}_{cu}).

The residual compressive strength of the cubic concrete specimen (edge length 100 mm) after it is cooled is expressed as f^{TR}_{cu}/f_{cu} and varies as the testing temperature changes.[1-4,9-12] The varied regularity (Fig. 1-8) can be simulated using an empirical formula as follows:

$$\frac{f^{TR}_{cu}}{f_{cu}} = \frac{1}{1+26\left(\frac{T}{1000}\right)^{6.5}}. \tag{1.3}$$

Comparing Fig. 1-8 with Fig. 1-7, it is found that the residual strength of the concrete after it has cooled has similar regularity but with a slightly lower value than the strength of the concrete at elevated temperatures with the same value, i.e., $f^{TR}_{cu}/f_{cu} \le f^{T}_{cu}/f_{cu}$. In the temperature range T = 500–800 °C, the difference between them reaches the maximum value, i.e., about 5–10%; in the other temperature ranges (T < 400 °C or T > 800 °C), the values are close to each other. The compressive strength of prism concrete specimens under these conditions behaves similarly, but the relative strength (f^{TR}_{cu}/f_{cu}) has a lower value.[9-12]

The test results show that the interior of concrete is damaged gradually when it is heated and a high temperature is maintained. When the concrete is cooling down, the temperature on its outer surface decreases quickly but the temperature in its interior remains high, so a nonuniform temperature field with opposite gradient is formed and new damage occurs in the interior of the concrete. This results in a further decrease in strength after the concrete is cooled compared with before cooling (i.e., at elevated temperatures).

Similar experimental investigations throughout the world have reached the same conclusion.[1-9,1-14,1-15] However, the values obtained for the residual strength of concrete after cooling are different. The minimum value is 20% less than the strength at elevated temperatures, probably because the specimen cooled down too quickly or due to difference in the testing methods. Some factors, such as the types of cement and aggregate, the water/cement ratio, and the age of the concrete, also influence the residual strength after cooling.[1-15]

Some investigations[1-9,1-15] show that if the specimen experiences heating and cooling and is kept in air, its compressive strength decreases continuously and the residual strength decreases with a higher maximum temperature. If after heating and cooling the specimen is immersed in water, the lost strength may partly recover and the lower the maximum temperature reached, the more the strength is recoverable. After several heating and cooling cycles, the strength of concrete decreases with time, but most of the strength appears already lost after the first cooling.

1.3 TENSILE STRENGTH AT ELEVATED TEMPERATURES

The splitting testing method of a cubic specimen (edge length 100 mm) is used to measure the tensile strength of concrete at elevated temperatures. The specimen is heated to a predetermined temperature and kept for 6 h in the preheating furnace. It is then moved into the strength test furnace, and is loaded after adjusting the temperature. Two square (5 mm × 5 mm) stainless steel bars are placed between the specimen and the upper and lower compression heads separately. The tensile (splitting) strength of the concrete at elevated temperatures is calculated[0-2] according to the load value when the specimen splits.

The tensile strength of concrete decreases monotonically as the testing temperature increases (Fig. 1-9). The range of its relative values (f_t^T / f_t) versus temperature is listed in Table 1-4. The tensile strength of concrete reduces quickly when T = 20–100 °C, but reduces slowly when T = 100–300 °C, and linearly when $T > 300$ °C. When T = 900 °C, the specimen approaches failure without further loading.

The splitting failure pattern of concrete at $T \leq$ 300 °C is the same as that at normal temperature.

TABLE 1-4 Tensile Strength of Concrete at Elevated Temperatures[1-4]

T (°C)	f_t^T / f_t
100	0.78–0.90
300	0.66–0.88
500	0.52–0.60
700	0.24–0.32
900	—

A splitting crack divides the specimen into two pieces along its central line and the boundaries on both sides of the crack are clear and approximate to a plane; little damage can be found on the boundaries but no crack is found on other outer surfaces. When the testing temperature $T >$ 500 °C, the failed specimen caves in locally at the positions of the steel bars, dregs appear on the crack boundaries, and wider cracks parallel with the crack boundary and irregular heating cracks appear on the outer surfaces of the specimen.

Comparing Fig. 1-9 with Fig. 1-7, the different deterioration regularities can be seen between the tensile and compressive strengths of concrete at elevated temperatures, especially within the range of T = 100–700 °C. The relative tensile strength is obviously lower than the relative compressive strength ($f_t^T / f_t < f_{cu}^T / f_{cu}$). This demonstrates that the internal damage in the concrete caused by thermal action has a stronger influence on its tensile strength. Therefore, the ratio between the tensile strength and the compressive strength of concrete varies as the testing temperature changes, but the value of the ratio is always less than that at normal temperatures, i.e., $f_t^T / f_{cu}^T < f_t / f_{cu}$.

The tensile strength of concrete has a small contribution to the bearing capacity of reinforced concrete structures and their members at elevated temperatures, the value of which can be calculated approximately according to a simple linear formula:

$$\frac{f_t^T}{f_t} = 1 - \frac{T}{1000} \tag{1.4}$$

A comparison between the theoretical and the experimental values is shown in Fig. 1-9.

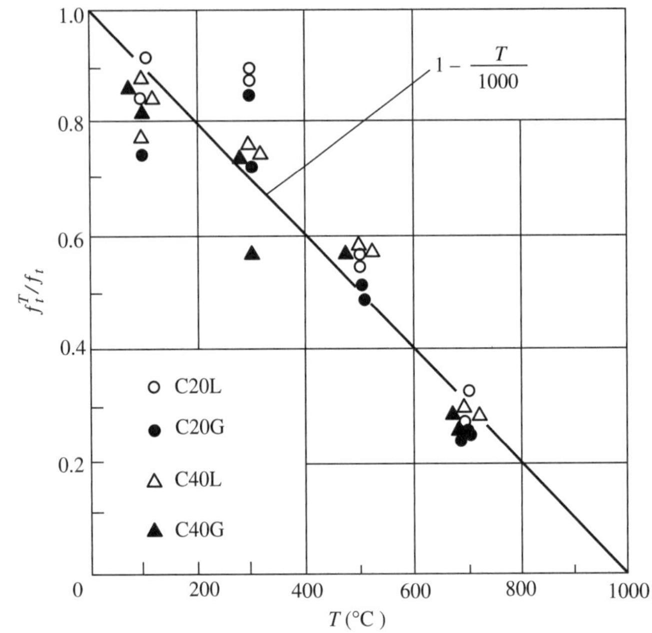

FIGURE 1-9 Tensile strength of concrete at elevated temperatures.[1-4]

The bond strength (τ_u) between the reinforcement and the concrete is the basis for using that combination in a reinforced concrete structure and is also an important factor influencing the behavior of the structures.[0-2] Some investigations[1-16] show that the bond strength between both materials decreases monotonically as the testing temperature increases, and the variation regularity is similar to that of the tensile strength of concrete. The reducing amplitude of the bond strength at elevated temperatures mainly depends on the surface shape and the corrosion level of the reinforcement. The relative bond strength of round-shaped reinforcements at elevated temperatures is even lower than the relative tensile strength of concrete, i.e., $\tau_u^T/\tau_u < f_t^T/f_t$. The design of the fire-resistant structures should take this into account.

CONCLUSIONS

Concrete is an artificial compound material. Cement is used as the main gluing substance and is blended proportionally with sand, stone aggregates, and water. After the process of mixing, casting in a mold, compacting, and curing successively, concrete coagulates and hardens gradually. The mechanical behavior of concrete varies considerably[0-2] because of the differences in the chemical components, behavior, and the proportion of the various composite materials, as well as the technological conditions, environment, and time factors during the production and hardening processes. At elevated temperatures, structural damage occurs in the interior of concrete material due to evaporation of water, differences in thermal behavior between aggregate and cement mortar, and expanding and breaking of aggregate particles. The internal damage in the concrete accumulates continuously as a high temperature is sustained, so the deterioration in the mechanical behavior of the concrete is accelerated.

The various mechanical behaviors of concrete at elevated temperatures arranged in order of the deterioration levels from lower to higher are as follows: weight lost, cubic compressive strength (f_{cu}^T), compressive strength after cooling (f_{cu}^{TR}), prism compressive strength (f_c^T), tensile strength (f_t^T), bond strength (τ_u^T), and elastic modulus

(E_0^T). The variation regularity of each behavior is also different as the testing temperature increases. However, the index for every behavior approaches exhaustion when $T \geq 800$ °C.

Normally, large quantities of local materials are used in concrete. Their behavior varies considerably and is influenced by many factors. In addition, there is no unified testing standard. Therefore, the existing experimental results show large deviations. Even the data, points of view, and conclusions derived by some researchers are contrary to one another. For example: Which type of aggregate has the highest or lowest strength at elevated temperatures? Should Poisson's ratio increase or decrease at elevated temperatures? Does the heating–loading path obviously influence the mechanical behavior? Furthermore, experimental data are still insufficient and the mechanism of some types of behavior require further investigations.

The chapters in this book introduce the various mechanical behaviors of concrete at elevated temperatures. The fundamental variation regularities are generally in agreement. However, for specific concretes, the technical index of the various types of behavior may still vary to some extent. Once the materials in the concrete to be used in an important structural engineering project are determined, a number of specimens should be specially manufactured and tested to accurately measure the mechanical behavior of the concrete at elevated temperatures.

NOTE

[1]The empirical formula suggested in the previous references is

$$\frac{f_{cu}^T}{f_{cu}} = \frac{1}{1+a(T-20)^b},$$

and the value of the parameters obtained in the references are listed in the table below.

Reference	a	b
[1-4]	2.4×10^{-17}	6
[1-11]	3.3×10^{-16}	5.5
[1-12]	1.183×10^{-20}	7.1
[9-3]	1.7×10^{-17}	6
[8-3]	8.24×10^{-15}	5

REFERENCES

[1-1] J.A. Purkiss, J.W. Dougill, Apparatus for compression test on concrete at high temperature, Magazine of Concrete Research (1973) 102–108.

[1-2] V.V. Bertero, M. Polivka, Instrumentation and techniques for study of concrete properties at elevated temperature, ACI Special Publication: Concrete for Nuclear Reactors, SP 34-64, Detroit (1972) 1377–1419.

[1-3] H. Niu, Z. Lu, L. Chen, Experimental investigation of constitutive relations of reinforcement and concrete under high temperature, Journal of Tongji University 18 (3) (1990) 287–297.

[1-4] W. Li, Experimental investigation on strength and deformation of concrete at high temperature. Masters dissertation, Tsinghua University, Beijing, 1991.

[1-5] W. Li, Z. Guo, Experimental investigation on strength and deformation behaviors of concrete at elevated temperature, Journal of Building Structures 14 (1) (1993) 8–16.

[1-6] J. Nan, Experimental investigation of mechanical behaviors of concrete considering temperature-stress coupling. Masters dissertation, Tsinghua University, Beijing, 1994.

[1-7] S.E. Philajavaara, An analysis of the factors exerting effect on strength and other properties of concrete at elevated temperature, ACI SP 34-19, Detroit (1972) 347–354.

[1-8] C. Carlos, A.J. Durrani, Effect of transient high temperature on high-strength concrete, ACI Materials Journal 87 (1) (1990) 47–53.

[1-9] T. Harada, et al., Strength, elasticity and thermal properties of concrete subjected to elevated temperature, ACI SP 34-21, Detroit (1972) 377–406.

[1-10] M.F. Kaplan, F.J.P. Roux, Effect of elevated temperature on the properties of concrete for the containment and shielding of nuclear reactors, ACI SP 34-24, Detroit (1972) 437–441.

[1-11] L. Jiang, Experimental investigation on strength and deformation of concrete with different temperature-stress paths. Masters dissertation, Tsinghua University, Beijing, 1992.

[1-12] X. Shi, Experimental investigation and non-linear finite element analysis of reinforced concrete beam-column structures under loading and elevated temperature. Doctoral thesis, Tsinghua University, Beijing, 1992.

[1-13] G.G. Carette, V.M. Malhotra, Performance of dolostone and limestone concrete at sustained high temperature, Special Technical Publication 858, ASTM, Philadelphia (1985) 38–67.

[1-14] Z. Qian, R. Chen, Z. Zhang, Physical and mechanical behaviors of concrete and reinforcement after fire. Symposium of the Second National Conference on Basic Theory and Application of Reinforced Concrete Structure, Beijing, Department of Civil Engineering, Tsinghua University (1990) 131–137.

[1-15] H. Weigler, R. Fisher, Influence of high temperature on strength and deformation of concrete, ACI SP 34-26, Detroit (1972) 481–493.

[1-16] U. Diederichs, U. Schneider, Bond strength at high temperature, Magazine of Concrete Research (6) (1981) 75–84.

CHAPTER 2

Deformation of Concrete at Elevated Temperature

2.1 DEFORMATION DURING HEATING AND COOLING

2.1.1 Deformation During Monotonic Heating and the Linear Expansion Coefficient

When a concrete specimen is heated or cooled freely (stress $\sigma = 0$), it is elongated or shortened and expanded or contracted. When the temperature $T > 400$ °C, the value of the thermal deformation of concrete is extremely high and far exceeds the value of the compressive peak strain (about 2×10^{-3}) of concrete at normal temperature. This thermal behavior has a considerable influence on the mechanical behavior of concrete material and structures at elevated temperatures.

The testing devices used for measuring the thermal deformation of concrete are the same as those described in Section 1.1. The testing procedure is as follows:

1. Put the specimen directly into the furnace for deformation testing.
2. Install the deformation transducer and recording instrument.
3. Switch on the electrical power.
4. Heat the specimen in a low velocity (2–5 °C/min) until the temperature reaches the predetermined value.
5. Maintain the same temperature for 30 min.
6. Switch off the power and take out the insulation material that is packed up between the chamber hole and the specimen; this makes the air blow through the chamber and the specimen cools naturally. The cooling velocity is –5–10 °C/min at the maximum test temperature (700 °C or 500 °C) and reduces gradually to –1 °C/min as the temperatures of the chamber and the specimen decrease.

The temperatures of the chamber and the interior of the specimen (T_a and T_c, respectively) are measured and recorded during the testing process. They vary with the heating time (t) and are shown in Fig. 2-1.

The thermal strain (ε_{th}) of the specimen is also measured during the testing process. The heating strain and temperature curves are drawn in Fig. 2-2.[1-4,1-6,1-11] As the predetermined temperature cannot be maintained for 6 h due to the limitations of the testing method, the internal temperature of the specimen distributes nonuniformly and the average value calculated by Eqn (3.3) is taken as the nominal temperature of the specimen.

According to the experimental results, the general regularity of the freely expanding strain of concrete can be analyzed as discussed below.

When the testing temperature $T < 200$ °C, the solid components of the concrete, including the coarse aggregate and cement mortar, expand due to elevated temperatures and simultaneously shrink due to water loss. Both factors compensate and cause smaller strain (elongation) at a slowly increasing rate. When $T = 200$ °C, the thermal strain reaches $\varepsilon_{th} \approx (0.8\text{–}1.5) \times 10^{-3}$.

When $T = 300\text{–}600$ °C, the solid components expand continuously as the temperature

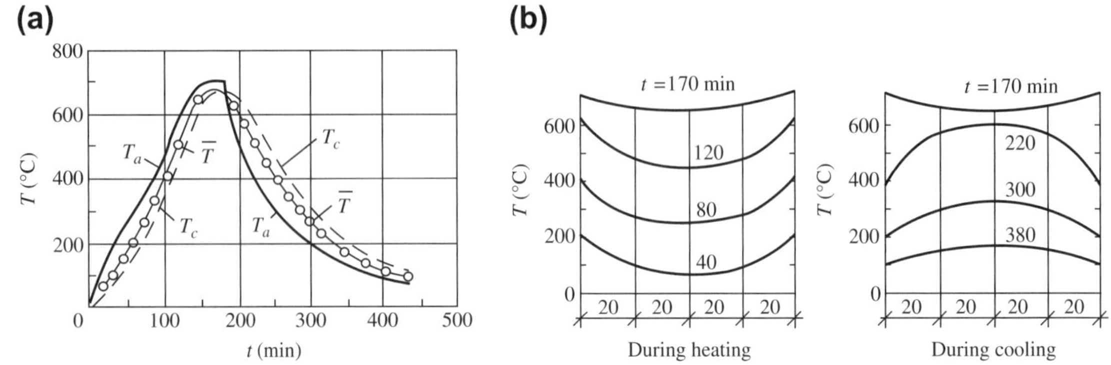

FIGURE 2-1 Heating and cooling curves for the chamber and the specimen. (a) Heating–cooling and time curves[1-6]; (b) temperature distribution in the interior of the specimen.

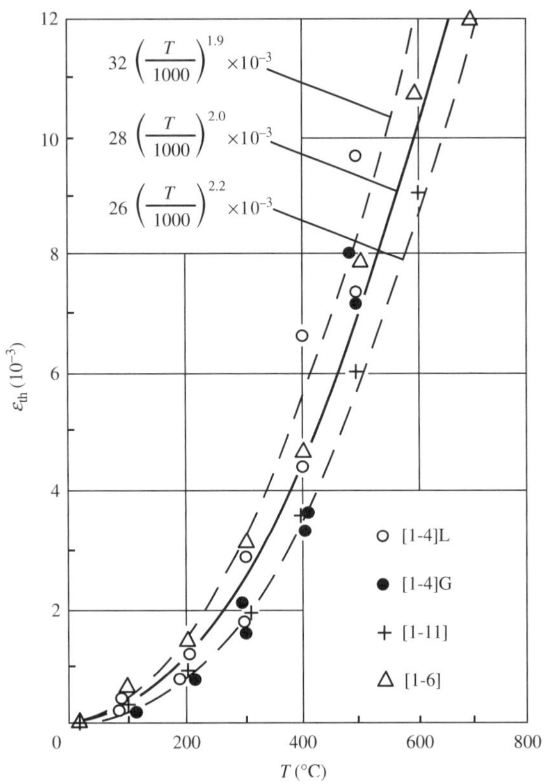

FIGURE 2-2 Free heating strain–temperature curve.

is elevated, and the cracks on the boundary of the aggregate appear and extend; the resulting strain increases quickly. The thermal strain reaches a high value, e.g., ε_{th} = (6–9) × 10^{-3} when T = 500 °C.

When T = 600–700 °C, the increasing rate of the thermal expansion strain slows down or even ceases.[1-11,0-6,2-1] The expansion strain is obstructed possibly because the crystal of the mineral component within the aggregates varies and the internal damage in the concrete accumulates. At this time, the strain ε_{th} is greater than 10 × 10^{-3}.

According to some investigations,[0-7,1-9,1-15] the main factors influencing the thermal expansion strain of concrete are: the types of mineral components in the aggregates, the mix and water content of the concrete, and the heating velocity of the specimen. In addition, differences in the testing methods and the measuring techniques cause greater deviation in the experimental data on the thermal strain of concrete.[2-2]

On the basis of the experimental data provided by Tsinghua University,[1-4,1-6,1-11] the relationship between the average value of freely expanding strain of concrete (ε_{th}) and the temperature (T, °C) can be represented by a simplified regression formula:

$$\varepsilon_{th} = 28\left(\frac{T}{1000}\right)^2 \times 10^{-3} \le 12 \times 10^{-3} \tag{2.1}$$

The upper and the lower bounds of the varying data are also shown in Fig. 2-2, and the bounds are within the varying scope of the experimental data in several countries.[2-2]

In accordance with the definition, the average linear expansion coefficient of concrete should be

$\varepsilon_{th} / (T - 20)$ at elevated temperatures; it can be expressed approximately as

$$\overline{\alpha}_c = \frac{\varepsilon_{th}}{T} = 28\left(\frac{T}{1000}\right) \times 10^{-6}/°C \qquad (2.2)$$

Its value is about (6–20) × 10^{-6}/°C (when <650 °C) and increases as the temperature is elevated.

2.1.2 Deformation During the Heating–Cooling Cycle

A building fire experiences at least one complete cycle of heating and cooling. If several instances of extinguishing and burning occur, the same number of irregular cycles of heating and cooling occur. Correspondingly, the temperatures on the surface and interior of the structural concrete increase and decrease alternately. This influences the different levels of deformation, internal forces (stress), bearing capacity, and the damage to the concrete materials and structures.

When the concrete specimen is heated freely ($\sigma = 0$) to different predetermined temperatures and is then cooled down naturally to room temperature, the thermal (expansion) strains measured are shown in Fig. 2-3.[1-6]

The expansion strains of specimens heated to different maximum testing temperatures increase with increasing acceleration during heating. These heating strain curves coincide and show good repeatability. The length of the specimen shortens during cooling, i.e., the thermal strain reduces, all the cooling strain curves are nearly parallel, and the cooling strain rate is closed (8–12) × 10^{-6}/°C. When the specimen cools down to room temperature, its expansion strain does not vanish completely and residual strain (elongation) exists. The residual strain is small when the maximum testing temperature T < 300 °C; it increases quickly when T > 500 °C, and it reaches 5.2 × 10^{-3} when T = 700 °C, which is about 40% of the expansion strain at the maximum temperature.

The expansion strain (ε_{th}) of concrete during the heating process is composed of four parts: heating expansion of the solid materials, shrinking due to water loss, appearance and extension of cracks on the boundary between the aggregate and cement mortar, and damage to the interior of the aggregates. However, during the cooling process, only the heating expansion of the solid materials may restore completely; the others remain unchanged. Therefore, all the cooling strain curves are parallel in the cycle of free heating and cooling, and the residual strain (elongation) of concrete after it is cooled increases considerably as the concrete reaches the maximum elevated temperature.

When concrete experiences multicycle testing of heating and cooling with the same maximum

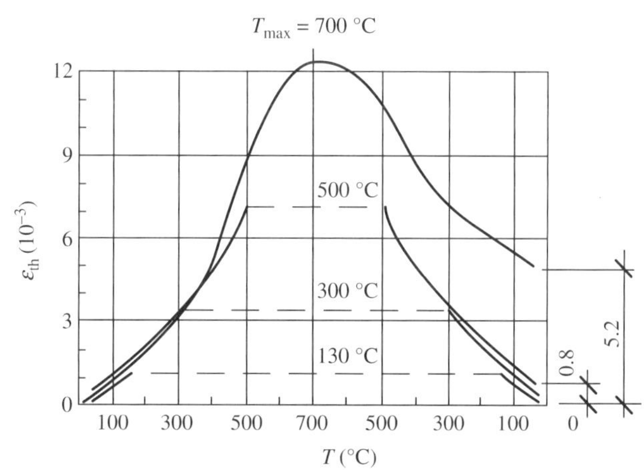

FIGURE 2-3 Thermal strain of concrete during a free heating and cooling cycle.[1-6]

temperature in each cycle, the total expansion strain at the maximum temperature and the residual strain after it is completely cooled increase but tend toward convergence with repeating cycles, and most parts of both strains appear after the first cycle. When concrete experiences multicycle testing of heating and cooling with different maximum temperature in each cycle, the strains measured in the process are shown in Fig. 2-4.

If the maximum temperature in the multicycle testing of heating and cooling increases after each cycle (300 °C to 500 °C to 700 °C), the values of the maximum expansion strain and the residual strain of the concrete specimen accumulate and increase after each cycle. On the contrary, if the maximum temperature in the cycles decreases after each cycle (700 °C to 500 °C to 300 °C), the absolute maximum expansion strain of the specimen occurs at the maximum temperature of the first cycle, and the strain at the maximum temperature of the other cycles decreases gradually. The residual strain of the specimen also appears after the first cooling and remains almost constant after the succeeding cycles.

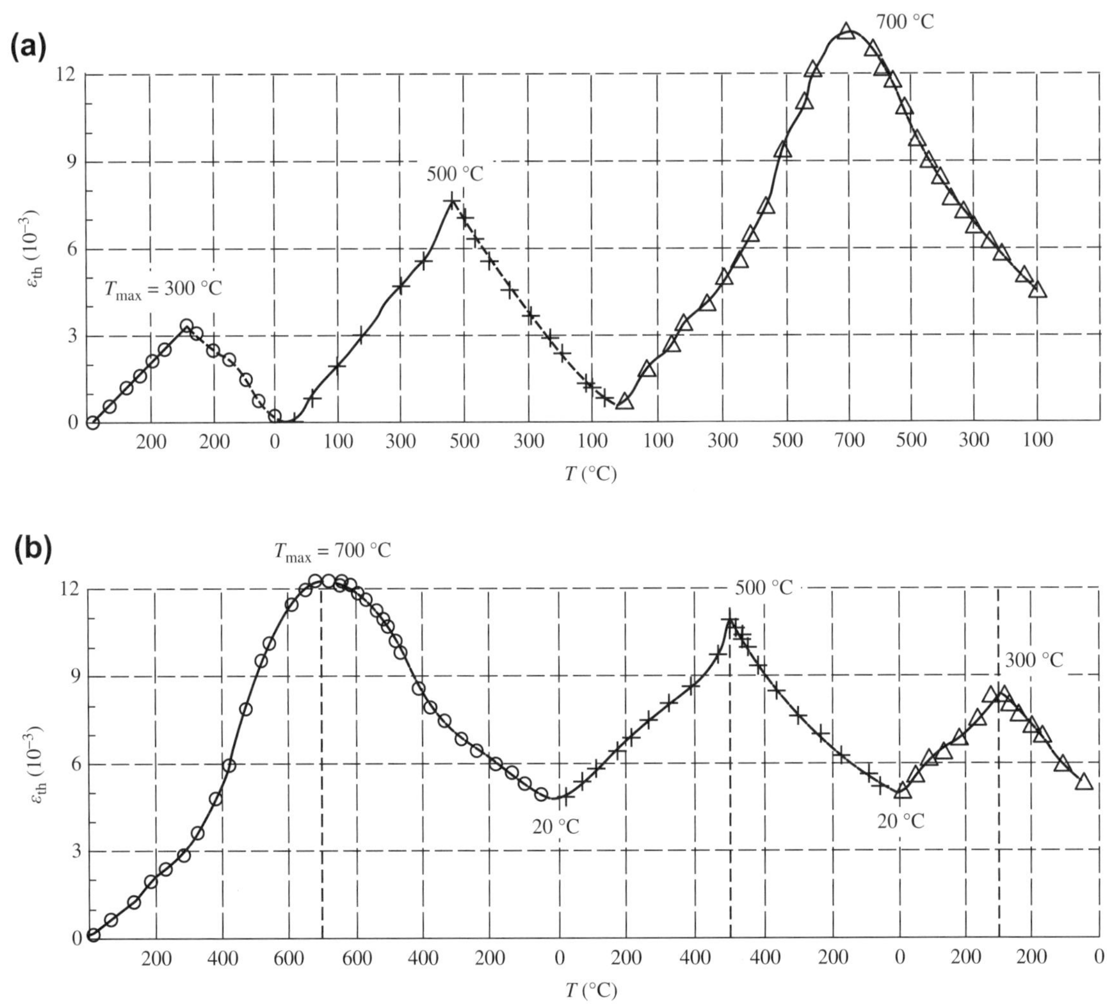

FIGURE 2-4 Thermal strains of concrete during multicycle testing of free heating and cooling[1-6] (a) T_{max}, 300 °C to 500 °C to 700 °C; (b) T_{max}, 700 °C to 500 °C to 300 °C.

These test results indicate an important characteristic of the thermal strain of concrete under free ($\sigma = 0$) conditions. For multicycle testing of heating and cooling, the heating and cooling strain curves of each cycle coincide with that of a single cycle test (Fig. 2-3) regardless of the maximum temperature and the order of every cycle. The residual strain of concrete after cooling depends mainly on the absolute maximum temperature reached in the multicycles, but is independent of the number and order of the multicycles. The mechanism of these phenomena is the same as described above.

2.2 COMPRESSIVE DEFORMATION AND THE STRESS–STRAIN CURVE AT ELEVATED TEMPERATURE

2.2.1 Characteristics of Compressive Deformation

The compressive deformation and the stress–strain relationship in concrete at normal temperatures have been investigated fully, and accurate formulas have been established[0-2,2-4] and can be used for structural analysis. Similar investigations of concrete at elevated temperatures have been conducted worldwide.[1-3,1-9,2-5,2-6] However, the descending branch of the stress–strain curve was not measured and the value of the peak strain was not given in some findings; they are not satisfactory for analyzing structures at elevated temperatures.

A prism specimen measuring 100 mm × 100 mm × 300 mm or 80 mm × 80 mm × 300 mm is used for the deformation test of concrete at elevated temperatures.[1-4,1-6] A set of specimens is put into the preheating furnace, heated, and maintained for 6 h at the predetermined temperature; each specimen is taken out individually and put into the furnace for the deformation test. Then, the deformation transducers are installed (Fig. 1-3) and various measuring instruments are connected, the test machine is operated, and the specimen is loaded with a constant strain velocity. The complete compressive stress–strain curve for concrete at elevated temperatures (Fig. 2-5) is drawn on the *X–Y* function recorder or using isolated measurement pods.

As the concrete at elevated temperatures has low strength and large deformation, the linear stiffness of the descending branch of its stress–strain curve is considerably less than that at normal temperatures and is generally smaller than the stiffness of the test machine. Thus, the complete compressive stress–strain curve is stable without any special stiffening construction attached to the specimen.[0-3]

The compressive stress–strain curve of concrete at elevated temperatures clearly shows that it tends to flatten and its peak obviously drops and moves toward the right-hand side as the testing temperature increases. This means that the compressive strength at elevated temperatures (f_c^T) decreases, the corresponding peak strain (ε_p^T)

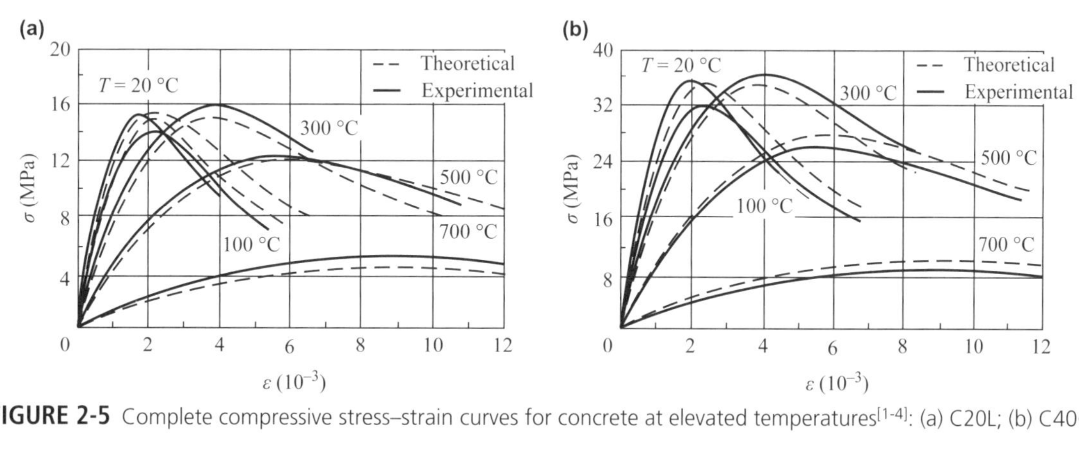

FIGURE 2-5 Complete compressive stress–strain curves for concrete at elevated temperatures[1-4]: (a) C20L; (b) C40G.

increases considerably, and the modulus of elasticity decreases sharply.

During heating and maintaining the temperature, before loading of the specimen, the initial microstress (strain) state and many cracks are formed in the interior of the concrete for various reasons, such as water evaporation, thermal behavior difference between the coarse aggregate and cement mortar, and the expansion and breaking of the aggregate. The deformation process from the start of loading until failure of the specimen can be divided into three stages:

1. When the stress of the specimen is low, e.g., $\sigma/f_c^T \leq 0.4$, the strain of the concrete increases approximately linearly with the stress. But, if the testing temperature is higher (e.g., $T >$ 500 °C), several cracks are formed in the specimen before loading and the stress scope of the initial straight line is reduced.
2. As the stress increases, the plastic deformation develops quickly and the slope of the stress–strain curve decreases gradually because of the cracks formed on the surface and in the interior of the heated specimen. When the stress is at a maximum, i.e., the prism compressive strength at elevated temperatures (f_c^T) is reached, the tangent of the curve is horizontal and the corresponding strain is called the compressive peak strain at elevated temperatures (ε_p^T). As the testing temperature increases, the peak part of the curve tends to flatten and the peak point is not clearly visible.
3. When the stress passes the peak point and enters the descending branch of the curve, the deformation of the specimen increases continuously ($\varepsilon > \varepsilon_y^T$) and internal cracks extend further, but no sudden breaking occurs in the concrete. The bearing capacity of the specimen reduces steadily.

The final failure pattern of the concrete specimen at elevated temperatures (Fig. 2-6) is similar to that at normal temperature. When the testing temperature $T \leq 300$ °C, an inclined main crack appears clearly on the surface of the specimen, the inclined angle is larger, the cracking and breaking area is longer, and irregular cracks are distributed on the other parts of the surface. When the testing temperature $T > 500$ °C, most of the specimens break into two pieces after failure and

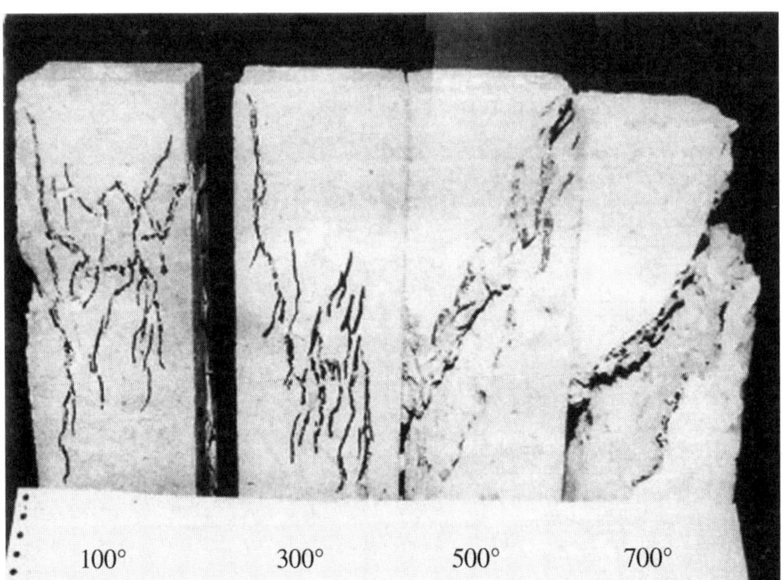

FIGURE 2-6 Compressive failure pattern of a prism specimen at elevated temperatures.[1-5]

cannot be taken out intact from the chamber of the furnace. The angle of the inclined main crack decreases but the inclined breaking belt is wider, and the dregs fall down from both sides of the crack. The irregular cracks on the other parts of the surface become wider.

An important phenomenon is shown in Fig. 2-5. When the strains of the specimens have the same value and are large enough, the specimen being tested at a lower temperature enters the stress descending branch, while the specimen being tested at a higher temperature is still located in the ascending branch, or the flatter part, of the descending branch. So, the bearing capacity of the latter may exceed that of the former.

2.2.2 Prismatic Compressive Strength and Corresponding Strain

The ordinate and abscissa of the peak point in the compressive stress–strain curve of concrete at elevated temperatures are the prismatic compressive strength (f_c^T) and corresponding strain (ε_p^T), respectively. All the experimental data[1-4,1-6,1-11,8-3,9-3] obtained from tests of concrete material under the same conditions and using the same facilities (Fig. 1-1), are shown in Figs 2-7 and 2-8.

Comparing Fig. 2-7 with Fig. 1-7, it is found that the variation regularity of the prismatic compressive strength of concrete is similar to that of the cubic compressive strength as the testing temperature increases. The prismatic compressive strength reduces when T = 20–100 °C, but increases slightly when T = 100–300 °C, and reduces monotonically and quickly when T > 400 °C.

Comparing the relative values of both strengths of concrete, the prismatic strength (f_c^T/f_c) is obviously lower than the cubic strength (f_{cu}^T/f_{cu}). The differences between them are 10–20% within T = 400–700 °C, and 2–10% within other temperature ranges (T < 400 °C and T > 700 °C). This shows that high temperatures cause internal damage in concrete and have a more prominent influence on the mechanical behavior of the prism specimen, resulting in more loss and deviation of the strength.

The prismatic compressive strength of concrete varies with temperature (T, °C) and can be expressed by the similar empirical formula

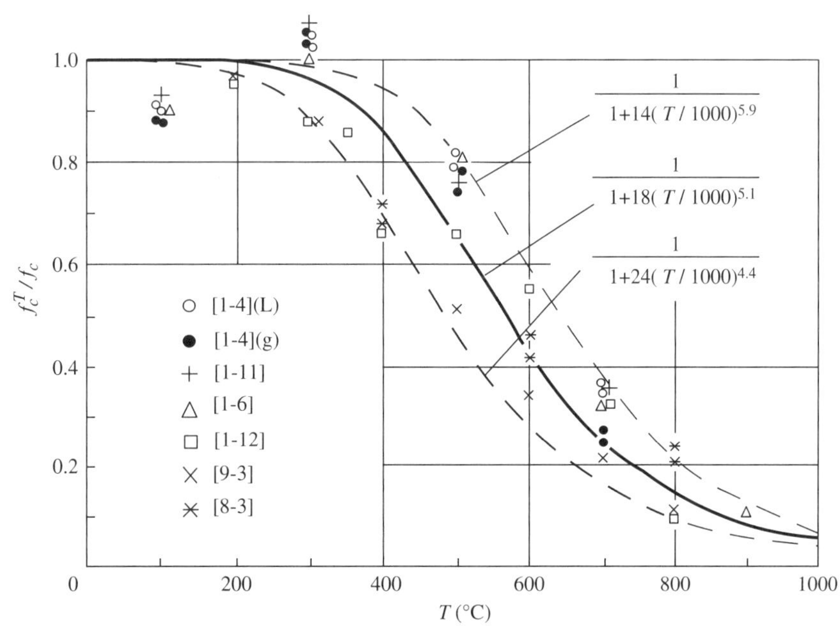

FIGURE 2-7 Prismatic compressive strength of concrete at elevated temperatures.

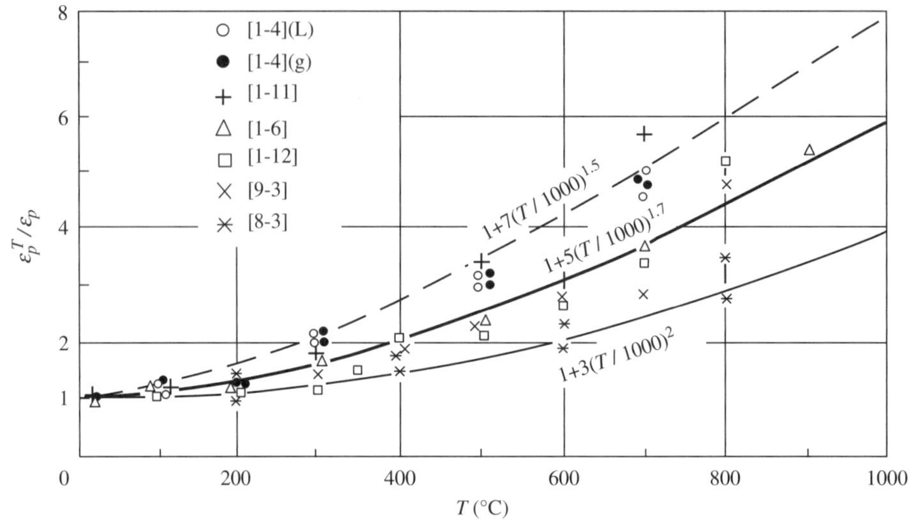

FIGURE 2-8 Compressive peak strain of concrete at elevated temperatures.

(Eqn (1.1)) used for cubic strength. The parameters in the formula can be calibrated from the experimental data and the formula is obtained as follows:

$$\frac{f_c^T}{f_c} = \frac{1}{1 + 18\left(\frac{T}{1000}\right)^{5.1}} \qquad (2.3)$$

The experimental data deviates and the envelopes of the upper and the lower bounds are also shown in Fig. 2-7. The formula used in the references listed in Fig. 2-7 is the same as the note after Eqn (1.1), and various theoretical curves fall inside both bounds.

The ratio between the prismatic and cubic compressive strength of concrete varies with temperature and its value can be derived from Eqns (1.1) and (2.3):

$$\frac{f_c^T}{f_{cu}^T} = \frac{1 + 16\left(\frac{T}{1000}\right)^{6.3}}{1 + 18\left(\frac{T}{1000}\right)^{5.1}} \cdot \frac{f_c}{f_{cu}} \le \frac{f_c}{f_{cu}} \qquad (2.4)$$

The value of the ratio fluctuates as the testing temperature increases, but is always less than that at normal temperatures.

The compressive peak strain of concrete at elevated temperatures accelerates with the testing temperature (T, °C). The ratio between it and the value at normal temperatures can be calculated from

$$\frac{\varepsilon_p^T}{\varepsilon_p} = 1 + 5\left(\frac{T}{1000}\right)^{1.7} \qquad (2.5)$$

The experimental data fluctuate and the upper and the lower bounds of the envelopes are shown in Fig. 2-8.

2.2.3 Equation of a Complete Stress–Strain Curve

The measured stress–strain curves of the specimens are modified from the standard curves, in which the relative peak stress and strain are $\sigma / f_c^T = \varepsilon / \varepsilon_p^T = 1$. The standard curves of the four types of concrete (Table 1-1) at different testing temperatures approach one another.[1-4] Therefore, the complete stress–strain curves of the concrete at elevated and normal temperatures can be expressed reasonably by the same equation.

In Guo,[0-2] the loading and deforming processes of the concrete at normal temperature are analyzed and the multinomial of the third power and the fractional expression are suggested, respectively, for the ascending and descending branches of the complete stress–strain curve, and both branches continue at the peak point:

$$\left.\begin{array}{l} x \le 1,\ y = ax + (3-2a)x^2 + (a-2)x^3 \\ x \ge 1,\ y = \dfrac{x}{\alpha(x-1)^2 + x} \end{array}\right\} \quad (2.6)$$

where $y = \sigma / f_c$ and $x = \varepsilon / \varepsilon_p$. This equation satisfies all the boundaries and geometrical conditions of the experimental curves:

- $x = 0,\ y = 0$.
- $d^2y/dx^2 < 0$ when $0 \le x \le 1$, which shows that the slope of the tangent dy/dx decreases monotonically and no point of inflection will appear on the ascending branch.
- single peak, i.e., $y = 1$ and $dy/dx = 0$ when $x = 1$.
- one point of inflection $d^2y/dx^2 = 0$ appears on the descending branch ($x > 1$), where the slope of the tangent has the maximum (absolute) value.
- $y = 0$ and $dy / dx = 0$ when $x \to \infty$.
- $0 \le y \le 1$ when $x \ge 0$.

In Eqn (2.6), there is one parameter for each branch, i.e., a and α for the formulas of the ascending and descending branches, respectively. The values of these parameters can be calibrated by the experimental data of concrete at elevated temperatures and the following formulas are obtained:

Let

$$y = \frac{\sigma}{f_c^T},\ x = \frac{\varepsilon}{\varepsilon_p^T} \quad (2.7)$$

when

$$\left.\begin{array}{l} x \le 1:\ y = 2.2x - 1.4x^2 + 0.2x^3 \\ x \ge 1:\ y = \dfrac{x}{0.8(x-1)^2 + x} \end{array}\right\} \quad (2.8)$$

where σ and ε are the stress and strain of concrete at elevated temperatures, respectively. f_c^T and ε_p^T are the compressive prismatic strength and corresponding peak strain of the concrete, respectively, at temperature T °C and they can be calculated from Eqns (2.3) and (2.5).

The theoretical curves calculated from these formulas fit well with the experimental results (Fig. 2-5).

2.2.4 Initial Elastic Modulus and Secant Modulus at Peak Stress

The initial elastic modulus of concrete at elevated temperatures (E_0^T), just like that at normal temperatures,[0-2] is defined as the ratio between the stress $\sigma = 0.4 f_c^T$ and the corresponding strain ε, or the secant slope at this point on a measured stress–strain curve. The secant modulus at peak stress is the ratio between the prismatic compressive strength and the corresponding peak strain, i.e., $E_p^T = f_c^T / \varepsilon_p^T$.

The initial elastic modulus and the secant modulus at peak stress of the concrete vary with the testing temperature and are shown in Fig. 2-9. The reducing amplitudes of the moduli of concrete obviously exceed that of the compressive strength of the concrete at the same temperature, i.e., $(E_0^T / E_0, E_p^T / E_p) < (f_{cu}^T / f_{cu}, f_c^T / f_c)$. The reason for this is that the decrease in the strength and the increase in the peak strain of concrete occur simultaneously, and the ratio between them has to reduce quickly.

Because the ascending branch of the compressive stress–strain curve of concrete at elevated temperatures adopts the same formula (Eqn (2.8)) as that used for concrete at normal temperature, the ratio between the initial elastic modulus and the secant modulus at peak stress is certainly a constant and is independent of temperature. The solution of Eqn (2.8) is $x = 0.209$ when $y = 0.4$, so

$$\frac{E_0^T}{E_p^T} = \frac{E_0}{E_p} = \frac{0.4}{0.209} = 1.914 = \text{const.} \quad (2.9)$$

This conforms to the experimental data 1.87–1.96[1-4] for various specimens.

Both the initial elastic modulus and the secant modulus at peak stress of concrete decrease as the testing temperature increases and they have the same variation regularity. After modifying Eqn (2.9), the following formula is derived:

$$\frac{E_0^T}{E_0} = \frac{E_p^T}{E_p} = \frac{f_c^T / \varepsilon_p^T}{f_c / \varepsilon_p} = \frac{f_c^T / f_c}{\varepsilon_p^T / \varepsilon_p} \quad (2.10)$$

When the numerator and the denominator of the formula are replaced by Eqns (2.3) and (2.5), respectively, the formula for both moduli

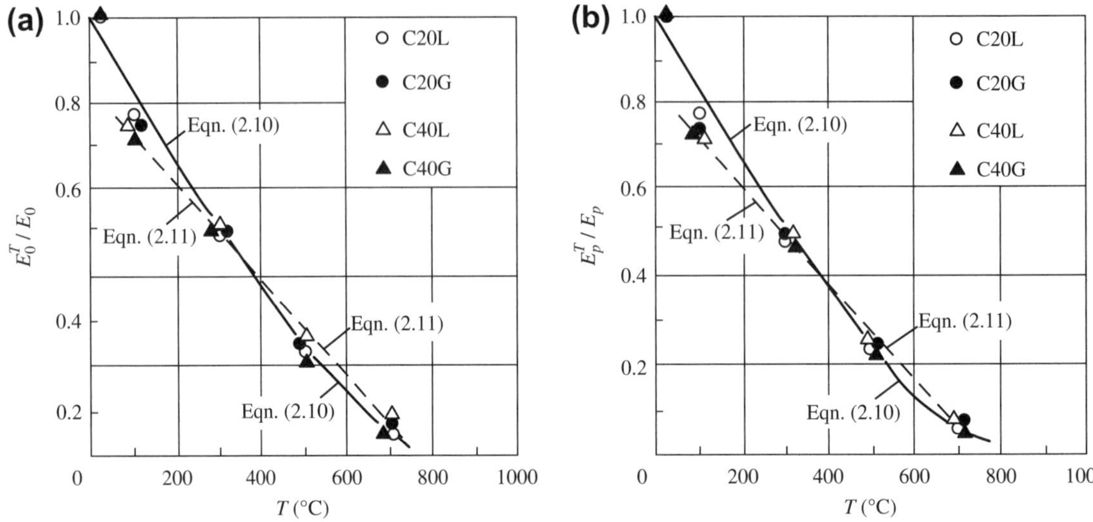

FIGURE 2-9 Initial elastic modulus and secant modulus at peak stress of concrete at elevated temperatures[1-5]: (a) E_0^T / E_0; (b) E_p^T / E_p.

is obtained and the corresponding theoretical curves are shown by the solid lines in Fig. 2-9.

In order to simplify the calculation, the variation in the initial elastic modulus and the secant modulus at peak stress of concrete at different temperature can be expressed by a linear equation:

$$60\ °\text{C} \le T \le 700\ °\text{C}: \frac{E_0^T}{E_0} = \frac{E_p^T}{E_p} = 0.83 - 0.0011T \quad (2.11)$$

The calculated results are shown by the dashed lines in Fig. 2-9 and approach the experimental results and Eqn (2.10).

During the cooling process, both moduli of the concrete basically maintain the values at elevated temperatures (E_0^T, E_p^T)[2-7] and may not recover. This is similar to the compressive strength of concrete and the cause is also the same.

2.3 STRESS–STRAIN CURVES UNDER REPEATED LOADING

2.3.1 Envelope and Loci of the Common Point and the Stability Point

The internal forces and stress redistribute extensively in a statically indeterminate structure at elevated temperatures (see Chapter 10); even the value of the direction (positive or negative) of the bending moment on a cross-section may change alternately and the stress of concrete may also alternate. The strength and deformation behavior of concrete at elevated temperatures and subjected to repeated load (compressive stress) is introduced in this section.

A prismatic concrete specimen is put into the preheating furnace and heated to the predetermined temperature, which is maintained for 6 h; then it is moved into the furnace for the deformation test (Fig. 1-1). Two types of repeated loading–unloading (axial compression) tests are performed after adjusting and maintaining the value of the temperature.

- One cycle of complete loading and unloading is conducted at the predetermined strain, which is increased by equal increments. When the predetermined strain value is reached after loading the specimen, the load (stress) is released to zero. Then the specimen is reloaded until the next predetermined strain on the envelope is reached.
- Multicycles of complete loading and unloading are conducted at the predetermined strain, which is increased by equal increments. In each

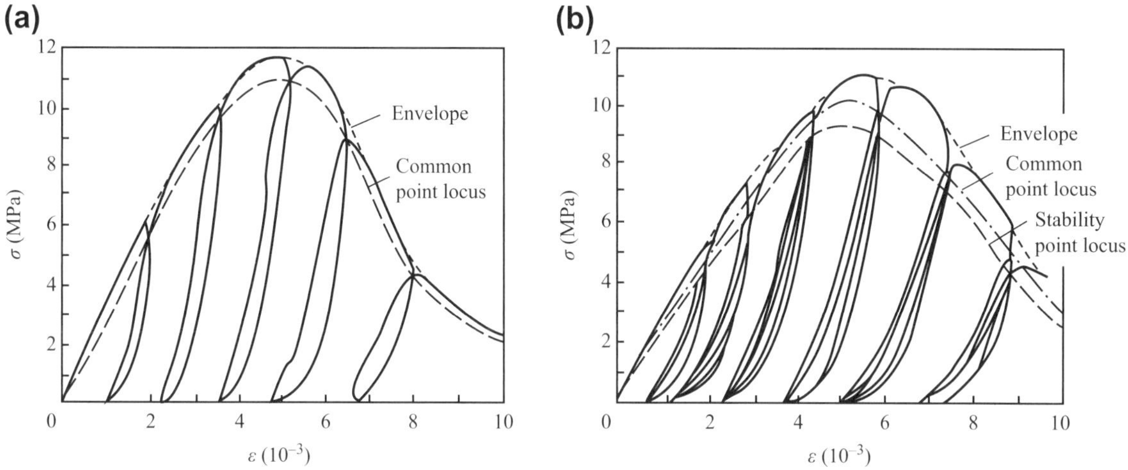

FIGURE 2-10 Compressive stress–strain curves of concrete at elevated temperature and under repeated loading.[1-6] (a) Loading and unloading with equal strain increment; (b) loading and unloading cycles with equal strain increment.

unloading, the stress is released to zero. When the value of the stress at the predetermined strain after each loading does not decrease, the residual strain after each unloading no longer increases, and the loading and unloading curves tend to be stable, no more cycles at this predetermined strain are conducted. The measured stress–strain process during testing ($T = 500$ °C) is shown in Fig. 2-10.

There are three loci found on the stress–strain curves of concrete specimens under repeated loading and they have definite physical meaning.

- The envelope is an outline along the outer boundary of all the curves. It shows the maximum stress or strength of the concrete at corresponding strain values and coincides in both shape and value with the complete stress–strain curve under monotonic loading (Fig. 2-5).
- The locus of the common point. When the specimen is unloaded until its stress is reduced to zero from any point on the envelope and is then reloaded, the intersecting point of the unloading and reloading curves is called the common point. The slope of the reloading curve passes the common point and reduces considerably and the strain increases quickly. This shows that new damage occurs in the interior of the concrete. All the common points of unloading–reloading curves are connected smoothly and the locus of the common point is composed.
- The locus of the stability point. After the specimen is unloaded and reloaded at the predetermined strain for several cycles, and when the stress (strength) of the concrete does not reduce and the residual strain does not increase, the unloading–reloading curves compose a stable closed ring, the top of which is called the stability point. All the stability points in the figure are connected and the locus of the stability point is obtained. It is equivalent to the ultimate envelope of fatigue strength of concrete under low cycles.

All the envelopes and the loci of the common point and stability point obtained from repeated loading–unloading testing of concrete at elevated temperatures are compared carefully with the complete stress–strain curve obtained from the monotonic loading (Fig. 2-5). They are similar in shape and the same formula (Eqn 2.8) can be used to calculate both. The similarity ratios are, respectively:

$$\begin{aligned} &\text{Envelope}: K_e = 1 \\ &\text{Locus of the common point}: K_c = 0.90\text{–}0.94 \\ &\text{Locus of the stability point}: K_s = 0.82\text{–}0.88 \end{aligned} \quad (2.12)$$

These phenomena and conclusions coincide with the repeated loading test of concrete at normal temperature. Under repeated loading, the relative strength and deformation behavior of concrete at elevated temperatures are similar to that at normal temperatures and no over-damage occurs in the concrete.

2.3.2 Formulas for the Unloading and Reloading Curves

There are two kinds of stress–strain curves, besides the envelope, for concrete under repeated loading: the unloading curve when the stress decreases to zero from the envelope and the reloading curve when the stress increases from zero to the tangent point on the envelope. The shapes and variation regularities of these curves are similar to that at normal temperatures,[2-6] but the parameters in the formulas are different.

1. Unloading curve

When the stress is decreased to zero from any point (ε_u and σ_u) on the envelope, the residual strain (ε_s) is obtained. The measured ε_u and ε_s from each unloading curve of the specimens are expressed as the relative strains and drawn in Fig. 2-11 with the coordinates:

$$x_u = \frac{\varepsilon_u}{\varepsilon_p} \quad \text{and} \quad y_s = \frac{\varepsilon_s}{\varepsilon_p} \tag{2.13}$$

where ε_p is the peak strain at the corresponding temperature. The residual strain ε_s increases monotonically with the unloading strain ε_u, and also increases as the testing temperature (T, °C) increases. The variation regularity can be represented by a simplified empirical regression formula:

$$\left.\begin{aligned} x_u \le 1: \; y_s &= \left[0.5\left(\frac{T}{1000}\right) + 0.44\right] x_u^{1.36} \\ x_u \ge 1: \; y_s &= x_u + 0.5\left(\frac{T}{1000}\right) - 0.56 \end{aligned}\right\} \tag{2.14}$$

When the unloading curves are expressed by the relative values of the stress and strain

$$\xi_u = \frac{\varepsilon - \varepsilon_s}{\varepsilon_u - \varepsilon_s} \quad \text{and} \quad \eta_u = \frac{\sigma}{\sigma_u} \tag{2.15}$$

the standardized curve, which starts from the point (1,1) and unloads at the point (0,0), is obtained. Analyzing and comparing the standardized unloading curves of the specimens at different testing temperatures, the curves can be compared using a power function:

$$\eta_u = \xi_u^n \quad \text{or} \quad \left(\frac{\sigma}{\sigma_u}\right) = \left(\frac{\varepsilon - \varepsilon_s}{\varepsilon_u - \varepsilon_s}\right)^n \tag{2.16}$$

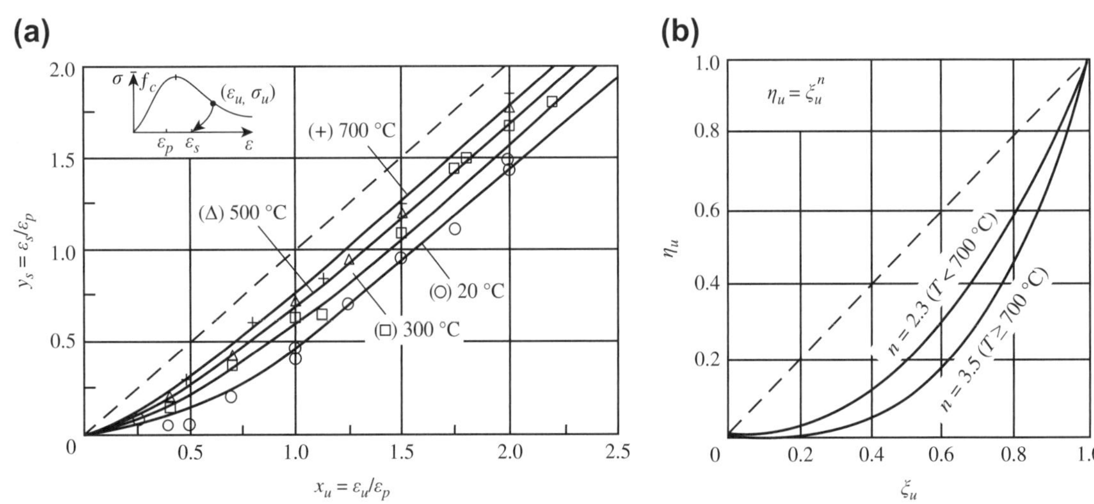

FIGURE 2-11 Unloading curve and its calculation parameters.[1-6] (a) Unloading strain and residual strain; (b) theoretical unloading curves.

where n is a parameter. It takes the value of $n = 2.3$ when $T < 700$ °C or $n = 3.5$ when $T \geq 700$ °C. The theoretical curves are shown in Fig. 2-11(b).

2. Reloading curve
Starting from any strain (ε_s) with zero stress, the specimen is reloaded until the reloading curve is tangential to and coincides with the envelope. The tangential point is the end of the reloading curve and its coordinates are (ε_r, σ_r). The measured ε_s and ε_r from each reloading curve of the specimens are expressed as the relative strains and are drawn in Fig. 2-12 with the coordinates:

$$x_s = \frac{\varepsilon_s}{\varepsilon_p}, \quad y_r = \frac{\varepsilon_r}{\varepsilon_p} \tag{2.17}$$

The end strain ε_r increases monotonically with the starting strain ε_s, but decreases as the testing temperature (T, °C) increases. The variation regularity can be represented by a simplified empirical regression formula:

$$\left.\begin{aligned} x_s \leq 0.4: \ y_r &= \left[2.124 - 1.2\left(\frac{T}{1000}\right)\right] x_s^{0.74} \\ x_s > 0.4: \ y_r &= 1.05x_s - 0.6\left(\frac{T}{1000}\right) + 0.66 \end{aligned}\right\} \tag{2.18}$$

The reloading curve is expressed by the relative values of the stress and strain

$$\xi_r = \frac{\varepsilon - \varepsilon_s}{\varepsilon_r - \varepsilon_s} \text{ and } \eta_r = \frac{\sigma}{\sigma_r} \tag{2.19}$$

and the standardized curve is obtained. The end point of the reloading stress–strain curve has to coincide with the envelope. When $\varepsilon_r < \varepsilon_p$, i.e., the curve is tangential to the ascending branch of the envelope, the slope at the end point is $d\eta_r / d\xi_r > 0$. When $\varepsilon_r > \varepsilon_p$, i.e., the curve is tangential to the descending branch of the envelope, the slope at the end point is $d\eta_r/d\xi_r < 0$. These two kinds of reloading curves must be distinguished and have different empirical formulas:

$$\left.\begin{aligned} \varepsilon_r \leq \varepsilon_p: \ \eta_r &= \xi_r^{1.4}[1 + 0.3\sin(\pi\xi_r)] \\ \varepsilon_r > \varepsilon_p: \ \eta_r &= \xi_r^{1.4}[1 + 0.6\sin(\pi\xi_r)] \end{aligned}\right\} \tag{2.20}$$

The theoretical curves are shown in Fig. 2-12.

2.4 SHORT TIME CREEP AT ELEVATED TEMPERATURE

2.4.1 Creep Under Constant Temperature and Stress

When concrete is stressed, the strain (ε_c) appears instantaneously. In addition, if the stress is sustained, the strain increases continuously for the duration and is called creep (ε_{cr}). Usually, the increasing rate of the compressive creep of concrete under ambient temperatures decreases gradually with for the time (t, day) that the stress (σ, MPa) is sustained. The creep develops quickly within the first 3 months then slowly after 2 or 3 years. However, a small increment in the creep can be measured even after 20–30 years. The ultimate convergence value of the creep is called the ultimate creep $\varepsilon_{cr}(\infty)$, and the ultimate creep under unit stress is called the specific ultimate creep $c_{cr}(\infty) = \varepsilon_{cr}(\infty)/\sigma(1/\text{MPa})$. The ratio between the ultimate creep and the instantaneous strain is called the ultimate creep coefficient $\phi(\infty) = \varepsilon_{cr}(\infty)/\varepsilon_c(t_0)$.

The value and variation regularity of concrete creep are influenced by the following main factors:

- Stress level (σ/f_c),
- Sustaining time (t) of the stress,
- Concrete age when loading,
- Kind and mix of the raw materials,
- Environmental temperature and humidity within the service period, and
- The shape and size of the cross-section of the member.[0-2]

It is found from the experimental data that concrete creep deviates considerably and the range of the specific ultimate creep $c_{cr}(\infty) = (10–140) \times 10^{-6}$/MPa, with an average value of about 70×10^{-6}/MPa. The ultimate creep coefficient is about $\phi(\infty) = 2–4$.

When a building fire occurs, the sustaining time is generally a few hours (1–4 h; see Section 5.1) depending on the conditions of combustion and the extinguishing method. Therefore, the concrete structure will support the load for a

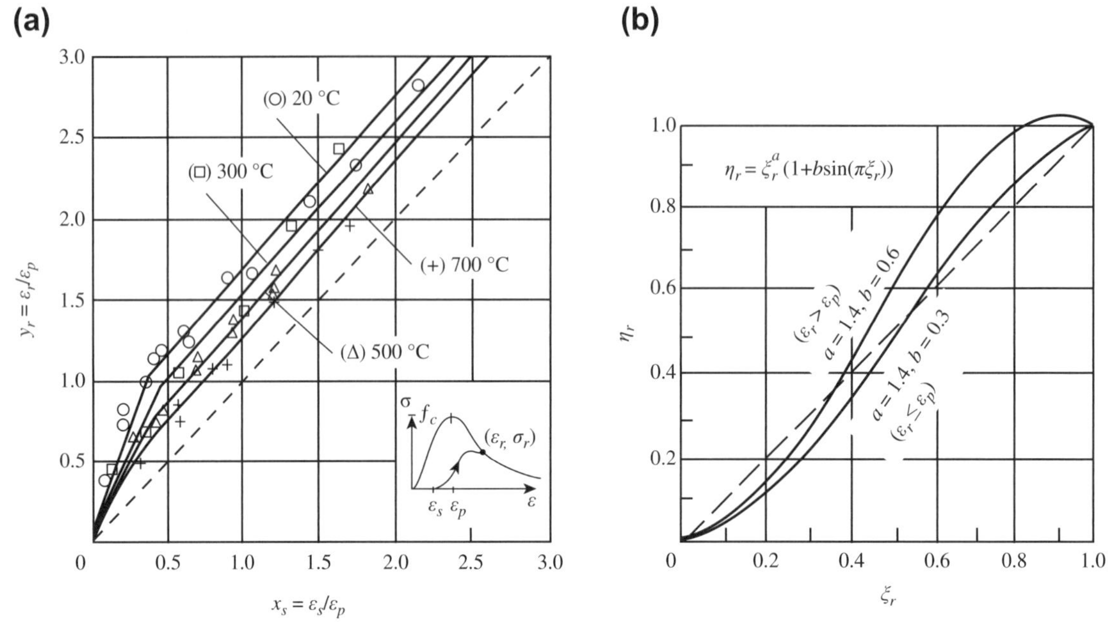

FIGURE 2-12 Reloading curve and its calculation parameters.[1-6] (a) Starting strain and end strain; (b) theoretical reloading curves.

few hours at elevated temperatures. Under these conditions, the short time creep of concrete at elevated temperatures is different from that at normal temperatures.

First, the short time creep test of concrete under basic conditions, i.e., constant temperature and stress, is introduced. The prismatic specimen is used for testing. It is put into the furnace for the deformation test after being heated to the predetermined temperature and maintained at that temperature for 6 h, and it is then loaded with the predetermined stress. The creep of the specimen is measured for 2–3 h under constant temperature and stress.[1-6] Some of the experimental results are shown in Fig. 2-13.

It is seen from the figures that the short time creep of concrete at elevated temperatures has a large value and is greater than that at normal temperatures for several decades. However, the variation regularities are similar. The creep at elevated temperatures develops quickly in the early stages of sustaining a load, and the increasing rate of creep decreases gradually afterward. The higher the temperature and stress of the specimen, the longer the convergence time needed for creep. At the same temperature, the value of the creep is approximately proportional to the stress level when $\sigma/f_c^T \le 0.6$. It is called linear creep, and the specific creep limit is a constant for the concrete. When the stress is higher ($\sigma > 0.6f_c^T$), the experimental data deviate seriously because of the difficulty in controlling the testing. The maximum stress that the specimen can bear at that temperature is the ultimate strength of concrete at elevated temperatures for a long period. The increase in the value of creep accelerates with the testing temperature at the same relative stress level (σ/f_c^T, rather than σ/f_c). If the temperature is even higher, the creep of the specimen will diverge and cause failure of the specimen.

The short time creep of concrete at elevated temperatures, especially when $T > 500$ °C, has considerable value and develops quickly, because the internal cracks expand and elongate enormously. The main factors influencing the creep are the temperature, the level and duration of the stress, and the types and mix of the aggregate.[2-9] These cause large deviations in the data measured

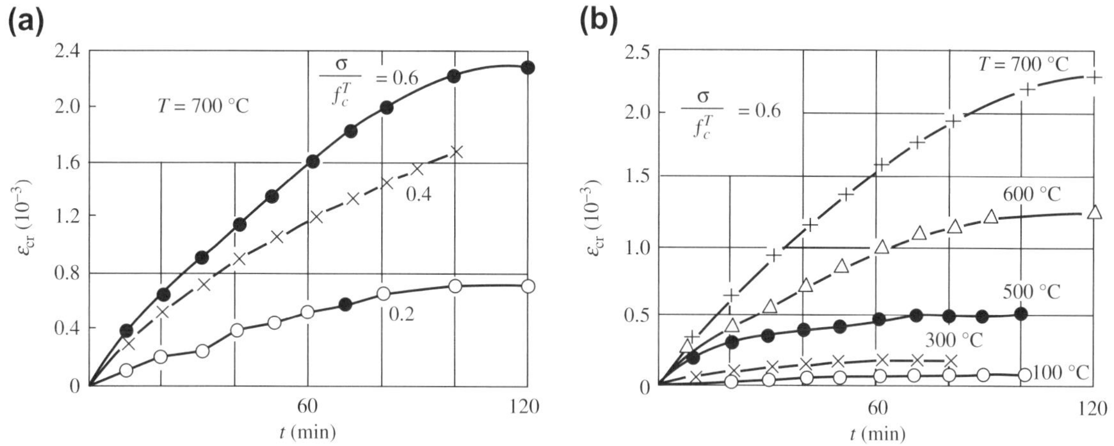

FIGURE 2-13 Short time creep of concrete at elevated temperatures.[1-6] (a) For different stress ratios; (b) at different temperatures.

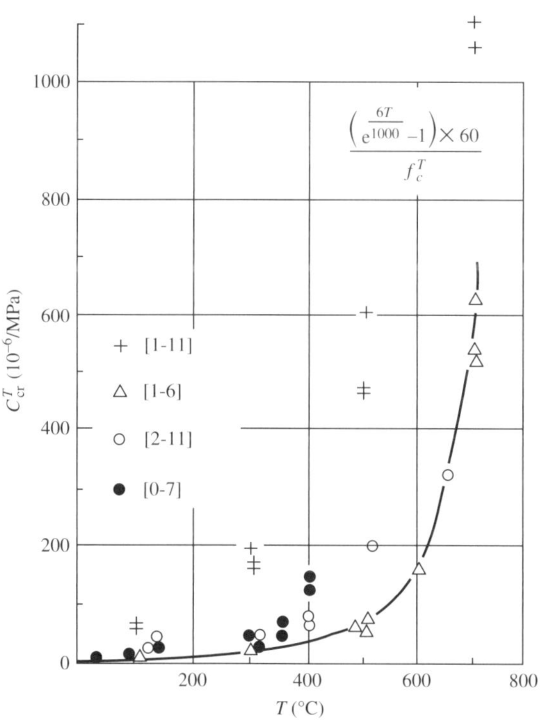

FIGURE 2-14 Relation between specific creep of short duration and temperature.

from creep tests, and the differences may be several times higher,[1-11,2-10,2-11] and about the same as that at normal temperatures. In addition, there are two other important reasons: no unified testing standard exists and measuring strain has technical difficulties for the testing of concrete at elevated temperatures.

The definition of the specific creep of concrete at elevated temperatures for a short time period is the creep value that occurs within 2 h when the specimen is subjected to unit stress (1 MPa) under constant temperature (T, °C):

$$c_{cr}^T = \frac{\varepsilon_{cr}(2)}{\sigma}(1\text{MPa}) \tag{2.21}$$

All the relevant experimental data are shown in Fig. 2-14. The specific creep of concrete at elevated temperatures for a short period of time is quite low when the testing temperature $T < 400$ °C; it then increases rapidly and is far greater than that at normal temperatures when $T > 400$ °C.

As the compressive strength of concrete at elevated temperatures varies considerably, the stress level σ/f_c^T can better represent the mechanical damage level of concrete at different temperatures. Therefore, a dimensionless parameter of specific creep is introduced and its definition is the creep value, which occurs within 2 h when the specimen is subjected to unit stress level at constant temperature:

$$c^T = \frac{\varepsilon_{cr}(2)}{\sigma/f_c^T} = c_{cr}^T f_c^T \tag{2.22}$$

A simplified regression formula is obtained for the parameter from the experimental data provided by Nan[1-6]:

$$c^T = (e^{6T/1000} - 1) \times 60 \times 10^{-6} \quad (2.23)$$

The short time creep of concrete varies with the temperature and stress duration (t) and is approximately proportional to $\sqrt{t/t_0}$, where t_0 equals 2 h or 120 minutes, i.e., the duration used to determine the parameter of short time creep. Then, the calculation formula for short time creep of concrete at elevated temperatures can be derived:

$$\frac{\sigma}{f_c^T} \leq 0.6, \varepsilon_{cr}(t) = \frac{\sigma}{f_c^T}\sqrt{\frac{t}{t_0}}(e^{T/1000} - 1) \times 60 \times 10^{-6} \quad (2.24)$$

The results calculated by this formula are compared with the experimental data in Fig. 2-14 and fit well with those provided by Nan,[1-6] but are lower than other data. If more reliable data are accumulated, the parameters in Eqn (2.24) may be modified to calculate the short time creep of concrete at elevated temperatures more accurately.

2.4.2 Creep Under Variable Temperature and Stress

When a structure in engineering practice is subjected to fire, the temperature and stress of the concrete within it vary with the duration of the fire; the variation in short time creep of concrete is complex.

The short time creep of concrete was investigated at a constant temperature but variable stress,[1-6] and some of the experimental results are shown in Fig. 2-15. When the stress acting on the concrete varies and the succeeding stress is increased, the increasing rate of creep accelerates suddenly and the creep curve turns away obviously. When the succeeding stress remains constant, the increasing rate of creep decreases gradually and the curve tends to flatten. The creep curve of concrete subjected to succeeding stress approaches that at the beginning for the same temperature and stress (see Fig. 2-13). When the stress decreases, the value of creep remains almost constant and the creep curve tends to be a horizontal line. Thus, the total creep of concrete is mainly controlled by the maximum temperature and stress experienced.

According to the conclusions of existing experimental investigations, the creep–time curve of concrete at constant temperature and stress can be assumed to be unique, i.e., it is not influenced by a different temperature–stress history, when the short time creep of concrete at elevated temperatures is calculated. Considering the deviation of the creep at elevated temperatures, the approximate theory of elastic

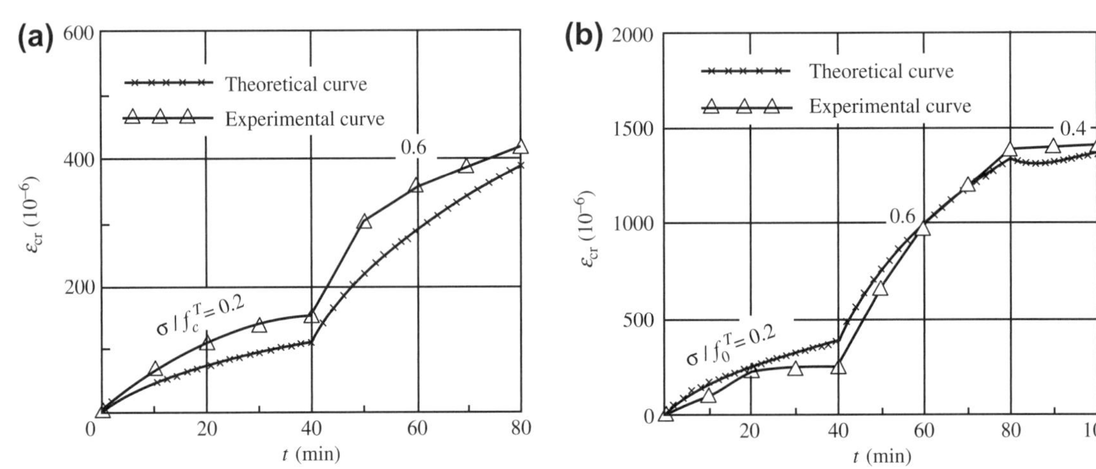

FIGURE 2-15 Short time creep of concrete under constant temperature and variable stress conditions[1-6]: (a) $T = 500$ °C; (b) $T = 700$ °C.

creep can be used for accumulative calculations. This means that the creep increment within every time period with the same temperature and stress can be calculated separately and then accumulated to obtain the total value of creep. An example is shown in Fig. 2-16 to explain this:

- The temperature–stress history experienced by the concrete is determined and divided into several time periods with the same temperature and stress.
- The unique creep–time (ε_{cr}–t) curve and the corresponding formula, e.g., Eqn (2.24), are decided for each time period of the temperature–stress (T, σ / f_c^T) condition.
- The creep ($\varepsilon_{cr,1}$) that appears within the first time period (t = 0–t_1) is calculated and the curve segment *oa* is obtained.
- The creep increment within the second time period ($t = t_1$–t_2) is calculated as follows. First, the equivalent time t_1^{eq} (point *a*′) corresponding to creep $\varepsilon_{cr,1}$ is found on the relevant creep–time curve (T_2, σ_2/f_{c2}^T). Then the creep increment within the time increment (t_2–t_1) and the total creep $\varepsilon_{cr,2}$ (point *b*′) at t_2 are calculated. The curve segment *a*′*b*′ is moved in a parallel fashion and creep segment *ab* is obtained.
- Similarly, the creep increment occurring within the third time period (t_2–t_3) and the total creep $\varepsilon_{cr,3}$ at t_3 are calculated, and the

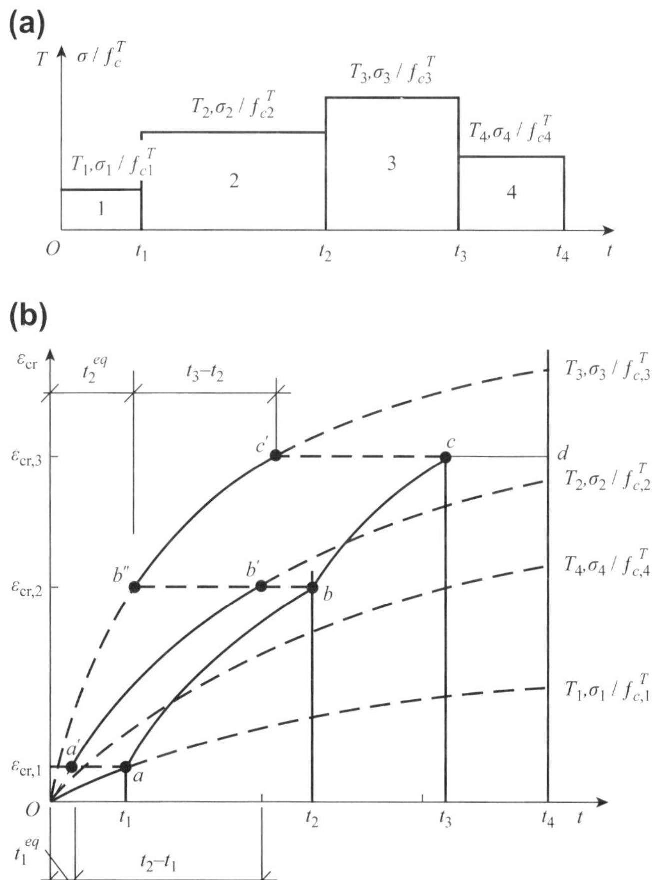

FIGURE 2-16 Short time creep of concrete under variable temperature and stress conditions. (a) Variations of temperature and stress with time; (b) calculation scheme.

curve segment $b''c'$ is obtained and moved in a parallel fashion to *bc*.

- Because the values of the temperature and stress within the fourth time period (t_3–t_4) are lower, the intersection of the equivalent time cannot be found on the corresponding creep–time curve $(T_4, \sigma_4/f_{c4}^T)$. It is assumed that the creep will not increase and a horizontal line *cd* is obtained.
- The segments of the creep–time curves within all time periods are connected one by one; the curve *abcd* of several segments, i.e., the total creep–time curve, is completed.

Obviously, the method and procedure described above can also be used for calculating the short time creep of concrete under other conditions, e.g., constant temperature and stress, constant temperature but variable stress, and variable temperature but constant stress. However, if the concrete under stress is heated and its temperature is elevated, the freely expanding strain (ε_{th}) and the transient thermal strain (ε_{tr}) (see Chapter 3) also have to be calculated.

CONCLUSIONS

The deformation of concrete at elevated temperatures includes the free heating strain, strain caused by the stress load, and short time creep. These strains accelerate with increasing temperature. The value of concrete deformation is considerable when the temperature $T > 400$–500 °C and is several times, even several tens of times, that at normal temperature. It is also far greater than the peak compressive strain (about 2×10^{-3}) of concrete at normal temperature.

Concrete at elevated temperatures shows large deformations mainly caused by cracks spreading and extending into the interior of the cement mortar and on the boundary of the aggregate and damage within the aggregate. Generally, this damage in the concrete does not recover during the cooling and unloading process, so the residual strain is still large after the concrete has cooled and unloaded.

When the concrete experiences several cycles of heating–cooling and loading–unloading at the same maximum temperature and stress, the maximum strain and residual strain appear in the first cycle and the strain increments occurring in the succeeding cycles decrease sharply. If the values of the maximum temperature and stress of the latter cycles are greater than in the previous cycles, the maximum strain and residual strain in this cycle increase notably. On the contrary, if the values of the maximum temperature and stress of the latter cycle are less, the maximum strain does not exceed the maximum value reached previously and the residual strain remains constant. Therefore, the maximum strain and residual strain of concrete depend on the maximum values of temperature and stress ever experienced within all the heating–cooling and loading–unloading cycles.

Many types of deformation behaviors of concrete at elevated temperatures have the same regularity as that at normal temperatures. Therefore, some important concepts and analysis methods, including the formulas, can be used for reference, e.g., the formulas for the complete compressive stress–strain curve, the formulas for the envelope, the loci of the common and stability points, the unloading and reloading curves of the specimen under repeated loading, the concepts and analyses of linear creep, specific creep, and the creep coefficient. However, the values of the parameters in these formulas should be determined from the corresponding characteristic values of the concrete behavior at elevated temperatures.

REFERENCES

[2-1] U. Schneider, K. Kordina, On the behavior of normal concrete under steady state and transient temperature conditions, Proceedings of the Third International Conference on Structural Mechanics in Reactor Technology (SMiRT), London (1975) H1/6.

[2-2] B. Bresler, R.H. Iding, Fire response of prestressed concrete members. Fire safety of concrete structures, ACI SP 80-2, Detroit (1983) 69–113.

[2-3] V.V. Bertero, M. Polivka, Influence of thermal exposure on mechanical characteristics of concrete, ACI SP 34-28, Detroit (1972) 505–531.
[2-4] Z. Guo, X. Zhang, Experimental investigation on complete stress–strain curve of concrete, Journal of Building Structures 3 (1) (1982) 1–12.
[2-5] R. Baldwin, M.A. North, A stress–strain relationship for concrete at high temperature, Magazine of Concrete Research 12 (1973) 208–211.
[2-6] Commission of the European Communities. Eurocode No. 2, Design of Concrete Structures. Part 10: Structural Fire Design; April 1990.
[2-7] J.C. Marechal, Variations in the modulus of elasticity and Poisson's ratio with temperature, ACI SP 34-27, Detroit (1972) 495–503.
[2-8] Z. Guo, X. Zhang, Experimental investigation on complete stress–strain curve of concrete under repeated load, Proceedings No.3, Aseismic Behaviors of Reinforced Concrete Structures Tsinghua University Press, Beijing (1981) 38–53.
[2-9] H.G. Geymayer, Effect of temperature on creep of concrete: a literature review, ACI SP 34-31, Detroit (1972) 565–589.
[2-10] E. Crispino, Studies on the technology of concrete under thermal conditions, ACI SP 34-25, Detroit (1972) 443–479.
[2-11] H. Gross, On high-temperature creep of concrete, Proceedings of the Second International Conference on Structural Mechanics in Reactor Technology (SMiRT), Berlin (1973) H6/5.

CHAPTER

3

Temperature–Stress Paths and Coupling Constitutive Relation of Concrete

3.1 TEMPERATURE–STRESS PATHS AND DEFORMATION COMPONENTS

3.1.1 Temperature–Stress Path and Its Resolution

In structural engineering, concrete that experiences successive casting, curing, and hardening reaches a certain value of strength and starts to bear load action. In the latter period of construction and placing into service, the structure carries various dead and live loads, and sustains frequent or occasional environmental temperature variations. This reflects a complicated long-term load (or internal forces)–temperature history; it also includes redistribution of internal forces of the structure during heating and cooling processes. The stress and temperature of the concrete in the structure vary in a more complicated fashion: both either increase or decrease simultaneously or alternately. Therefore, every point of the concrete has a particular temperature–stress path.

When the temperature and stress of the concrete vary from the original condition to certain values, there may be many different paths; generally it should be an irregular path (e.g., *OCP* in Fig. 3-1). If the temperature and stress increase proportionally, this results in a special path, *OP*, which is seldom found in engineering practice. Moreover, there are two extreme but basic paths as follows:

- Path (*OAP* in Fig. 3-1) of loading under constant temperature (T–σ). The concrete is heated to and maintained at a certain value at elevated temperatures, and is then loaded. The various tests on concrete behavior at elevated temperatures introduced in the previous chapters involve this path.
- Path (*OBP* in Fig. 3-1) of heating under constant load (σ–T). The concrete is loaded to and maintained at a certain value of stress, and is then heated. For example, in a building sustaining a fire accident or in a workshop with high temperature, the structure carries various dead and live loads from an early stage, and then experiences thermal action.

Any arbitrary temperature–stress path can be simulated by finite steps of temperature and stress increments.[3-1] All the ordinate increments are

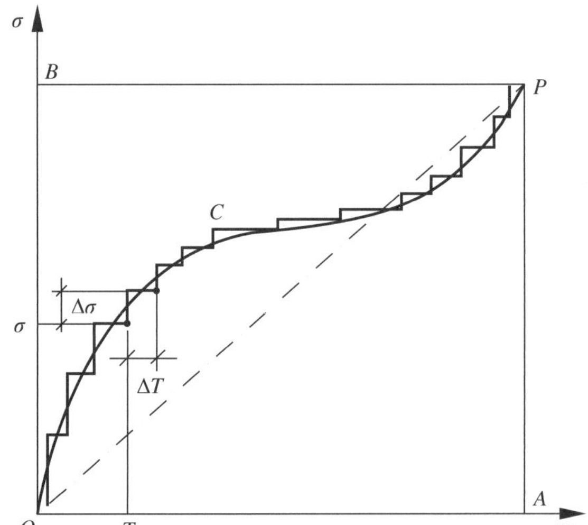

FIGURE 3-1 Different temperature–stress paths.[3-1]

the paths of loading ($\Delta\sigma$) under constant temperature (T), and all the abscissa increments are the paths of heating (ΔT) under constant loading (σ). The number of the steps and the length of the increments can be determined to satisfy the requirement of calculation accuracy.

The strength and deformation behavior of concrete at elevated temperatures vary considerably with the different temperature–stress paths, so it is necessary to first investigate the mechanical behavior of concrete under these two basic paths. As the behavior of concrete under the path of loading under constant temperature (T–σ) has been introduced comprehensively in the previous chapters, the behavior of concrete under the path of heating under constant load (σ–T) is discussed in this chapter. The strength and deformation behavior of concrete under an arbitrary temperature–stress path and their calculations are also investigated.

3.1.2 Composite Components of Deformation at Elevated Temperatures

At any point in the concrete in a structure, the temperature–stress condition starts with (T, σ) and reaches another condition ($T + \Delta T$, $\sigma + \Delta\sigma$) after sustaining a time increment (Δt) and following a known temperature–stress path, yields the corresponding strain increment ($\Delta\varepsilon$). As the temperature–stress path is resolved according to the method shown in Fig. 3-1, the strain increment of concrete in each step is composed of three deformation components[3-2]:

- The strain increment under the path of loading under constant temperature (T–σ). When the temperature T = constant, the strain increment is produced instantaneously by the stress increment ($\Delta\sigma$). It depends on the current values of the temperature and stress and is written as $\Delta\varepsilon_\sigma(T,\sigma)$;
- The strain increment under the path of heating under constant load (σ–T). When the stress σ = constant, the strain increment is given by the temperature increment (ΔT). It also depends on the current values of the temperature and stress and is written as $\Delta\varepsilon_T(T,\sigma)$;
- Short time creep at elevated temperatures. Under the condition of constant temperature (T) and stress (σ), the creep increment is induced after the time increment (Δt) and is written as $\Delta\varepsilon_{cr}(T, \sigma/f_c^T, t)$.

If the signs of the compressive stress and the elongated strain of concrete are taken as positive, the total strain increment of each step is

$$\Delta\varepsilon = -\Delta\varepsilon_\sigma(T,\sigma) + \Delta\varepsilon_T(T,\sigma) - \Delta\varepsilon_{cr}(T,\sigma/f_c^T,t) \quad (3.1)$$

The total strain of concrete is obtained by accumulating all the increments of the temperature–stress-time steps, i.e.,

$$\varepsilon = -\sum\Delta\varepsilon_\sigma(T,\sigma) + \sum\Delta\varepsilon_T(T,\sigma) - \sum\Delta\varepsilon_{cr}(T,\sigma/f_c^T,t) \quad (3.2)$$

The loading strain under constant temperature (ε_σ) and the short time creep (ε_{cr}) among the deformation components are investigated separately in Sections 2.2 and 2.4. The heating strain of concrete under constant loading (ε_T) is introduced in Section 3.3.

3.1.3 Testing Method and Average Temperature of the Specimen

The differences in the strength and deformation behavior of concrete under different temperature–stress paths can be found by corresponding experiments. Therefore, the experiments under different temperature–stress paths must be performed under identical or approximate testing conditions in order to ensure that the experimental results are comparable.

Concrete is a material of thermal inertia as mentioned earlier. If the path of loading under constant temperature (T–σ) is performed to measure its material behavior at elevated temperatures, the specimen experiences a longer heating time (about 1–2 h) and a constant temperature in the furnace chamber for 6 h (Fig. 1-2) to make the temperature in the interior of the specimen approach that outside it. In the succeeding loading stage, only a short time (10–20 min) is needed

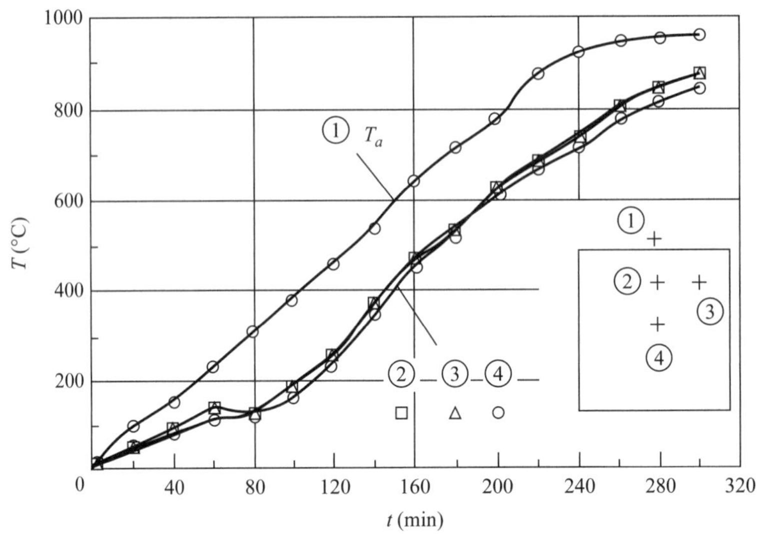

FIGURE 3-2 Heating–time curve for the furnace chamber and the temperature in the interior of the specimen.[1-6]

for the specimen to load until failure. In contrast, in another extreme path, i.e., the path of heating under constant loading (σ–T), the specimen is loaded first and the stress is kept constant at normal temperature. This takes quite a short time. Then, the specimen is heated until failure and the total time needed is 1–2 h. Within the process of the latter path, there is no time to achieve a uniform temperature in the specimen, so the temperatures in the interior and exterior of the specimen are not uniform. Obviously, the time at high temperature and the internal temperature distribution of the specimen under this path are quite different from that under the path of loading at constant temperature. Similarly, other temperature–stress paths (e.g., increasing proportionally or increasing in several steps) also have different testing conditions. Therefore, it is doubtful whether the strength and deformation values measured in these experiments are comparable.

In order to create identical or approximate test conditions for experiments with different temperature–stress paths, the following test and analysis methods are adopted:

- The same temperature–time curve is used for all the comparative experiments. The specimen is loaded soon after the predetermined temperature is reached, and the constant temperature stage is canceled.
- The heating velocity is reduced to 2–5 °C/min, in order to decrease the temperature difference between the inside and outside of the specimen.
- The average temperature at the central section of the specimen is calculated and used as the nominal value of the testing temperature.

Thermocouples are set in the interior of the cubic specimen (edge length 100 mm) and used to measure the temperature and its variation during heating (Fig. 3-2). The maximum difference in the temperatures at the center and the outside of the specimen is less than 200 °C, but most of the difference (gradient) is concentrated within the outer layer (20–25 mm) of the specimen. The temperature difference within the central part is less than 30 °C.

The temperature distribution in the middle section of the specimen is assumed as shown in Fig. 3-3. The temperature on the surface of the specimen is taken as that of the furnace chamber (T_a); the temperature varies linearly within the outer layer (s, mm), and the constant temperature (T_c) is taken within the central part ($b - 2s$, mm) of the section. The formula for the average

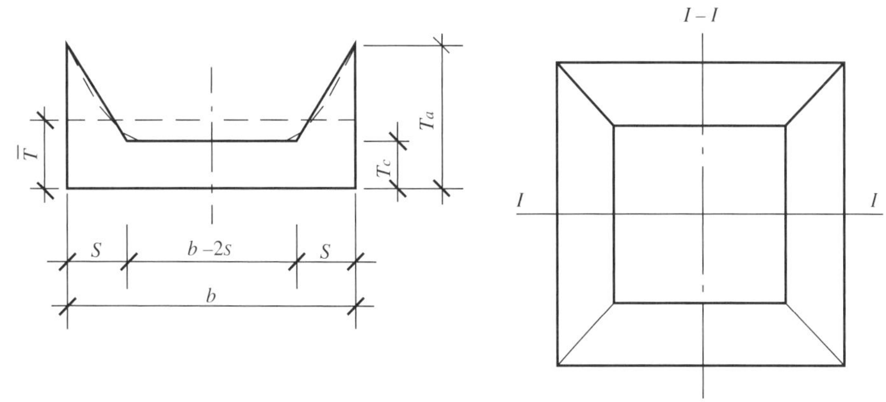

FIGURE 3-3 Temperature distribution and average temperature of the middle section of the specimen where b is the edge length of the specimen and s is the thickness of the area with variable temperature and may be taken as 20 mm.

temperature of the specimen can be derived according to the volume of the geometric figure:

$$\begin{aligned}\overline{T} &= T_a - (T_a - T_c)\left[1 - 2\left(\frac{s}{b}\right) + \frac{4}{3}\left(\frac{s}{b}\right)^2\right] \\ &= T_c + (T_a - T_c)\left[2\left(\frac{s}{b}\right) - \frac{4}{3}\left(\frac{s}{b}\right)^2\right]\end{aligned} \tag{3.3}$$

The experimental data, which are obtained from the tests with a heating velocity of 5 °C/min, are entered into the formula, and the ratio between the average temperature of the specimen and the temperature of the furnace is derived:

$$T \leq 300\ ^\circ\text{C}: \overline{T}/T_a = 0.5 - 0.7$$

$$T \geq 400\ ^\circ\text{C}: \overline{T}/T_a = 0.8 - 0.9$$

If the heating velocity is slower (e.g., 2 °C/min) during testing, the value of $\overline{T}/T_a$ is slightly larger.

The compressive strength of concrete at the average temperature measured and calculated using the method is the average strength of the specimen with nonuniform temperature distribution in the section. Comparing this with the test results of the specimens loaded after keeping the temperature constant for 6 h (f_{cu}^T, Chapter 1), Fig. 3-4 shows that both have the same variation regularity and approximate values, and may be replaced by each other. This demonstrates that the method can be used for testing different temperature–stress paths.

3.2 COMPRESSIVE STRENGTH OF CONCRETE UNDER DIFFERENT TEMPERATURE–STRESS PATHS

3.2.1 Upper and Lower Bounds of Compressive Strength

The procedure for testing a cubic concrete specimen under the path of heating under constant load (σ–T) is:

1. Apply compressive stress (σ) under normal temperature and keep the stress constant.
2. Then, heat the specimen until failure when the stress cannot be sustained and decreases.
3. Obtain the corresponding ultimate temperature.

All the experimental data are drawn and connected smoothly in Fig. 3-5 and the variation regularity is presented.

In the figure, the curve of the path of heating under constant load (σ–T) is obviously above that of the path of loading under constant temperature (T–σ). The ultimate temperature and stress values of the concrete obtained from testing various other temperature–stress paths (see Section 3.2.2) fall between these two curves. These tests prove that the connecting line of the experimental data measured from the path of loading under constant temperature (T–σ) is the lower envelope among different temperature–stress paths, i.e., the lower bound of the

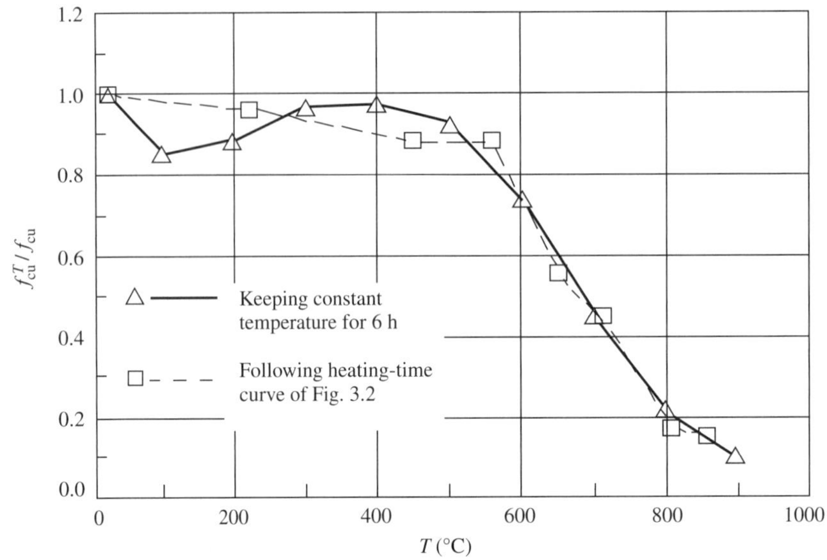

FIGURE 3-4 Comparison of compressive strength of concrete at elevated temperatures measured from different testing methods.

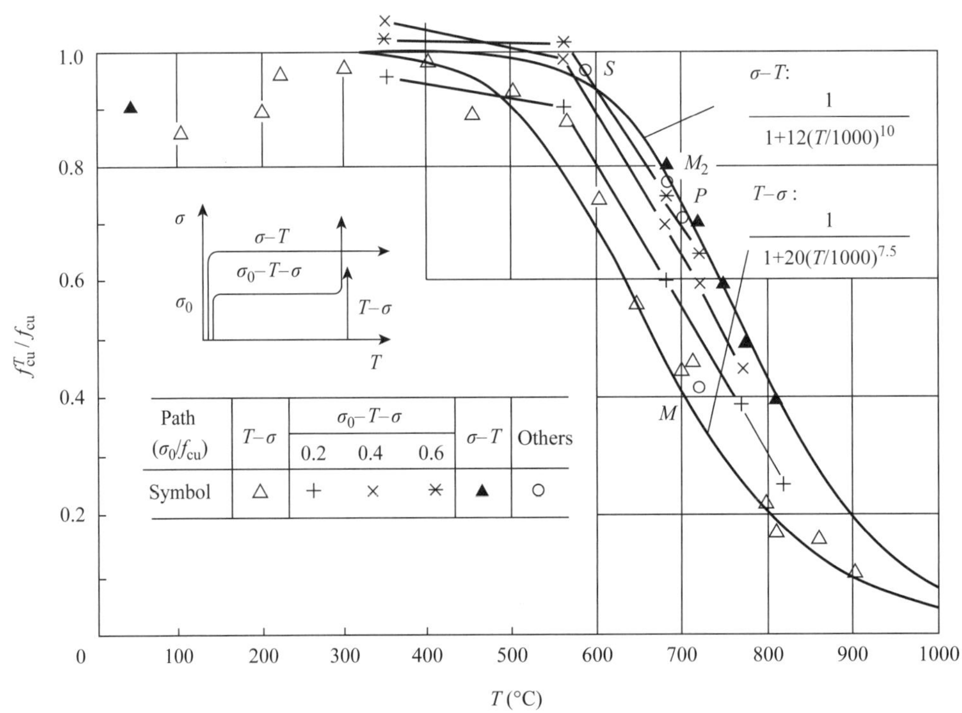

FIGURE 3-5 Upper and lower bounds of compressive strength of concrete at elevated temperatures.[1-6]

compressive strength of concrete at elevated temperatures. Correspondingly, the connecting line of the experimental data measured from the path of heating under constant load (σ–T) is the upper envelope, i.e., the upper bound of the compressive strength of concrete at elevated temperatures.

The difference between the upper bound (f_{cu}^{Tu}) and the lower bound (f_{cu}^{T}) of the compressive strength of cubic concrete at elevated temperatures reaches the maximum value within the temperature range T = 600–800 °C. The absolute difference and the ratio between them are (0.20–0.32)f_{cu} and $f_{cu}^{Tu}/f_{cu}^{T} = 1.4$–$2.0$, respectively. It is apparent that different temperature–stress paths greatly influence the strength value of concrete.

The compressive strength of concrete at elevated temperatures is increased for the specimen under the path of heating under constant load (σ–T) and other paths with compressive stress before heating, compared with that under the path of loading under constant temperature (T–σ). This has been demonstrated by many experiments[1-11,1-15,3-3,3-4] and is generally agreed. The main reasons for this are that the compressive stress acting on the concrete in advance effectively confines the freely expanding deformation of concrete during the heating process and restricts extension of the crack perpendicular to the stress direction. In the meantime, the transient thermal strain (ε_{tr}, see Section 3.3) appears as a large quantity, induces the relaxing and releasing of the internal stress of the concrete, and mitigates the failure process on the boundary between the cement mortar and the aggregate. Also, the compressive stress reduces the volume expansion resulting from the crystallization of the aggregate and dehydration of the cement hydration products at elevated temperatures.

Corresponding to the two extreme temperature–stress paths, the upper and the lower bounds of the compressive strength of concrete at elevated temperatures vary with the temperature and can be expressed, respectively, by the following formulas:

$$\text{Upper bound (path } \sigma - T): \frac{f_{cu}^{Tu}}{f_{cu}} = \frac{1}{1+12(T/1000)^{10}} \tag{3.4}$$

$$\text{Lower bound (path } T - \sigma): \frac{f_{cu}^{T}}{f_{cu}} = \frac{1}{1+20(T/1000)^{7.5}} \tag{3.5}$$

3.2.2 Influence of Different Temperature–Stress Paths

Numerous different temperature–stress paths may appear in concrete in practical structural engineering and not all of them can be tested and measured. Only several typical temperature–stress path tests can be tested to investigate the general regularity of the concrete strength and demonstrate the upper and the lower bounds.

The experimental data of concrete strength under various temperature–stress paths are provided in references [1-6] and [1-11].

- In the series of experiments under the path of previous stress–heating–loading (σ_0–T–σ, Fig. 3-6(a)), the level of previous compressive stress is selected as σ_0/f_{cu} = 0.2, 0.4, and 0.6 and the testing temperature is T = 350–820 °C.
- In the series of experiments under the path of (1) heating–prestressing, (2) heating–reloading (T_1–σ_0–T–σ, Fig. 3-6(b)), the parameters selected are T_1 = 20–500 °C, σ_0/f_{cu} = 0.2–0.6, and T = 270–630 °C.
- Other complicated paths (Fig. 3-6(c)) include the path of temperature and stress increasing proportionally (path P), multisteps increasing alternately (path S), and prestressing–unloading and heating–reloading (path M_1, M_2). All the experimental results are shown in Fig. 3-5 and fall between the curves of the upper and the lower bounds of the concrete strength at elevated temperatures.

The compressive strength of concrete ($f_{cu}^{T\sigma}$) under the path of prestressing–heating–loading (σ_0–T–σ) is related to the value of the prestress (σ_0) and is shown in Fig. 3-7. When the testing temperature $T \le 300$ °C, the prestress has only a minor influence on the compressive strength, i.e.,

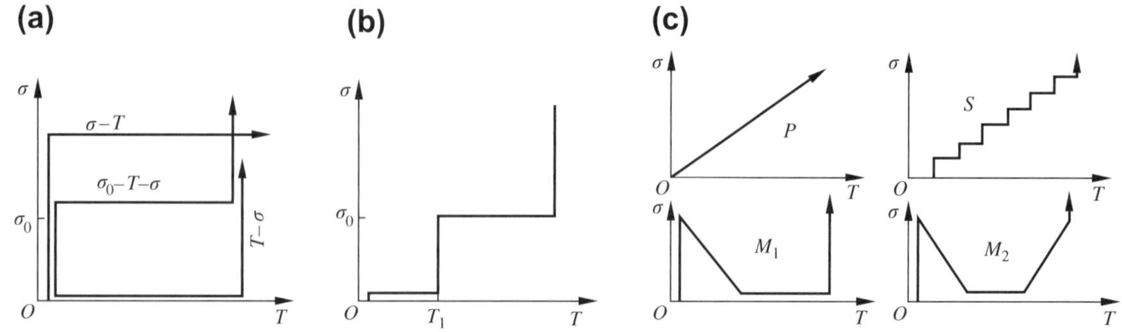

FIGURE 3-6 Various experiments with complicated temperature–stress paths. (a) σ_0–T–σ path; (b) T_1–σ_0–T–σ path; (c) four other paths.

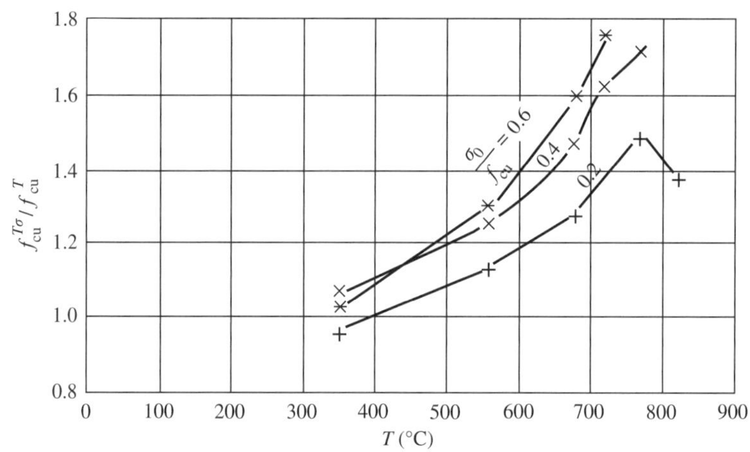

FIGURE 3-7 Influence of prestress on the compressive strength of concrete at elevated temperatures.[1-6]

$f_{cu}^{T\sigma}/f_{cu}^{T} \approx 1.0$, where f_{cu}^{T} is the lower bound strength corresponding to the path of loading under constant temperature (T–σ). When $T \geq 400$ °C, the relative compressive strength of concrete increases with the temperature ($f_{cu}^{T\sigma}/f_{cu} > 1.0$) and the rate of increase accelerates gradually. When the testing temperature is identical, the relative compressive strength increases with the level of prestress (when $\sigma_0/f_{cu} \leq 0.6$) but the rate of increase reduces gradually.

When testing under the path of twice heating–twice loading (T_1–σ_0–T–σ), the first heating and the prestress have an influence on the compressive strength of concrete ($f_{cu}^{T\sigma}$) and the results measured from the tests at temperature $T = 630$ °C are shown in Fig. 3-8. But, when the temperature of the first heating $T_1 = 20$–500 °C, the compressive strength ($f_{cu}^{T\sigma}$) of the specimens varies slightly with temperature T_1. It is concluded that the compressive strength of concrete at elevated temperatures increases greatly and only depends on the level of prestress (σ_0/f_{cu}) when the difference between the heating $T - T_1 > 130$ °C or 150 °C, but it increases slightly when $T - T_1 < 130$ °C.

According to the available data from experimental investigations, the compressive strength of concrete under different temperature–stress paths with the action of prestress can be calculated approximately by the linear interpolation:

$$f_{cu}^{T\sigma} = f_{cu}^{T} + \left(f_{cu}^{Tu} - f_{cu}^{T}\right)\frac{\sigma_0}{f_{cu}^{Tu}} \tag{3.6}$$

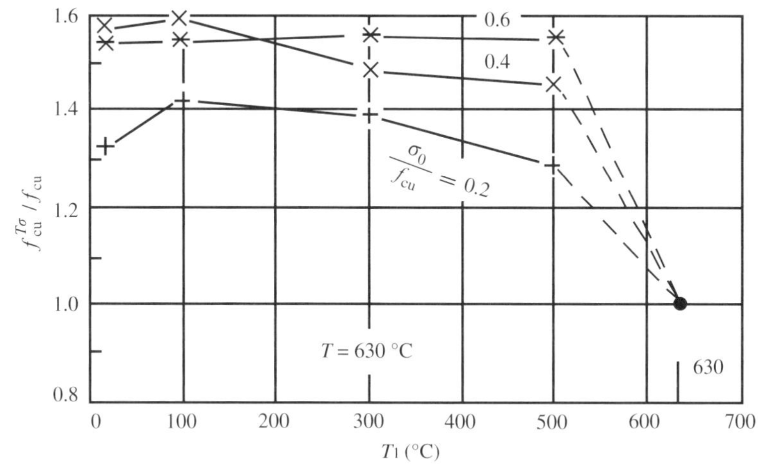

FIGURE 3-8 Influence of the first heating on the compressive strength of concrete at elevated temperatures.[1-11]

where f_{cu}^{T} and f_{cu}^{Tu} are the lower and the upper bounds of compressive strength of concrete when the maximum temperature is T, respectively, and σ_0 is the value of prestress before the temperature reaches T − 150 °C. If the stress is not a constant within the temperature range from 20 °C to T − 150 °C, the average stress weighted by the temperature range (ΔT) is used.

The compressive strength of a concrete prism at elevated temperatures also varies with the different temperature–stress paths. It is demonstrated in the existing experiments[1-11] that the values of the lower and the upper bounds of a concrete prism sample (f_c^T and f_c^{Tu}) can also be obtained from the experiments under the paths of loading under constant temperature (T–σ) and of heating under constant load (σ–T), respectively. The variation regularity and the difference between both values are similar to that of the compressive strength of a concrete cube (Fig. 3-5), but the relative compressive strength of a concrete prism is slightly lower than that of a cube, i.e., $f_c^T/f_c < f_{cu}^T/f_{cu}^T, f_c^{Tu}/f_c < f_{cu}^{Tu}/f_c$. In the experiments under other temperature–stress paths, the prestress (σ_0) and multiheating also influence the compressive strength of the concrete prism, and the regularity and the level of influence are similar to that of a concrete cube.

3.3 THERMAL STRAIN UNDER STRESS AND TRANSIENT THERMAL STRAIN

3.3.1 Thermal Strain Under Stress

The procedure of the deformation test of concrete under the paths of heating (or cooling) under constant load (σ–T) is as follows:

1. The prism specimen is put into the furnace chamber for the deformation test, and is loaded to and maintained at the predetermined stress level (σ/f_c = 0–0.6) at normal temperature.
2. The chamber is heated at a velocity of 2–5 °C/min until it reaches the predetermined temperature (500 °C or 700 °C), and the same temperature is maintained for 30 min.
3. Then, the electrical power is turned off, the upper and the lower gaps between the chamber and the specimen are cleared, and the specimen is cooled down naturally.

The temperature and deformation of the specimen are measured in the testing process and are shown in Fig. 3-9, and the strain–temperature curves of the concrete are obtained. The abscissa in the figure is the average temperature (Eqn 3.3) of the specimen.

The heating and cooling strain curve of the specimen with the stress, $\sigma = 0$, is the same as that

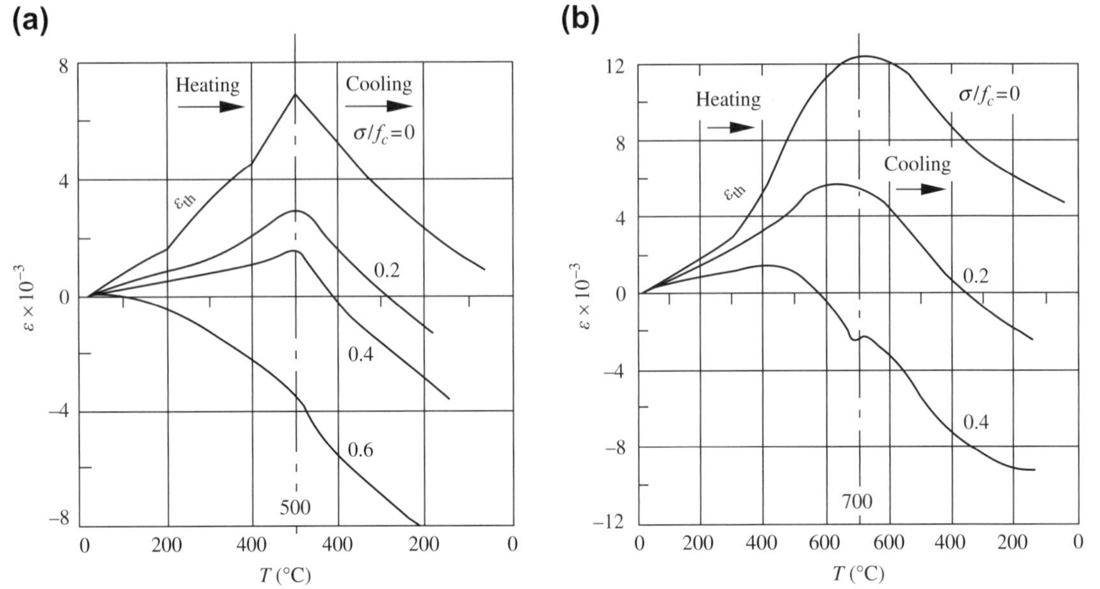

FIGURE 3-9 Heating and cooling strains of concrete under different stress levels[3-5]: (a) T_{max} = 500 °C; (b) T_{max} = 700 °C.

shown in Fig. 2-3. The expansion strain (taken as a positive value) of the specimen increases quickly with temperature; it may reach $\varepsilon_{th} = +(7\text{–}12) \times 10^{-3}$ when T = 500–700 °C. The specimen contracts almost linearly during the cooling process. The residual strain (elongation) of the specimen after it cools down to the normal temperature is about $+(1\text{–}5) \times 10^{-3}$, depending on the maximum temperature reached before cooling.

The compressive strain (taken as a negative value) is produced instantaneously when the specimen is loaded at normal temperature, and its value is rather small, i.e., $|\varepsilon_\sigma| < 0.7 \times 10^{-3}$. In the successive heating process, the thermal strain of the specimen varies considerably with the stress level (σ/f_c):

- When $\sigma/f_c \leq 0.4$ and $T < 500$ °C, the thermal strain of the specimen is expansive and increases with the elevating temperature. However, the higher the stress level of the specimen, the smaller the expansive strain.
- When $\sigma/f_c \geq 0.6$ and $T > 100$ °C, or $\sigma/f_c = 0.4$ and $T > 500$ °C, the specimen is contracted with the elevating temperature and its strain is taken as a negative value that develops quickly.

Therefore, the total thermal strain of the concrete heated from the beginning to a predetermined temperature may be elongation or contraction, depending mainly on the stress level acted before heating.

In the cooling process, all the specimens under different stress levels contract and the temperature–strain curves parallel approximately with that under free cooling ($\sigma = 0$). This means that the values of the total contracted strain approach one another at about $-(5\text{–}7) \times 10^{-3}$, when the specimens are cooled down to normal temperature from 500 °C or 700 °C. Thus, the values of the residual strain of the specimens are considerably different because of the different stress levels acted on the concrete specimen previously. When the specimen is loaded with the stress $\sigma/f_c \geq 0.2$ before heating, the residual strain after cooling is generally contracted (negative value) and is opposite to that of the specimen without stress ($\sigma = 0$).

3.3.2 Transient Thermal Strain

In the heating–cooling process, the thermal (expansive) strain of the concrete specimen without stressing ($\sigma = 0$) is ε_{th} (curve *oa* in Fig. 3-10).

For the specimen under the path of heating under constant load, the strain $\varepsilon\sigma$ (oo') appears instantaneously when the stress is acted before heating, and the strain after heating is ε_T (curve $o'b'$ or ob after moving in a parallel fashion). Therefore, the thermal strains of the concrete under different stresses (σ) differ greatly and the difference between them is called the transient thermal strain (contraction) (in Khoury et al.,[3-6,3-7] the difference is called load-induced thermal strain (LITS) and is considered to be composed of two parts: "transient thermal strain" and "basic creep," but the former is the main part):

$$\varepsilon_{tr}\ (T, \sigma / f_c) = \varepsilon_{th}\ (T) - \varepsilon_T\ (T, \sigma) \tag{3.7}$$

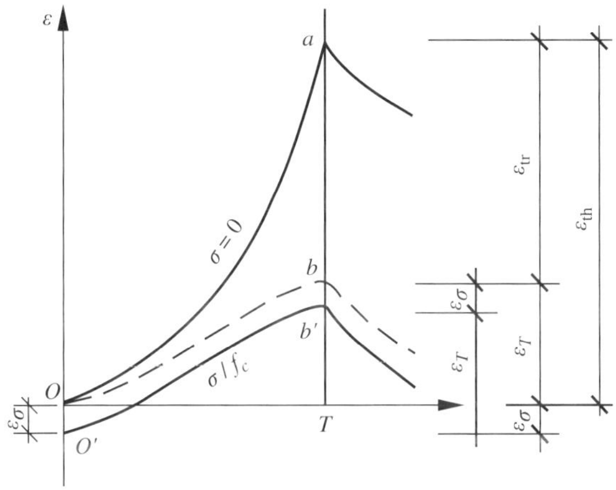

FIGURE 3-10 Thermal deformation and transient thermal strain of concrete.

According to the measured thermal strains ε_{th} and ε_T from the experiments, the transient thermal strain occurring during one cycle of heating–cooling can be calculated and is shown in Fig. 3-11. It is a general regularity that the transient thermal strain increases quickly as the temperature increases during the heating process and its value is approximately proportional to the stress level (σ/f_c). However, it varies less during the cooling process and maintains almost the maximum value at the maximum temperature. Therefore, transient thermal strain of concrete occurs only during heating and is not restored during cooling.

If the concrete specimen experiences multi-cycles of heating–cooling under constant stress, the thermal strain is measured from the testing (Fig. 3-12). In the first set of the two cycles of heating–cooling, the maximum temperature in the first cycle is higher than that in the second cycle (700 °C > 500 °C, Fig. 3-12(a)). The thermal strains of the specimen within the first cycle of heating–cooling are the same as those shown in Fig. 3-9(b). Within the second cycle

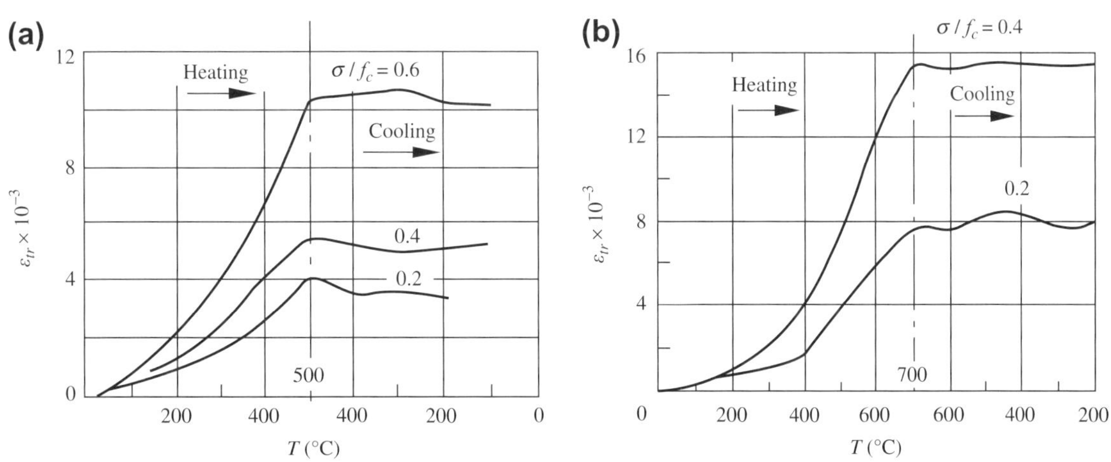

FIGURE 3-11 Transient thermal strain of concrete during one cycle of heating–cooling[3-5]: (a) T_{max} = 500 °C; (b) T_{max} = 700 °C.

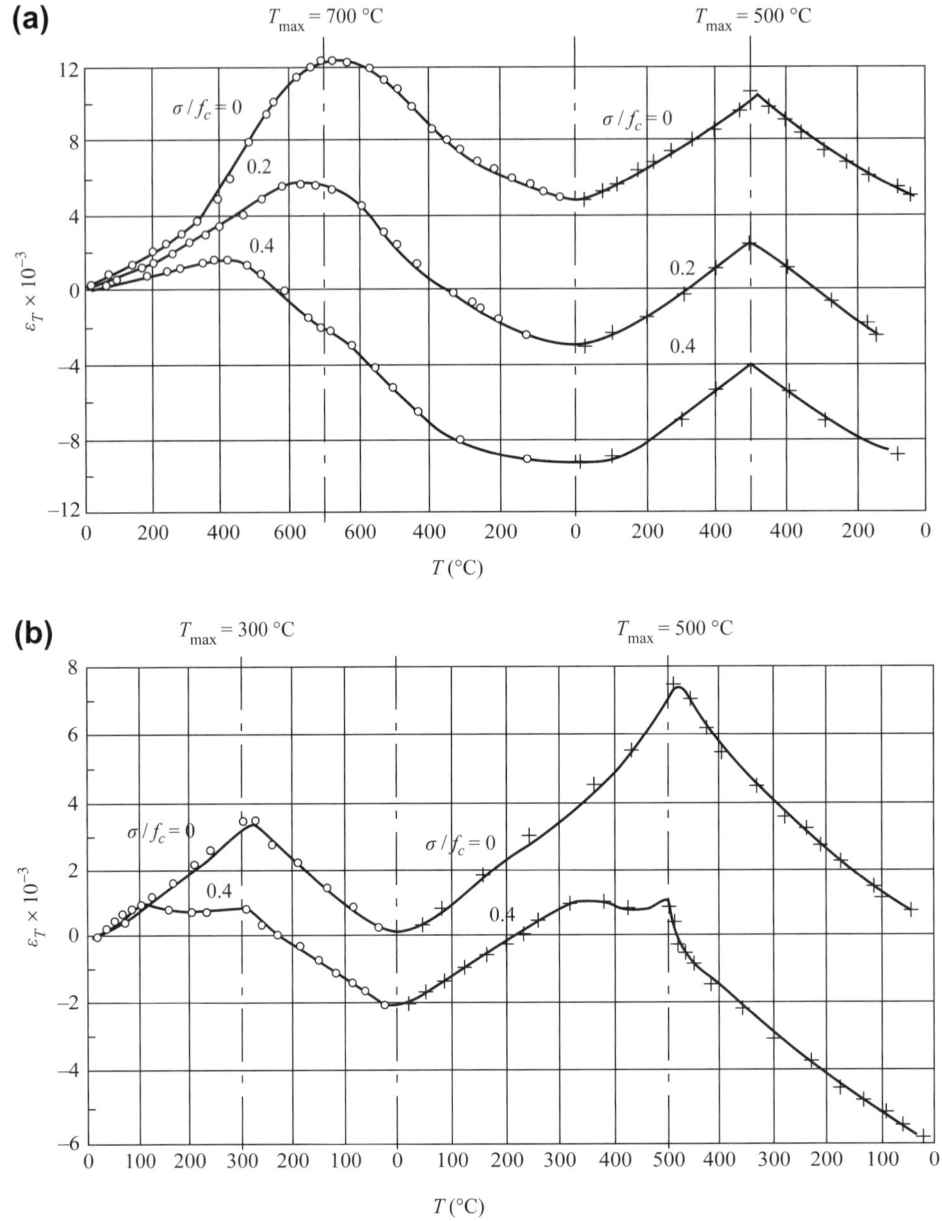

FIGURE 3-12 Thermal strain of concrete during multicycles of heating–cooling[3-5]: (a) 700–500 °C; (b) 500–700 °C.

of heating–cooling, the total strain–temperature curves of the specimens without stress ($\sigma = 0$) and with constant compressive stress ($\sigma/f_c = 0.2$ and 0.4) are basically parallel, and the transient thermal strains (Eqn 3.7) approach a constant, with an increment that is nearly zero. The expanding strain of the specimen occurring during the second heating restores during cooling and the residual strain does not increase (refer to Fig. 2-4(b)).

In the second set of the two cycles of heating–cooling, the maximum temperature in the first cycle is lower than that in the second cycle (300 °C < 500 °C, Fig. 3-12(b)). The strains in the specimen within the first cycle of heating–cooling are similar to that shown in Fig. 3-9. Within the second cycle of heating–cooling, the thermal strain of the specimen with no stress ($\sigma = 0$) is the same as those in Fig. 2-4(a); the thermal strain of the specimen with stress ($\sigma/f_c = 0.4$) increases nearly linearly with increasing temperature before $T \leq 300$ °C, but stagnates when the temperature exceeds the maximum temperature in the first heating, i.e., $T = 300$ °C or 500 °C. This means that the transient thermal strain appears again. During the second cooling, the thermal strain curve of the specimen with stress is parallel to the corresponding curve of the specimen without stress.

It is demonstrated by all the experiments under the path of multicycles of heating–cooling that the transient thermal strain (contraction) of concrete with compressive stress appears only during the first heating or when the temperature during the succeeding heating exceeds the maximum temperature experienced in previous cycles. The transient thermal strain keeps a constant value, the maximum value reached, when cooling or when the temperature during heating does not exceed the maximum temperature experienced in the previous cycles of heating–cooling.

The transient thermal strain of concrete appears instantaneously during heating, when it is acted on with compressive stress. The value of transient thermal strain is far greater than that of the short-term creep (see Section 2.4) and the strain induced by the stress (see Section 2.2) at elevated temperatures. Thus, the transient thermal strain is the main component of thermal strain of concrete at elevated temperatures and influences considerably the deformation and stress relaxation (or redistribution) of the structure at elevated temperatures, and it should be taken into account in the analysis of the structure at elevated temperatures.

The mechanism of transient thermal strain in concrete is not yet clear. Possibly, when concrete is first heated, a chemical reaction occurs in the cement hydrates causing variation in the microstructure of the cement mortar and a change in the volume of the internal holes, resulting in remarkable compressive strain of the concrete under stress and at elevated temperatures.

It is convenient to introduce a dimensionless parameter of transient thermal strain, $\beta(T)$. This parameter is defined as the value under a specific stress level (σ/f_c):

$$\beta(T) = \frac{\varepsilon_{tr}}{\sigma / f_c} \tag{3.8}$$

The experimental data from various tests were introduced into the formula, and a series of β–T curves (Fig. 3-13) were obtained after calculation. The curves for the tests at different stress levels ($\sigma/f_c = 0$–0.6) and heating velocity 2–5 °C/min are quite close together, and a simplified formula is obtained after regression analysis:

$$\beta(T) = \left[72\left(\frac{T}{1000}\right)^2 - \left(\frac{T}{1000}\right)\right] \times 10^{-3} \tag{3.9}$$

Therefore, the transient thermal strain of concrete at the first heating can be calculated by the following formula:

$$\varepsilon_{tr} = \frac{\sigma}{f_c}\left[72\left(\frac{T}{1000}\right)^2 - \left(\frac{T}{1000}\right)\right] \times 10^{-3} \tag{3.10}$$

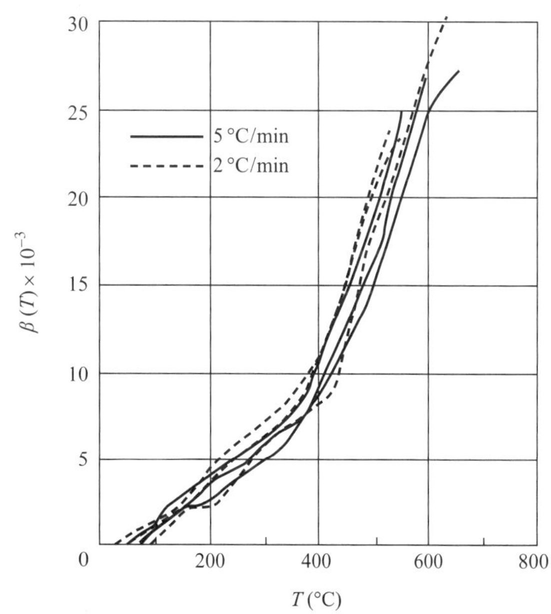

FIGURE 3-13 Relation between parameters of transient thermal strain and temperature.

3.4 COUPLING TEMPERATURE–STRESS CONSTITUTIVE RELATION

3.4.1 Comparison of Strains Under Different Temperature–Stress Paths

The difference between the strain values of concrete experiencing different temperature–stress paths can be represented by the comparison of the experimental results of the two extreme paths (Fig. 3-14). When the concrete (C20L, see Table 1-1) reaches $T = 500$ °C and $\sigma = 0.6f_c$ from the original conditions (T = 20 °C and $\sigma = 0$), the specimen tested under the path of loading under constant temperature (OAP) experiences successively free heating strain ($\varepsilon_{th} = +7.4 \times 10^{-3}$) and instantaneous strain $\varepsilon_\sigma = -2.65 \times 10^{-3}$ under loading at constant temperature (T = 500 °C). The total strain is $\varepsilon_{T-\sigma} = +4.75 \times 10^{-3}$ ($\overline{PR}$, elongation). The specimen tested under the path of heating under constant load (OBQ) experiences successively the strain ($\varepsilon_\sigma = -0.6 \times 10^{-3}$) under loading at normal temperature and thermal strain ($\varepsilon_T = +0.2 \times 10^{-3}$) during heating under constant load, and the total strain is $\varepsilon_{\sigma-T} = -0.4 \times 10^{-3}$ ($\overline{RQ}$, contraction).

In this example, the strains of concrete under the two extreme paths ($\varepsilon_{T-\sigma}$ and $\varepsilon_{\sigma-T}$) have opposite signs, and the difference between them is large $\Delta\varepsilon = 5.15 \times 10^{-3}$ (PQ). The total strains under both paths with other values of stress ($\sigma < 0.6f_c$) are presented separately as lines AP ($\varepsilon_{T-\sigma}$) and AQ ($\varepsilon_{\sigma-T}$) in Fig. 3-14, and the variation of the difference between them ($\Delta\varepsilon$) can be seen in the figure. If the concrete reaches $T = 500$ °C and $\sigma \le 0.6f_c$ following other various temperature–stress paths (e.g., Fig. 3-1 and Fig. 3-6), the total strain of the specimen should be situated between the curves AP and AQ.

The total strain values of concrete under different temperature–stress paths are considerably different, mainly because the transient thermal strain of concrete under compressive stress is large during the heating process and is not restored after it appears. Therefore, it should be carefully taken into account when the coupling temperature–stress constitutive relation of concrete is established.

3.4.2 Basic Formulas of Coupling Constitutive Relation

Following an arbitrary temperature–stress path, the concrete experiences a time increment (Δt, min) and reaches the temperature $T + \Delta T$ and stress $\sigma + \Delta\sigma$ from the temperature–stress condition (T,σ); the total strain increment produced

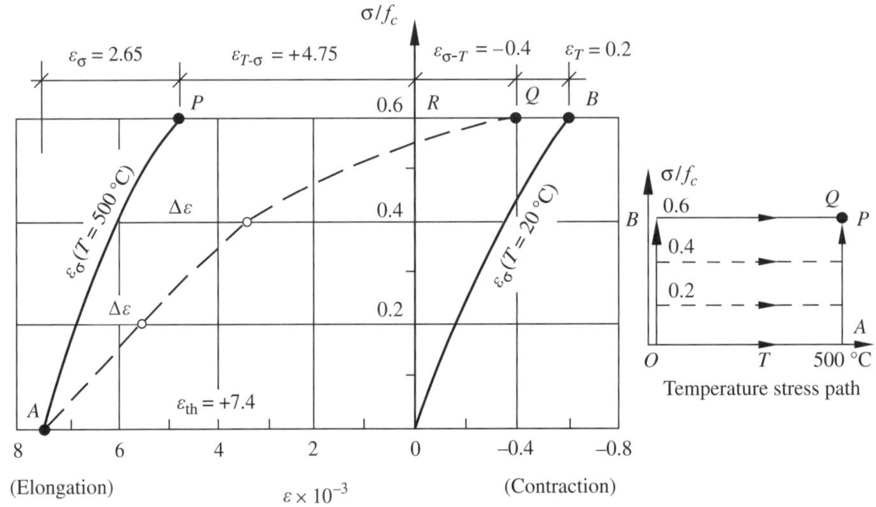

FIGURE 3-14 Strains of concrete experienced with different temperature–stress paths.[3-1]

can be calculated by Eqn (3.1), in which the strain increment due to heating under constant stress $\sigma(\Delta\varepsilon_T)$ can be replaced by Eqn (3.7). The basic formula to calculate the total strain increment is:

$$\Delta\varepsilon = -\Delta\varepsilon_\sigma(T,\sigma) + \Delta\varepsilon_{th}(T) - \Delta\varepsilon_{tr}(T,\sigma/f_c) - \Delta\varepsilon_{cr}(T,\sigma/f_c^T,t) \quad (3.11)$$

The definition and formula for each strain component are as follows:

$\Delta\varepsilon_\sigma(T,\sigma)$: the strain increment produced at constant temperature T when the stress increases from σ to $\sigma+\Delta\sigma$:

$$\Delta\varepsilon_\sigma(T,\sigma) = \frac{1}{d\sigma/d\varepsilon}\Delta\sigma \quad (3.12)$$

where $d\sigma/d\varepsilon$ is the first derivation at σ on the stress–strain curve (Eqn (2.8));

$\Delta\varepsilon_{th}(T)$: the strain increment of freely expanding strain (Eqn (2.1)) is produced when the temperature rises to $T+\Delta T$ from T.

$$\Delta\varepsilon_{th}(T) = \frac{d\varepsilon_{th}(T)}{dT}\Delta T = 56\left(\frac{T}{1000}\right)\times 10^{-6}\Delta T \quad (3.13)$$

$\Delta\varepsilon_{tr}(T,\sigma/f_c)$: the increment of transient thermal strain (Eqn (3.10)) appears when the temperature rises to $T+\Delta T$ from T and under stress:

$$\Delta\varepsilon_{tr}(T,\sigma/f_c) = \frac{\partial\varepsilon_{tr}(T,\sigma/f_c)}{\partial T}\Delta T = \frac{\sigma}{f_c}\left[144\left(\frac{T}{1000}\right)-1\right]\times 10^{-6}\Delta T \quad (3.14)$$

$\Delta\varepsilon_{cr}(T,\sigma/f_c^T,t)$: the increment of short time creep (Eqn (2.24)) is produced in the time from t to $t+\Delta t$ under the actions of temperature T and stress σ:

$$\Delta\varepsilon_{cr}(T,\sigma/f_c^T,t) = \frac{\partial\varepsilon_{cr}(T,\sigma/f_c^T,t)}{\partial t}\Delta t = \frac{\sigma}{f_c^T}\left(e^{6T/1000}-1\right)\frac{30\times 10^{-6}}{\sqrt{t^{eq}t_0}}\Delta t \quad (3.15)$$

t^{eq} is the equivalent time (see Fig. 2-16) and can be calculated by modifying Eqn (2.24), in which ε_{cr} is the total creep at T and σ, and t_0 = 120 min.

$$t^{eq} = \left[\frac{\varepsilon_{cr}\times 10^6}{(\sigma/f_c^T)\left(e^{6T/1000}-1\right)\times 60}\right]^2 t_0 \quad (3.16)$$

Equations (3.11)–(3.16) comprise the coupling thermal–mechanical constitutive relation of concrete and include the main influencing factors and reveal their variation regularities. They show the basic characteristics of the coupling actions of temperature and stress, and improve on comparisons with some available coupling constitutive relations.[3-1,3-8,3-9]

3.4.3 Calculation Rules for Strain Increments

The coupling constitutive relation of concrete introduced above is expressed in terms of the increments and fits for the analysis of the structure under complicated temperature–stress paths when the corresponding computer program is developed. The smaller increments in temperature and stress can be used to increase computing accuracy. As for a simple temperature–stress path, the corresponding formulas for strains are used to calculate step by step, and accurate results can be achieved.

An arbitrary temperature–stress path in concrete can be resolved into multisteps of loading under constant temperature and heating under constant load (Fig. 3-1). When the increments of various strain components are calculated, the following rules can be used for simplification, if variable temperature or stress occurs.

1. Stress-induced strain ε_σ

When the concrete is loaded under temperature T_1 and its compressive stress reaches σ_1, the strain $\varepsilon_{\sigma1}(T_1)$ induced can be calculated by Eqn (2.8) (Fig. 3-15), in which the prismatic compressive strength and corresponding strain are $f_c(T_1)$ (Eqn (2.3)) and $\varepsilon_p(T_1)$ (Eqn (2.5)), respectively.

After the concrete had been heated under constant load (σ_1) and reaches T_2, it is loaded again and the compressive stress increases to σ_2 under constant temperature T_2. The strain increment induced by the stress increment should be the difference between the strains corresponding to σ_2 and σ_1 on the stress–strain curve referring to T_2, i.e.,

$$\Delta\varepsilon_\sigma = \varepsilon_{\sigma2}(T_2) - \varepsilon_{\sigma1}(T_2)$$

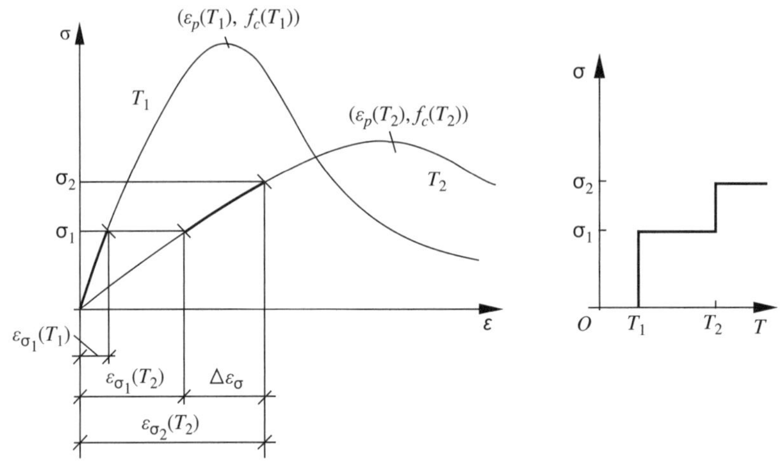

FIGURE 3-15 Calculation of the strain increment induced by stress.

The values of $\varepsilon_{\sigma 2}$ and $\varepsilon_{\sigma 1}$ can be calculated from Eqn (2.8), in which the prismatic strength $f_c(T_2)$ and corresponding strain $\varepsilon_p(T_2)$ are referred to temperature T_2.

2. Freely expanding strain ε_{th}
When the concrete is heated first to T_1, the freely expanding strain $\varepsilon_{th}(T_1)$ can be calculated from Eqn (2.1). When the concrete is loaded and reaches σ_1 under constant temperature T_1, compressive strain $\varepsilon_{\sigma 1}(T_1)$ is produced. Then it is heated again and reaches T_2 under constant load (σ_1); the increment of the freely expanding strain (Fig. 3-16) is

$$\Delta \varepsilon_{th} = \varepsilon_{th}(T_2) - \varepsilon_{th}(T_1)$$

The expanding strain $\varepsilon_{th}(T_2)$ and $\varepsilon_{th}(T_1)$ are also calculated using Eqn (2.1).

3. Transient thermal strain ε_{tr}
When the concrete is heated first to T_1, the transient thermal strain is zero, $\varepsilon_{tr} = 0$, as no stress is acted. When the concrete is heated to T_2 from T_1 under the action of stress σ_1, the transient thermal strains at T_2 and T_1 can be calculated individually by Eqn (3.10) and the difference between them is the increment of transient thermal strain, i.e., $\Delta\varepsilon_{tr,1} = \varepsilon_{tr}(T_2) - \varepsilon_{tr}(T_1)$ (Fig. 3-17). When the concrete is heated to T_3 from T_2 under the action of stress σ_2, the increment in the transient thermal strain can be calculated similarly.

4. Short time creep ε_{cr}
The calculation of the increment in short time creep is presented in Section 2.4.2 (Fig. 2-16).

3.4.4 Example and Experimental Demonstration

1. Example
If the cubic and prismatic compressive strengths of concrete C40 under normal temperature are f_{cu} = 40 MPa and f_c = 32 MPa, respectively, and the corresponding strain is $\varepsilon_p = 2 \times 10^{-3}$, the strains at T = 500 °C and $\sigma = 0.5f_c$ for various temperature–stress paths (Fig. 3-18(a)) are calculated, and the strain under the path of temperature and stress increasing proportionally are estimated (without considering short time creep at elevated temperatures).
Solution:
Path 1: OAP, i.e., loading after heating

OA ($\sigma = 0$, T = 20–500 °C):

$$\varepsilon_{th} = 28\left(\frac{500}{1000}\right)^2 \times 10^{-3} = 7.0 \times 10^{-3}$$

AP (T = 500°C, σ = 0–0.5f_c): $f_c^T = 0.656 f_c$, $y = \dfrac{0.5f_c}{0.656f_c} = 0.762$

Substituting into Eqn (2.8) and obtaining x = 0.487, we get,

$$\varepsilon_\sigma = 0.487\left[1 + 5\left(\frac{500}{1000}\right)^{1.7}\right] \times 2 \times 10^{-3} = 2.47 \times 10^{-3}$$

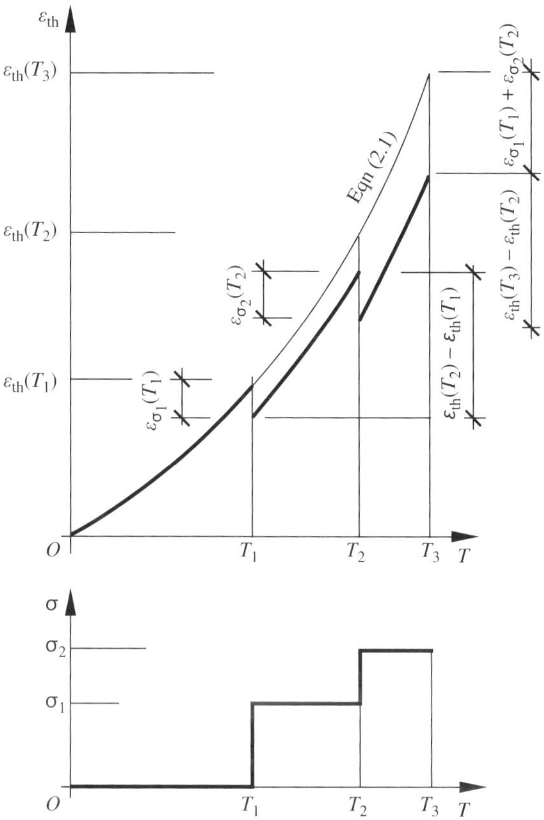

FIGURE 3-16 Calculation of the increment of freely expanding strain.

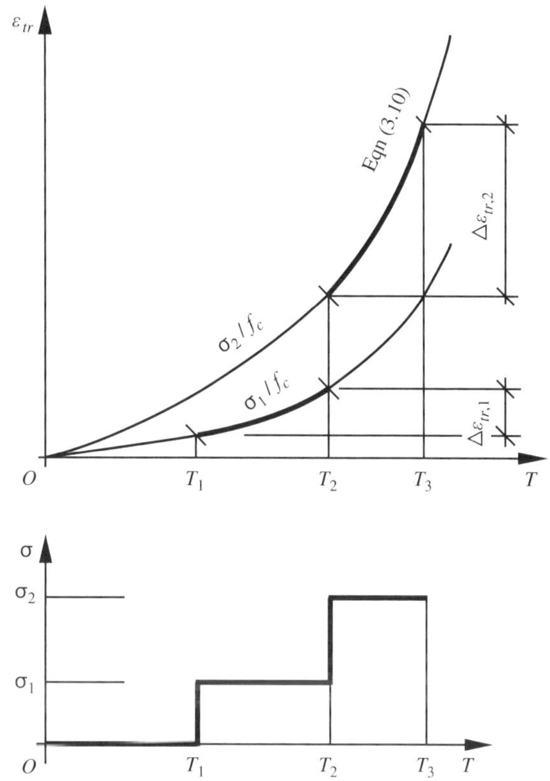

FIGURE 3-17 Calculation of the increment of transient thermal strain.

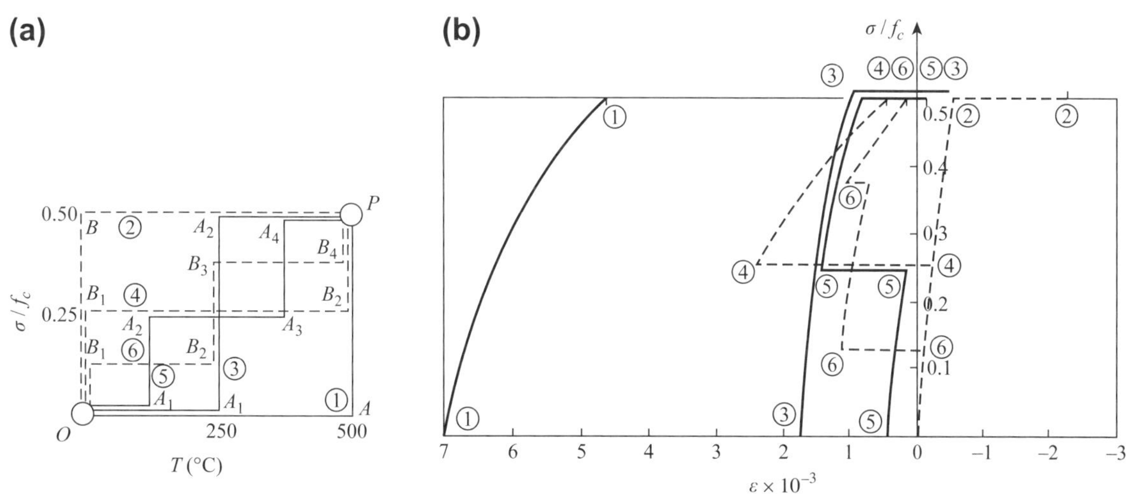

FIGURE 3-18 Comparison of strains of concrete under various temperature–stress paths (example): (a) various temperature–stress paths; (b) comparison of strains.

$$\sum \sigma = \sigma_{th} - \varepsilon_\sigma = +4.53 \times 10^{-3}$$

Path 2: *OBP*, i.e., heating after loading

$$OB\ (T=20°\ C,\ \sigma=0.5f_c):\ \varepsilon_\sigma = 0.27\times 2\times 10^{-3} = 0.54\times 10^{-3}$$

$$BP\ (\sigma=0.5f_c,\ T=20\text{–}500\ °C):\ \varepsilon_{th} = 7\times 10^{-3}$$

$$\varepsilon_{tr} = 0.5\left[72\left(\frac{500}{1000}\right)^2 - \frac{500}{1000}\right]\times 10^{-3} = 8.75\times 10^{-3}$$

$$\therefore \sum \varepsilon = -\varepsilon_\sigma + \varepsilon_{th} - \varepsilon_{tr} = -2.29\times 10^{-3}$$

The calculations for paths 3–8 are listed in Table 3-1.

The variation processes of the strains of concrete under various temperature–stress paths are shown in Fig. 3-18(b) and the total strains are listed in Table 3-2. According to the results, the strain of concrete at $T = 500$ °C and $\sigma = 0.5f_c$ experiencing the path of temperature–stress increasing proportionally should be estimated between $+(0.009–0.076) \times 10^{-3}$ (paths 8 and 7). If a smaller increment step is used for the calculations, a more accurate value of the strain can be achieved.

2. Experimental demonstration

The coupling temperature–stress relation for concrete introduced above is established based on the experimental data, which are measured individually in the various arbitrary tests for the strain components of ε_σ, ε_{th}, ε_{tr}, and ε_{cr}, and derived after analysis. In order to confirm the reasonableness and accuracy of the constitutive relation, a comprehensive experiment should be performed.

Therefore, a short strut of constant height (confined) is designed for the heating experiment (Fig. 3-19).[1-6] The prismatic concrete specimen (80 mm × 80 mm × 300 mm, f_c = 23.4 MPa) is put into the furnace and a load transducer is put on top. Both are placed on the working table of a hydraulic testing machine and the upper cross head is moved to make contact with the transducer. When the furnace is heated at a velocity of 5 °C/min after turning on the electrical power, the concrete specimen expands and presses the testing machine during heating, and the compressive force and the deformation are measured simultaneously. The hydraulic valve of the testing machine is adjusted continuously during the testing to keep the height (strain) of the specimen constant. Actually, the error in the practical operation is small and the strain fluctuates at $<\pm 50 \times 10^{-6}$. The compressive force (stress σ)–temperature curve of the specimen is recorded (Fig. 3-19(b)). Obviously, this is a complicated temperature–stress path.

According to the traditional method of structural mechanics, the thermal stress of the specimen is analyzed using the following principle: the freely expanding strain of the specimen during heating is compensated fully by the strain induced by compressive stress, without considering transient thermal strain and short time creep. Therefore the compressive stress of the specimen increases quickly and reaches the corresponding strength f_c^T at temperature $T = 380$ °C; this means that the specimen fails. If the specimen is heated further, the stress of the specimen enters the descending branch. Obviously, the calculation curve obtained is considerably different from the experimental one.

When the coupling temperature–stress constitutive relation of concrete introduced above is used, the example can be calculated as follows:

1. The behavior parameters of concrete material are determined.
2. The temperature increment for every calculation step is given, the corresponding time increment is obtained, and the average temperature of the specimen is calculated.
3. The stress value obtained in the last step is used as a parameter and placed in the constitutive formulas, and the increments for freely expanding strain, transient thermal strain, and short time creep ($\Delta\varepsilon_{th}$, $\Delta\varepsilon_{tr}$, and $\Delta\varepsilon_{cr}$) occurring in the current temperature step are calculated individually.
4. The condition of confined deformation of the specimen is established as

$$\Delta\varepsilon = -\Delta\varepsilon_\sigma + \Delta\varepsilon_{th} - \Delta\varepsilon_{tr} - \Delta\varepsilon_{cr} = 0$$

The strain $\Delta\varepsilon_\sigma$ induced by the stress is then obtained. The stress increment $\Delta\sigma$ and total stress σ can then be calculated from Eqn (2.8),

TABLE 3-1 Example of the Calculations for the Coupling Constitutive Relation of Concrete ($\varepsilon \times 10^{-3}$)

Path		T (°C)	σ/f_c	ε_σ	ε_{th}	ε_{tr}	Accumulation
3	OA_1	20→250	0		+1.75	0	+1.75
	A_1A_2	250	0→0.5	−0.82			+0.93
	A_2P	250→500	0.5		+5.25	−6.624	−0.444
4	OB_1	20	0→0.25	−0.25			−0.25
	B_1B_2	20→500	0.25		+7.0	−4.375	+2.375
	B_2P	500	0.25→0.5	−1.468			+0.907
5	OA_1	20→125	0		+0.438		+0.438
	A_1A_2	125	0→0.25	−0.281			+0.157
	A_2A_3	125→375	0.25		+3.50	−2.188	+1.469
	A_3A_4	375	0.25→0.5	−0.680			+0.789
	A_4P	375→500	0.5		+3.06	−3.875	−0.026
6	OB_1	20	0→0.125	−0.118			−0.118
	B_1B_2	20→250	0.125		+1.75	−0.53	+1.102
	B_2B_3	250	0.125→0.375	−0.404			+0.698
	B_3B_4	250→500	0.375		+5.25	−4.969	+0.979
	B_4P	500	0.375→0.50	−0.828			+0.151
7	OA_1	20→100	0		+0.28		+0.28
	A_1A_2	100	0→0.2	−0.213			+0.067
	A_2A_3	100→300	0.2		+2.24	−1.112	+1.195
	A_3A_4	300	0.2→0.4	−0.385			+0.810
	A_4A_5	300→500	0.4		+4.48	−4.528	+0.762
	A_5P	500	0.4→0.5	−0.686			+0.076
8	OB_1	20	0→0.1	−0.094			−0.094
	B_1B_2	20→200	0.1		+1.12	−0.268	+0.758
	B_2B_3	200	0.1→0.3	−0.275			+0.483
	B_3B_4	200→400	0.3		+3.36	−2.532	+1.311
	B_4B_5	400	0.3→0.5	−0.632			+0.679
	B_5P	400→500	0.5		+2.52	−3.19	+0.009

TABLE 3-2 Comparison of Strains of Concrete Under Various Temperature–Stress Paths

	Path							
	1	4	6	7	8	5	3	2
ε (10^{-3})	+4.53	+0.907	+0.151	+0.076	+0.009	−0.026	−0.444	−2.29
Paths compared	1 vs 2		4 vs 3		6 vs 5		7 vs 8	
Strain difference ε (10^{-3})	6.82		1.351		0.177		0.067	

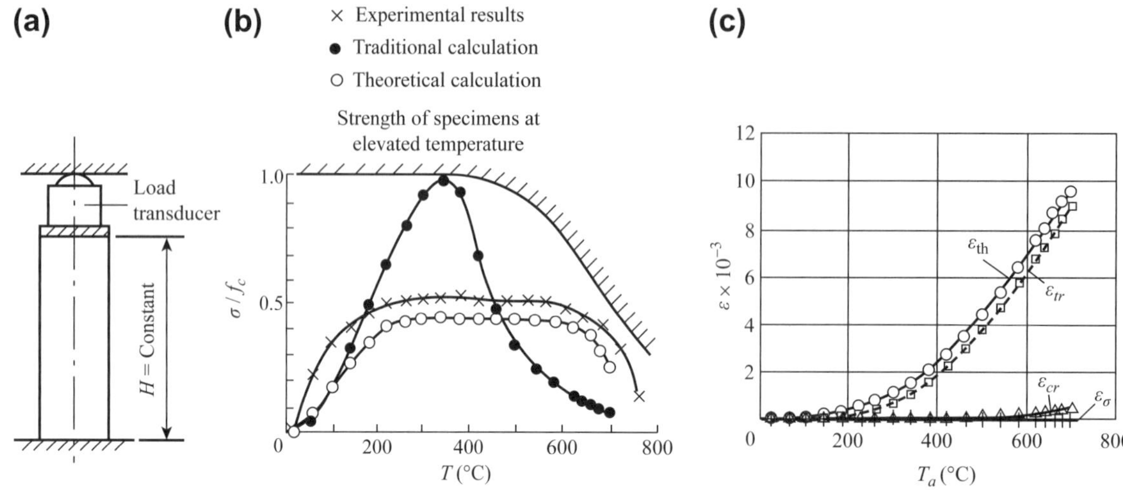

FIGURE 3-19 Heating experiment on a concrete strut of constant height[1-6]: (a) specimen; (b) temperature–stress curve; (c) strain components.

considering the values of temperature and stress in the current step.

5. The variation in the specimen stress throughout the heating process can be obtained by repeating steps 2–4.

A computer program for the calculation procedure has been compiled,[1-6] and the temperature–stress curve of the specimen and the variations in the strain components during heating are computed and shown in Fig. 3-19(b) and (c), respectively. Their variation regularities coincide with and their values approach that of the experimental results.

When the temperature is lower ($T < 200$ °C), the values of the transient thermal strain and creep of concrete are small and the freely expanding strain is compensated mainly by the strain induced by the compressive stress, satisfying the condition $\varepsilon = 0$. As the value of the elastic modulus of concrete is still high at this time, the stress increases quickly. Afterward, the concrete is heated continuously under the action of the compressive stress, the transient thermal strain increases sharply and has a value approximating the freely expanding strain, so the stress of the specimen varies less. When the temperature $T >$ 600 °C, the freely expanding strain of concrete increases gently, but its transient thermal strain and short time creep develop more quickly and exceed the increment in the expanding strain, so stress relaxation occurs and the stress decreases in the specimen, satisfying the condition $\varepsilon = 0$.

CONCLUSIONS

The compressive strength and deformation behavior of concrete at elevated temperatures varies considerably with the temperature–stress path experienced. Among various paths, the path of loading under constant temperature (T–σ) refers to the lower bound, and the path of heating under constant load (σ–T) refers to the upper bound; other arbitrary paths are located between the upper and the lower bounds.

The strain of concrete at elevated temperatures is composed of four components. Among them, the freely expanding strain (ε_{th}) and the transient thermal strain (ε_{tr}) have considerable values but with opposite signs. The others are the stress-induced strain (ε_{σ}) and short time creep (ε_{cr}), both of which have relatively smaller values. These strain components vary with temperature, stress, and time, but have different variation regularities, so the complicated temperature–stress–strain–time (T–σ–ε–t) coupling constitutive relation of concrete is formed. When a concrete structure and its members at elevated temperatures are analyzed, the coupling constitutive relation of concrete should be introduced carefully in order to obtain reasonable and accurate results in the theoretical calculation.

The mechanical behavior of concrete at elevated temperatures is influenced by many factors and varies considerably. There is no unified standard and the measuring technique is not sufficient for testing at elevated temperatures, resulting in large deviation from the experimental data. In addition, the experimental data in some aspects is still not sufficient and the mechanisms of some physical phenomena are not very clear; they cause more difficulty in developing the coupling constitutive relation with high accuracy. However, the effect of a fire accident on the structure has more severe probability and deviation (see Chapter 5). Thus, it is possible to make some necessary simplifications and approximate assumptions for developing the coupling constitutive relation of concrete.

The coupling constitutive relation of concrete suggested in this chapter includes the main influential factors and their variation regularities, and reflects the basic characteristics of the coupling action between temperature and stress. Some approximate assumptions are introduced, e.g., the uniqueness of the compressive stress–strain curve of concrete at elevated temperatures, simplified treatments of transient thermal strain, and short time creep. Better coupling constitutive relations require further experimental and theoretical research.

REFERENCES

[3-1] Z. Guo, W. Li, Deformation behavior and constitutive relation of concrete under different stress-temperature paths, China Civil Engineering Journal 26 (5) (1993) 58–69.

[3-2] J. Nan, Z. Guo, X. Shi, Temperature-stress coupling constitutive relationship of concrete, Journal of Tsinghua University (Science and Technology) 37 (6) (1997) 87–90.

[3-3] G.A. Khoury, B.N. Grainger, P.J.E. Sullivan, Transient thermal strain of concrete: literature review. Conditions within specimen and behavior of individual constituents, Magazine of Concrete Research 37 (1985) 131–143.

[3-4] M.S. Abrams, Compressive strength of concrete at temperature to 1600 °F, ACI SP-25, Detroit (1971).

[3-5] J. Nan, Z. Guo, X. Shi, Experimental investigation of deformation behavior of concrete under heating-cooling cycles, Building Science 52 (2) (1997) 16–21.

[3-6] G.A. Khoury, B.N. Grainger, P.J.E. Sullivan, Strain of concrete during first heating to 600 °C under load, Magazine of Concrete Research 37 (133) (1985) 195–215.

[3-7] G.A. Khoury, B.N. Grainger, P.J.E. Sullivan, Strain of concrete during first cooling from 600 °C under load, Magazine of Concrete Research 38 (134) (1986) 3–12.

[3-8] L. Jiang, Z. Guo, Experimental investigation on strength and deformation of concrete under actions of stress and temperature, Symposium of the 7th National Conference on Structural Mechanics in Reactor Technology, Shanghai (1992) 442–447.

[3-9] Y. Anderberg, Predicted fire behavior of steel and concretestructures,InternationalSeminaronThreeDecades of Structural Fire Safety (1983) 115–137.

CHAPTER 4

Mechanical Behavior and Constitutive Relation of Reinforcement at Elevated Temperatures

4.1 TESTING METHOD AND DEVICE

4.1.1 Testing Program

Devices for testing the mechanical behavior of reinforcement at elevated temperatures include three systems: a loading system, a system for heating and controlling temperature, and a system for measuring and recording data. Such a testing device has been developed and manufactured in house in the Structural Engineering Laboratory of Tsinghua University (Fig. 4-1).

1. The loading system

The existing lever-type tension–compression testing machine (100 kN) is used as the main loading device. After the reinforcement specimen is clamped down, the tensile force can be applied and controlled by hand or electricity

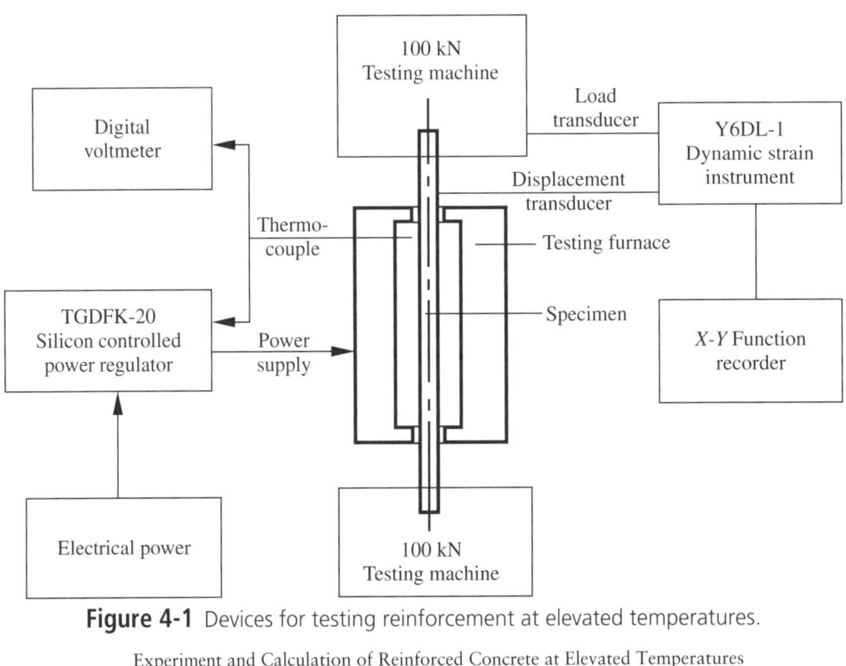

Figure 4-1 Devices for testing reinforcement at elevated temperatures.

Experiment and Calculation of Reinforced Concrete at Elevated Temperatures

and its signal is transmitted through the load transducer.

In order to suspend the testing furnace, a special horizontal slipping track is installed perpendicular to the plane of the loading screws in the testing machine. After the furnace is moved along the track and into the loading position in the testing machine, testing at elevated temperatures can be started. Usually, the furnace is moved away from the loading position, and the regular use of the testing machine is not affected.

2. The system for heating and controlling temperature

A furnace for testing reinforcement at elevated temperatures cannot be purchased on the market and must be developed by researchers. Reinforcement with a small diameter, generally $d \leq 32$ mm, is used in structural engineering practice. The value of the thermal conductivity of steel is high (see Section 5.2.2) and heat conducts quickly through it. Therefore, only one testing furnace is used for the heating operation, and the preheating furnace is not necessary.

The testing furnace is composed mainly of chamber brick, opening brick, heating wire, an insulation layer, and an outer wall (Fig. 4-2). The chamber brick is tubular in shape with a height of 200 mm, and external and internal diameters of 88 mm and 52 mm, respectively; the tube wall contains 12 longitudinal holes. The heating wires (4 kW) pass through these holes. Two opening bricks protect both ends of the chamber brick, and each has a central hole 40 mm in diameter. The insulation layers are composed of refractory siliceous aluminum felts, which are wrapped around the chamber brick. The outer wall of the furnace is cylindrical in shape and is manufactured from stainless steel plate; the external diameter is 200 mm.

The outer ends of the heating wires in the furnace are connected to a silicon controlled power regulator (type TGDFK-20, same as in Fig. 1-1), which supplies electrical energy for heating. The thermocouple is extended into the chamber and

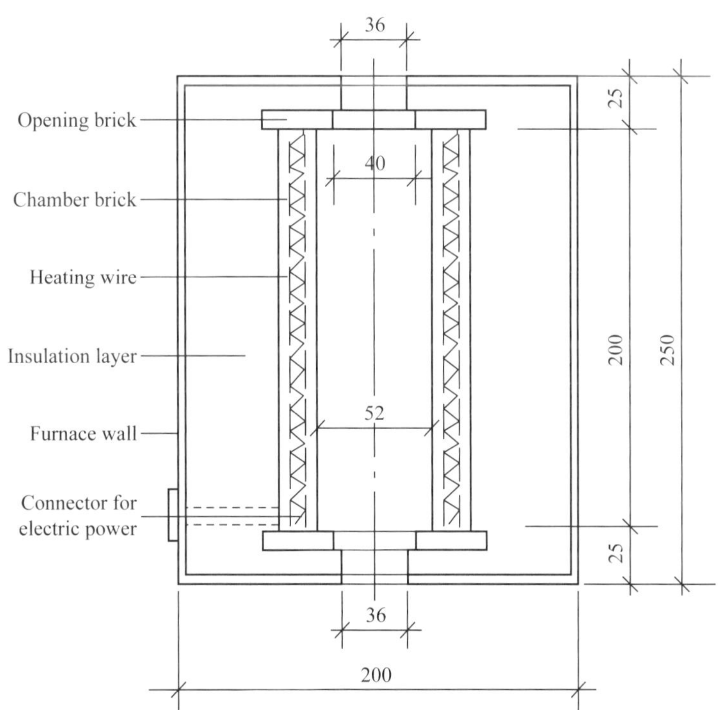

Figure 4-2 Construction of the furnace for testing reinforcement at elevated temperatures.[4-1]

the outer ends are connected to the regulator, which controls the temperature in the chamber. A digital voltmeter is also connected to the thermocouple, parallel to the regulator, and displays the temperature (voltage) in the chamber instantaneously.

This device setup has been used many times for testing reinforcement at elevated temperatures. The maximum temperature reaches 1000 °C, the maximum heating velocity is 50 °C/min, the temperature control is accurate within ±0.5 °C, and the temperature on the outer wall of the furnace is <60 °C after being heated for several hours. The testing furnace is thermally insulated and displays stable thermal behavior. It provides a longer area of uniform temperature in the middle and is convenient to use.

3. System for measuring and recording data
The factors that need to be measured during the testing of reinforcement at elevated temperatures are temperature, stress (tensile force), and deformation. The thermocouple in the furnace and the load transducer in the testing machine supply the information on temperature and tensile force, respectively.

The strain value of the reinforcement is difficult to measure directly because of the high temperature in the furnace, but can be transmitted outside the furnace and measured indirectly. Two pairs of stretched bars, specially designed and manufactured, are clamped using strong springs on the top and bottom of the reinforcement, and extend out of the furnace. Two displacement transducers are set up between the ends of the corresponding stretched bars (Fig. 4-3(a)) and

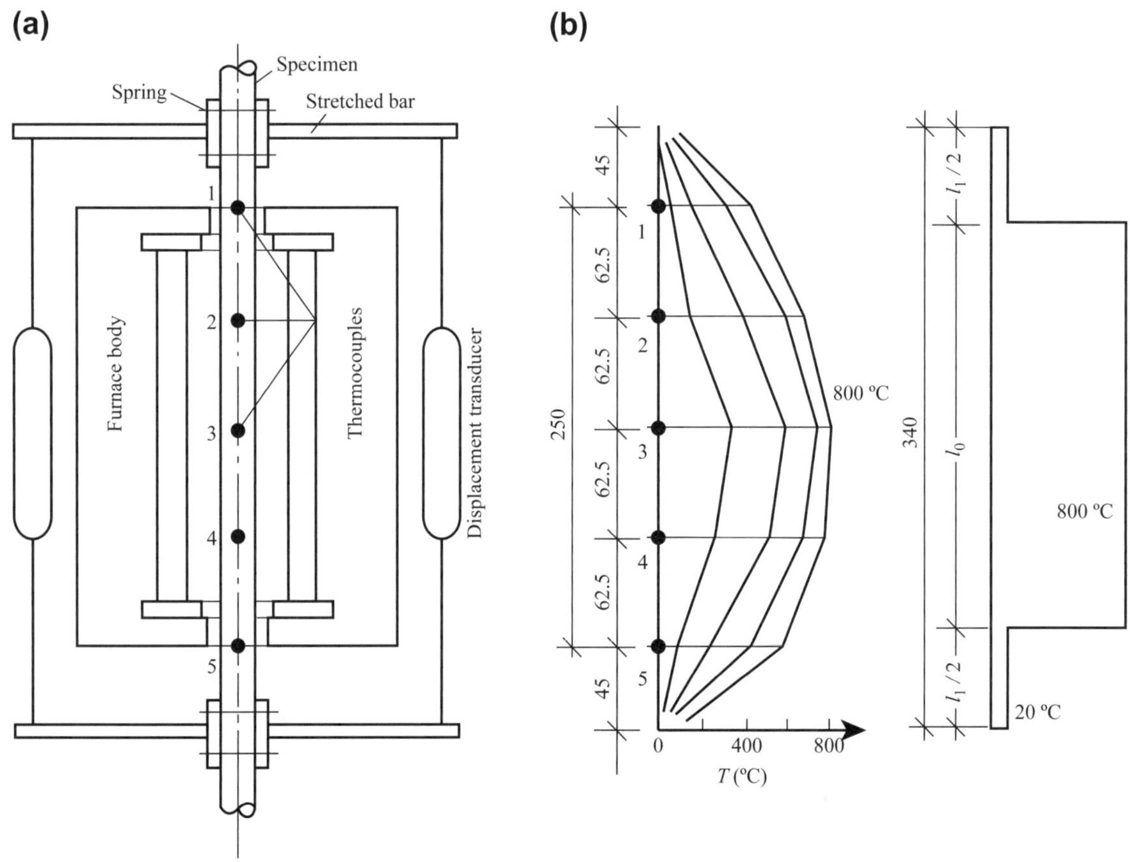

Figure 4-3 Measurement and calculation of the strain of a specimen: (a) measuring method; (b) temperature distribution.

connected in series to form a complete bridge circuit for measuring the average strain of the reinforcement.

4.1.2 Test Contents and Data Processing

The basic mechanical behavior of reinforcement at elevated temperatures includes yield strength, ultimate strength, stress–strain curve, and elastic modulus at different temperatures. In addition, the strength and deformation of reinforcement under different temperature–stress paths and after cooling and creep at elevated temperatures greatly influence the behavior of the structure at elevated temperatures and after a fire accident.

There are five grades of strength of reinforcement used in concrete structural engineering. Generally, the diameter of the reinforcement is not greater than 32 mm. The diameter, the yield, and the ultimate strength at normal temperature of the reinforcement specimens selected for testing are listed in Table 4-1. The testing temperature is in the range of 20–800 °C and not less than three specimens are used for each testing condition.

The general testing process for reinforcement at elevated temperatures is as follows. The testing furnace is moved into the loading position of the testing machine, and the reinforcement specimen is passed through the chamber; its upper end is clamped to the upper head of the machine and the lower end is free. After the gaps between the specimen and the furnace opening are blocked with refractory fiber, the testing furnace is heated to the predetermined temperature, which is maintained for 5–10 min. The displacement transducers are set up and the measuring and recording instruments are adjusted. Finally, the lower end of the specimen is clamped to the lower head of the machine and the specimen is loaded at the velocity of 0.5 MPa/s until failure.

Various types of data are measured and recorded during the testing process and values for the temperature, stress, and strain of the reinforcement can be obtained using the methods described below.

The temperature in the middle part of the furnace chamber is taken to be the approximate temperature of the specimen, because its diameter is small, heat conducts quickly through it, and the temperature distributes uniformly on its section.

The average stress calculated from the original area of the specimen is taken as its nominal stress, without considering the reduction in tensile area at elevated temperatures and after yielding.

The strain value of the reinforcement at elevated temperatures has to be calculated and modified according to the deformation value (δ) measured by the displacement transducers. The temperature distribution in the chamber is measured during testing and it varies with time as shown in Fig. 4-3(b). The temperature distributes quite nonuniformly within the length (l = 340 mm) of deformation measured. The temperature in the middle part of the chamber reaches the maximum and distributes uniformly, but the temperature near the opening of the chamber reduces obviously, and the temperature at the fixed ends of the stretched bars, which extend out of the furnace, approaches room temperature. In order to simplify the calculation, the measuring length is divided approximately into two parts, i.e., the testing part and the part at room temperatures, according to the principle that the area of the measured temperature distribution is equal to that of the equivalent temperature distribution. Then, the length of the two parts (l_t and l_0) are determined and the strain value of the specimen

TABLE 4-1 Types of Reinforcement Tested at Elevated Temperatures[4-1]

Strength grade	I	II	III	IV	V
Diameter (mm)	12	12, 25	12	12	5
f_y (MPa)	288	426	504	579	1274
f_u (MPa)	376	577	633	930	1681

at testing temperature can be calculated using the following formula:

$$\varepsilon_s(T) = \frac{\delta - (\sigma / E_S)\, l_t}{l_0} \tag{4.1}$$

where δ is the deformation value (mm) measured by the displacement transducers, σ is the stress of the reinforcement, E_s is elastic modulus of the reinforcement at room temperature, and l_0 and l_1 are the length at the testing and normal temperatures, respectively, based on the equivalent figure of temperature distribution.

4.2 TENSILE STRENGTH AT ELEVATED TEMPERATURE

4.2.1 Characteristics and Ultimate Tensile Strength

The path of loading at constant temperature (T–σ) is mainly used for testing the tensile strength of reinforcement at elevated temperatures. The reinforcement specimen is put into the testing furnace, heated to the predetermined temperature which is maintained for several minutes. Then, the specimen is loaded until failure and the deformation process and the strength value are measured simultaneously.

The color on the surface of the reinforcement varies gradually during heating. The color is gray-black and the same as that at room temperature when $T \leq 300$ °C; it becomes darker when $T > 400$ °C, slightly red when $T = 600$ °C, and red-black after $T > 700$–800 °C.

The specimen is taken out of the furnace at the end of the test, and its failure mode is observed after it is cooled. The specimen is broken into two and the fracture section is obviously necked down when the testing temperature $T \leq 300$ °C. The failure mode of the specimen tested when $T \leq 300$ °C is same as that tested at room temperature. When $T = 400$–600 °C, the failed specimen still has local neck and its surface layer is lost, but the length of the neck increases and the diameter decreases as the testing temperature increases. When $T = 700$–800 °C, the local neck of the specimen is hardly seen, the middle part elongates uniformly at higher temperature, and the diameter clearly decreases; the loss of the surface layer is significant. However, the length and diameter of both ends of the specimen at lower temperature vary considerably less compared to those before testing.

The ultimate tensile strength of the reinforcement at different temperatures (f_u^T) is taken as the quotient between the maximum tensile force during testing and the original area of the section, or as the stress value at the highest point of the stress–strain curve. The experimental results of the reinforcements for the five strength grades are represented by the relative values (f_u^T / f_u) and shown in Fig. 4-4,

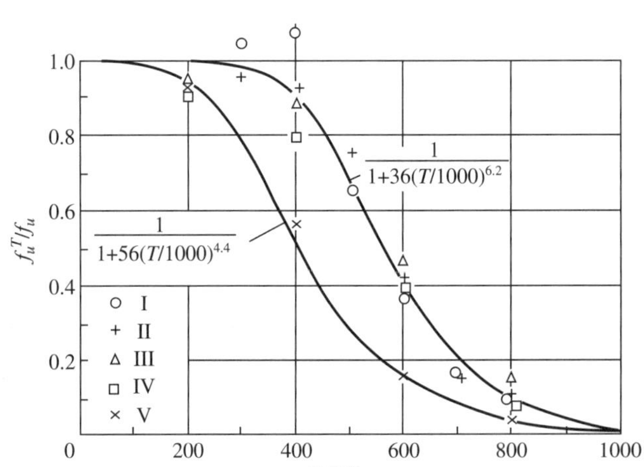

Figure 4-4 Ultimate tensile strength of reinforcement at elevated temperatures.[4-2]

where f_u is the ultimate tensile strength of the reinforcement of the same grade at room temperature.

The ultimate tensile strength of the reinforcements of all grades decreases slightly ($f_u^T/f_u > 0.9$) when the testing temperature $T \leq$ 300 °C, but decreases sharply within the range $T =$ 300–800 °C, and reaches a rather low value ($f_u^T/f_u \approx 0.04$–0.12) at $T =$ 800 °C. However, the ultimate strength of the reinforcements of different types and strength grades varies.

The ultimate strength of grade I reinforcement increases slightly ($1.0 < f_u^T/f_u < 1.08$) when $T =$ 300–400 °C, but it decreases when $T >$ 500 °C and varies similarly to grades II, III, and IV.

The grade V specimen is heat-treated high-strength wire ($\phi 5$). Distortion of the metal crystal construction induced during heat-treating is relieved gradually during heating and the heat-treated function is basically lost when the temperature exceeds 400 °C.[4-3,4-4] Therefore, the relative strength at elevated temperatures decreases sharply, and it is about $f_u^T/f_u \approx 0.04$ at $T =$ 800 °C and is far smaller than that of grades I–IV (Fig. 4-4). However, the grade V reinforcement contains more alloy elements and the absolute value of its ultimate strength (f_u^T) is greater than that of grades II–IV when $T \leq$ 600 °C, but approaches that of grades I–IV when $T >$ 600 °C.

According to the above experimental results, the ultimate tensile strength at elevated temperatures of the reinforcements of the five strength grades can be summarized into two simplified regression formulas:

$$\text{Steel grades I, II, III, IV:} \frac{f_u^T}{f_u} = \frac{1}{1 + 36\,(T/1000)^{6.2}} \tag{4.2a}$$

$$\text{Steel grade V:} \frac{f_u^T}{f_u} = \frac{1}{1 + 56\,(T/1000)^{4.4}} \tag{4.2b}$$

Experimental investigations[2-6] also show that the relative ultimate strength (f_u^T/f_u) of hard wire (grade V) is lower than that of mild steel (other grades), and the difference between them may reach 25–35% within $T =$ 400–600 °C. Grade V wire is mainly used in prestressed concrete structures and it loses strength considerably at elevated temperatures, so the fire-resistant behavior of the corresponding structural member should be emphasized.

4.2.2 Yield Strength at Elevated Temperatures

1. Method for determining the value of yield strength

The yield strength (f_y) of reinforcement is the main factor that decides the bearing capacity and ductility of reinforced concrete structures, and is the strength limit used in the design of the structure. Although the ultimate tensile strength of reinforcement is much higher than the yield strength ($f_u/f_y \approx 1.5$), the strain corresponding to the ultimate strength is several times the yielding strain. Normally, when the strain at the ultimate strength is reached, the structure fails because of too much deformation or loss of stability. Therefore, the ultimate strength of reinforcement cannot be used in the structural design, but can be used as an additional safety reserve.

The tensile–elongation (σ–ε) curve of mild steel (grades I–IV) has an apparent yielding step, and the lower yielding point on the curve is taken as the yield strength (f_y) of the steel.[0-2] But the tensile–elongation curve of hard steel (grade V) has no yielding step and the stress corresponding to the residual strain $\varepsilon = 2 \times 10^{-3}$ is taken as the nominal yield strength of the steel. However, the tensile–elongation curve of the reinforcement at elevated temperatures (Fig. 4-9) also has no apparent yielding step, and if the nominal yield strength corresponding to the residual strain $\varepsilon = 2 \times 10^{-3}$ is used, it does not fit well with the trend of the elongation curve. Therefore, another method has to be found to determine the yield strength of the reinforcement at elevated temperatures.

The following methods have been used and compared[4-1] to determine the yielding points of the materials or structures:

- Residual strain: described earlier.
- Energy: determined according to the principle that the areas under the experimental and nominal stress–strain (or load–displacement for structure) curves should be equal.

- Ductility ratio: determined by calculating the yielding strain ($\varepsilon_y = \varepsilon_u/\mu$) according to the strain at ultimate strength (ε_u) and the ductility ratio required (μ).
- Yield/ultimate strength ratio: determined by calculating ($f_y^T = f_u^T f_y / f_u$) from the ultimate strength at elevated temperatures (f_y^T) and the yield/ultimate strength ratio of the reinforcement at room temperature (f_y/f_u).

All the yield points on the tensile–elongation curve of reinforcement at elevated temperatures determined by these methods are not quite correct. The yield strains of the reinforcement (ε_y^T) obtained are too high and the structure may fail because the corresponding displacement exceeds the limit, although the material still retains some strength reserve.

From the physical concept of the yielding point, it is reasonable to define it as follows: plastic strain of the material starts to develop or the strain rate of the material increases suddenly, and a turning point forms on the tensile–elongation (stress–strain) curve. The practical method and criteria are described below.

The stress–strain curve of the reinforcement measured during testing is divided into finite stress increments ($\Delta\sigma$), and the corresponding strain increments ($\Delta\varepsilon_i$) are calculated one by one. When the strain increasing rate of the (n + 1)th increment exceeds several times (a) that of the nth increment, the slope of the curve decreases obviously and a turning point is formed on the curve. Then, this point is defined as the theoretical yielding point. This means that when

$$\frac{\Delta\varepsilon_{n+1}/\Delta\sigma}{\Delta\varepsilon_n/\Delta\sigma} = \frac{\Delta\varepsilon_{n+1}}{\Delta\varepsilon_n} \geq a \tag{4.3a}$$

the yield strength and the corresponding strain of the reinforcement at elevated temperatures are

$$f_y^T = n\cdot\Delta\sigma,\ \varepsilon_y^T = \sum_{i=1}^{n}\Delta\varepsilon_i \tag{4.3b}$$

If the deformation processes (e.g., bending moment–curvature, load–displacement) of the concrete structures and their members have no apparent yielding step, the same method can also be used for defining the unified standard to determine the yield bending moment, and yield bearing capacity, and then to calculate the ductility ratios. Of course, the magnitude of the stress (or force) increment and the bound (a) of the deformation rate can be adjusted properly.

2. Yield strength at elevated temperatures

The stress increment is taken as one-tenth of the ultimate tensile strength at T = 800 °C of the reinforcement for each grade, i.e., $\Delta\sigma \approx 0.1 f_u^{800} = (3.3 \sim 9.6)\text{MPa} \approx (0.005 \sim 0.020) f_y$, and the bound of the strain increasing rate is taken as a = 2. Then, the yield strengths of the reinforcements at different temperatures are calculated and the relative values (f_y^T / f_y) are shown in Fig. 4-5.

The yield strength of grade I–IV reinforcements (f_y^T) decreases monotonically as the testing temperature increases; the general variation regularity is similar to that of the ultimate tensile strength (f_u^T). However, there are some differences between them. The yield strength of the reinforcement decreases by about 10–15% at T= 200 °C and has lower relative value (f_y^T / f_y) when T = 200–500 °C, so the yield/ultimate strength ratio (f_y^T / f_u^T) (Fig. 4-6) reduces gradually within T = 20–500 °C, but increases again gradually within the range T = 600–800 °C. However, the variable amplitude of the ratio is not large and most of the ratios are within the range of $f_y^T / f_u^T = 0.6$–0.8.

The relative yield strength (f_y^T / f_y) of grade V reinforcement is lower than that of grades I–IV and the variation regularity is similar to that of its relative ultimate strength (f_u^T / f_u) but with a slightly lower value. Therefore, the yield/ultimate strength ratio (f_y^T / f_u^T) approximates to a constant, possibly because the obvious yielding point does not appear originally for grade V reinforcement at room temperature.

According to the above experimental results, at elevated temperatures the yield strength of reinforcements of all the five grades can also

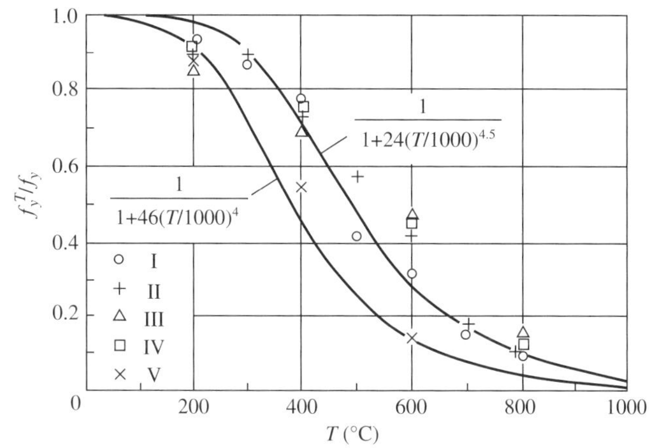

Figure 4-5 Yield strength of reinforcement at elevated temperatures.[4-1]

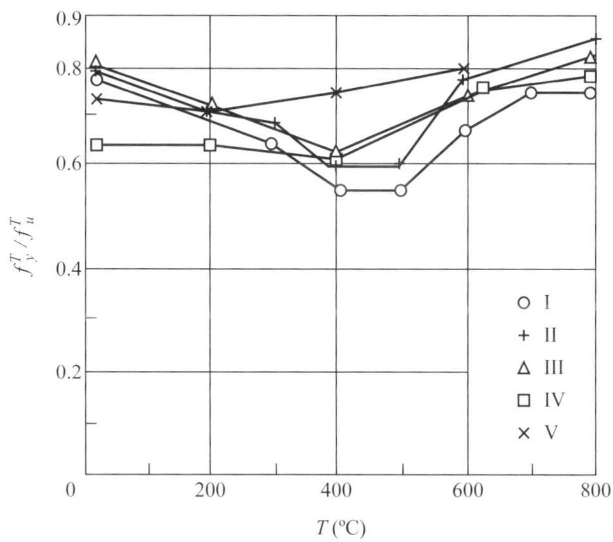

Figure 4-6 Yield/ultimate strength ratio of reinforcement at elevated temperatures.[4-1]

be summarized in two simplified regression formulas:

$$\text{Steel grades I, II, III, IV:} \quad \frac{f_y^T}{f_y} = \frac{1}{1+24\,(T/1000)^{4.5}} \tag{4.4a}$$

$$\text{Steel grade V:} \quad \frac{f_y^T}{f_y} = \frac{1}{1+46\,(T/1000)^{4}} \tag{4.4b}$$

The theoretical curve is compared with the experimental data in Fig. 4-5.

Comparing the grade II reinforcement specimens of diameter $\varphi 25$ mm $\varphi 12$ mm, there is no apparent difference between the ultimate tensile strengths measured during the testing temperature range $T = 600$–800 °C.

4.2.3 Influence of Temperature–Stress Path

The reinforcement, like the concrete, in a structure subjected to a fire accident has to experience a complicated long-term temperature–load (stress) history. The mechanical behavior and the variation of the reinforcement under different

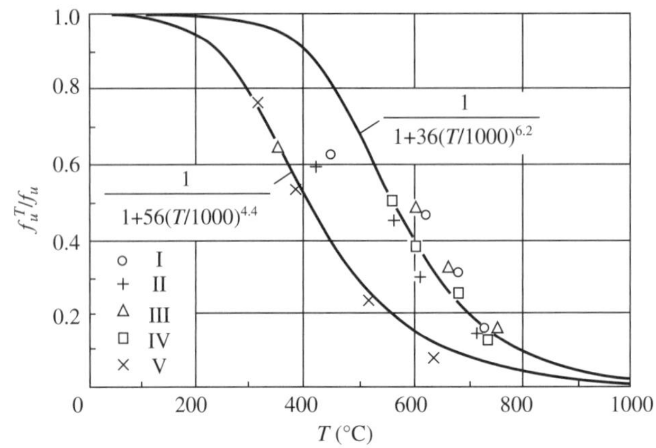

Figure 4-7 Ultimate temperature/tensile strength of reinforcement under the path of heating under constant load.[4-1]

temperature–stress paths must be investigated by proper testing.

The testing methods for reinforcements at elevated temperatures introduced above correspond to the path of loading at constant temperature (T–σ). Another extreme path is that of heating under constant load (σ–T, Fig. 3-1) and its testing procedure is as follows. The specimen is placed inside the testing furnace and both its ends are clamped by the testing machine. It is then loaded and reaches the predetermined stress level (σ/f_u) under ambient temperature. After the gaps near the furnace openings are obstructed, the furnace is supplied with electrical power and heated. In the meantime the testing machine is operated and the specimen is elongated simultaneously in order to keep the stress constant. The specimen is heated continuously until failure and the ultimate temperature is measured.

The ultimate temperatures of the five grades of reinforcements under different stress levels ($\sigma/f_u = f_u^T/f_u$) are presented in Fig. 4-7. Compared with the theoretical curves of the ultimate tensile strengths of the reinforcements under the path of loading at constant temperature (T–σ, Eqn (4.2)), the strengths of the reinforcements of grades I–IV (mild steel) under the path (σ–T) is lower than that under the path (T–σ) when $T < 500$ °C, but the strength under both paths is not obviously different when $T > 500$ °C. The strength of grade V reinforcements (hard steel) under both extreme paths is not obviously different throughout the temperature range.

Grade II reinforcement diameter ϕ12 mm is used for all the tests under various complicated temperature–stress paths. The ultimate strengths at elevated temperatures (f_u^T) of the reinforcements are tested and measured under five temperature–stress paths (paths 3–7 in Fig. 4-8(b)), and they are compared in Table 4-2 with the experimental results under two extreme paths (1 and 2).

The average ultimate tensile strength and the temperature of the specimens are 271 MPa and 598 °C, respectively, under all seven paths. The fluctuation of the tensile strength and temperature is ≤±11.8%, which is within the reasonable error range of testing at elevated temperatures. Therefore, the experimental results demonstrate that the temperature–stress path does not obviously influence the ultimate tensile strength of the reinforcement.

Some experiments[1-14,9-12] have shown that the ultimate tensile strength and yield strength of a reinforcement after heated to 900 °C and then cooled down (naturally or in water) are respectively approaching, or even slightly higher than, that of the reinforcement at room temperature without heating-cooling.

Therefore, according to the existing experimental results, it is considered that the different

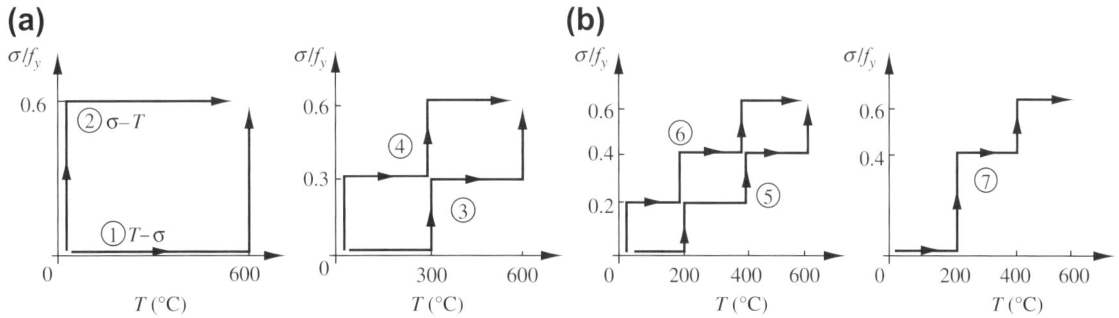

Figure 4-8 Various temperature–stress paths for testing reinforcement[4-1]: (a) extreme paths; (b) complicated paths.

TABLE 4-2 Comparison of the Ultimate Strength of Reinforcements Under Different Temperature–Stress Paths

	Temperature–stress path							
	1	2	3	4	5	6	7	Average
T_{max} (°C)	580	563	572	638	625	590	618	598
Relative difference (%)	−3.0	−5.9	−4.3	+6.7	+4.5	−1.3	+3.3	
f_u^T (MPa)	249	268	303	270	270	266	274	271
Relative difference (%)	−8.1	−1.1	+11.8	−0.4	−0.4	−1.8	+1.1	

temperature–stress paths and the heating–cooling cycles do not obviously influence the strength of the reinforcement. This characteristic of steel is caused by its properties, manufacture, and working technology and is totally different from that of concrete.

4.3 TENSILE STRAIN AND STRESS–STRAIN CURVE AT ELEVATED TEMPERATURE

4.3.1 Characteristics of Tensile Strain

Tension testing and strain calculations of reinforcement at constant temperature follow the methods introduced in Section 4.1.2 and the stress–strain curves of the reinforcements of the five strength grades under different temperature (T_{max} = 800 °C) are obtained (Fig. 4-9). The shapes and the variations in these curves illustrate the characteristics of the reinforcement deformation at elevated temperatures.

The stress–strain curve of grade I reinforcement at room temperature has an apparent and long (about 5×10^{-3}) horizontal yielding step, which is followed by a hardening part of stress increase and the peak point; the specimen is then necked and the stress descending branch is formed. When the testing temperature increases but $T < 400$ °C, the yield strength of the reinforcement decreases, the turning at the yield point is blurred gradually, the yielding step contracts and slopes gradually, and the hardening part and the descending branch still follow obviously. When $T > 500$ °C, the yielding step disappears, the ultimate strength decreases quickly, and the stress–strain curve is composed of the ascending branch with the slope decreasing monotonically and the descending branch after the peak point. The higher the testing temperature ($T > 700$ °C), the flatter the curve top and the less clear the descending branch. When the test temperature ranges from T = 20 °C – 200 °C, the values of the ultimate elongation of the specimen (reinforcement) measured after failure are less different. The ultimate elongation decreases when $T > 300$ °C and the minimum value of the

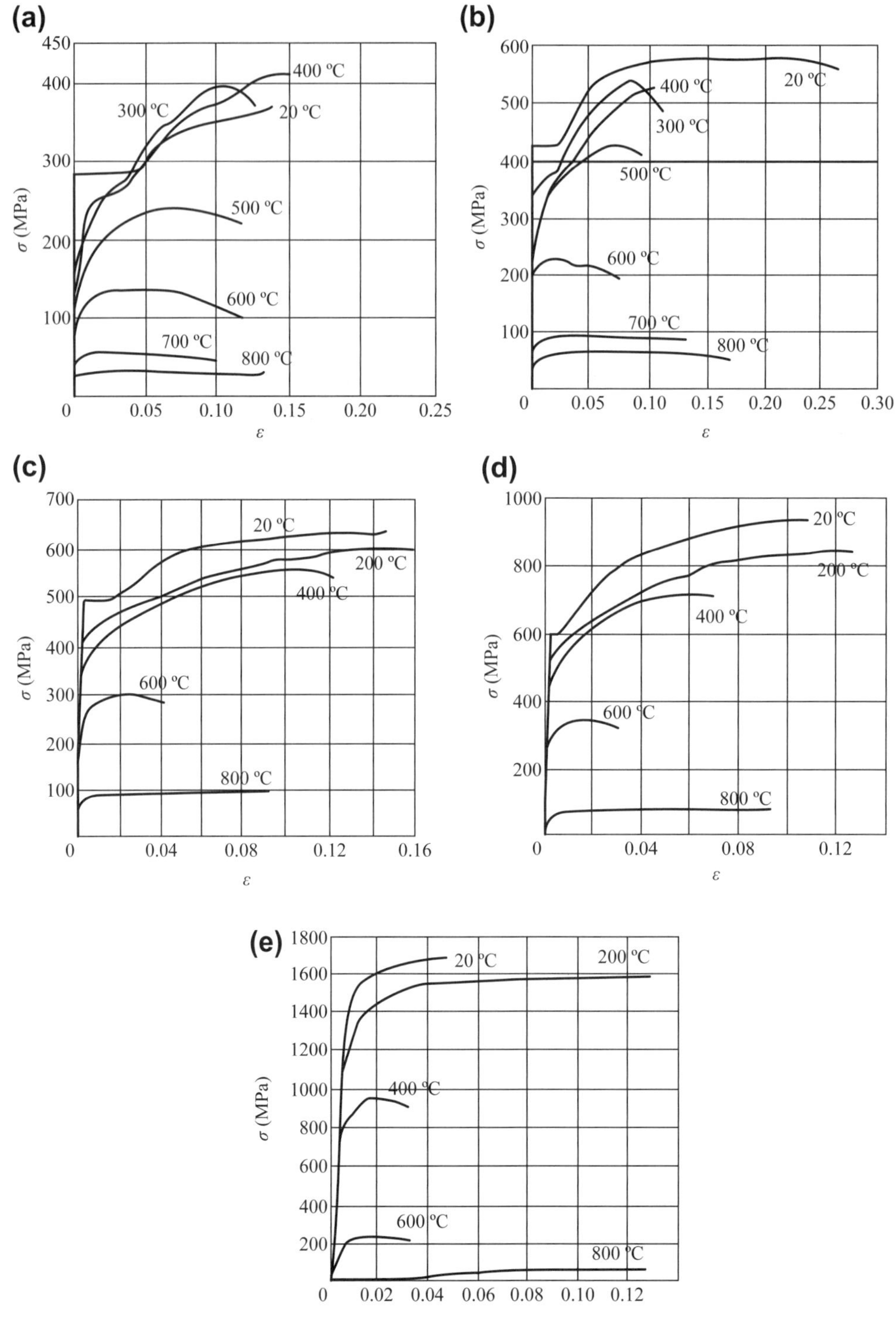

Figure 4-9 Stress–strain curves for reinforcements at different temperatures[4-2]: (a) steel grade I; (b) steel grade II; (c) steel grade III; (d) steel grade IV; (e) steel grade V.

ultimate elongation is measured at about T = 600–700 °C. When T > 800 °C, the ultimate elongation increases because the reinforcement softens and stable flow occurs.

The deformation behavior of grade II–IV reinforcements at elevated temperatures is similar to that of grade I, but varies gradually. When the strength grade of the reinforcement increases, the yielding step of the stress–strain curve at room temperature is clear, but the length contracts to (20–10) × 10^{-3}, and the temperature corresponding to the disappearance of the yielding point is reduced to 200 °C from 400 °C. The ultimate strength decreases sharply when T > 400 °C. The ultimate elongation of the specimen after failure decreases gradually when T > 300 °C and reaches the minimum value when T = 600 °C, which is only about one-fourth of that at room temperature. However, the ultimate elongation increases again when T > 700 °C.

The stress–strain relationship for grade V reinforcement at room temperature, and also at elevated temperatures, is a continuous curve without the yielding step. The ultimate tensile strength of the reinforcement decreases quickly when T > 200 °C, and the ultimate elongation of the failed specimen reaches the minimum value within the range of T = 400–600 °C and increases greatly when T > 800 °C.

In the stress–strain curves of reinforcement under different temperatures, the shapes of the top part and the descending branch can be divided into two categories. One has an apparent peak and a descending branch. When the testing temperature is not too high ($T \le 600$ °C), the specimen is necked down locally soon after the peak tensile load (or f_u^T) is reached, then the area of its critical section decreases and the load reduces, and finally the specimen suddenly breaks in two. The stress–strain curve is obtained when the stress (or strength) of the specimen is calculated by using its original section area. However, if the practical area of the critical section is used in the calculation, the stress after the peak load (f_u^T) still rises slowly. Another category of the stress–strain curve has a gentle top part and the descending branch is not clear. For this category, the specimen is tested at a higher temperature (T > 700 °C) and is softened with a much lower elastic modulus. The part of the specimen with high temperature elongates uniformly and the diameter decreases uniformly without a local neck, so the specimen has a large ultimate elongation.

4.3.2 Equation of Stress–Strain Curve

The stress–strain relationships (curves) of reinforcement at room temperature are usually divided into two types. For grade I–IV reinforcements, the curve has an apparent yielding step and is simulated by an ideal elastoplastic model, i.e., the stress and strain increase proportionally and the elastic modulus (E_s) is a constant when $\sigma < f_y$, and the stress is taken as a constant (f_y) when the strain is greater than the yield strain ($\varepsilon \ge \varepsilon_y$). For grade V reinforcement, the relation is usually simulated by a continuous equation.[0-2]

The stress–strain curve of the reinforcement at elevated temperatures is different from that at room temperature. Observing and comparing the stress–strain curves of the reinforcements of all strength grades, the shapes of the curves at different temperatures when T > 200 °C are similar, i.e., the strain before yielding ($\sigma < f_y^T$) is relatively small and the yielding step is not clear. The strain increase accelerates after yielding ($\sigma \ge f_y^T$) and the curve tends to be horizontal when the ultimate tensile strength is reached. Therefore, a unified mathematical model (Fig. 4-10) is used for simulating the stress–strain curve of the reinforcement at elevated temperatures to simplify the calculation with sufficient accuracy, and it can satisfy the analysis requirement for structures at elevated temperatures.

The stress–strain curve for reinforcements at high temperatures (T > 200 °C) can be divided into two parts, the elastic branch before yielding and the hardening branch after yielding, and the equations of both branches are presented separately:

$$\varepsilon \le \varepsilon_y^T, \ \sigma = \frac{f_y^T}{\varepsilon_y^T} \cdot \varepsilon = E_s^T \cdot \varepsilon \qquad (4.5a)$$

$$\varepsilon_y^T \le \varepsilon \le \varepsilon_u^T, \quad \sigma = f_y^T + (f_u^T - f_y^T) \cdot \eta, \tag{4.5b}$$

where

$$\eta = (1.5\xi - 0.5\xi^3)^{0.62} \tag{4.6}$$

The relative coordinates used for the hardening branch are

$$\xi = \frac{\varepsilon - \varepsilon_y^T}{\varepsilon_u^T - \varepsilon_y^T}, \quad \eta = \frac{\sigma - f_y^T}{f_u^T - f_y^T} \tag{4.7}$$

Equation (4.6) is a simplified regression formula obtained from the analysis of the partial experimental data (Fig. 4-10(b)) and it satisfies the boundary conditions, i.e., $\sigma = f_u^T$ and $d\sigma/d\varepsilon = 0$, when $\varepsilon = \varepsilon_u^T$.

The ultimate strain (ε_u^T) of the reinforcement at elevated temperatures has a high value, which is hardly exceeded in a practical structure during a fire accident, so the equations above can satisfy the requirement for thermal analysis of the structure. If necessary, a plastic horizontal line can be attached after $\varepsilon \ge \varepsilon_u^T$ when $T > 700$ °C, i.e., the stress is taken as a constant $\sigma = f_u^T$. Or a descending branch may be attached when $T < 700$ °C.

The yield strength f_y^T and the ultimate strength f_u^T of reinforcement in these constitutive relations can be calculated using the formulas presented earlier. The experimental data for the corresponding yield and ultimate strains (ε_y^T and ε_u^T) are shown in Fig. 4-11.[4-1]

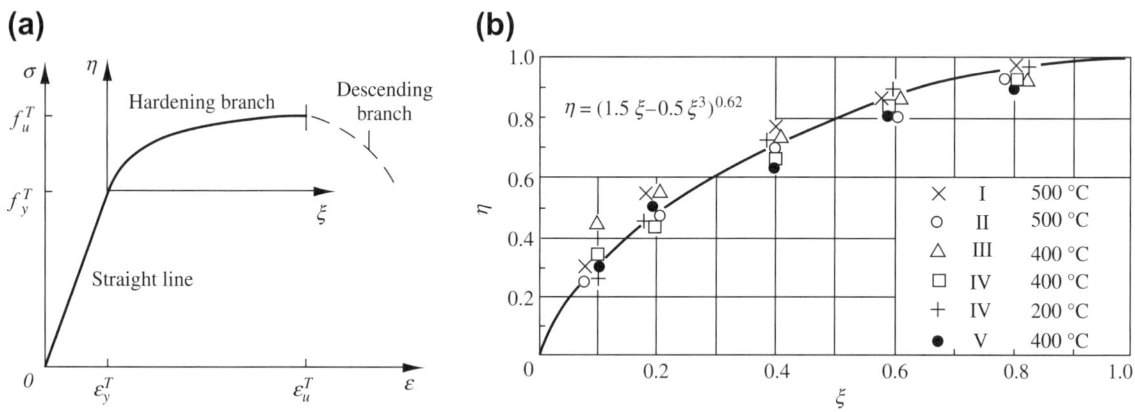

Figure 4-10 Model of the stress–strain curve for reinforcement at elevated temperatures: (a) composition of the curve; (b) the hardening branch of the curve.

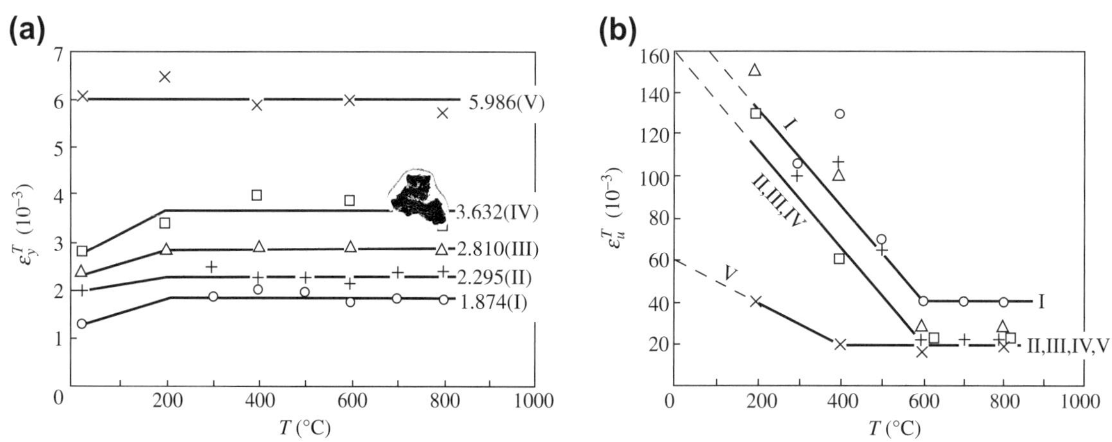

Figure 4-11 Yield and ultimate strains of reinforcement at elevated temperatures[4-1]: (a) ε_y^T; (b) ε_u^T.

The yield strain of each grade of reinforcement varies irregularly with temperature when $T \geq 200$ °C, but the difference between them is less. Therefore, the yield strain of each grade of reinforcement can be taken approximately as a constant when the structure and its members at elevated temperatures are analyzed. The yield strains of the reinforcements of various strength grades at elevated temperatures can be calculated by a unified formula:

$$\varepsilon_y^T = \alpha \cdot \varepsilon_y \tag{4.8}$$

where $\varepsilon_y = f_y/E_s$ is the yield strain of the reinforcement at room temperature. The strain ratio a suggested in Table 4-3 can be used in the analysis.

The ultimate strains ε_u^T of the various strength grades of reinforcements at elevated temperatures decrease as the testing temperature increases and reach minimum values at $T = 600$ °C, i.e., 40×10^{-3} for grade I or 20×10^{-3} for grades II–V. The minimum values are still greater than the permissible maximum strain value (10×10^{-3}) of steel (reinforcement), which is listed in the design codes for steel and reinforced concrete structures[0-1] and is used for preventing failure of structures because of excessive deformation. Although the ultimate strain increases when $T > 700$ °C, the corresponding stress–strain curve is rather flat (Fig. 4-9). Therefore, if the minimum value at $T = 600$ °C is taken, the analysis of the structure at elevated temperatures will not be influenced. The ultimate strains for all five grades of reinforcements at elevated temperatures ($T > 200$ °C) can be calculated separately as follows:

$$\begin{aligned} &\text{Grade I:}\ \varepsilon_u^T = 0.18 - 0.23\left(\frac{T}{1000}\right) \geq 0.04 \\ &\text{Grades II, III, IV:}\ \varepsilon_u^T = 0.16 - 0.23\left(\frac{T}{1000}\right) \geq 0.02 \\ &\text{Grade V:}\ \varepsilon_u^T = 0.06 - 0.1\left(\frac{T}{1000}\right) \geq 0.02 \end{aligned} \tag{4.9}$$

4.3.3 Elastic Modulus

The stress–strain relation ($\sigma < f_y^T$) before the reinforcement yields at elevated temperatures is simulated approximately to a straight line and its slope is the corresponding elastic modulus:

$$E_s^T = f_y^T / \varepsilon_y^T \tag{4.10}$$

The experimental data[4-1] for various grades of reinforcements tested at different temperatures are used and the values of their elastic modulus can be calculated (Fig. 4-12).

The value of the elastic modulus of reinforcement decreases as the temperature increases, and its variation is similar to that of the yield and ultimate strengths, except that its decreasing amplitude is greater. The elastic modulus of grade I–IV reinforcement starts to decrease more when $T \geq 200$ °C.

When Eqns (4.4) and (4.8) are substituted into Eqn (4.10), formulas for the elastic modulus of various strength grades of reinforcement at elevated temperatures are obtained, and the theoretical curves are compared with the experimental results in Fig. 4-12.

TABLE 4-3 Experimental and Suggested Values of Yield Strain of Reinforcement at Elevated Temperatures ($T \geq 200$ °C)

Grade of reinforcement	Yield strain at normal temperature ε_y (10^{-3})	Experimental value ε_y^T (10^{-6})		$\alpha = \varepsilon_y^T / \varepsilon_y$	
		Range	Average	Average	Suggested
I	1370	1759–2003	1874	1.368	1.36
II	2029	2127–2500	2295	1.131	
III	2381	2796–2821	2810	1.180	1.20
IV	2843	3295–3962	3632	1.278	
V	6067	5700–6407	5986	0.987	1.0

4.4 THERMAL STRAIN UNDER STRESS

The thermal deformation of concrete under the action of stress includes freely expanding strain (ε_{th}) and transient thermal strain (ε_{tr}), and its value is considerable, so it is the main part of the total deformation of concrete at elevated temperatures and is also a key factor in the coupling constitutive relation. This behavior of reinforcement has not been investigated fully up to now, and there is no final conclusion. Some experimental research and analyses have revealed a few important phenomena that are worth investigating further.

4.4.1 Freely Expanding Strain

The reinforcement specimen is heated to the predetermined temperature and is maintained at the same temperature under free conditions ($\sigma = 0$) until it does not deform any more, then the strain, the freely expanding strain (ε_{th}) of the reinforcement, is recorded. When the testing temperature is $T = 800$ °C, the time needed for heating is 20 min and the temperature is maintained for 40 min. The measured expanding strains for the five grades of reinforcements at temperature T = 20–800 °C are shown in Fig. 4-13.

Although the chemical components are different and the strength differences are considerable for the five grades of reinforcements, the expanding strains are similar at the same temperature. The expanding strains of all the reinforcements are less but the increases accelerate when $T \leq 200$ °C, and are almost proportional when $T \geq 300$ °C.

The freely expanding strain of reinforcement can be calculated following the empirical regression formula:

$$\varepsilon_{th} = 16\left(\frac{T}{1000}\right)^{1.5} \times 10^{-3} \tag{4.11}$$

Correspondingly, the linear expansion coefficient is

$$\overline{\alpha}_s = \frac{\varepsilon_{th}}{T} = 0.5\sqrt{T} \times 10^{-6} \tag{4.12}$$

where T is the temperature of the reinforcement (in °C).

4.4.2 Thermal Strain Under Constant Stress

The testing procedure for measuring the thermal strain of reinforcement under constant stress is as follows. The testing furnace is set up and the reinforcement specimen is clamped down and loaded under tension at room temperature. After

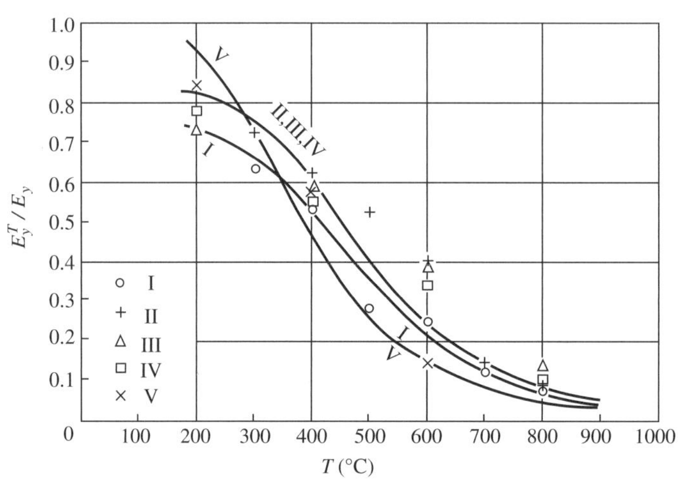

Figure 4-12 Elastic modulus of reinforcement at elevated temperatures.

the tensile stress reaches the predetermined level ($\sigma/f_y = 0.2$, 0.4, or 0.6), the furnace is heated by electrical power and the testing machine is operated simultaneously to keep the tension constant; the temperature and elongation deformation of the specimen are measured and recorded. To limit the influence of short time creep of the reinforcement at elevated temperatures, the specimen is heated continuously without the constant temperature stage. This takes about 20 min when the temperature reaches $T = 800$ °C.

The thermal strains of the five grades of reinforcements under different stress levels have similar variation regularity (Fig. 4-14), i.e., the strain of the reinforcement accelerates as the temperature increases continuously. The thermal strain is less when $T \leq 200$ °C and increases quickly when $T \geq 300$ °C, because the elastic modulus of the reinforcement at elevated temperatures decreases considerably and the creep appears unavoidably. The specimen at high stress level ($\sigma/f_y = 0.6$) fails because its strain diverges when the ultimate temperature (Fig. 4-7) approaches.

The thermal strains of the reinforcements at the same temperature increase approximately linearly with the stress. The values of the thermal strains of grade I–IV reinforcements (mild steel) under the same conditions are similar, but that of grade V is obviously larger. Empirical formulas for thermal strain are given by Lu.[4-1]

Compared with the freely expanding strain (Fig. 4-13), the thermal strain of reinforcement under the action of stress is less when the temperature $T < 500$–650 °C because the specimen is heated quickly without the constant temperature stages and the strain does not develop fully. However, the thermal strain is obviously larger when the temperature is even higher, because the elastic modulus decreases at elevated temperatures and the plastic strain increases quickly.

The total strain of the reinforcement under the path of heating under constant load (σ–T) includes the load causing strain at normal temperature and the heating causing strain under constant stress. The total strain of the reinforcement under the path of loading at constant temperature (T–σ) includes the freely expanding strain and the load (stress) causing strain at elevated temperatures. The values of both total strains are not equal, but there is a lack of detailed comparison and analyses.

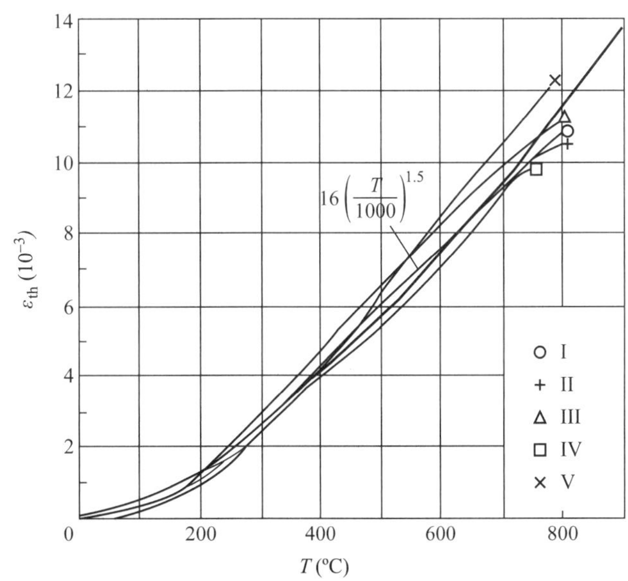

Figure 4-13 Freely expanding strain of reinforcement.[4-1]

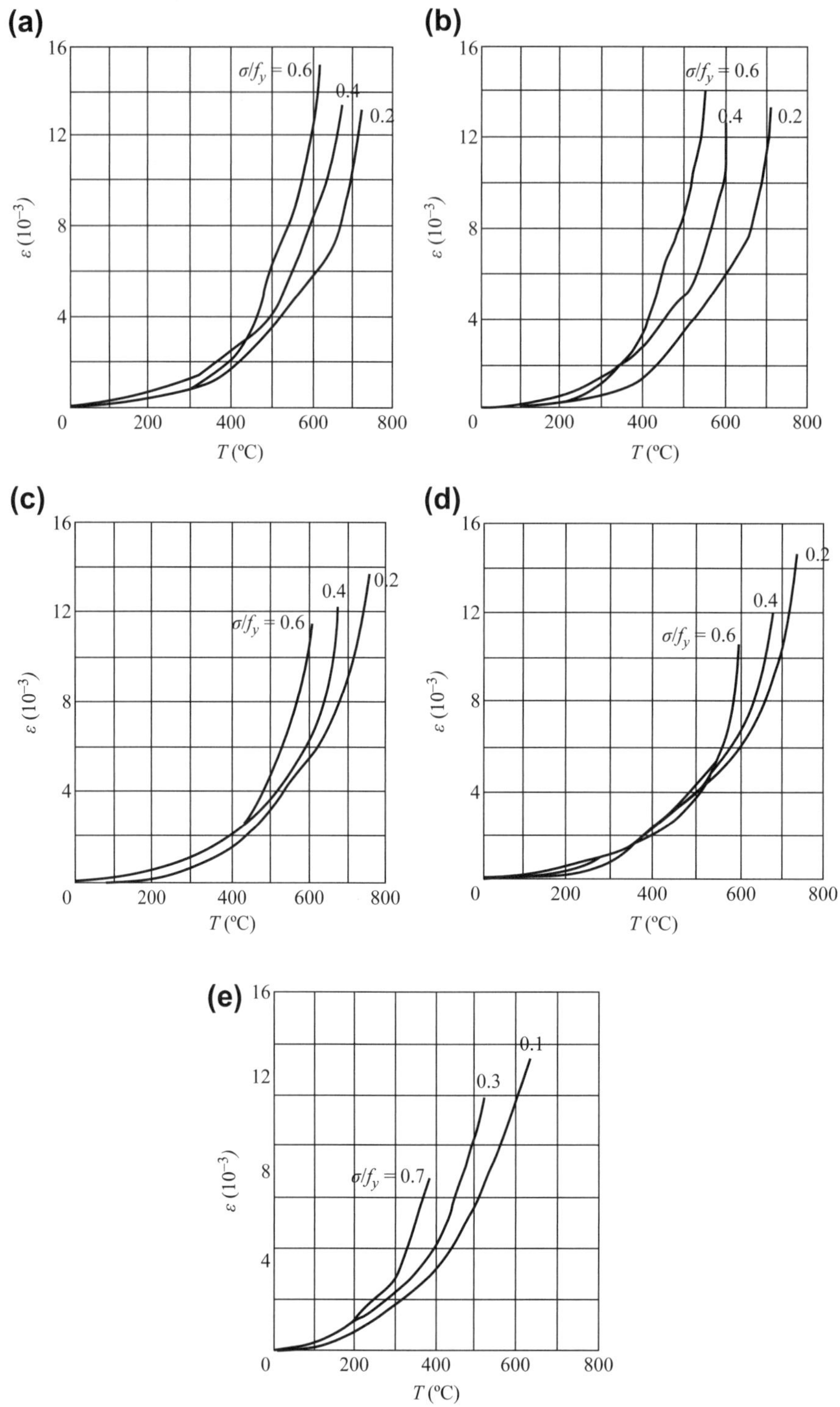

Figure 4-14 Thermal strain of reinforcement under the action of stress[4-1]: (a) steel grade I; (b) steel grade II; (c) steel grade III; (d) steel grade IV; (e) steel grade V.

4.4.3 Short Time Creep at Elevated Temperatures

Many experimental investigations have been conducted for the creep of steel at normal temperature and identical conclusions are drawn.[0-2] For mild steel (grades I–IV) with an obvious yield step, when a specimen is loaded and sustained for a long time or loaded–unloaded repeatedly, no creep and residual strain appears if the stress applied is within the elasticity limit, which is slightly lower than the yield strength. The high strength reinforcement (grade V or hard steel) shows plastic strain under high stress (σ/f_u>0.6), so the creep (or relaxation) occurs when the stress is sustained for a long time. The value of the creep is about 3–8% of the corresponding elastic strain when the designed stress acts for 1000 h, and the creep tends to be stable afterward.

Grade II reinforcement is used for the creep test at elevated temperatures. The specimen is heated to the predetermined temperature (200–600 °C) in the testing furnace and expands fully when the constant temperature is maintained. It is then loaded to $\sigma/f_y = 0.2$–0.8. The creep of the specimen is then measured under the conditions of constant temperature (±0.5 °C) and stress (±0.5 MPa) and it develops within 2 h as shown in Fig. 4-15.

When the testing temperature $T = 200$ °C, the creep of the reinforcement developed within 2 h is very small and is far less than the yield strain at normal temperature ($\varepsilon_{cr} = \varepsilon_y$), and the stress level reaches $\sigma/f_y = 0.8$. When the testing temperature $T = 600$ °C, the creep of the reinforcement under low stress level ($\sigma/f_y = 0.2$) increases proportionally with time and reaches 16×10^{-3} after 2 h and the increasing rate does not reduce. When the stress level is higher ($\sigma/f_y \geq 0.4$) and exceeds the yield strength at elevated temperatures (f_y^T), the strain of the specimen increases quickly with time (t) and the specimen then fails.

When the reinforcements are tested at the same temperature $T = 400$ °C but under different stress levels, the creeps developed within 2 h are recorded completely. The creep increases quickly when the stress is applied, and the value of the creep yielded within the first 10 min is about one-half of that yielded within 2 h; the increasing rate is reduced gradually. For specimens with stress level $\sigma/f_y \leq 0.6$, the creeps tend to steady when $t = 80$ min. For specimens with stress level $\sigma/f_y = 0.8$, the creep still increases obviously without convergence after 2 h.

These experimental results demonstrate that if the stress of the reinforcement is less than the yield strength at elevated temperatures ($\sigma < f_y^T$), the creep value will converge within 2 h and is approximately proportional to the stress

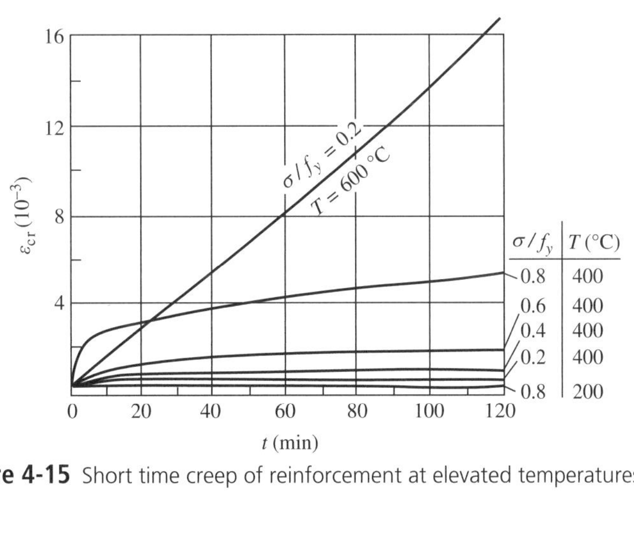

Figure 4-15 Short time creep of reinforcement at elevated temperatures.[4-1]

value, but the increase accelerates with the temperature.

CONCLUSIONS

Steel (reinforcement) is a product manufactured by modern industry and has uniform properties, stable behavior, and guaranteed quality. The mechanical behavior mainly depends on the composition of the alloys and production technology, including thermal treatment. The metal crystal structure in the interior of steel transforms at elevated temperatures and this causes the corresponding variations in its mechanical behavior.

The experimental data provided from testing reinforcement at elevated temperatures demonstrate that the strength and deformation behavior deteriorate gradually as the temperature increases. In the order of decreasing amplitude from smaller to greater, the behavior indices are ultimate tensile strength (f_u^T / f_u), yield strength (f_y^T / f_y), and elastic modulus (E_s^T / E_s). The ultimate elongation of reinforcement also decreases continuously when the temperature $T = 20$–600 °C. The various strength and deformation indices (f_u^T, f_y^T, E_s^T) for all strength grades of reinforcement reach much lowers value at temperature $T = 800$ °C, which is only about 10% of that at normal temperature.

At the same temperature, the decreasing amplitudes of grade I–IV reinforcement (mild steel) approach one another, but that of grade V (hard steel) is larger.

The strength and deformation behavior of reinforcement vary under different temperature–stress paths, and this is revealed experimentally. However, the experimental data and theoretical analysis available are still not sufficient to draw a quantitative conclusion, and more comprehensive experimental investigations are needed.

REFERENCES

[4-1] T. Lu, Experimental investigation of strength and deformation of reinforcing bars under high temperature, Masters dissertation, Tsinghua University, Beijing (1996).

[4-2] T. Lu, X. Shi, Z. Guo, Experimental investigation of strength and deformation of reinforcing bars of grades I–IV under high temperature, Journal of Fuzhou University (Science and Technology) (Supplementary Issue) 24 (1996) 11–17.

[4-3] Y. Yang, T. Wu, J. Zhu, Strength of Metal under High Temperature and its Experiment, Metallurgical Industry Press, Beijing (1979).

[4-4] M. Huang, Z. Jin, et al., Mechanical Behaviors of Metal, Metallurgical Industry Press, Beijing (1986).

PART II

Temperature Field on a Cross Section of a Structural Member

CHAPTER 5

Temperature–Time Curve of Fire and the Equation of Heat Conduction

5.1 TEMPERATURE–TIME CURVE OF FIRE

5.1.1 Characteristics of the Temperature Change of Fire

Fire is a burning phenomenon caused after a combustible substance is ignited and reacts intensely with oxygen in the air. During the burning process of the combustible material, enormous heat is produced and spread out in various ways. The temperature of the surrounding air and materials rises quickly, causes new burning, and further spread of heat. If burning gets out of control, as the flame and smoke at a high temperature flow and heat spreads, more materials nearby burn and a fire accident quickly occurs.

When a fire accident occurs in a building or a local space within a building, the fire generally involves three stages: the start of the fire, the spread of fire, and the decline and stop of the fire. It is also possible, under some conditions, that the fire declines and burns again repeatedly. Finally, when all the combustibles burn out, the surrounding oxygen is exhausted, or the fire-fighters fight the fire successfully, the fire declines and goes out.

Reinforcement concrete is not a combustible material and will not burn and produce heat when it is exposed to a fire. A fire on a concrete structure only elevates the temperature and forms a nonuniform temperature distribution in the interior of the concrete, because the concrete is under the action of the surrounding air at a high temperature and absorbs heat gradually. Therefore, when analyzing the temperature fields of a structure and its members, the first thing to find out is how the temperature varies on the periphery of the structure.

The typical temperature–time curve of a fire accident is shown in Fig. 5-1. In practice, the temperatures at various points in the space sustaining the fire are quite different; the maximum temperature appears near the flame surrounding the combustibles and the temperature outside the flame reduces gradually. Due to the heat flow floating up, an air layer at high temperature is formed near the bottom of the horizontal members, such as the beams and slabs above a room, and along the side surfaces of the vertical members, such as the walls and columns. The value of the temperature and its uniformity of distribution in the air layer surrounding the structural members depend on the position and distribution of

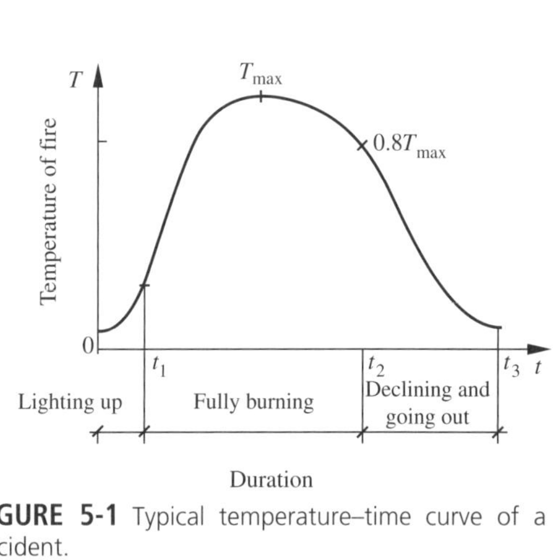

FIGURE 5-1 Typical temperature–time curve of a fire accident.

the combustibles, the burning time, and the direction of the air flow in the room. The temperature of a fire accident is referred to the temperature of the air layer, and it is normally assumed that the temperature distributes uniformly within the layer and is only the function of the burning time.

When a fire occurs in a room, only a few combustibles are burning and limited heat is sent out at the beginning. Although the local temperature nearby elevates and heat spreads quickly, the temperature of the surroundings is still lower and the temperature in the room elevates slowly. This is the first stage of a fire accident. As the temperature in the room is low and the temperature in the interior of the structural concrete is even lower (<300 °C), the mechanical behavior of the material loses less and has little influence on the behavior of the structure, so no obvious damage is caused.

As the burning time continues and the burning combustibles increase, the released heat increases quickly and accumulates in the room, causing the temperature of the fire to increase gradually; the fire is now in the second stage. When the combustibles burn fully, the temperature of the fire approaches and reaches the maximum temperature (T_{max}). The temperature of the fire then decreases slowly, due to reduction of the combustibles or oxygen provided. When the fire temperature elevates, the temperatures of the structural members and the architectural finishing layers surrounding them, such as ceilings and plastering, are lower than that of the air flow layer near the boundary. But the temperature of the interiors of these materials increases because heat is absorbed. When the fire temperature reduces to about 80% of the maximum temperature, the temperature of the air flow layer is lower than that on the surfaces of the structure and architectural layers, so heat conducts in the opposite way, i.e., heat spreads from the building materials to the air layer. This indicates the end of the second stage of a fire. The temperature of the structure increases quickly during the second stage; the temperature on its surface reaches the maximum value, and the temperature in its interior elevates continuously. Because of the heat inertia of concrete, the temperature on a section of the structural member decreases from the surface to the interior and distributes rather nonuniformly, and a considerable temperature gradient is formed on the outer layer of the section. Correspondingly, the bearing capacity and deformation behavior of the structure and its members deteriorate rapidly and different levels of damage result. This is the most dangerous period during a fire accident.

At the beginning of the third stage of a fire, the temperature on the surface layer of the structure will not increase but its absolute value is still high, although the temperature of the air flow layer has decreased gradually. As the structure has experienced a period under sustained high temperature, its behavior may deteriorate further and the damage tends to be more serious. During the later period of the third stage, the fire temperature decreases rapidly, the heat absorbed in the interior of the structure is released gradually to the outer layer and the surrounding air, and a temperature field with the opposite gradient and a corresponding stress field are formed on the section. When the fire goes out and the normal temperature is restored in the room, the damage that occurred in the structural concrete at elevated temperature will not recover, and the strength of the concrete even decreases slightly (Fig. 1-8). The deformation of the reinforcement that occurred at elevated temperature will also not recover, although the yield strength will recover. Therefore, after the fire goes out, serious damage in the structure, such as considerable residual deformation, cracks, and local spall, is visible, and the residual bearing capacity is far lower than the initial value at normal temperature.

5.1.2 Factors That Influence Fire Temperature

The typical temperature–time curve of a fire is shown in Fig. 5-1, but the parameters, such as the duration of each stage and the total process, the maximum temperature reached (T_{max}), the duration at high temperature (e.g., $T > 600$ °C), and the shape of the curve, may differ considerably for any particular fire.

According to the statistics from China,[0-12] 80% and 95% of the building fires on the ground are extinguished within 1 h and 2 h, respectively, i.e., the total duration of most fires is no more than 2 h. However, some serious fires occurring in underground spaces (e.g., in a building, tunnel, or cave) are sustained more than 10 h, or even several days (see Section 14.2.1), usually because of difficulties in extinguishing the fire.

The temperature–time curves of building fires vary considerably; the main factors influencing the duration and the maximum temperature of a fire are discussed below.

1. Properties, quantity, and distribution of the combustibles in a room

   The types and properties of various combustibles are different, so the burning behavior, i.e., the temperature of the ignition point, burning velocity, and the quantity of heat from a specific combustible mass, are also different.

   The quantity of combustibles, also called the fire load, determines the total quantity of heat produced during burning, the maximum temperature, and the duration of the fire.

   The distribution conditions of the combustibles, which include concentrated, continuous, or separate distribution, in a compact fashion or scattered, and the height and area of the stack, influence the burning velocity, concentration level of the flame, and the spread of the fire.

2. Area and shape of the room, and the area and position of windows and doors

   A room with a large area may contain more combustibles and the temperatures occurring in the center and near the window or door in the room make a great difference. The shape of the room influences the direction and velocity of the heat smoke flow.

   The area and position of the windows and doors in the room determine the direction and quantity of the air flow. The air flowing into the room supports continuous burning and production of heat from the combustibles; the heat smoke flowing away from the room possibly causes the fire to spread into other rooms.

3. Thermal behavior of building material

   If the building materials in and around the room are combustible, they support and develop the fire. If the mass heat capacity of the combustibles is small, they absorb less heat but the temperature increases quickly. If the value of the thermal conductivity (see Section 5.2) is large, the building materials transfer heat quickly and the temperature elevates quickly, which may cause the fire to spread, because the temperature on the outer surface of the room elevates too much.

   The actual variation in fire temperature can be illustrated by a series of experimental results (Fig. 5-2) on wood burning in a full-sized room. The quantities of the wood piled up in the room were 15, 30, and 60 kg/m^2 and the area of the window and door holes in the walls as a percentage of the gross area of the walls were 12.5%, 25%, and 50%, respectively. The relationship between the average temperature in the room and time was measured under various conditions (Fig. 5-2).

   The temperature quickly reaches the maximum value (within 10–40 min) after the wood is lit under all the test conditions; there is a longer stage of decline and the fire goes out.

   When the area of the holes on the wall is the same, the more wood (or fire load) in the room, the higher the maximum temperature reached and the longer the duration of the fire.

   When the quantity of the wood in the room is the same, the room with a smaller area of holes on the wall takes longer to reach the maximum temperature, so the total duration of the fire is longer. If the area of holes in the wall is increased, the burning process is accelerated, the maximum temperature of the fire is reached earlier and the duration is shortened. On the other hand, when the area of the holes on the wall is larger (50%), the cool air flowing into the room and the hot air flowing outside increase, causing a great quantity of heat loss, the maximum temperature in the room decreases, and the duration of the fire is shortened.

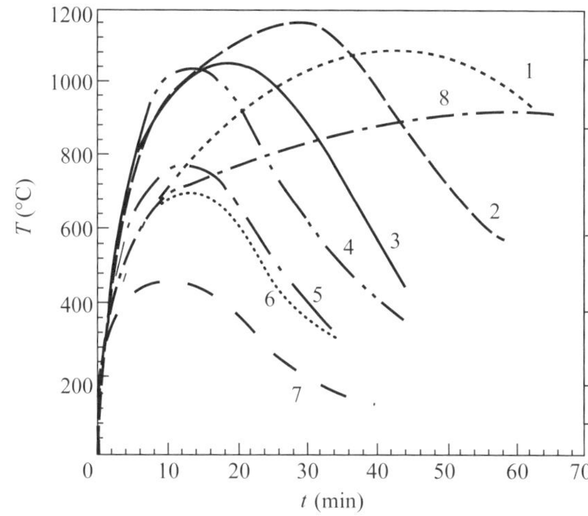

Curve No.	Wood (kg/m²)	Window hole (%)
1	60	12.5
2	60	25
3	60	50
4	30	25
5	30	50
6	15	25
7	15	50
8	ISO	

FIGURE 5-2 Experimental temperature–time curves of wood burning.[0-13]

5.1.3 Standard Temperature–Time Curve of Fire

For many years, scientists have conducted a great deal of research on the variation regularity of the temperature in a building fire, which has included surveying and gathering statistics from practical fire fields, burning tests in analogous rooms, and various analyses of burning theory. Many experimental and theoretical research results have been published up to now. In a simple building space, a complete temperature–time curve of the fire can be analyzed theoretically and accurately[0-7,0-12,0-14] according to the theorem of energy (heat) conservation and the principle of heat conduction, if the values of the parameters of the various influencing factors are determined. The dynamic process of a given fire can even be described by a visualization on the computer.

However, an actual fire in a building is far more complicated than that in a simple room. Many rooms with different area, height, and shape are connected in a building and there are many holes for windows and doors, so the burning and spread of the fire are influenced by adjacent rooms. These are difficult to describe accurately using a simple mathematical model. In addition, various combustibles are mixed in every room of the building and they are of different types, quantity, and distribution, with no unified regularity. As the functions of the rooms may change and the users of the rooms have different interests, these are very difficult to predict. Furthermore, the cause of the fire and the material and location where the fire started are obviously undefined. Therefore, the fire temperature–time curve of a practical building presents considerable randomness.

In view of the complexity of a building fire, the relevant research and academic organizations in various countries have worked out standard temperature–time curves of fire (Fig. 5-3(a)) to provide unified fire resistance requirements for building structures and for use as the basis for fire resistance experiments on the building members. These curves illustrate the monotonic heating process. The temperature elevates very quickly within the first 30 min after the fire starts and the increasing velocity then reduces gradually but without reaching the cooling stage. These standard curves of elevating temperature are similar, with a few exceptions.

The fire resistance curve suggested by the International Standardization Organization (ISO 834) for experiments on building members is the

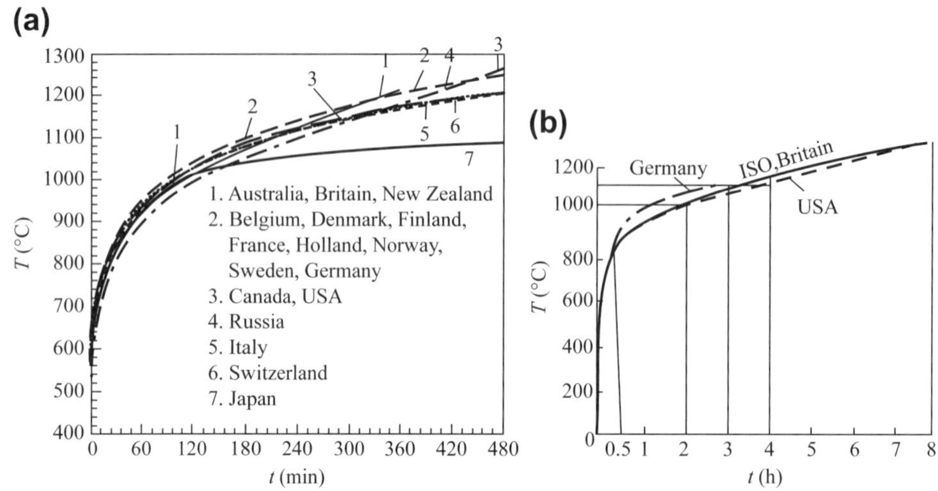

FIGURE 5-3 Standard temperature–time curves of fire: (a) fire resistance test standards in various countries; (b) ISO standard curve.

same as that suggested in the United Kingdom; the formula for the curve is

$$T = T_0 + 345\log(8t+1) \tag{5.1}$$

where T_0 is the initial temperature in the testing furnace (°C), and T is the average temperature of the air in the furnace at t minutes after the fire starts.

The ISO standard temperature–time curve is also a unified process of a monotonic increase regardless of how long the fire lasts, and the temperature increases continuously without the decreasing and extinguishing process. It cannot replace the variable temperature of a real fire occurring in a building (similar to Fig. 5-1), and it is considerably different from the experimental results for wood burning in a room (Fig. 5-2). However, the curve can be used as a standard to conduct fire resistance tests, to analyze behavior at elevated temperature, to check the fire endurance limit of structural members, to ensure an identical fire resistance behavior for structures, or to offer comparable fire resistance safety in different structures.

A recent contentious consideration is that the actual temperature–time curve of a possible fire occurring in a building may be used for the fire resistance design or fire endurance check of its structure, if the curve is far lower than the standard (e.g., ISO) curve and is confirmed in various ways.

5.2 THERMAL BEHAVIOR OF MATERIALS

The mechanical behavior of concrete and steel (reinforcement) materials at elevated temperature has been introduced fully in the previous four chapters. The temperature distribution and its variation in the interior of the structure and its members at elevated temperatures (e.g., under fire) depend only on the thermal behavior of the structural materials, in addition to the temperature conditions surrounding the structure. The temperature distribution is unrelated to the stress (strain) state and the mechanical behavior of the materials.

When the temperature field of a structure is analyzed theoretically and the basic equation of heat conduction (Eqn (5.4)) is established, the relevant thermal behavior of the material is described using three terms. Correspondingly, there are three basic parameters: coefficient of heat conductivity, specific heat capacity, and mass density. Other thermal parameters can be derived from these parameters.

Another basic thermal parameter of the material is the linear expansion coefficient (see Sections

2.1.1 and 4.4.1), which influences only the thermal strain and stress of the material and structure, but is unrelated to the analysis of the temperature field.

5.2.1 Thermal Parameters of Concrete

Concrete is a man-made compound material and is composed of cement, water, and fine (sand) and coarse (gravel) aggregates, and is sometimes mixed with some additional materials. They are mixed following predetermined ratios and then undergo casting, compacting, and curing. Concrete is formed in the coagulation and hardening process due to the adhesive action of cement. The various raw materials in concrete have very different chemical mineral compositions and textures, and their original thermal parameters are different. The values of the thermal parameters of the concrete formed are different and the experimental data vary due to the differences in the mixing ratio, water content, age, and production technique. The experimental results and data listed below give general values for the thermal parameters of concrete.

1. Thermal conductivity or coefficient of heat conduction (λ_c)

   The coefficient of heat conduction of a material is defined as the quantity of heat (J) passing through per unit area (m^2) with uniform temperature within a unit of time (h) and per unit temperature gradient (K/m). Its units are W/(m K) or W/(m °C).

   The coarse aggregate has the highest fraction of the total volume of concrete and has the predominant influence on its thermal behavior. The coarse aggregate of concrete of normal weight is alluvial gravel or crushed stone, which is broken from igneous or aqueous rock. The coefficient of heat conduction of the aggregate depends mainly on the mineral composition, crystal character, and structure of the granules because of small porosity (normally <5%) in the interior.

   The coefficients of heat conduction of various rocks and their variations with temperature are given in Fig. 5-4.[5-1] The values of the coefficients of heat conduction of these rocks at normal temperature are considerably different (may exceed 300%), but they tend to be similar at elevated temperature (T > 200 °C). Also, they vary differently as the temperature increases. The coefficient of heat conduction of siliceous sandstone, dolomite, and limestone decreases quickly, that of granite and gneiss decreases slowly, but that of diabase and calcareous feldspar increases slowly when the temperature increases.

   The coefficient of heat conduction of hardened cement mortar fluctuates slightly as the temperature increases (Fig. 5-5). The water/cement ratio (W/C) of the mortar during mixing has some influence on the coefficient of heat conduction. A specimen with a large water/cement ratio contains more water and more microporosity is formed after the water is lost during hardening. This causes a decrease in the coefficient of heat conduction.

   The coefficient of heat conduction of concrete composed of an aggregate of different kinds of rock varies with temperature and is shown in Fig. 5-6. The coefficient of ordinary

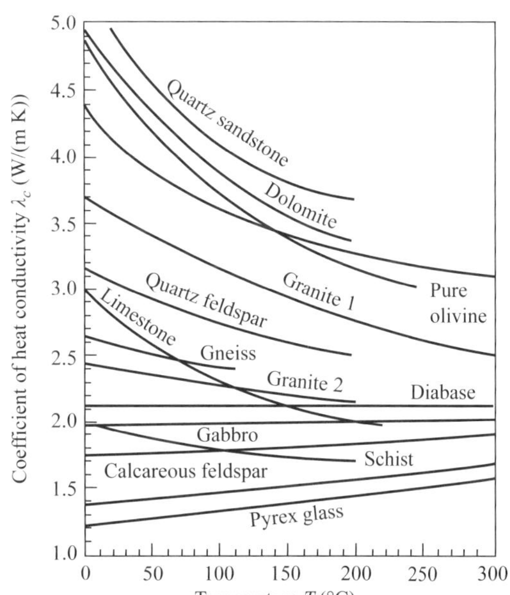

FIGURE 5-4 Coefficient of heat conduction of rock.[5-1]

concrete with siliceous aggregate is slightly higher than that of calcareous aggregate, but both reduce less as the temperature increases and are similar at very high temperature (e.g., >800 °C).

Coarse aggregate can also be made of various porous materials, such as pumice, slag, expanded clay, or shale, and it can replace ordinary rock aggregate and be mixed to produce light-weight concrete. The granules of light-weight aggregate contain many interior

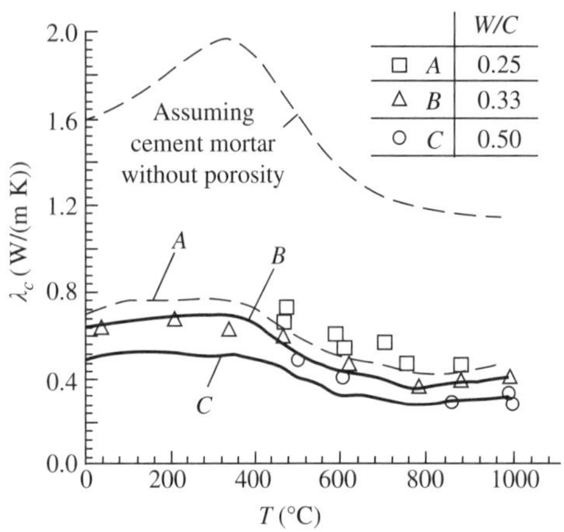

FIGURE 5-5 Coefficient of heat conduction of cement mortar.[5-2]

pores, and the efficiency of heat conduction reduces considerably. Therefore, the coefficient of heat conduction of light-weight concrete is far lower than that of ordinary concrete, and its variable amplitude decreases as the temperature increases.

The coefficient of heat conduction of concrete shows larger variability and scatter because of the influence of various factors. In order to simplify the calculation, concrete is divided into three categories according to the different types of aggregate, and the formulas for the coefficients of heat conduction (W/(m K)) are given individually in the design code.[2-6]

Siliceous aggregate (20 °C ≤ T ≤ 1200 °C)

$$\lambda_c = 2 - 0.24\left(\frac{T}{120}\right) + 0.012\left(\frac{T}{120}\right)^2 \tag{5.2a}$$

Calcareous aggregate (20 °C ≤ T ≤ 1200 °C)

$$\lambda_c = 1.6 - 0.16\left(\frac{T}{120}\right) + 0.008\left(\frac{T}{120}\right)^2 \tag{5.2b}$$

Light-weight aggregate

$$20\ ^\circ\mathrm{C} \le T < 800\ ^\circ\mathrm{C}: \lambda_c = 1.0 - \frac{T}{1600}$$

$$800\ ^\circ\mathrm{C} \le T < 1200\ ^\circ\mathrm{C}: \lambda_c = 0.5 \tag{5.2c}$$

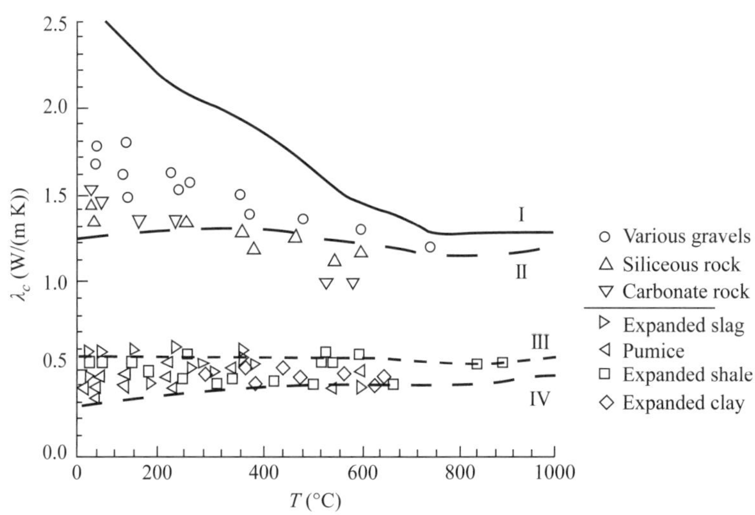

FIGURE 5-6 Coefficient of heat conductivity of concrete with various aggregates[5-2]: siliceous aggregate (20 °C ≤ T ≤ 1200 °C).

The corresponding relationship between the coefficient of heat conduction and temperature is shown in Fig. 5-7. Different calculations for the coefficient of heat conduction are also suggested in other references,[5-3] and can be compared and consulted.

2. Mass heat capacity or specific heat capacity (C_c)

The specific heat capacity is defined as the quantity of heat (J) absorbed per unit mass (kg) of the material when its temperature increases 1 K (or 1 °C), and its units are J/(kg K) or J/(kg °C).

The values of the specific heat capacity of concretes with different aggregates are measured from the experiments and are shown in Fig. 5-8. They increase gradually with temperature and tend to be stable when $T >$ 600 °C. However, a sharp peak appears near $T = 100$ °C, because the water contained in the interior of concrete evaporates and a lot of steaming heat is absorbed.

The type of aggregate has an influence, although not much, on the specific heat capacity of concrete. The specific heat capacity of concrete with siliceous (quartz) aggregate is slightly larger than that of calcareous (limestone) aggregate, and that for various light-weight aggregates is slightly smaller than that of ordinary concrete. Other factors, such as the mixing ratio, water content, and age, have less influence on specific heat capacity. A standard formula (in J/(kg K)) is suggested for various concretes by the Commission of the European Communities[2-6]:

$$20\ °C \le T < 1200\ °C: C_c = 900 + 80\left(\frac{T}{120}\right) - 4\left(\frac{T}{120}\right)^2 \quad (5.3)$$

Other references (e.g., Lie[5-3]) have different suggestions.

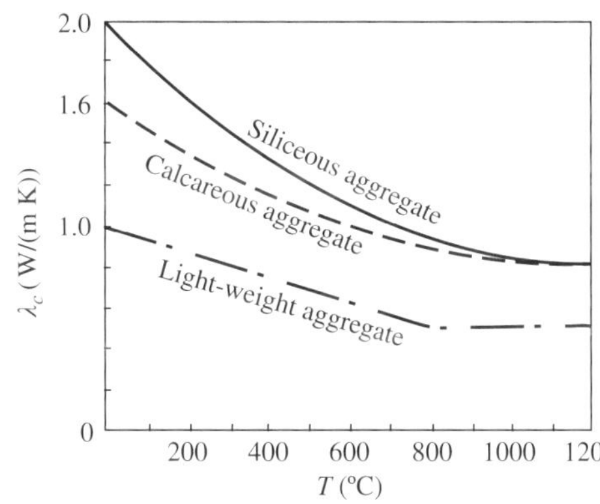

FIGURE 5-7 Calculated value of the coefficient of heat conduction of concrete.[2-6]

3. Mass density (ϱ_c)

The mass density, also called the volume density, is defined as the mass of the material per unit volume, and its units are kg/m³.

The mass density of concrete changes continuously during heating (Fig. 5-9). It reduces

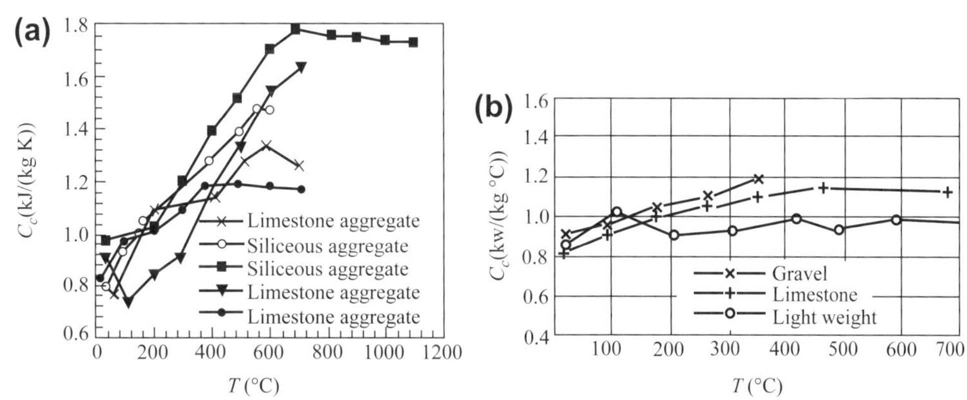

FIGURE 5-8 Specific heat capacity of concrete with different aggregates: (a) from Schneider[5-4] and Harmathy TZ, Allen[5-5]; (b) from Federation International de la Précontrainte.[0-7]

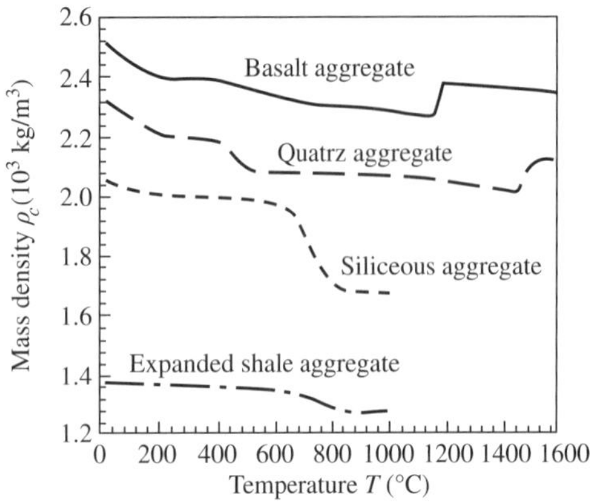

FIGURE 5-9 Mass density of concrete with different aggregates.[5-4,5-5]

obviously during the initial period, because the water content evaporates and overflows. The solid components, i.e., aggregate and cement, expand after heating, the volume increases, and the mass density decreases. This phenomenon exists throughout the heating process, and its influence increases gradually at high temperature.

In addition, some types of rock aggregate with different mineral components have special properties at elevated temperature, which influence the mass density. For example, siliceous aggregate dissolves and forms crystals at $T = 600$–800 °C, and is accompanied by considerable volume expansion and sudden decrease in the mass density. Basalt and quartz are melted and sintered when $T = 1200$–1400 °C, and then the mass density of the concrete increases suddenly.

The mass density of concrete composed of various light-weight aggregates varies with temperature similar to that of ordinary concrete, but the amplitude of the variation is smaller (Fig. 5-9).

The mass density of concrete does not vary sharply with temperature, and the influence on the temperature in the interior of the structure is smaller than that of the other main thermal parameters. To simplify the calculation during analysis of the temperature field of the structure, the mass density of concrete is normally taken as a constant (2200–2400 kg/m^3) irrespective of temperature. Or the mass density is combined with the specific heat capacity and the value of $(C_c\varrho_c)$[5-3] is given and introduced into the heat conduction equation (Eqn (5.4)).

The basic thermal behavior of concrete depends not only on the thermal behavior of the coarse and fine aggregates and the hardened cement mortar but also the composition, water content, age, casting and compacting technique, and compactness of the concrete. Thus, the thermal parameters present large variation and scatter. If accurate thermal parameters are required for analysis of a large engineering project, the specimens should be manufactured and tested specially and then the thermal parameters can be measured. As far as general structural engineering is concerned, no special requirement of accuracy is needed for the analyses of fire resistance and the temperature field. The simplified values of the thermal parameters suggested in the relevant code can be used for practical engineering for the randomness and scatter of the temperature variation in a fire accident.

5.2.2 Thermal Parameters of Reinforcement

A limited quantity of reinforcement (or wire) spreads in the interior of reinforced and prestressed concrete structures, and it generally makes up only a small percentage (<3%) of the total volume. The existing reinforcement has little influence on the temperature distribution in the interior of a structure under fire (high temperature) conditions. When the temperature field of a structure is analyzed, taking the required accuracy of the calculations into account, the structure is assumed to be composed of homogeneous concrete material and the reinforcement can be ignored. The main thermal behavior and the indices of the steel used in building are introduced briefly below.

The main constituents of the steel used in building structures are iron and carbon, and the steel is divided into low, medium, and high carbon-steels according to the carbon content. To improve the mechanical and working behavior of steel, a few alloy elements, such as manganese, silicon, niobium, vanadium, and titanium, are mixed in during the smelting process, to make low-alloy steel. The different element components and their content in the steel and the working and heat-treatment processes of the steel influence the indices of its thermal behavior.

The coefficient of heat conduction of various steels varies with temperature (Fig. 5-10). Pure iron has the highest coefficient of conduction and it decreases gradually as the content of carbon and alloy in the steel increases. The coefficients of heat conduction of carbon steel and low-alloy steel decrease monotonically as the temperature increases, but the variable rates are reduced gradually. However, some steels containing more alloy and the coefficient of heat conduction increases slowly with temperature.

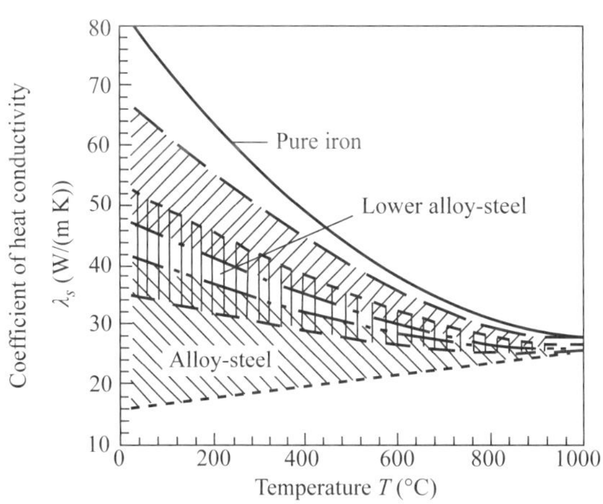

FIGURE 5-10 Coefficient of heat conduction of various steels.[5-5]

Because iron, carbon, and other alloy elements have different values of specific heat capacity, the steels composed of different types and content of alloy elements have corresponding values of heat capacity (C_s). The value increases slightly and gradually with temperature, but the variation is small.

The mass density (ϱ_s) of steel also varies slightly because of the different types and content of alloy elements in steel. The mass density of pure iron is high and reaches 7871 kg/m^3; for carbon and low-alloy steels, the mass density is 7850 kg/m^3. The volume of the steel expands (Fig. 4-13) and the mass density decreases slightly as the temperature increases, but it is generally taken as a constant during analysis of the temperature field.

General variable ranges for the thermal parameters of steel are given by Federation International de la Précontrainte,[0-7] and are listed in Table 5-1 and compared with concrete.

Comparing the data listed in Table 5-1, it is found that steel is a good heat conductor and concrete is a heat inertia material. The ratio of the coefficients of heat conduction is enormous. The mass heat capacity of steel is obviously smaller than that of concrete, because it is defined by the mass (kg) of the material. However, if per unit volume of material is considered, the $C_s\varrho_s$ (kJ/(m^3 K)) of steel is about double the $C_c\varrho_c$ of concrete.

This thermal behavior and the values of the parameters for concrete and steel have an obvious impact on the value and distribution of temperature in the structure at elevated temperatures.

TABLE 5-1 General Range of Thermal Parameters of Steel and Concrete

Material	Coefficient of heat conduction λ (W/(m K))	Mass heat capacity C (kJ/(kg K))	Mass density ϱ (kg/m^3)
Steel	55–28	0.42–0.84	7850
Concrete	1.6–0.6	0.84–1.26	2300

5.3 EQUATION OF HEAT CONDUCTION

5.3.1 Basic Equation of Heat Conduction

When a structure sustains a fire, the temperature of the heat flow surrounding it increases rapidly and varies continuously. Then, the structure is heated, the temperature on its surface elevates quickly, and heat penetrates gradually into the interior of the structure through conduction action. As concrete is a material of heat inertia, nonuniform temperature distribution is formed in the interior of the structure and varies continuously while the fire lasts. Thus, this is a problem of a dynamic or transient temperature field.

When a structure is heated, the mechanical behavior of the concrete and steel (reinforcement) deteriorates (see Part 1), the deformation of the structural member increases, and the bearing capacity decreases (see Part 3). In the meantime, the nonuniform temperature distribution of the structure causes thermal stress and stress redistribution on its section and redistribution of the internal forces in the statically indeterminate structure (see Chapter 10). Therefore, the mechanical responses of the structure at elevated temperature, including internal forces, deformation, and bearing capacity, depend on the temperature fields and their variations in the structure and structural members.

On the other hand, the mechanical responses of a structure at elevated temperature do not change the existing temperature distribution in general cases. This only occurs when very wide cracks appear in the structural concrete and heat flow penetrates its interior. The temperature may vary locally within a small area nearby.

Therefore, when the mechanical analysis of a structure at elevated temperature and the fire resistance design or checks on a structure are performed, the temperature field of the structure must first be analyzed, and the internal forces and bearing capacity (or fire endurance) must be checked.

The analysis of the temperature field of a structure is a based on heat conduction of solid matter. The basic differential equation of heat conduction can be developed and the solution found.

The concrete in a structure is assumed to be an isotropic material, and the thermal parameters, i.e., λ, c, and ϱ, are known and are functions of temperature. A microcuboid $dx\ dy\ dz$ is taken from the structure near an arbitrary point (x, y, z) in Cartesian coordinates (Fig. 5-11); its internal temperature is assumed to be uniformly distributed and is $T(x, y, z, t)$ at time t.

The microcuboid experiences heat exchange under the thermal action on its surfaces. If the X direction is considered first, the quantities of heat flowing into and out of the microcuboid, through unit area within unit time, are q_x and q_{x+dx}, respectively. They can be obtained as follows, according to the definition of the coefficient of heat conduction:

$$q_x = -\lambda \frac{\partial T}{\partial x}$$

and

$$q_{x+dx} = q_x + \frac{\partial q_x}{\partial x} dx = -\lambda \frac{\partial T}{\partial x} - \frac{\partial}{\partial x}\left(\lambda \frac{\partial T}{\partial x}\right) dx$$

The area of the microcuboid perpendicular to the X direction is $dy\ dz$, so the increase in the quantity of heat in the cuboid per unit time should be

$$(q_x - q_{x+dx})\, dy\, dz = \frac{\partial}{\partial x}\left(\lambda \frac{\partial T}{\partial x}\right) dx\, dy\, dz$$

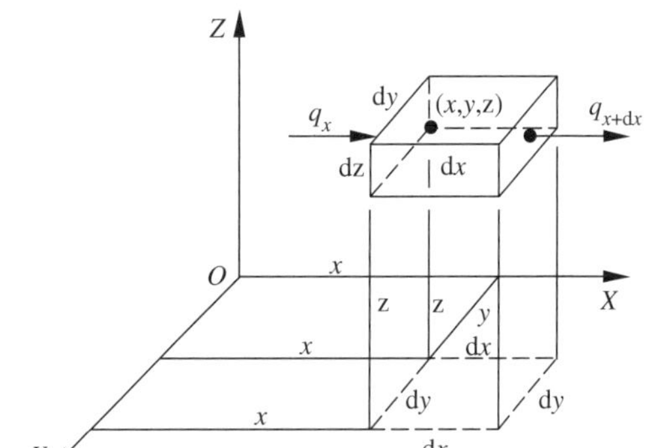

FIGURE 5-11 Analysis of the heat flow of a microcuboid.

Similarly, the quantities of heat obtained in the Y and the Z directions are

$$\frac{\partial}{\partial y}\left(\lambda\frac{\partial T}{\partial y}\right)\mathrm{d}x\,\mathrm{d}y\,\mathrm{d}z$$

and

$$\frac{\partial}{\partial z}\left(\lambda\frac{\partial T}{\partial z}\right)\mathrm{d}x\,\mathrm{d}y\,\mathrm{d}z$$

respectively. Therefore, the increase in the total quantity of heat in the microcuboid within unit time is

$$\left[\frac{\partial}{\partial x}\left(\lambda\frac{\partial T}{\partial x}\right)+\frac{\partial}{\partial y}\left(\lambda\frac{\partial T}{\partial y}\right)+\frac{\partial}{\partial z}\left(\lambda\frac{\partial T}{\partial z}\right)\right]\mathrm{d}x\,\mathrm{d}y\,\mathrm{d}z \tag{i}$$

In addition, if the material can generate heat itself, e.g., hydration heat of cement or burning of combustibles mixed in concrete, and the quantity of heat generated per unit volume of the cuboid within unit time is q_d, then the quantity of heat generated in the cuboid within unit time should be

$$q_d\,\mathrm{d}x\,\mathrm{d}y\,\mathrm{d}z \tag{ii}$$

As the heat is absorbed into the microcuboid, the temperature increases. If the temperature increment per unit time is $\partial T/\partial t$, the total quantity of heat absorbed in the cuboid within unit time should be

$$C\rho\,\mathrm{d}x\,\mathrm{d}y\,\mathrm{d}z\frac{\partial T}{\partial t} \tag{iii}$$

according to the definition of specific heat capacity.

According to the principles of energy conservation, the sum of the quantities of heat that enter or leave the microcuboid through its surfaces and generated in its interior should be equal to the quantity of heat absorbed because of increased temperature or released because of reduction in temperature, respectively, in the microcuboid. Using equations (i), (ii), and (iii):

$$\left[\frac{\partial}{\partial x}\left(\lambda\frac{\partial T}{\partial x}\right)+\frac{\partial}{\partial y}\left(\lambda\frac{\partial T}{\partial y}\right)+\frac{\partial}{\partial z}\left(\lambda\frac{\partial T}{\partial z}\right)\right]\mathrm{d}x\,\mathrm{d}y\,\mathrm{d}z + q_d\,\mathrm{d}x\,\mathrm{d}y\,\mathrm{d}z = C\rho\frac{\partial T}{\partial t}\mathrm{d}x\,\mathrm{d}y\,\mathrm{d}z$$

or

$$\frac{\partial T}{\partial t}=\frac{1}{C\rho}\left[\frac{\partial}{\partial x}\left(\lambda\frac{\partial T}{\partial x}\right)+\frac{\partial}{\partial y}\left(\lambda\frac{\partial T}{\partial y}\right)+\frac{\partial}{\partial z}\left(\lambda\frac{\partial T}{\partial z}\right)\right]+\frac{q_\mathrm{d}}{C\rho} \tag{5.4}$$

This is the basic differential equation of the transient heat conduction.

In the general condition, concrete is not considered to generate heat itself and $q_d = 0$ when a structure at elevated temperature (or fire resistance) is analyzed.

If the temperature of the environment surrounding the structure is not variable, e.g., the second category of thermal problem in structural engineering (see Section 0.1), the internal temperature of the structure is also not variable as time continues. This means $\partial T/\partial t=0$, and when $q_d = 0$, Eqn (5.4) can be simplified:

$$\frac{\partial}{\partial x}\left(\lambda\frac{\partial T}{\partial x}\right)+\frac{\partial}{\partial y}\left(\lambda\frac{\partial T}{\partial y}\right)+\frac{\partial}{\partial z}\left(\lambda\frac{\partial T}{\partial z}\right)=0 \tag{5.5}$$

This is called the basic equation of stable heat conduction.

The equations above can be used to analyze any three-dimensional structure. The linear members, such as a beam and a column, are most commonly used in practice; it is generally assumed that the temperatures along the axis line are identical, and the temperature field is simplified into a two-dimensional field on its section. For planar members, such as a wall and a slab, the temperature field is simplified further into a one-dimensional field along the direction of its thickness. The corresponding basic equations of heat conduction are listed in Table 5-2.

If the coefficient of heat conduction (λ) is taken as a constant and unrelated to temperature, and when

$$d=\frac{\lambda}{C\rho} \tag{5.6}$$

is introduced into Eqns (5.4) and (5.5), the basic equations of heat conduction can be simplified and modified respectively as

$$\frac{\partial T}{\partial t}=\mathrm{d}\left(\frac{\partial^2 T}{\partial x^2}+\frac{\partial^2 T}{\partial y^2}+\frac{\partial^2 T}{\partial z^2}\right)+\frac{q_d}{C\rho} \tag{5.4a}$$

TABLE 5-2 Two- and One-Dimensional Basic Equation of Heat Conduction

Member	Transient	Stable
Linear member (two-dimensional)	$\frac{\partial T}{\partial t}=\frac{1}{C\rho}\left[\frac{\partial}{\partial x}\left(\lambda\frac{\partial T}{\partial x}\right)+\frac{\partial}{\partial y}\left(\lambda\frac{\partial T}{\partial y}\right)\right]$	$\frac{\partial}{\partial x}\left(\lambda\frac{\partial T}{\partial x}\right)+\frac{\partial}{\partial y}\left(\lambda\frac{\partial T}{\partial y}\right)=0$
Planar member (one-dimensional)	$\frac{\partial T}{\partial t}=\frac{1}{C\rho}\frac{\partial}{\partial z}\left(\lambda\frac{\partial T}{\partial z}\right)$	$\frac{\partial}{\partial z}\left(\lambda\frac{\partial T}{\partial z}\right)=0$

and
$$\frac{\partial^2 T}{\partial x^2}+\frac{\partial^2 T}{\partial y^2}+\frac{\partial^2 T}{\partial z^2}=0 \tag{5.5a}$$

d is called the heat diffusivity of the material and is a physical quantity derived from the basic thermal parameters, with units of m^2/h or m^2/s.

5.3.2 Conditions and Methods for Finding a Solution

In order to find the solution of the equation of heat conduction (Eqn (5.4) or (5.5)), the initial and boundary thermal conditions of the structure have to be determined in advance, and the thermal parameters of the material must be known.

A building structure is normally at the environmental temperature before a fire accident occurs; the temperature of the whole structure is usually assumed to be uniformly distributed and is equal to the environmental temperature (T_0). Therefore, the initial thermal condition of the structure can be written as

$$T_{(x,y,z,t=0)}=T_0 \tag{5.7}$$

The boundary thermal condition of the structure is generally divided into four categories,[0-9] depending on the different conditions of the environment surrounding the structure and heat exchanging with the surrounding medium:

1. It is known that the temperature on the boundary (l_1) of the structure is a function of time (t), e.g.,
$$T_{(x,y,z,t)}\mid_{l_1}=T_f(t) \tag{5.8}$$
2. It is known that the quantity of the heat flow
$$\left(-\lambda\frac{\partial T}{\partial n}\mid_{l_2},\ n \text{ is normal direction}\right)$$
on the boundary (l_2) of the structure is a function of time (t).
3. It is known that the temperature of the fluid medium in contact with the structure is T_a and the quantity of the heat flow passing through the boundary (l_3) is
$$\beta_T\left[T_{(x,y,z,t)}-T_a\right]\mid_{l_3} \tag{5.9}$$
where β_T is the coefficient of heat transfer between the boundary of the structure and the surrounding fluid medium, and it is defined as the quantity of heat passing through unit area within unit time and with unit temperature difference. The units of β_T are W/(m^2 K).
4. The condition of the heat exchange on the boundary (l_4) of the structure is known when it is in contact with other solid material.

For the basic methods for solving the problems of various thermal boundaries, consult relevant monographs (e.g., Zhu et al.[0-9]).

When a structure sustains a fire accident, the surface exposed to the fire is normally taken as the third category of the boundary condition. As the fire continues and the temperature increases, the temperature on the surface exposed to the fire also increases and gradually approaches that of the heat flow of the fire; it can then be considered as the first category of the boundary condition. The surface not exposed to fire and located far away from the fire (e.g., the back surface of a thick slab) may be treated as the first category of the boundary condition.

When the transient temperature field of a structure under a fire accident or at elevated temperature is analyzed, the heating processes are variable, the nonlinear thermal parameters of the material vary with the temperature, and the boundary conditions are complex. Therefore, an accurate solution to the partial differential equation of heat conduction, which is nonlinear and parabolic in shape, is very difficult to determine.

The equation of heat conduction is almost impossible to solve by the analytical method. The approximate analytical solution can be obtained only if the temperature field of the structure is simplified into a stable problem and constant values are used for the thermal parameters of the material.

In order to deal with the nonlinear equation of transient heat conduction, the most effective and successful choice is to use the numerical method, i.e., finite difference or finite element analysis, and a high-speed computer to calculate the solution.

The finite difference method can be used to find the solution for the temperature field for some simple structures. This method has advantages; the calculation format is simple and flexible, and the solution is reached quickly. However, the method fits only the difference network of regular shapes (square, rectangular, and equilateral triangle); it is quite inconvenient for complicated geometrical shapes. Another disadvantage is that the method considers only the action of the nodes, without taking into account the characteristics of the elements that are linked to the node.

The mixed finite element-difference method is now popular. It means that the space region of the structure is discreted into finite elements and the time (or temperature) region is analyzed recurrently step by step by using the difference method. The finite element method used in the space region can easily separate the structure of various complicated shapes, and the integration performed over the element can introduce the contribution of the element to the parameters of the node. This may compensate for the disadvantage of the finite difference method. On the other hand, the transient temperature field of a structure depends only on the previous temperature variation and current boundary conditions, and is not related to the condition occurring afterward. So, this is a step by step problem in the time coordinate and the solutions at different times need not be found simultaneously. Therefore, the calculation can be started from the initial value in the time region and is performed step by step with the time increment until the predetermined time.

Because the temperature variation of the concrete structure during a fire accident is influenced by many factors and deviates considerably, and accurate values for the relevant thermal parameters are not available so far, it is necessary to use some simplifications and assumptions in order to achieve more simple, clear, and practical calculations of the temperature field analysis, when the accuracy required for engineering practice is satisfied. The actual methods for finding the solution of the equation of heat conduction and the analysis of the temperature field (two- and one-dimensional) of the structure are introduced in detail in the next chapter.

CONCLUSIONS

The content of this chapter is the preparation for the analysis of the temperature field of a structure. The temperature–time curve of a fire accident is a concentrated representation of its action on the structure, and is the main basis for the analysis of the temperature field. Based on the principle of energy conservation, the differential equation of heat conduction is established and can be used to find the solution. It also represents, mathematically, the regularity of the temperature field. The thermal parameters of concrete and steel (reinforcement) reflect the characteristics of the structural materials and are the physical parameters necessary for the analysis of the temperature field.

Generally, a fire accident in a building experiences three stages: the start of the fire, the spread of the fire, and the decline and end of the fire. The corresponding temperature–time curve can reflect the main action of the fire on the structure. The practical temperature–time curve of a fire varies considerably with many factors, such as the combustibles, the shape of room, the air flow, and the thermal behavior of the building materials. In order to unify the fire resistance requirements of a structure, the standard (ISO) temperature–time curve can be used.

When a building sustains a fire, the structure is heated, its temperature increases, and a non-uniform temperature field is formed in its interior because concrete is a material of heat inertia. Correspondingly, the mechanical behavior of concrete and reinforcement deteriorates, the stress on the section and the internal forces of the structure redistribute, the behavior indices of the structure reduce, and damage at different levels appears. On the other hand, the internal forces, deformation, and various damage to the structure do not generally have an influence on the temperature distribution. Therefore, the temperature field of the structure can be analyzed first, i.e., by establishing the basic equation of heat conduction, introducing the values of the thermal parameters of concrete, determining the initial and boundary thermal conditions, and then finding the solution.

As the temperature of a fire accident varies with time and the thermal parameters of the materials vary with the temperature (time), the analysis of the temperature field of a structure is a nonlinear problem of transient heat conduction. The basic equation is a second order partial differential equation and it is difficult to determine an accurate analytical solution. Numerical methods, i.e., finite difference and finite element analyses, are normally used to find the solution.

The values of the thermal parameters of concrete vary and deviate considerably because of differences in factors such as the raw materials, mixing ratio, casting and compacting techniques, and age. The values of the thermal parameters suggested in the relevant code can be used for analysis of the temperature field in an ordinary building. When dealing with some important structural engineering, special experiments should be performed, if necessary, to accurately measure the values of the thermal parameters of the concrete, which are then used for the analysis of the temperature field.

REFERENCES

[5-1] H. Birch, H. Clark, The thermal conductivity of rocks and its dependence on temperature and composition, American Journal of Science 238 (1940) 529–558.

[5-2] T.Z. Harmathy, Thermal properties of concrete at elevated temperature, Journal of Materials 5 (1) (1970) 47–74.

[5-3] T.T. Lie, A procedure to calculate fire resistance of structural members, International Seminar on Three Decades of Structural Fire Safety, London (February 1983) 139–153.

[5-4] U. Schneider, Behavior of concrete at high temperature, Report to RILEM Committee, 44-PHT, The Hague (1982) 72.

[5-5] T.Z. Harmathy, L.W. Allen, Thermal properties of selected masonry unit concretes, Journal of the American Concrete Institute 70 (2) (1973) 132–142.

CHAPTER

6

Theoretical Analysis of the Temperature Field

6.1 DIFFERENCE ANALYSIS

The basic equation of heat conduction, i.e., Eqn (5.4) or Eqn (5.5), is a second order partial differential equation and can be solved approximately by the difference method. As for a concrete member of a rectangular section, the network of the section can be divided and the boundary condition can be represented easily, so the difference method is a simple and practical method for the analysis of a temperature field.

6.1.1 Discretization Method and Difference Format

In order to solve the two-dimensional temperature distribution on a section of the structural member, the section can be divided into a certain number of planar networks. It is assumed that the temperature and thermal behavior within each network are invariable and can be represented by the corresponding values at the center of the network. Therefore, the temperature distribution on the section can be represented by the temperatures at finite points.

The network can be divided in various ways. The simplest way is to divide the section into a network of the same size using the straight lines parallel to the coordinates. If the temperature on the section changes sharply, networks of different size can be divided. In order to represent the practical distribution of temperature, the finer network on the section should be divided, and then more temperature points are obtained, and more calculations result.

In order to solve the second order equation of heat conduction, the first and second differentials of temperature (e.g., $\partial T/\partial x$, $\partial T/\partial y$, $\partial^2 T/\partial x^2$, $\partial^2 T/\partial y^2$, and $\partial T/\partial t$) should be discreted. According to the requirements of calculation accuracy and stability for the solution, different difference formats (e.g., two or three points, forward, backward, or central difference) may be selected and obtained by expanding into a Taylor series.

If the coordinate is x_i and the temperature is T_i at a network point i, then the adjacent points are $i + 3$, $i + 2$, $i + 1$, $i - 1$, $i - 2$, and $i - 3$, the corresponding coordinates are $x_i + 3\Delta x$, $x_i + 2\Delta x$, $x_i + \Delta x$, $x_i - \Delta x$, $x_i - 2\Delta x$, and $x_i - 3\Delta x$, respectively, and corresponding temperatures are T_{i+3}, T_{i+2}, T_{i+1}, T_{i-1}, T_{i-2}, and T_{i-3}, respectively. Then, the difference formats for two and three points can be written for the first and second partial differentials ($\partial T/\partial x$ and $\partial^2 T/\partial x^2$) and are shown in Table 6-1.

Similarly, the corresponding difference formats can be obtained for more difference points and for the first and second partial differentials of temperature with respect to coordinate y and time t, respectively.

After the difference formula is substituted into Eqn (5.4) for heat conduction, an algebraic equation of heat equilibrium is obtained for each network point. The corresponding equations for all the network points on the section make up a series of algebraic equations, and the unknowns are the temperatures at every network

TABLE 6-1 Difference Formats for the First and Second Differentials of Temperature

Difference format»		$\partial T/\partial x$	$\partial^2 T/\partial x^2$
Two-point formula	Forward difference	$\frac{\partial T}{\partial x}\Big\vert_i = \frac{T_{i+1}-T_i}{\Delta x} + O(\Delta x)$	$\frac{\partial^2 T}{\partial x^2}\Big\vert_i = \frac{T_i - 2T_{i+1} + T_{i+2}}{\Delta x^2} + O(\Delta x)$
	Backward difference	$\frac{\partial T}{\partial x}\Big\vert_i = \frac{T_i - T_{i-1}}{\Delta x} + O(\Delta x)$	$\frac{\partial^2 T}{\partial x^2}\Big\vert_i = \frac{T_{i-2} - 2T_{i-1} + T_i}{\Delta x^2} + O(\Delta x)$
	Central difference	$\frac{\partial T}{\partial x}\Big\vert_i = \frac{T_{i+1} - T_{i-1}}{2\Delta x} + O(\Delta x^2)$	$\frac{\partial^2 T}{\partial x^2}\Big\vert_i = \frac{T_{i-1} - 2T_i + T_{i+1}}{\Delta x^2} + O(\Delta x^2)$
Three-point formula	Forward difference	$\frac{\partial T}{\partial x}\Big\vert_i = \frac{-3T_i + 4T_{i+1} - T_{i+2}}{2\Delta x} + O(\Delta x^2)$	$\frac{\partial^2 T}{\partial x^2}\Big\vert_i = \frac{2T_i - 5T_{i+1} + 4T_{i+2} - T_{i+3}}{\Delta x^2} + O(\Delta x^2)$
	Backward difference	$\frac{\partial T}{\partial x}\Big\vert_i = \frac{T_{i-2} - 4T_{i-1} + 3T_i}{2\Delta x} + O(\Delta x^2)$	$\frac{\partial^2 T}{\partial x^2}\Big\vert_i = \frac{-T_{i-3} + 4T_{i-2} - 5T_{i-1} + 2T_i}{\Delta x^2} + O(\Delta x^2)$

$O(\Delta x)$ and $O(\Delta x^2)$ are the first and second cut-off errors, respectively.

point. When the temperature condition on the boundary is introduced and the equation series is solved, the temperature distribution on the section is obtained.

6.1.2 Stable Heat Conduction Problem

The equation of stable heat conduction in a two-dimensional temperature field (Table 5-2) on the section is

$$\frac{\partial}{\partial x}\left(\lambda \frac{\partial T}{\partial x}\right) + \frac{\partial}{\partial y}\left(\lambda \frac{\partial T}{\partial y}\right) = 0 \tag{6.1}$$

The network of the section shown in Fig. 6-1 is divided into equal sizes along the X and Y axes. As an example, the algebraic equations for the central difference format for two points are derived below. Other difference formats can be derived similarly.

As far as an internal network point on the section is concerned, if the number and temperature are $P_{i,j}$ and $T_{i,j}$, respectively, the numbers and temperatures of the surrounding network points are, respectively:

number: $P_{i-1,j}$, $P_{i+1,j}$, $P_{i,j-1}$, $P_{i,j+1}$
temperature: $T_{i-1,j}$, $T_{i+1,j}$, $T_{i,j-1}$, $T_{i,j+1}$

When the central difference format of two points is used and substituted into Eqn (6.1), we obtain:

$$\lambda_{i-1,\,j:i,j}\frac{T_{i-1,j} - T_{i,j}}{\Delta x^2} + \lambda_{i+1,\,j:i,j}\frac{T_{i+1,j} - T_{i,j}}{\Delta x^2} + \lambda_{i,\,j-1:i,j}\frac{T_{i,j-1} - T_{i,j}}{\Delta y^2} + \lambda_{i,j+1:i,j}\frac{T_{i,j+1} - T_{i,j}}{\Delta y^2} = 0 \tag{6.2}$$

where Δx and Δy are the edge lengths of a network along the X and Y axes, respectively, $\lambda_{i-1,j:i,j}$ is the coefficient of heat conduction between the network points $P_{i-1,j}$ and $P_{i,j}$, and so on.

As the edge length of the network is far smaller than that of the section, the coefficient between the adjacent network points can be taken as the average value of both points:

$$\lambda_{i-1,\,j:i,j} = \frac{1}{2}\left(\lambda_{i-1,j} + \lambda_{i,j}\right) \tag{6.3}$$

where $\lambda_{i-1,j}$ and $\lambda_{i,j}$ are the coefficients of heat conduction of the network points $P_{i-1,j}$ and $P_{i,j}$, respectively.

For the network points along the periphery, including the corner, of the section, it is not necessary to establish the algebraic equations of heat equilibrium if the boundary condition is in the first category, i.e., the temperature on the boundary is already known. However, if the

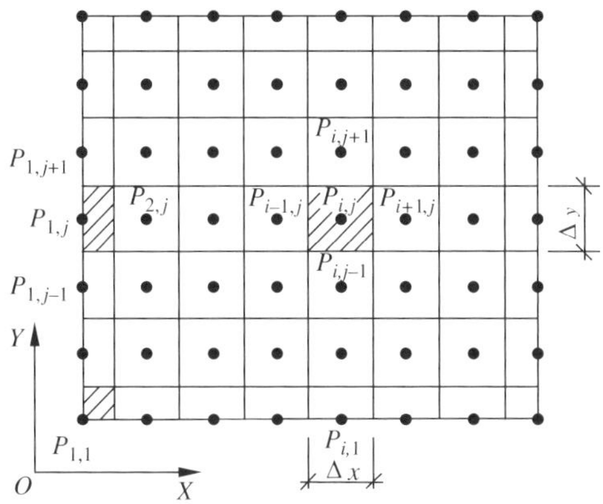

FIGURE 6-1 Divided network of the section.

boundary condition is in the third category, the temperature at the network point on the peripheral is unknown and the algebraic equation of heat equilibrium has to be given. For example, the following equation is given for the network $P_{1,j}$ in Fig. 6-1, according to Eqn (6.1):

$$\beta_T^1 \frac{T_a^1 - T_{1,j}}{\Delta x} + \frac{\lambda_{2,j} + \lambda_{1,j}}{2} \frac{T_{2,j} - T_{1,j}}{\Delta x^2} + \frac{\lambda_{1,j-1} + \lambda_{1,j}}{4} \frac{T_{1,j-1} - T_{1,j}}{\Delta y^2} + \frac{\lambda_{1,j+1} + \lambda_{1,j}}{4} \frac{T_{1,j+1} - T_{1,j}}{\Delta y^2} = 0 \tag{6.4}$$

where β_T^1 and T_a^1 are the coefficient of heat transfer and the environmental temperature on the left boundary of the network $P_{1,j}$, respectively.

The definitions of the other symbols, i.e., P, T, and λ, are the same as in Eqns (6.2) and (6.3).

If the boundary condition of the corner network of the section is in the third category, i.e., $P_{1,1}$ in Fig. 6-1, the algebraic equation of heat equilibrium can also be obtained from Eqn (6.1):

$$\beta_T^1 \frac{T_a^1 - T_{1,1}}{\Delta x} + \frac{\lambda_{2,1} + \lambda_{1,1}}{4} \frac{T_{2,1} - T_{1,1}}{\Delta x^2} + \beta_T^b \frac{T_a^b - T_{1,1}}{\Delta y} + \frac{\lambda_{1,2} + \lambda_{1,1}}{4} \frac{T_{1,2} - T_{1,1}}{\Delta y^2} = 0 \tag{6.5}$$

where β_T^b and T_a^b are the coefficient of heat transfer and the environmental temperature on the bottom boundary of the network $P_{1,1}$, respectively.

If the two outside boundaries of the corner network are in different categories, e.g., the left boundary of $P_{1,1}$ in Fig. 6-1 is in the third category but the bottom boundary is in the first category, the temperature of the network is still unknown and the corresponding algebraic equation of heat equilibrium is:

$$\beta_T^1 \frac{T_a^1 - T_{1,1}}{\Delta x} + \frac{\lambda_{2,1} + \lambda_{1,1}}{4} \frac{T_{2,1} - T_{1,1}}{\Delta x^2} + \frac{\lambda_{1,2} + \lambda_{1,1}}{4} \frac{T_{1,2} - T_{1,1}}{\Delta y^2} = 0 \tag{6.6}$$

Referring to these equations, the algebraic equation of heat equilibrium can be established individually for every network on the section including the internal, edge, and corner networks. The series of equations is compiled and can be solved, and then the temperature distribution on the section under a given boundary condition is obtained.

6.1.3 Transient Heat Conduction Problem

The method for solving the transient heat conduction problem is basically the same as that for the stable heat conduction problem, except one term of $\partial T/\partial t$ is added to consider the temperature varying with time. It is assumed that the section of the structural member is also divided into the network on the space field as shown in Fig. 6-1. On the time field, the total time is divided into n increments and the duration of the kth increment is Δt_k. Using the difference format given above, the transient heat conduction equation (Table 5-2),

$$\frac{\partial}{\partial x}\left(\lambda \frac{\partial T}{\partial x}\right) + \frac{\partial}{\partial y}\left(\lambda \frac{\partial T}{\partial y}\right) = \rho c \frac{\partial T}{\partial t} \tag{6.7}$$

can be changed into an algebraic equation. The central difference format of two points is used on the space field here, and the forward difference format of two points is used on the time field.

When $t = 0$, the initial condition is the value of the environmental temperature of every network point on the section:

$$T^0_{i,j} = T_0 \quad (i = 1, 2, \cdots, m;\ j = 1, 2, \cdots, n) \tag{6.8}$$

When the total time $t = \sum_{i=1}^{k} \Delta t_i$, i.e., at the kth incremental time step, the temperature of the internal network point, $P_{i,j}$, of the section can be written from Eqn (6.7):

$$T^k_{i,j} = \frac{\Delta t_k}{2(\rho C)^{k-1}_{i,j}\Delta x^2 \Delta y^2}\begin{bmatrix} \left(\lambda^{k-1}_{i,1,j} + \lambda^{k-1}_{i,j}\right)\Delta x^2 T^{k-1}_{i-1,j} \\ +\left(\lambda^{k-1}_{i+1,j} + \lambda^{k-1}_{i,j}\right)\Delta y^2 T^{k-1}_{i+1,j} \\ +\left(\lambda^{k-1}_{i,j-1} + \lambda^{k-1}_{i,j}\right)\Delta x^2 T^{k-1}_{i,j-1} \\ +\left(\lambda^{k-1}_{i,j+1} + \lambda^{k-1}_{i,j}\right)\Delta x^2 T^{k-1}_{i,j+1} \end{bmatrix} + \left\{\begin{matrix} 1 - \dfrac{\Delta t_k}{2(\rho C)^{k-1}_{i,j}\Delta x^2 \Delta y^2} \\ \begin{bmatrix} \left(\lambda^{k-1}_{i-1,j} + 2\lambda^{k-1}_{i,j} + \lambda^{k-1}_{i+1,j}\right)\Delta y^2 \\ +\left(\lambda^{k-1}_{i,j-1} + 2\lambda^{k-1}_{i,j} + \lambda^{k-1}_{i,j+1}\right)\Delta x^2 \end{bmatrix}\end{matrix}\right\} T^{k-1}_{i,j} \tag{6.9}$$

where $T^k_{i,j}$ and $T^{k-1}_{i,j}$ are the temperature values of the network point $P_{i,j}$ at the kth and $(k-1)$th incremental time steps, respectively, and $(\rho c)^{k-1}_{i,j}$ is the product of the mass density and the specific heat capacity of the network point $P_{i,j}$ at the $(k-1)$th incremental time step. The definitions of other symbols are listed in Table 6-2.

When the boundary of the peripheral network point of the section is in the third category, the temperature value of the point $(P_{1,j})$ at the $k$wth incremental time step has to be solved using the formula:

$$T^k_{1,j} = \frac{\Delta t_k}{2(\rho C)^{k-1}_{1,j}}\left[\frac{2\left(\beta^1_T\right)^{k-1}\left(T^1_a\right)^{k-1}\Delta x + \left(\lambda^{k-1}_{1,j} + \lambda^{k-1}_{2,j}\right)T^{k-1}_{2,j}}{\Delta x^2} + \frac{\left(\lambda^{k-1}_{1,j-1} + \lambda^{k-1}_{1,j}\right)T^{k-1}_{1,j-1} + \left(\lambda^{k-1}_{1,j+1} + \lambda^{k-1}_{1,j}\right)T^{k-1}_{1,j+1}}{2\Delta y^2}\right] + \left\{1 - \frac{\Delta t_k}{2(\rho C)^{k-1}_{1,j}}\left[\frac{2\left(\beta^1_T\right)^{k-1}\Delta x + \lambda^{k-1}_{1,j} + \lambda^{k-1}_{2,j}}{\Delta x^2} + \frac{\lambda^{k-1}_{1,j-1} + 2\lambda^{k-1}_{1,j} + \lambda^{k-1}_{1,j+1}}{2\Delta y^2}\right]\right\} T^{k-1}_{1,j} \tag{6.10}$$

where $(\beta^1_T)^{k-1}$ and $(T^1_a)^{k-1}$ are the coefficient of heat transfer and the environmental temperature on the left side boundary of the section at the $(k-1)$th incremental time step. The definitions of the other symbols are the same as in the previous formulas.

If the boundary condition of the corner network point (e.g., $P_{1,1}$ in Fig. 6-1) is in the third category, the temperature value $T^k_{1,1}$ can be calculated using the following formula:

$$T^k_{1,1} = \frac{\Delta t_k}{(\rho C)^{k-1}_{1,1}}\left[\frac{4\left(\beta^1_T\right)^{k-1}\left(T^1_a\right)^{k-1}\Delta x + \left(\lambda^{k-1}_{1,1} + \lambda^{k-1}_{2,1}\right)T^{k-1}_{2,1}}{4\Delta x^2} + \frac{4\left(\beta^b_T\right)^{k-1}\left(T^b_a\right)^{k-1}\Delta y + \left(\lambda^{k-1}_{1,1} + \lambda^{k-1}_{1,2}\right)T^{k-1}_{1,2}}{4\Delta y^2}\right] + \left\{1 - \frac{\Delta t_k}{(\rho C)^{k-1}_{1,1}}\left[\frac{4\left(\beta^1_T\right)^{k-1}\Delta x + \lambda^{k-1}_{1,1} + \lambda^{k-1}_{2,1}}{4\Delta x^2} + \frac{4\left(\beta^b_T\right)^{k-1}\Delta y + \lambda^{k-1}_{1,1} + \lambda^{k-1}_{1,2}}{4\Delta y^2}\right]\right\}^{-1}_{11} \tag{6.11}$$

TABLE 6-2 Definitions of Other Symbols

Network point	$P_{i-1,j}$	$P_{i,j}$	$P_{i+1,j}$	$P_{i,j-1}$	$P_{i,j+1}$
Temperature value at $(k-1)$th incremental time step	$T^{k-1}_{i-1,j}$	$T^{k-1}_{i,j}$	$T^{k-1}_{i+1,j}$	$T^{k-1}_{i,j-1}$	$T^{k-1}_{i,j+1}$
Coefficient of heat conduction	$\lambda^{k-1}_{i-1,j}$	$\lambda^{k-1}_{i,j}$	$\lambda^{k-1}_{i+1,j}$	$\lambda^{k-1}_{i,j-1}$	$\lambda^{k-1}_{i,j+1}$

where $(\beta_T^b)^{k-1}$ and $(T_a^b)^{k-1}$ are the coefficient of heat transfer and the environmental temperature on the bottom boundary of the section at the $(k-1)$th incremental time step, respectively. The definitions of other symbols are the same as before.

If the left boundary condition is in the third category, but the bottom boundary condition is in the first category for the corner network point, the temperature value can be calculated using the following formula:

$$T_{1,1}^k = \frac{\Delta t_k}{4(\rho C)_{1,1}^{k-1}}\left[\frac{4(\beta_T^1)^{k-1}(T_a^1)^{k-1}\Delta x + (\lambda_{1,1}^{k-1}+\lambda_{2,1}^{k-1})T_{2,1}^{k-1}}{\Delta x^2} + \frac{(\lambda_{1,1}^{k-1}+\lambda_{1,2}^{k-1})T_{1,2}^{k-1}}{\Delta y^2}\right] + \left\{1-\frac{\Delta t_k}{4(\rho C)_{1,1}^{k-1}}\left[\frac{4(\beta_T^1)^{k-1}\Delta x + \lambda_{1,1}^{k-1}+\lambda_2^{k-1}}{\Delta x^2} + \frac{\lambda_{1,1}^{k-1}+\lambda_{1,2}^{k-1}}{\Delta y^2}\right]\right\}T_{1,1}^{k-1} \tag{6.12}$$

Referring to these equations, the algebraic equation of the heat equilibrium can be established individually for the temperature value of every network on the section, including the internal, edge, and corner networks, at the kth incremental time step. Then the series of equations is compiled and can be solved, and the temperature of every network point $T_{i,j}^k$ is obtained. When the time increment reaches the sth step, the temperature distribution and its variation on the section are obtained within the whole time field.

The stability of the calculation above is conditional. When the series analysis method is used, the thermal parameters are assumed to be constants, and the maximum value of the coefficient of heat conduction λ and the minimum values of the specific heat capacity C and mass density ϱ are taken, the condition of stability is obtained to be the limits of the incremental time step:

$$\Delta t \le \frac{(\Delta x)^2(\Delta y)^2}{2(\lambda/\rho C)\left[(\Delta x)^2+(\Delta y)^2\right]} \tag{6.13}$$

If the boundary condition of the section is in the third category and the value of the coefficient of heat conduction is high, the limits of the incremental time step should be reduced further.

6.1.4 Examples

Example 1 There is an infinite plate of thickness l = 0.2 m (Fig. 6-2) and the coefficient of heat conduction is λ = 1 W/(m °C). The temperatures of the media on the outer side of both sides are T_{a1} = 100 °C and T_{a2} = 0 °C, and the coefficient of heat transfer of the media to the plate is β = 20 W/(m² °C). The temperature distribution is calculated by difference method.

Solution This is a problem of a stable temperature field and its boundary condition is in the third category. The theoretical solution is:

$$T = \frac{\beta(T_{a2}-T_{a1})}{2(\lambda+l\beta)}x + \frac{1}{2}(T_{a1}+T_{a2})$$

The plate is divided into five networks along its thickness direction, and the length of the network is Δx = 0.05 m. The algebraic equations of heat equilibrium are established successively

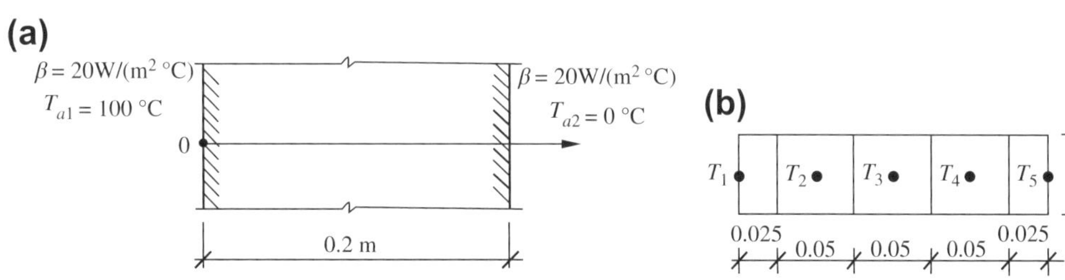

FIGURE 6-2 Analysis of the stable temperature field of an infinite plate: (a) boundary condition; (b) network division.

for every network point according to Eqns (6.2) and (6.4):

$$2T_1 - T_2 - 100 = 0$$

$$T_1 - 2T_2 + T_3 = 0$$

$$T_2 - 2T_3 + T_4 = 0$$

$$T_3 - 2T_4 + T_5 = 0$$

$$T_4 - 2T_5 = 0$$

The series of equations is solved and T_1 = 83.33 °C, T_2 = 66.67 °C, T_3 = 50 °C, T_4 = 33.33 °C, and T_5 = 16.67 °C. The solution is identical with the accurate theoretical solution.

Example 2 There is a semi-infinite uniform substance and its initial temperature is zero. When $t = 0$, a constant and uniform heat flow applies suddenly on its left surface as a thermal load (Fig. 6-3). The coefficient of heat conduction is $\lambda = 1.0$ W/(m °C), the mass density is $\varrho =$ 1.0 kg/m^3, the specific heat capacity is $C = 10$ J/(kg °C), and the heat diffusivity is $d = \lambda/\varrho C = 10$ m^2/s correspondingly for the substance. The temperature of the medium on the boundary is $T_a =$ 200 °C, and the coefficient of heat transfer is $\beta =$ 1.0 W/(m^2 °C). The temperature distribution is calculated using the difference method.

Solution This is a problem of transient heat conduction and the boundary condition is in the third category. The theoretical solution[6-1] is:

$$T(x,t) = T_\infty \left[\operatorname{erfc}\left(\frac{x}{2\sqrt{\alpha t}}\right) - \exp\left(\frac{\beta x}{\lambda} + \frac{\beta^2 \alpha t}{\lambda^2}\right) \operatorname{erfc}\left(\frac{x}{2\sqrt{\alpha t}} + \frac{\beta\sqrt{\alpha t}}{\lambda}\right) \right]$$

where the complementary error function is $\operatorname{erfc}(x) = 1 - \operatorname{erf}(x)$ and the Gauss error function is $\operatorname{erf}(x) = (2/\sqrt{\pi}) \int_0^x e^{-y^2} dy$.

The difference method is used to find the solution. The semi-infinite substance is divided into networks along the perpendicular direction of the boundary, and the lengths of the network and the time step are taken as $\Delta x = 0.1$ m and $\Delta t = 0.001$ s, respectively. The algebraic equations of heat equilibrium are established according to Eqns (6.9) and (6.10), and the temperature value at arbitrary time and position can be obtained after the equations are solved. Part of the calculation results and the curve of the accurate theoretical solution are drawn together in Fig. 6-4.

The temperatures varying with time at points located at $x = 0.5$ m and 2 m from the boundary are given in Fig. 6-4(a), and the temperature variation with the position, i.e., temperature distribution on the section, at times $t = 0.1$ s, 0.5 s, and 1.0 s are presented in Fig. 6-4(b). It is seen that the calculation results for the difference method agree well with the accurate theoretical solution.

These examples show that the difference method used for analyzing the temperature field of a regular section of a structure is accurate. However, when the difference method is used for analyzing the transient temperature field, the accuracy of the calculation result reduces and depends not only on the difference format selected but is also closely related to the divisions of the network and the time step.

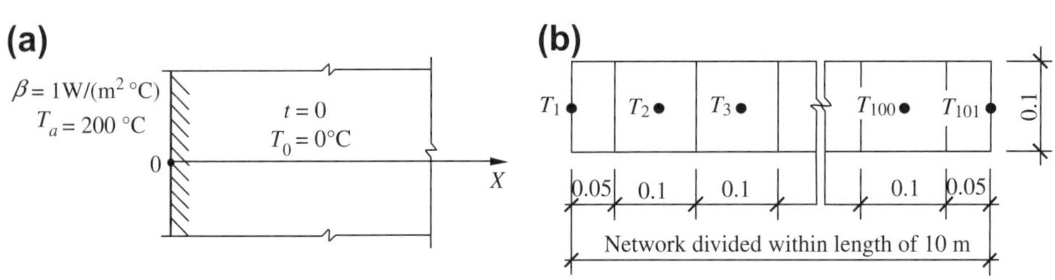

FIGURE 6-3 Analysis of the transient temperature field of a semi-infinite substance: (a) boundary condition; (b) network division.

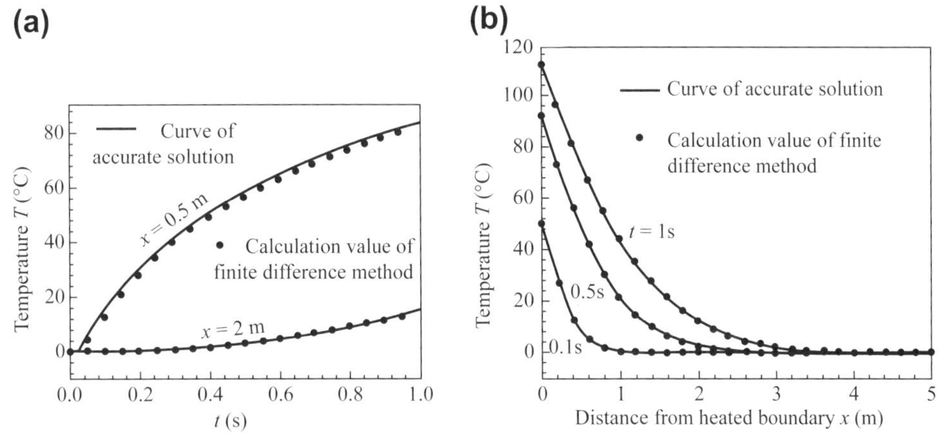

FIGURE 6-4 Calculated results of the temperature field of a semi-infinite substance: (a) temperature–time curve; (b) temperature distribution on the section.

6.2 COMBINED FINITE ELEMENT AND DIFFERENCE ANALYSIS

The combined finite element and difference method is now commonly used to analyze the transient temperature field. It uses the finite element method to discrete the space field and the difference method to discrete the time field, and then the finite element is calculated successively along the time progress.[6-2–6-7]

This method also has some disadvantages. Some researchers argue that the calculation value is stable and does not fluctuate, but with lower accuracy, if the backward difference format of two points is used. However, if the Crank–Nicolson format, the Galerkin format, and the backward difference format of three points are used, a stable calculation with high accuracy is achieved, but the size of the finite element and the incremental time step have to be selected reasonably. Otherwise, fluctuation occurs most likely during the calculation process.[6-8–6-10] Generally, variable step length can be used to deal with this (see Section 6.2.5).

6.2.1 Heat Conduction Equation for an Element

As well as the difference method, the extreme value method of the pan function and the weighted complementary method are also used to solve the partial differential equation. However, generally, the pan function of a partial differential equation of nonlinear parabolic type is complicated.[6-11] Among the weighted complementary methods, the Galerkin method is used for finite element analysis. In this chapter it is used to derive the equation of heat conduction of an element.

According to the Galerkin method, the basic equation (Eqn (6.7)) of transient heat conduction can be expressed as an integral:

$$\int_{A^e}\left[\frac{\partial}{\partial x}\left(\lambda(T)\frac{\partial T}{\partial x}\right)+\frac{\partial}{\partial y}\left(\lambda(T)\frac{\partial T}{\partial y}\right)-\rho(T)\cdot C(T)\frac{\partial T}{\partial t}N_i\,\mathrm{d}A\right]=0$$

$$\int_{A^e}\left[\frac{\partial}{\partial x}\left(\lambda(T)\frac{\partial T}{\partial x}\right)+\frac{\partial}{\partial y}\left(\lambda(T)\frac{\partial T}{\partial y}\right)\right]N_i\,\mathrm{d}A = \int_{A^e}\rho(T)\cdot C(T)\frac{\partial T}{\partial t}N_i\,\mathrm{d}A \tag{6.14}$$

where N_i is the weighted function and A^e is the area of the element.

The left of Eqn (6.14) is partially integrated and gives:

$$\int_{A^e}\rho(T)\cdot C(T)\frac{\partial T}{\partial t}N_i\,\mathrm{d}A+\int_{A^e}\begin{pmatrix}\dfrac{\partial N_i}{\partial x} & \dfrac{\partial N_i}{\partial y}\end{pmatrix}\begin{Bmatrix}\lambda(T)\dfrac{\partial T}{\partial x}\\ \lambda(T)\dfrac{\partial T}{\partial y}\end{Bmatrix}\mathrm{d}A + \int_{l_1^e}(\boldsymbol{q}\cdot\boldsymbol{n})N_i\,\mathrm{d}l+\int_{l_3^e}\beta_T(T-T_a)N_i\,\mathrm{d}l=0 \tag{6.15}$$

where l_1^e and l_3^e are the element boundaries in the first and third categories, respectively, q is the density vector of heat flow, and $\boldsymbol{n}$ is the direction cosine vector outside the normal of the boundary.

The temperature at an arbitrary point within the element can be represented by the interpolation function:

$$T(x,y,t) = [N(x,y)]\{T(t)\}^e \tag{6.16a}$$

where the matrix of the shape function and the temperature vector of the element node are, respectively,

$$[N(x,y)] = [N_1(x,y)\ N_2(x,y)\ \ldots\ N_n(x,y)] \tag{6.16b}$$

$$\{T(t)\}^e = [T_1(t)\ \ T_2(t)\ \ T_n(t)]^T \tag{6.16c}$$

The partial derivatives of the temperature $T(x,y,t)$ with respect to coordinates x and y are found:

$$\begin{Bmatrix} \frac{\partial T}{\partial x} \\ \frac{\partial T}{\partial y} \end{Bmatrix} = \begin{bmatrix} \frac{\partial N_1}{\partial x} & \frac{\partial N_2}{\partial x} & \cdots & \frac{\partial N_n}{\partial x} \\ \frac{\partial N_1}{\partial y} & \frac{\partial N_2}{\partial y} & \cdots & \frac{\partial N_n}{\partial y} \end{bmatrix}$$

$$\{T(t)\}^e = [B]\{T(t)\}^e \tag{6.17}$$

Equations (6.16) and (6.17) are substituted into Eqn (6.15) to obtain:

$$[C]^e \frac{\partial}{\partial t}\{T(t)\}^e + [K]^e\{T(t)\}^e = \{P\}^e \tag{6.18a}$$

where

$$[K]^e = [K_T]^e + [K_\beta]^e \tag{6.18b}$$

$$\{P\}^e = \{P_T\}^e + \{P_\beta\}^e \tag{6.18c}$$

where the matrix for the specific heat capacity of the element is:

$$[C]^e = \int_{A^e} \rho(t)\, C(t) [N]^T [N]\, dA \tag{6.19a}$$

the matrix for heat conduction of the element is:

$$[K_T]^e = \int_{A^e} \lambda(T)[B]^T[B]\, dA \tag{6.19b}$$

the matrix for heat transfer of the element is:

$$[K_\beta]^e = \int_{A^e} \beta_T [N]^T [N]\, dA \tag{6.19c}$$

the vector for the heat load of the element is:

$$\{P_T\}^e = \int_{l_1^e} (\boldsymbol{q},\boldsymbol{n})[N]^T\, dl \tag{6.19d}$$

and the vector for the heat transfer load of the element is:

$$\{P_\beta\}^e = \int_{l_3^e} \beta_T T_a [N]^T\, dl \tag{6.19e}$$

When the behavior of the concrete structure at elevated temperature (fire resistance) is analyzed, considerable variation and deviation appear because of the influences of many factors. To simplify the calculation, a simple element can be used. A rectangular element of four nodes (Fig. 6-5) is selected here and the temperature at any point within the element can be represented by the temperatures at the four nodes (1, 2, 3, and 4). The shape function is

$$N_i = \frac{1}{4}\left(1+\xi_i \frac{x}{a}\right)\left(1+\eta_i \frac{y}{b}\right), (i=1,2,3,4) \tag{6.20}$$

where $\xi_1 = -1$, $\eta_1 = -1$; $\xi_2 = +1$, $\eta_2 = -1$; $\xi_3 = -1$, $\eta_3 = +1$; $\xi_4 = +1$, $\eta_4 = +1$.

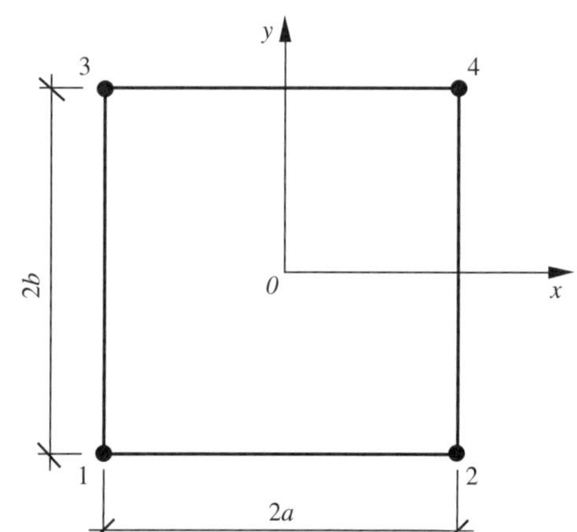

FIGURE 6-5 Rectangular element of four nodes.

The values of the thermal parameters λ, C, and ϱ of concrete vary with temperature (see Section 5.2). The temperature value at every point of the section is different and is a function of the coordinates x and y. The functions to be integrated in Eqns (6.19a) and (6.19b) are very complicated and their integrations are difficult to find directly. Thus, a digital method, such as integrating using the Gauss formula, has to be used to find the solution.

In order to simplify the calculation but to maintain the necessary accuracy of the calculation, the values of the thermal parameters of concrete are simplified. When the divided element is small, the temperature gradient approaches zero. The experimental result shows that the values of the thermal parameters do not vary sharply with temperature. So, it is assumed that the values of the thermal parameters in an element are constants and are calculated by the average temperature of its four nodes. Therefore, the thermal parameters λ, C, and ϱ are only functions of the temperature at the nodes of the element and are not related to the coordinates. Therefore, these parameters can be removed from the integration and Eqns (6.19a) and (6.19b) become:

$$[C]^e = \rho\left(\{T(t)\}^e\right)C\left(\{T(t)\}^e\right) \int_{A^e}[N]^T[N]dA \tag{6.21a}$$

$$[K_T]^e = \lambda\left(T(t)^e\right) \int_{A^e}[B]^T[B]dA \tag{6.21b}$$

The following formulas are obtained after integration:

$$C_{ij}^e = \frac{\rho(\{T(t)\}^e)C(\{T(t)\}^e)ab}{36}(3+\xi_i\xi_j)(3+\eta_i\eta_j) \tag{6.22a}$$

$$K_{ij}^e = \frac{\lambda(\{T(t)\}^e)}{12}\left[\frac{b}{a}\xi_i\xi_j\left(3+\xi_i\xi_j\right)+\frac{a}{b}\eta_i\eta_j\left(3+\eta_i\eta_j\right)\right] \tag{6.22b}$$

(i,j = 1, 2, 3, 4)

Therefore, the matrixes for the specific heat capacity and the heat conduction of the rectangular element of four nodes are written as:

$$[C]^e = \frac{\rho\left(\{T(t)\}^e\right)C\left(\{T(t)\}^e\right)ab}{9}\begin{bmatrix}4 & 2 & 2 & 1\\ 2 & 4 & 1 & 2\\ 2 & 1 & 4 & 2\\ 1 & 2 & 2 & 4\end{bmatrix} \tag{6.23}$$

$$[K_T]^e = \frac{\rho\lambda\left(\{T(t)\}^e\right)}{6}\begin{bmatrix}2\frac{b}{a}+2\frac{a}{b} & -2\frac{b}{a}+\frac{a}{b} & \frac{b}{a}-2\frac{a}{b} & -\frac{b}{a}-\frac{a}{b}\\ -2\frac{b}{a}+\frac{a}{b} & 2\frac{b}{a}+2\frac{a}{b} & -\frac{b}{a}-\frac{a}{b} & \frac{b}{a}-2\frac{a}{b}\\ \frac{b}{a}-2\frac{a}{b} & -\frac{b}{a}-\frac{a}{b} & 2\frac{b}{a}+2\frac{a}{b} & -2\frac{b}{a}+\frac{a}{b}\\ -\frac{b}{a}-\frac{a}{b} & \frac{b}{a}-2\frac{a}{b} & -2\frac{b}{a}+\frac{a}{b} & 2\frac{b}{a}+2\frac{a}{b}\end{bmatrix} \tag{6.24}$$

6.2.2 Total Collective Equation and Boundary Condition

The total collective equation of heat conduction is obtained for finite element analysis from Eqn (6.18):

$$[C]\frac{\partial}{\partial t}\{T(t)\} + [K]T(t) = \{P\} \tag{6.25}$$

where $[K] = [K_T] + [K_\beta]$, $\{P\} = \{P_T\} + \{P_\beta\}$, and $[C]$ is the total matrix of specific heat capacity, $[K_T]$ is the total matrix of heat conduction, $[K_\beta]$ is the total matrix of heat transfer, $\{P_T\}$ is the total vector of heat load, and $\{P_\beta\}$ is the total vector of heat transfer load.

$[C]$ and $[K_T]$ can be collected directly from Eqns (6.23) and (6.24), respectively. $\{P_T\}$ is the vector of the summation of the heat quantity flowing into (or out of) the node. The formula is given separately for elements at different positions.

1. Internal element

An internal element (i, j) has four nodes located not on the boundary of the section. The total heat load $(P_T)_l$ at its node l relates only to the four connected elements, i.e., (i – 1, j – 1), (i, j – 1), (i – 1, j), and (i, j) (Fig. 6-6). It is known from Eqns (6.19d) and (6.20) that $(P_T)_l$ relates only to the corresponding boundaries of the connected elements. Therefore,

$$\begin{aligned}(P_T)_l = {} & (P_T)_{34}^{i-1,j-1} + (P_T)_{24}^{i-1,j-1} + (P_T)_{13}^{i,j-1}\\ & + (P_T)_{34}^{i,j-1} + (P_T)_{12}^{i-1,j} + (P_T)_{24}^{i-1,j}\\ & + (P_T)_{12}^{i,j} + (P_T)_{13}^{i,j}\end{aligned} \tag{6.26}$$

where $(P_T)_{34}^{i-1,j-1}$ is the contribution of boundary 34 of the element (i –1,j –1) to the heat load at node l. Other symbols are the same as before.

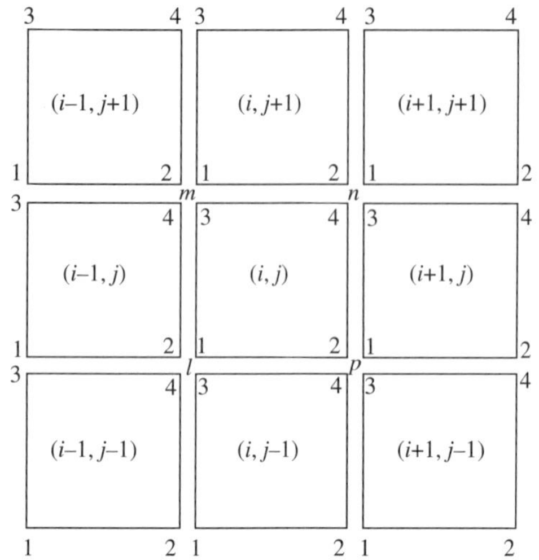

FIGURE 6-6 Internal element and its adjacent elements.

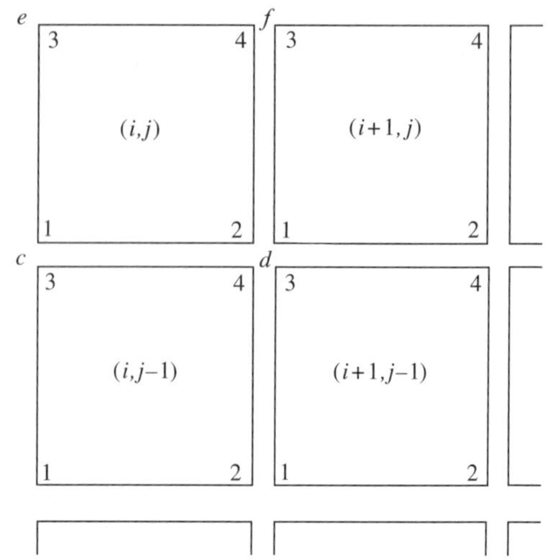

FIGURE 6-7 Boundary element and its adjacent elements.

The temperature and the density of heat flow should be continuous on the connected boundary between two adjacent elements. The continuity of the temperature on the boundary is ensured as the shape function of Eqn (6.20) is used. However, the normal density of heat flow is not continuous, so its continuous condition has to be added into the total collective equation of heat conduction for finite element analysis. For example, the density of heat flow on boundary 34 of element $(i-1, j-1)$ should be equal to that on boundary 12 of element $(i-1, j)$. So,

$$\left(-\lambda\frac{\partial T}{\partial y}\right)_{34}^{i-1,j-1} = \left(-\lambda\frac{\partial T}{\partial y}\right)_{12}^{i-1,j} \tag{6.27}$$

is obtained. The contributions of both boundaries to the heat load of node l can be obtained separately from Eqns (6.19d) and (6.20):

$$(P_T)_{34}^{i-1,j-1} = -\int_{-a}^{a}\frac{1}{2}\left(1+\frac{x}{a}\right)\left(\lambda\frac{\partial T}{\partial y}\right)_{34}^{i-1,j-1}dx \tag{6.28a}$$

$$(P_T)_{12}^{i-1,j} = \int_{-a}^{a}\frac{1}{2}\left(1+\frac{x}{a}\right)\left(\lambda\frac{\partial T}{\partial y}\right)_{12}^{i-1,j}dx \tag{6.28b}$$

Comparing Eqn (6.27) with Eqn (6.28), we have:

$$(P_T)_{34}^{i-1,j-1} + (P_T)_{12}^{i-1,j} = 0 \tag{6.29a}$$

Similarly,

$$(P_T)_{24}^{i-1,j-1} + (P_T)_{13}^{i-1,j} = 0 \tag{6.29b}$$

$$(P_T)_{34}^{i,j-1} + (P_T)_{12}^{i,j} = 0 \tag{6.29c}$$

$$(P_T)_{24}^{i-1,j} + (P_T)_{13}^{i,j} = 0 \tag{6.29d}$$

Substituting Eqn (6.29) into Eqn (6.26), we have:

$$(P_T)_l = 0 \tag{6.30a}$$

Similarly, for the nodes m, n, and p:

$$(P_T)_m = (P_T)_n = (P_T)_p = 0 \tag{6.30b}$$

2. Peripheral element

A peripheral element (i, j) has at least one node on the periphery of the section (Fig. 6-7). When the boundary condition is in the third category, $(P_T)_e = 0$ at node e. The heat load at node c is contributed only by boundary 34 of element $(i, j-1)$ and boundary 12 of element (i, j). Both boundaries are internal boundaries and also have

$$(P_T)_c = (P_T)_{34}^{i,j-1} + (P_T)_{12}^{i,j} = 0 \tag{6.31}$$

Similarly, the heat load at node f is $(P_T)_f = 0$. However, node d is in the interior of the section and $(P_T)_d = 0$ is also obtained according to the derivation above (Eqn (6.29)).

When the boundary condition is in the first category,

$$\left(\frac{\partial T}{\partial y}\right)_{13}^{i,j-1} = \left(\frac{\partial T}{\partial y}\right)_{13}^{i,j} = \left(\frac{\partial T}{\partial y}\right)_{34}^{i,j} \tag{6.32}$$

is obtained from Eqn (5.8). Substituting this formula and Eqn (6.20) into Eqn (6.19b),

$$(P_T)_{13}^{i,j-1} = (P_T)_{13}^{i,j} + (P_T)_{34}^{i,j} = 0 \tag{6.33}$$

is obtained. The heat loads at nodes e and c are, respectively:

$$(P_T)_e = (P_T)_{13}^{i,j} + (P_T)_{34}^{i,j} \tag{6.34a}$$

and

$$(P_T)_c = (P_T)_{13}^{i,j-1} + (P_T)_{34}^{i,j-1} + (P_T)_{12}^{i,j} + (P_T)_{13}^{i,j} \tag{6.34b}$$

Obviously, both are zero. Similarly, the heat load at node f ($(P_T)_f$) is also zero. The heat load at node d is the same as that under the boundary condition in the third category, i.e., $(P_T)_d = 0$.

Summarizing, we have

$$\{P_T\} = 0 \tag{6.35}$$

Then, Eqn (6.25) can be written as:

$$[C]\frac{\partial}{\partial t}\{T(t)\} + [K]\{T(t)\} = \{P_\beta\} \tag{6.36}$$

$[K_\beta]$ and $\{P_\beta\}$ occur on the boundary transferring heat. Generally, the coefficient of heat transfer β_T and the temperature of the flow medium T_a vary slightly along the boundary of the section of the structural member and vary even less along the edge length of the element. It can be assumed that β_T and T_a do not vary along the edge of the element and are only functions of the fire temperature. Therefore, β_T and T_a can be removed from the integration, and Eqns (6.19c) and (6.19e) can be written as:

$$[K_\beta]^e = \sum_i \beta_{Ti} \int_{l_{3i}^e} [N]^T [N]\, dl \tag{6.37a}$$

$$\{P_\beta\}^e = \sum_i \beta_{Ti} T_{ai} \int_{l_{3i}^e} [N]^T dl \tag{6.37b}$$

where l_{3i}^e is the length of the ith boundary in the third category of the element, β_{Ti} is the coefficient of heat transfer on the ith boundary of the element, and T_{ai} is the temperature of the flow medium on the ith boundary of the element.

If boundary 12 of the element is in the third category, substituting Eqn (6.20) into Eqn (6.37) gives:

$$[K_\beta]_{12}^e = \frac{a\beta_{T12}}{3}\begin{bmatrix} 2 & 1 & 0 & 0 \\ 1 & 2 & 0 & 0 \\ 0 & 0 & 0 & 0 \\ 0 & 0 & 0 & 0 \end{bmatrix}, \quad \{P_\beta\}_{12}^e = a\beta_{T12} T_{a12} \begin{Bmatrix} 1 \\ 1 \\ 0 \\ 0 \end{Bmatrix} \tag{6.38a}$$

where β_{T12} and T_{a12} are the coefficient of heat transfer and the temperature of the flow medium, respectively, on boundary 12 of the element.

Similarly, if boundaries 24, 13, and 34 of the element are in the third category, we have:

$$[K_\beta]_{24}^e = \frac{b\beta_{T24}}{3}\begin{bmatrix} 0 & 0 & 0 & 0 \\ 0 & 2 & 0 & 1 \\ 0 & 0 & 0 & 0 \\ 0 & 1 & 0 & 2 \end{bmatrix}, \quad \{P_\beta\}_{24}^e = b\beta_{T24} T_{a24} \begin{Bmatrix} 0 \\ 1 \\ 0 \\ 1 \end{Bmatrix} \tag{6.38b}$$

$$[K_\beta]_{13}^e = \frac{b\beta_{T13}}{3}\begin{bmatrix} 2 & 0 & 1 & 0 \\ 0 & 0 & 0 & 0 \\ 1 & 0 & 2 & 0 \\ 0 & 0 & 0 & 0 \end{bmatrix}, \quad \{P_\beta\}_{13}^e = b\beta_{T13} T_{a13} \begin{Bmatrix} 1 \\ 0 \\ 1 \\ 0 \end{Bmatrix} \tag{6.38c}$$

$$[K_\beta]_{34}^e = \frac{a\beta_{T34}}{3}\begin{bmatrix} 0 & 0 & 0 & 0 \\ 0 & 0 & 0 & 0 \\ 0 & 0 & 2 & 1 \\ 0 & 0 & 1 & 2 \end{bmatrix}, \quad \{P_\beta\}_{34}^e = a\beta_{T34} T_{a34} \begin{Bmatrix} 0 \\ 0 \\ 1 \\ 1 \end{Bmatrix} \tag{6.38d}$$

Collecting Eqn (6.38) for every element, the total matrix of heat transfer $[K_\beta]$ and the total vector of heat transfer load $\{P_\beta\}$ of the section of the structural member are obtained.

6.2.3 Equation of Stable Heat Conduction and Its Calculation

Equation (6.1) of stable heat conduction of a two-dimensional temperature field of a section is a partial differential equation of ellipse type, and the finite element equation can be derived using the method described in Section 6.2.2.

The problem of stable heat conduction can also be considered as a special example of the problem of transient heat conduction, in which the time tends to be infinite and the temperature does not vary. When $\partial\{T(t)\}/\partial t|_\infty = 0$, the finite element equation (Eqn (6.25)) of transient heat conduction is changed into that of stable heat conduction:

$$[K]\{T_\infty\} = \{P_\beta\} \tag{6.39}$$

where $\{T_\infty\}$ is stable temperature vector on the section.

Eqn (6.39) is a nonlinear equation series because $[K]$ is related to $\{T_\infty\}$. Many methods[6-12–6-18] can be used to solve this type of equation series. Usually, the solution of a set of a series of linear equations is used to approach the real solution; the Newton–Raphson and modified Newton–Raphson methods are commonly used. To increase the accuracy and to accelerate the convergence of the calculation, various calibration methods based on both methods can be used. Therefore, it is not difficult to find the solution of Eqn (6.39).

6.2.4 Difference Format of the Transient Heat Conduction Equation

The time field is divided into a certain number of time elements. Within each time element Δt_k, the vector of temperature $[T]$, the total vector of heat transfer load $\{P_\beta\}$, the total matrix of specific heat capacity $[C]$, and the total matrix of heat conduction $[K]$ are composed of the simplest linear equations (Fig. 6-8):

$$\{T\} = N_k\{T\}_k + N_{k+1}\{T\}_{k+1} \tag{6.40a}$$

$$\{P_\beta\} = N_k\{P_\beta\} + N_{k+1}\{P_\beta\}_{k+1} \tag{6.40b}$$

$$[C] = N_k[C]_k + N_{k+1}[C]_{k+1} \tag{6.40c}$$

$$[K] = N_k[K]_k + N_{k+1}[K]_{k+1} \tag{6.40d}$$

The partial derivative of Eqn (6.40a) with respect to time t is:

$$\frac{\partial}{\partial t}T = \dot{N}_k\{T\}_k = \dot{N}_{k+1}\{T\}_{k+1} \tag{6.41}$$

where the symbols for the various vectors and matrixes corresponding to time are:

when $t = \sum_{r=1}^{k} \Delta t_r$ $\quad \{T\}_k \quad \{P_\beta\}_k \quad [C]_k \quad [K]_k$

when $t = \sum_{r=1}^{k+1} \Delta t_r$ $\quad \{T\}_{k+1} \quad \{P_\beta\}_{k+1} \quad [C]_{k+1} \quad [K]_{k+1}$

In addition:

$$N_k = 1-\xi,\ N_{k+1} = \xi$$
$$\dot{N}_k = -1/\Delta t_k,\ \dot{N}_{k+1} = 1/\Delta t_k$$
$$\xi = \tau/\Delta t_k,\ 0 \le \xi \le 1$$

Using the method of weighted magnitude, Eqn (6.36) of a time element can be described as:

$$\int_0^1 w_j\left([C]\frac{\partial}{\partial t}\{T\} + [K]\{T\} - \{P_\beta\}\right)d\xi \tag{6.42}$$

where w_j is the weighted function.

Substituting Eqns (6.40) and (6.41) into Eqn (6.42),

$$\begin{bmatrix} \frac{1-\theta_1}{\Delta t_k}[C]_k + \frac{\theta_1}{\Delta t_k}[C]_{k+1} + (\theta_1-\theta_2)[K]_k \\ +\theta_2[K]_{k+1} \end{bmatrix}\{T\}_{k+1} + \begin{bmatrix} \frac{-1+\theta_1}{\Delta t_k}[C]_k - \frac{\theta_1}{\Delta t_k}[C]_{k+1} + (1-2\theta_1+\theta_2)[K]_k \\ +(\theta_1-\theta_2)[K]_{k+1} \end{bmatrix}$$
$$\{T\}_k - (1-\theta_1)\{P_\beta\}_k - \theta_1\{P_\beta\}_{k+1} = 0 \tag{6.43}$$

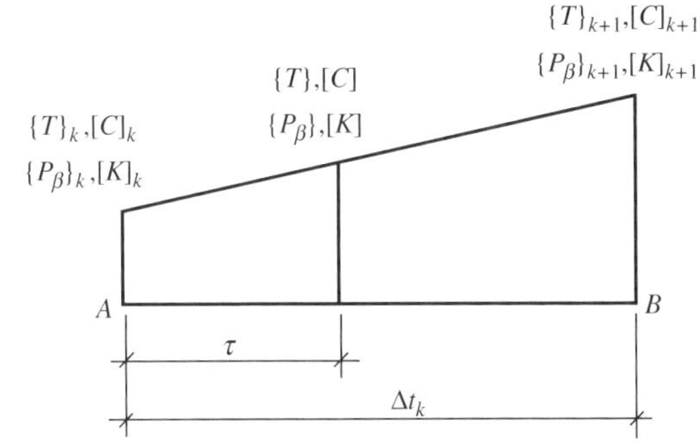

FIGURE 6-8 Linear variation within time element.

is obtained, where

$$\theta_1 = \frac{\int_0^1 w_j \xi \, d\xi}{\int_0^1 w_j \, d\xi}, \quad \theta_2 = \frac{\int_0^1 w_j \xi^2 \, d\xi}{\int_0^1 w_j \, d\xi}$$

There are various difference formats for different weighted functions. According to the Galerkin method, the weighted function is taken as:

$$w_i = \frac{\partial}{\partial \{T\}_{k+1}} \{T\} = \xi$$

When $\theta_1 = 2/3$, $\theta_2 = 1/2$ are substituted into Eqn (6.43), we obtain:

$$\begin{bmatrix} \frac{1}{\Delta t_k}\left(\frac{1}{3}[C]_k + \frac{2}{3}[C]_{k+1}\right) \\ + \frac{1}{6}[K]_k + \frac{1}{2}[K]_{k+1} \end{bmatrix} \{T\}_{k+1} + \begin{bmatrix} -\frac{1}{\Delta t_k}\left(\frac{1}{3}[C]_k + \frac{2}{3}[C]_{k+1}\right)[C] \\ + \frac{1}{6}[K]_k + \frac{1}{6}[K]_{k+1} \end{bmatrix} \{T\}_k - \frac{1}{3}\{P_\beta\}_k - \frac{2}{3}\{P_\beta\}_{k+1} = 0 \tag{6.44}$$

If $[C]_k = [C]_{k+1}$, $[K]_k = [K]_{k+1}$, and $\{P_\beta\}_k = \{P_\beta\}_{k+1}$, this is the Galerkin format.

If the matrixes $[C]$ and $[K]$ and the vector $\{P_\beta\}$ in Eqn (6.43) are not related to temperature, and when $\theta_1 = 0$, 1/2, and 1 are taken separately, the forward difference, the Crank–Nicolson formula, and the backward difference formats can be obtained. Therefore, Galerkin and these three formats are only the special cases of the difference format of Eqn (6.43).

Equation (6.44) is an implicit format and the iteration has to be performed within each incremental time step. The iteration formulas are:

$$\{T\}^0_{k+1} = \left(\frac{1}{\Delta t_k}[C]_k + \frac{2}{3}[K]_k\right)^{-1} \left(\frac{1}{\Delta t_k}[C]_k - \frac{1}{3}[K]_k\right)\{T\}_k + \left(\frac{1}{\Delta t_k}[C]_k + \frac{2}{3}[K]_k\right)^{-1}\{P_\beta\}_k \tag{6.45a}$$

and

$$\{T\}^i_{k+1} = \begin{bmatrix} \frac{1}{\Delta t_k}\left(\frac{1}{3}[C]_k + \frac{2}{3}[C]^{i-1}_{k+1}\right) \\ + \frac{1}{6}[K]_k + \frac{1}{2}[K]^{i-1}_{k+1} \end{bmatrix}^{-1} \times \begin{bmatrix} \frac{1}{\Delta t_k}\left(\frac{1}{3}[C]_k + \frac{2}{3}[C]^{i-1}_{k+1}\right)[C] \\ - \frac{1}{6}[K]_k - \frac{1}{6}[K]^{i-1}_{k+1} \end{bmatrix} \{T\}_k + \begin{bmatrix} \frac{1}{\Delta t_k}\left(\frac{1}{3}[C]_k + \frac{2}{3}[C]^{i-1}_{k+1}\right) \\ + \frac{1}{6}[K]_k + \frac{1}{2}[K]^{i-1}_{k+1} \end{bmatrix}^{-1} \left(\frac{1}{3}\{P_\beta\}_k + \frac{2}{3}\{P_\beta\}^{i-1}_{k+1}\right) \tag{6.45b}$$

The iterating procedure will not stop until

$$\| T^i_{k+1} - T^{i+1}_{k+1} \| \le T_g \text{ or } i = n_{\max} \tag{6.46}$$

where ‖•‖ is the norm with respect to •, T_g is the given convergence tolerance, and $n_{\max}$ is a positive integer for controlling the maximum number of iterations.

6.2.5 Stability of the Numerical Solution

In order to examine the stability of the transient heat conduction calculations, the decomposing vibration pattern method, which is popular in the dynamic analysis of structures, is used here. There are m coupling algebraic equations for the system of m degrees of freedom (temperature at node), and they are transformed into a series of m noncoupling equations. The temperature of the component of any vibration pattern $\{\hat{T}\}_i$ is given by the product of the vector of the vibration pattern $\{\phi\}_i$ and the amplitude of the vibration pattern Y_i, i.e.:

$$\{\hat{T}\}_i = \{\phi\}_i Y_i$$

The summation of the components of all vibration patterns is the total temperature vector of all the nodes:

$$\{T\} = \{\phi\}_1 Y_1 + \{\phi\}_2 Y_2 + \ldots + \{\phi\}_m Y_m = \sum_{i=1}^{m} \{\phi\}_i Y_i \tag{6.47}$$

Equation (6.47) is substituted into Eqn (6.43), and both sides of the equation are multiplied by $\{\phi\}_i^T$ $(i = 1,2,\ldots,m)$. Corresponding to free vibration, $\{P_\beta\} = \{0\}$ and

$$\begin{bmatrix} \frac{1-\theta_1}{\Delta t_k}(C_i)_k + \frac{\theta_1}{\Delta t_k}(C_i)_{k+1} \\ + (\theta_1 - \theta_2)(\lambda_i)_k + \theta_2(\lambda_i)_{k+1} \end{bmatrix} (Y_i)_{k+1} + \begin{bmatrix} \frac{-1+\theta_1}{\Delta t_k}(C_i)_k - \frac{\theta_1}{\Delta t_k}(C_i)_{k+1} \\ + (1 - 2\theta_1 + \theta_2)(\lambda_i)_k \\ + (\theta_1 - \theta_2)(\lambda_i)_{k+1} \end{bmatrix} (Y_i)_k = 0 \tag{6.48a}$$

Let $\delta = (Y_i)_{k+1}/(Y_i)_k$; the equation becomes:

$$\delta \begin{bmatrix} \frac{1-\theta_1}{\Delta t_k}(C_i)_k + \frac{\theta_1}{\Delta t_k}(C_i)_{k+1} \\ + (\theta_1 - \theta_2)(\lambda_i)_k + \theta_2(\lambda_i)_{k+1} \end{bmatrix} + \begin{bmatrix} \frac{-1+\theta_1}{\Delta t_k}(C_i)_k - \frac{\theta_1}{\Delta t_k}(C_i)_{k+1} \\ + (1 - 2\theta_1 + \theta_2)(\lambda_i)_k + (\theta_1 - \theta_2)(\lambda_i)_{k+1} \end{bmatrix} = 0 \tag{6.48b}$$

If $|\delta| > 1$, the instability (i.e., unlimited response) appears in the calculation, and the oscillation condition occurs when $|\delta| < 1$. Obviously, both cannot achieve the required solution.

In order to make the analysis convenient, the method of freezing coefficient is used to analyze the stability of the difference format with variable coefficient.[6-5] Equation (6.48b) is written as:

$$|\delta| = \frac{1 - (1 - \theta_i)\,\omega_i\,\Delta t_k}{1 + \theta_1 \omega_i\,\Delta t_k} \tag{6.49}$$

where $\omega_i = \lambda_i/C_i$ is the characteristic value of the difference format and is always larger than zero.

It is known from Eqn (6.49) that $\delta < 1$, because $0 < \theta_1 \le 1$. If $|\delta| < 1$, it must be $\delta > -1$. From Eqn (6.49):

$$(1 - 2\theta_1)\,\omega_i\,\Delta t_k < 2 \tag{6.50}$$

When $\theta_1 \ge 1/2$, $|\delta| < 1$ is always satisfied. These types of formats, such as Crank–Nicolson, Galerkin, and the backward difference format, satisfy this condition and are unconditionally stable. When $0 < \theta_1 < 1/2$, the stability of the format is conditional and it has to satisfy:

$$\omega_i\,\Delta t_k < \frac{2}{1 - 2\theta_1} \tag{6.51}$$

When a different format (i.e., a different value of θ_1) is used, the relation between δ and $\omega_i \Delta t_k$ is shown in Fig. 6-9. It is seen that $\delta > 0$ is ensured only when $\theta_1 = 1$, and the format is unconditionally stable and no oscillation will occur, but its accuracy is low and of first order. However, if $\theta_1 < 1$, oscillation occurs when $\delta < 0$, i.e., when $\omega_i \Delta t_k$ is greater than a certain value. The value of $\omega_i \Delta t_k$ reduces correspondingly with θ_1, so oscillation is more possible. Therefore, because it prevents oscillation, the Galerkin format ($\theta_1 = 2/3$) is better, but the calculation accuracy is slightly worse than that of the Crank–Nicolson format ($\theta_1 = 1/2$).

It is seen from Eqn (6.22) that c_{ij}^e is related to the size (edge lengths a and b) of the element, and K_{ij}^e is related to the size ratio (a/b) of the element. As ω_i

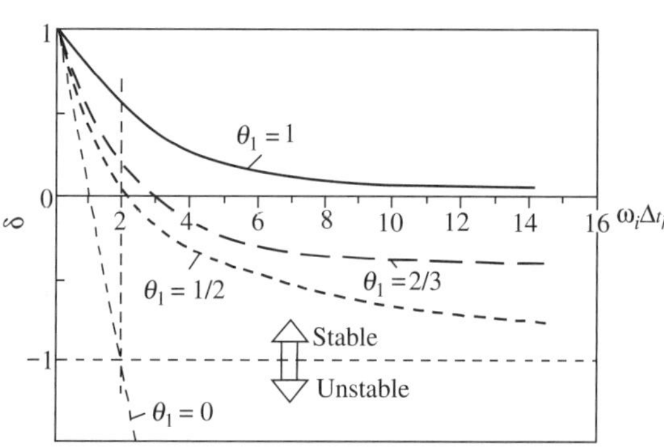

FIGURE 6-9 Relation between δ and $\omega_i \Delta t_k$ for different formats.

is approximately proportional to $1/(ab)$, it increases sharply when the size of the element decreases. In order to let the value of $\omega_i \Delta t_k$ be less than a certain limit, Δt_k has to be reduced considerably. This shows that the determination of the incremental time step is limited not only by the difference format but also by the step length in the space field (the size of the element). If the step in the time field is not matched properly with that in the space field, oscillation will also occur. Generally, oscillation reduces with Δt_k. However, when the edge length of the element remains constant and Δt_k is smaller than the limit value, new oscillation may occur because the accumulative and rounding off errors increase obviously. The limit value is given by Kong[6-8]:

$$\frac{d \Delta t_k}{\Delta x^2} > 0.01\text{–}0.1 \tag{6.52}$$

where d is the heat diffusivity and Δx is the edge length of the element.

The standard temperature–time curve of fire is a logarithmic function (Eqn (5.1)); the temperature elevates very quickly at the beginning and then becomes slow and gradual later. Correspondingly, the temperature field of a structural member also varies with time and shows similar variation regularity. The larger the rate of increase in temperature, the more serious the oscillation during calculation. So, the oscillation reduces gradually with time and the time step can be increased. Therefore, a constant temperature increment is used in this chapter, which can be calculated from the standard temperature–time curve of fire, and the corresponding variable time step increment is obtained; integration is performed with respect to the time step.

6.3 COMPUTING PROGRAM AND AN EXPERIMENTAL DEMONSTRATION

6.3.1 Compiling the Computer Program

According to the method and equations derived in Section 6.2, a computer program for the temperature field on a section of a concrete structure is easily compiled. The program is compiled using FORTRAN 90 and is called HTARC (heat transfer analysis of reinforced concrete). Considering the characteristics of concrete structures and fire accidents, the measures used for the program are as follows.

1. Network division

The shape of the section of the concrete structural member is mostly regular, and a simple rectangular element of four nodes is suitable for use. The short and long edges of the divided element can be selected arbitrarily, but the ratio between them should be >0.1 in order to avoid too short an incremental time step and too much computing time.

The network can be divided automatically by the program. If both the shape and the boundary condition of the member section are symmetrical, only one-half or one-fourth of the section needs to be analyzed for the condition of one or two symmetrical axes, respectively. A network of equal distance is divided separately along the width and height directions of the section, but there is no finer division locally. When the temperature gradient on the section is considerable, a smaller and equal sized network can be used in order to avoid complicated division in the program. Although the total number of networks in the section increases, this does not cause any problems for the computer.

2. Boundary condition and thermal parameters of the material

Generally, the thermal boundary condition of a concrete structural member during a fire is in the third category. Some boundary, which is far away from the surface exposed to fire, can be considered as being in the first category, and the symmetrical axis (boundary) of the section is in the second category. These three categories of boundary conditions are included in the program, and the same boundary condition category, but with different temperature parameters for each boundary of the section, is also considered.

The values of the thermal parameters of concrete can be determined by the related formulas (Eqns (5.2a) and (5.3)) in Section 5.2.

3. Time of temperature increase and incremental time step

This program can be used for various temperature–time curves. For the third category of the boundary condition, the ISO standard temperature–time curve and some existing experimental temperature–time curves have been deposited in subprograms, and various other temperature–time curves can also be imported as required.

The velocity of the increase in fire temperature varies considerably with time. If the incremental time step given is a constant, the temperature increments differ greatly, and oscillation may occur during the calculation process. Therefore, equal increments for the temperature of the medium surrounding the section are given automatically in this program, according to the edge length of the divided element of the section. Then the variable time step (Δt_k) can be calculated from the temperature–time curve and is used as the basis for successive analysis along the time direction.

4. Flow chart for the program

The flow chart for the HTARC program is shown in Fig. 6-10.

The program to compute the temperature field in a section of a structure (HTARC) must be calibrated and demonstrated. The simple linear problem of transient thermal conduction can be compared with the accurate analytical solution, but a nonlinear problem can only be measured and demonstrated by testing a structural member at elevated temperature, because the analytical solution is not available.

6.3.2 Calculated Examples and Demonstration

Example 1 The initial temperature ($T_0 = 800$ °C) is distributed uniformly in a concrete slab of thickness 10 m (Fig. 6-11), and then put in a medium at constant temperature ($T = 0$ °C). This is a thermal problem with the boundary condition in the first category, and the theoretical solution[0-9] of the temperature field of the slab is:

$$T = \frac{4T_0}{\pi}\sum_{k=0}\frac{1}{2k+1}\exp\left(\frac{d(2k+1)^2\pi^2 t}{l^2}\right)\sin\left(\frac{2k+1}{l}\pi x\right) \quad (6.53)$$

where T is the temperature (°C) at the point located x (m) from the left surface after time t (days), d is the heat diffusivity (m^2/s), and l is the thickness of the slab (m). The temperature field of the slab is analyzed.

Solution This is a problem of a one-dimensional temperature field. The temperature of the slab varies along the direction of its thickness but does not vary along the direction of its plane. The temperature field of the slab is analyzed using the HTARC program and the slab is divided into 20 networks along its thickness, with a network thickness of 0.5 m. The length of the network can be taken as 0.5 m. The heat diffusivity on the boundary is taken as $d = 1.157 \times 10^{-6}$ m^2/s.

The results computed by the program and the theoretical curves of Eqn (6.53) are shown in Fig. 6-12. The temperatures varying with time at the points located at $x = 2.0$, 3.5, and 5.0 m from the surface of the slab are shown in Fig. 6-12(a), and the temperature distribution along its thickness, when the time $t = 1$–300 days, is shown in Fig. 6-12(b).

Example 2 This is a concrete slab. The temperature of the medium on its left surface is $T = 0$ °C and the coefficient of heat transfer there is β_1; the temperature of the medium on its right surface is $T = A_0\cos\omega t$, i.e., it varies harmonically, and the coefficient of heat transfer there is β_2. The theoretical solution of the temperature field[0-9] is

$$T(x,t) = A_0\bar{C}\left(\varphi\cos\varphi x + \frac{\beta_1}{\lambda}\mathrm{sh}\varphi x\right)e^{i\omega t} \quad (6.54)$$

where

$$\bar{C} = \frac{\beta_2/\lambda}{\left(\varphi^2 + \left(\beta_1\beta_2/\lambda^2\right)\right)\mathrm{sh}\phi x_0 + \varphi\left(\beta_1\beta_2/\lambda\right)\mathrm{ch}\varphi x_0}$$

$$\text{and } \varphi = (1+i)\sqrt{\frac{\omega}{2a}}$$

The temperature field of the slab is analyzed.

Solution This is also a problem of a one-dimensional transient temperature field, but with different boundary temperatures and initial conditions. The values of the parameters used in computing are: the thickness of the slab

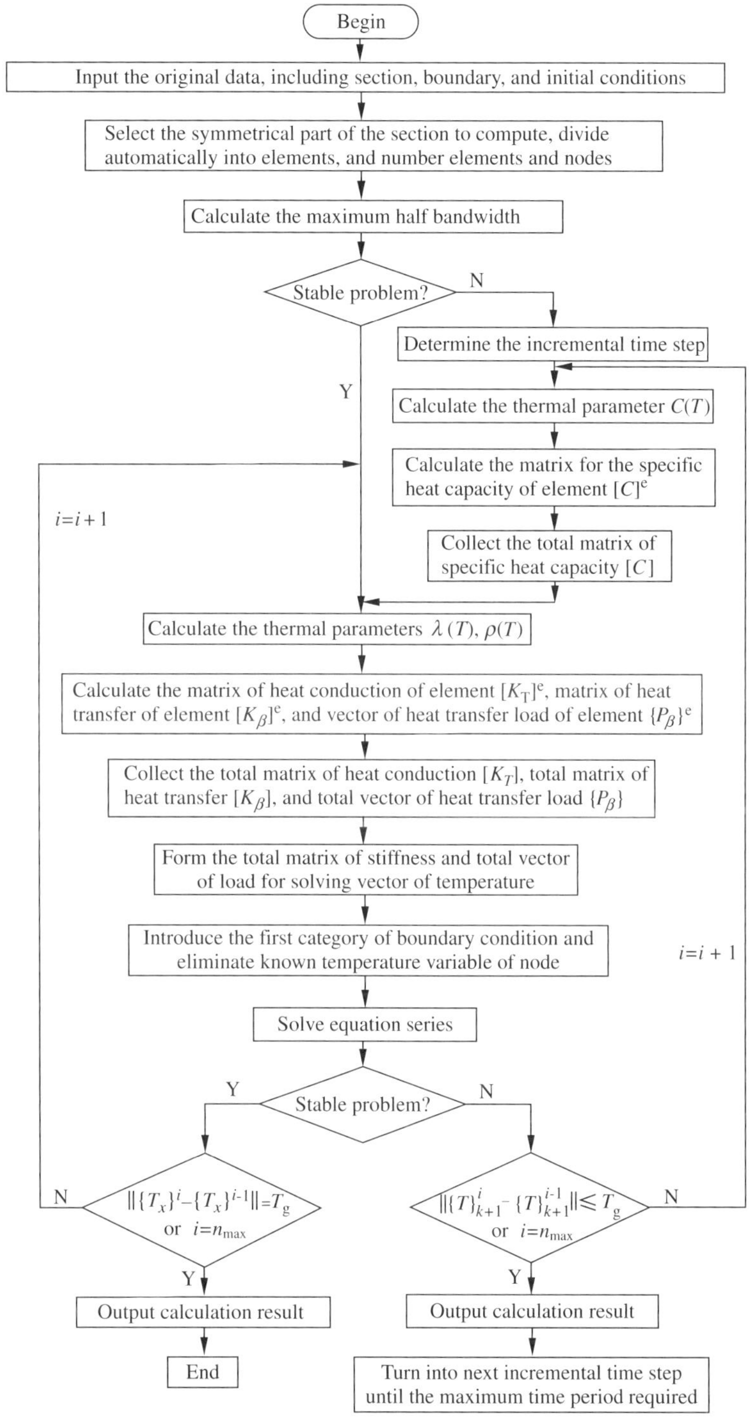

FIGURE 6-10 Flow chart for the HTARC program.

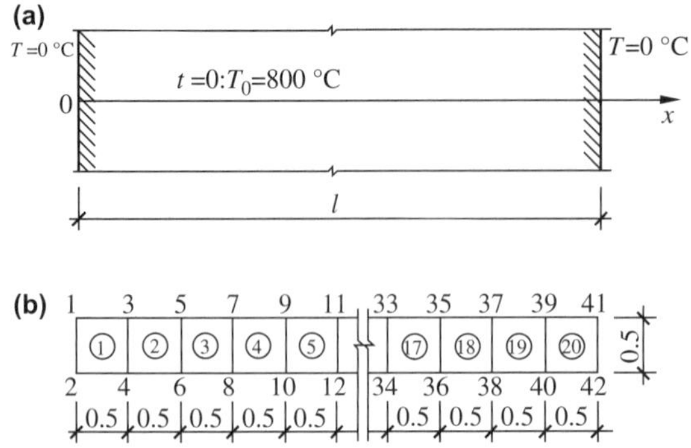

FIGURE 6-11 Analysis of the transient temperature field of a slab: (a) initial and boundary conditions; (b) division of the network.

$l = 2.6$ m, the coefficient of heat conduction of the material $\lambda = 0.00145$ W/(m^2 °C), the heat diffusivity $d = \lambda / C\rho = 6.028 \times 10^{-7}$ m^2/s, the coefficient of heat transfer on both surfaces $\beta_1 = \beta_2 = 0.00678$ W/(m^2 °C), the frequency of the temperature variation $\omega = 2\pi / r$, and $A_0 = 1$ °C.

The values of these parameters are substituted into Eqn (6.54) and the theoretical solution is obtained:

$$T(x,t) = \begin{pmatrix} 0.0619\text{ch}\xi\cos\xi - 0.0299\text{sh}\xi\sin\xi \\ +0.189\text{sh}\xi\cos\xi + 0.542\text{ch}\xi\sin\xi \end{pmatrix} \times \cos\frac{2\pi t}{r} + \begin{pmatrix} 0.029\,9\text{ch}\xi\cos\xi - 0.061\,9\text{sh}\xi\sin\xi \\ -0.189\text{ch}\xi\sin\xi + 0.542\text{sh}\xi\cos\xi \end{pmatrix} \times \sin\frac{2\pi t}{r} \tag{6.55}$$

where $\xi = 0.4062x$, t is time in months, and $r = 12$.

The results computed by the program and the theoretical curve of Eqn (6.54) are drawn together in Fig. 6-13. The temperatures varying with time at the points located at $x = 0.65$, 1.30, and 1.95 m from the left surface of the slab are shown in Fig. 6-13(a), and the temperature distribution along its thickness when time $t = 1$–11 months is shown in Fig. 6-13(b).

Examples 1 and 2 are in the first and third categories of boundary conditions, respectively. The results computed by the HTARC program agree well with the theoretical solutions. This demonstrates that the program is reliable and accurate for analysis of linear transient heat conduction.

6.3.3 Experimental Demonstration

When the mechanical behavior of fire resistance (or at elevated temperature) of a structure and its members is tested, certain thermocouples can be mounted in advance on the surface and in the interior to measure the temperature and its variation in the interior of the section during the process of heating (or cooling) and loading.

Generally, the thermal-electrical transducer is made of different alloy materials, and the thermocouple is composed of nickel–chromium and nickel–silicon wires. One end of fine wires of two different materials are welded together. When the temperature at this end varies, the potential difference between the other ends of both wires occurs correspondingly. After the value of the potential difference is measured by some instrument, the temperature at the welded end can be calculated.

When the temperature of the air surrounding the structural member or the temperatures on the surface and in the interior of the member need to be measured, the thermocouples should be positioned and mounted on the necessary places. There are two methods for mounting the thermocouples to measure the temperature in the interior of a concrete structural member. In the first method, the welded end of the thermocouple is fixed in position in the mold of the member before the concrete is cast, and the two wires pass through two insulation tubes separately and extend out of the mold. For the second method, a hole is left in the concrete in advance or is drilled after the concrete has hardened, then the thermocouple is put into the hole and cement mortar is poured in and compacted afterward. Both methods have advantages and disadvantages.

Experiment 1 A concrete member, length 1300 mm and cross section 100 mm × 180 mm (see Fig. 8-5), is heated in an electrical furnace (see Fig. 8-3). Only three surfaces of the member, i.e., both side surfaces and the bottom, are heated; the top surface is exposed to the environmental air in the laboratory. Eight thermocouples are placed on the central section of the member (Fig. 6-14(a)) and are used to measure the temperature at the corresponding points during the heating

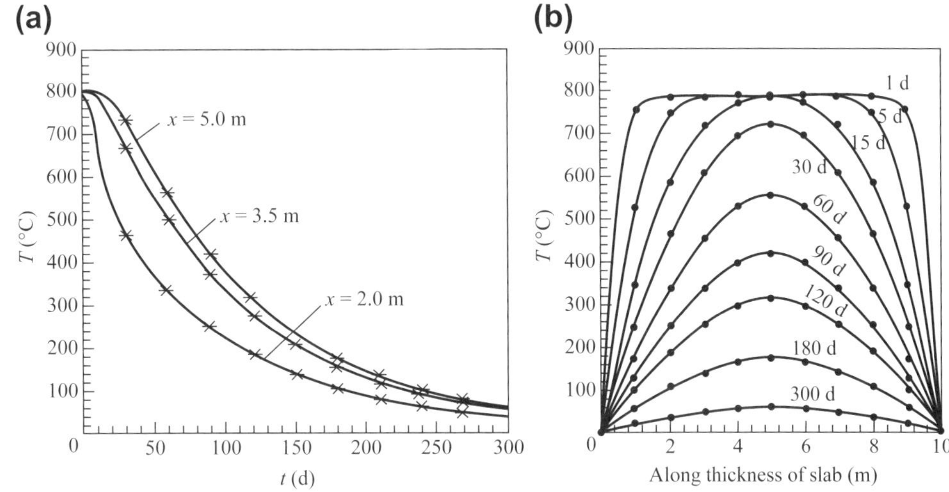

FIGURE 6-12 Analysis results for the temperature field of the slab (example 1): (a) temperature–time curves at various positions. Solid line, accurate solution; *, value computed from the program. (b) Temperature distribution varying with time. Solid line, accurate solution; •, value computed from the program.

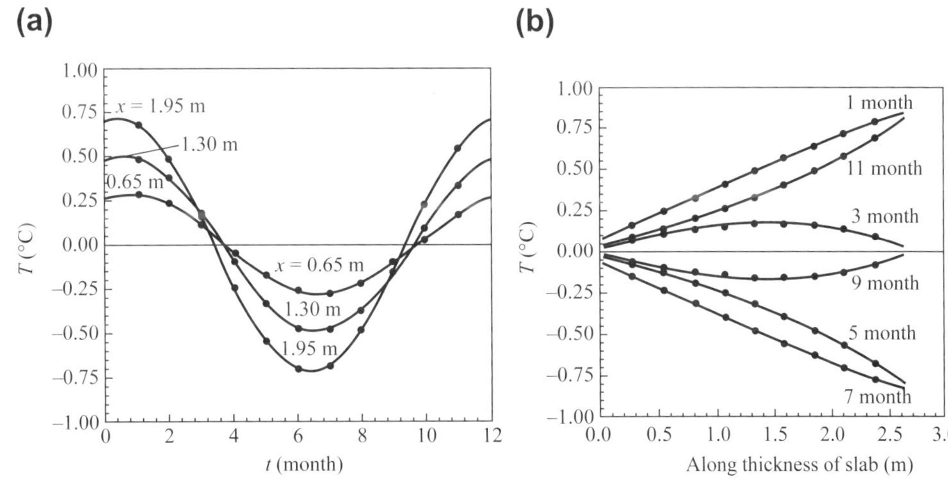

FIGURE 6-13 Analysis results for the temperature field of the slab (example 2): (a) temperature–time curves at various positions; (b) temperature distribution varying with time. Solid line, accurate solution; •, value computed from the program.

process. The measured temperature–time curves are shown in Fig. 6-14(c); the temperature distribution on the section can also be obtained and is shown in Fig. 8-6.

The temperature field on the section of the member is computed using the HTARC program. Because the temperatures on the section and the boundary are symmetrical about the vertical axis, only the left half of the section is analyzed and the network as divided is shown in Fig. 6-14(b). The thermal parameters λ and C of the concrete are calculated using Eqns (5.2a) and (5.3), respectively. The mass density is taken as a constant $\varrho = 2400$ kg/m^3. The unheated top surface of the member is in the first category of boundary conditions and

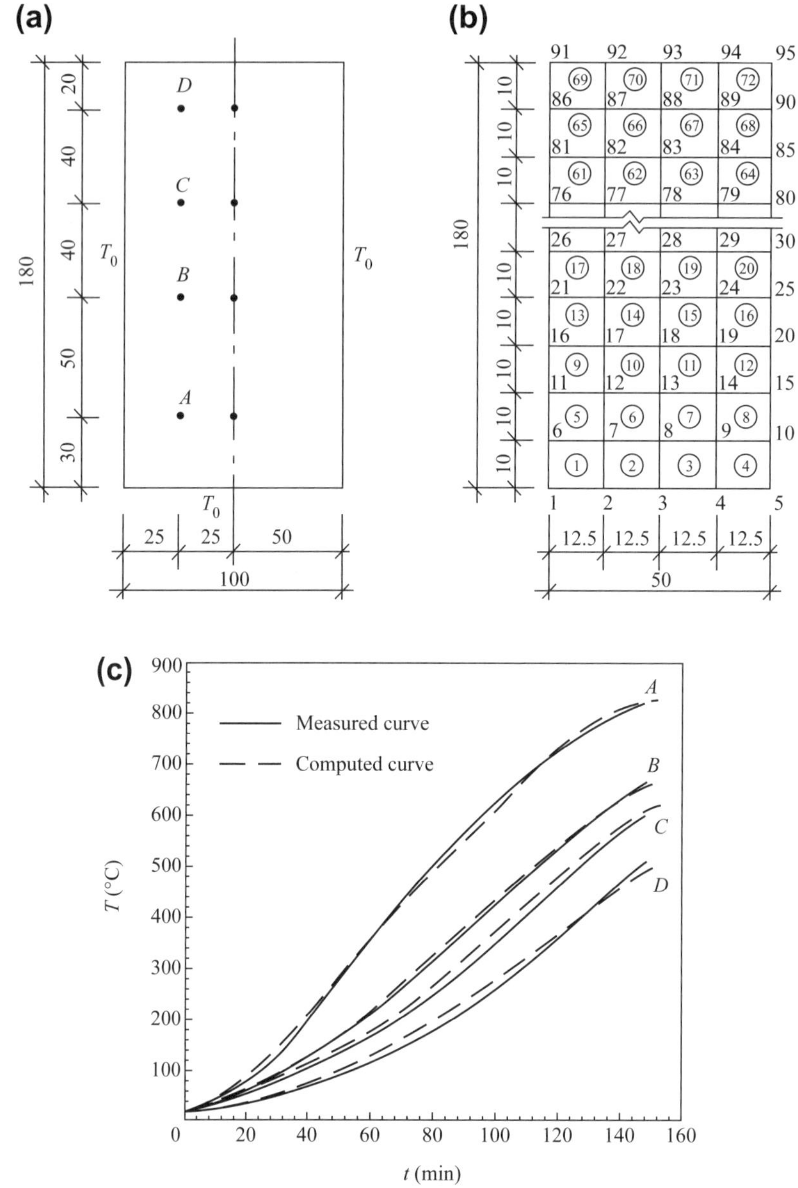

FIGURE 6-14 Experimental and calculated results for the temperature field of a member heated on three surfaces[1-12]: (a) section and measuring points for temperature; (b) network division; (c) temperature–time curves for the measuring points.

the temperature there is taken as 20 °C; the other heated sides and bottom surface are in the third category of boundary conditions, and the coefficient of heat transfer is taken as $\beta = 5.5\ \text{W/(m}^2\ {}^\circ\text{C)}$. The temperature of the medium surrounding the heated surfaces of the member is taken as the temperature measured (see Fig. 8-4) in the chamber of the furnace during testing. The temperatures at the measuring points and their variation with time obtained from the

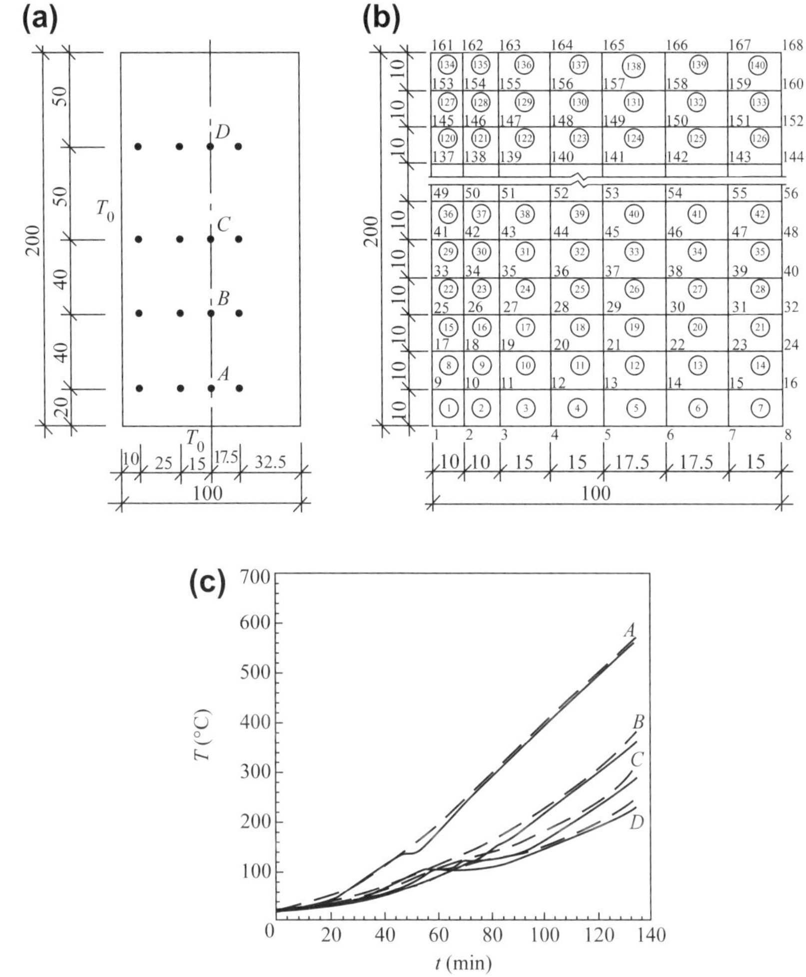

FIGURE 6-15 Experimental and calculated result for the temperature field of a member heated on two adjacent surfaces[9-6]: (a) section and measuring points for temperature; (b) network division; (c) temperature–time curves for the measuring points. Solid line, measured curve; dashed line, computed curve.

program agree well with the measured curves (Fig. 6-14(c)).

Experiment 2 A concrete member, length 1440 mm and cross section 100 mm × 200 mm (see Fig. 9-8). is heated in an electrical furnace (see Fig. 9-19). Only the two adjacent surfaces of the member, i.e., the left side and bottom surface, are heated; the right side and top surface are exposed to the air in the laboratory. Sixteen thermocouples are placed on the central section of the member (Fig. 6-15(a)), and are used to measure the temperatures at the corresponding points during the heating process. The measured temperature–time curves are shown in Fig. 6-15(c), and the temperature distribution on the section can be seen in Fig. 9-22.

The temperature field on the section of the member is computed using the HTARC program. The whole section is divided into networks, shown in Fig. 6-15(b), and the thermal parameters of the concrete and the parameter value of the boundary condition of the section are the same as in Experiment 1. The temperature of the medium near the heated boundary of the member is taken as the temperature program in the chamber of the test furnace (see Fig. 9-20). The temperatures at the measuring points and their variation with time obtained from the program agree well with the measured curves (Fig. 6-15(c)).

These two experiments and other experiments, which are not reported, show that the combined finite element difference method and the HTARC program are suitable for analyzing the nonlinear transient temperature field on a section of a concrete structure under various boundary conditions.

6.3.4 General Regularity of the Temperature Field on a Section

Using the HTARC computer program described above, the temperature field on the section of a structural member of various shapes and sizes can be computed in series for different types of temperature–time curves for a fire accident and different boundary condition. In addition, the temperature field on the section can be measured from a large number of experiments on the structural member at elevated temperatures. Then, the temperature distribution on the section of the member and its variation with time under various conditions can be obtained. Using these theoretical and experimental research results, the general regularity of the temperature field on the section of a concrete member under monotonic heating can be summarized as follows:

- The temperature field on a section of a structural member and its variation with time depend mainly on the temperature–time curve of the medium (air flow layer) surrounding it and the heating time.
- The temperature on the surface of a member heated or exposed to fire is always lower than that of the surrounding medium. The temperature difference may reach 100–200 °C at the beginning of heating, but reduces gradually to 20–30 °C as the heating time and the absolute value of the temperature increases.
- The one-dimensional temperature field of a planar member heated on one surface is the most fundamental temperature distribution. The temperature on its section decreases monotonically and the temperature gradient reduces as well along the normal line of the heated surface. The maximum temperature occurs on the outer surface, the temperature fall or gradient within the surface layer (about 20–40 mm) decreases approximately linearly, and the temperature at the furthest distance remains basically constant, i.e., the temperature gradient is nearly zero.
- When more than one peripheral surface of the section of the structural member, e.g., both surfaces of a planar structure, two adjacent surfaces of a beam or column, or three or four surfaces of a rectangular section, are heated or exposed to fire, the temperature field on the section is approximately the sum of the one-dimensional temperature fields from various directions.
- The thermal behavior of the material of a structural member has an obvious influence on the temperature field on its section. With smaller coefficients of heat conduction (λ), a greater specific heat capacity (C) and mass density (ϱ), i.e., smaller heat diffusivity ($d = \lambda / C\varrho$), the greater the temperature gradient on the surface layer of the section and the smaller the temperature in the interior.

These regularities apply to the temperature field on a rectangular section of an ordinary concrete structural member under monotonic heating conditions. As far as a member of a nonrectangular section (e.g., T, I, and circular sections), or special heating (and cooling) conditions, or thermal parameters of material is concerned, the corresponding temperature field on the section has to be computed. However, the regularities described above still apply to the qualitative analysis of a temperature field.

CONCLUSIONS

Variation in the temperature conditions on the boundary and the thermal parameters of the material of a concrete structural member with time (or temperature) is complicated. The transient temperature field on its section can only be obtained after solving the nonlinear second partial differential equation of heat conduction. As an accurate analytical solution is not achieved because of the mathematical difficulties, a numerical method of analysis is used and the solution can be obtained with enough accuracy, relying on numerous but fast calculations on the computer.

The difference method is suitable for analyzing the temperature field of a simple structural member with regular sections. After reasonably dividing the network and selecting the difference format and incremental time step, the differential equation of heat conduction is changed into an algebraic series. The approximate solution with certain accuracy of the temperature field is obtained after solving the series of equations.

The combined finite element difference analysis method uses the finite element method to discrete the space field and the difference method to discrete the time field, and the finite element analysis is performed successively with time. This method is now commonly used for analyzing the temperature field. Among the weighted complementary methods, the Galerkin method is used in this chapter to derive the heat conduction equation of the element in the finite element analysis and the difference format in the time field. In addition, the values of the thermal parameters of the element are simplified and the limiting condition is determined to satisfy the stability of the digital solution; the analysis of the nonlinear transient temperature field is then achieved successfully.

The HTARC computer program was developed for analysis of a nonlinear transient temperature field on the section of a concrete structural member based on the theoretical analysis described above. The variable incremental time step, which is equivalent to equal increments of temperature of the surrounding medium, is used in the program, and it is helpful to ensure the stability of the calculation process. This program can be used widely for the analysis of a temperature field on the section of a structural member under various conditions of the temperature–time curve of fire and the temperature on the boundary. The computed results are compared with the accurate theoretical solutions of some special examples and many experimental results for various structural members at elevated temperature. The results from this program show that the calculation method is correct, the computing process is stable, and the results computed are accurate.

REFERENCES

[6-1] F.P. Incropead, D.P. Dewitt, Basic Principle of Thermal Conduction, Translated by Ge X, Wang Y, Guo K, Education Press of Anhui Province, Hefei (1985).

[6-2] W.L. Wood, R.W. Lewis, A comparison of time marching schemes for the transient heat conduction equation, International Journal for Numerical Methods in Engineering 9 (1975) 679–689.

[6-3] A. Cardona, S. Idelsohn, Solution of non-linear thermal transient problems by a reduction method, International Journal for Numerical Methods in Engineering 23 (1986) 1023–1024.

[6-4] J.F. Stelzer, R. Welzel, Experiences in non-linear analysis of temperature field with finite elements, International Journal for Numerical Methods in Engineering 24 (1987) 59–73.

[6-5] J. Lu, Z. Guan, Numerical Solution of Partial Differential Equation, Tsinghua University Press, Beijing (1985).

[6-6] H. Tang, J. Hu, Numerical Method of Differential Equation, Nankai University Press, Tianjin (1990).

[6-7] X. Shi, Z. Guo, Temperature field of reinforced concrete structure, Engineering Mechanics 13 (1) (1996) 35–43.

[6-8] X. Kong, Application of Finite-Element Method to Thermal Conduction, Science Press, Beijing (1998).

[6-9] Research Group 515, Finite element solution and principle of the maximum norm for transient thermal problem, Calculation Mathematics (2) (1982) 114–120.

[6-10] K. Guo, Numerical Calculation for Heat Conduction, Science and Technology Press of Anhui Province, Hefei (1987).

[6-11] Shi G, Non-linear finite element analyses of temperature field and thermal stress in reinforced concrete structures. Doctoral thesis, Tsinghua University, Beijing (1990).

[6-12] O.P. Mondkar, G.H. Powell, Evaluation of solution schemes for nonlinear structures, Computers and Structures 9 (1978) 223–236.

[6-13] W.F. Schmit, Extending the convergence domain of the Newton-Raphson method in structural analysis, Computers and Structures 9 (1978) 265–272.

[6-14] M.A. Crinfield, A faster modified Newton–Raphson iteration, Computer Methods in Applied Mechanics and Engineering 20 (1979) 267–278.

[6-15] H. Mathies, G. Strong, The solution of nonlinear finite element equations, International Journal for Numerical Methods in Engineering 14 (1979) 1613–1626.

[6-16] K.J. Bathe, A.P. Cimento, Some practical procedures for the solution of nonlinear finite element equations, Computer Methods in Applied Mechanics and Engineering 22 (1980) 59–85.

[6-17] G.H. Powell, J. Simons, Improved iteration strategy for nonlinear structures, International Journal for Numerical Methods in Engineering 17 (1981) 1455–1467.

[6-18] K.J. Bathe, E.N. Dvorkin, On the automatic solution of nonlinear finite element equations, Computer and Structures 17 (1983) 871–879.

CHAPTER

7

Calculation Charts for a Temperature Field on a Cross Section

7.1 BASIC ASSUMPTIONS AND APPLICATION CONDITIONS

The temperature field on the cross section of a structural member is an important premise for analyzing the mechanical behavior of the structure at elevated temperatures (fire resistance). Although various theoretical methods and computer programs are available to analyze the individual temperature field, which is a complex problem of nonlinear transient heat conduction, a large number of complicated computing work has to be performed. The most relevant problems in engineering practice are evaluation of the endurance of fire resistance, the bearing capacity at elevated temperature, and the residual bearing capacity after a fire in the main members in a structure. Therefore, it is necessary to provide the temperature field for a section of a structural member, quickly and conveniently, and with sufficient accuracy.

Based on the theoretical analysis of the finite element difference method and the HTARC computer program described in Chapter 6, a large amount of computing work has been performed. Two practical methods and corresponding auxiliary means, i.e., calculation tables and temperature contour lines, are given in this chapter for the temperature field on a section of a concrete structural member to satisfy the requirements of engineering practice.

In addition to the basic assumptions and simplifications given for the theoretical analysis and computer program (see Chapter 6), the assumptions and application conditions for all the tables, contours, calculation formulas, and various data shown in this chapter are as follows:

1. The ISO standard temperature–time curve (Eqn (5.1)) is used for the process of a fire accident and the duration of the fire is divided into six grades, i.e., 30, 60, 90, 120, 150, and 180 min.
2. It is assumed that a section of the structural member is composed of homogeneous and continuous concrete material. Neither the influence of the reinforcement area of the section nor the temperature redistribution caused by local changes in the section area due to cracking and surface spalling of concrete are taken into account.
3. The values of the thermal parameters of concrete material vary with temperature. The coefficient of heat conduction (λ) and the specific heat capacity (C) are calculated using Eqns (5.2a) and (5.3), respectively; the mass density is taken as a constant $\varrho = 2400\ \text{kg/m}^3$. The coefficient of heat transfer ($\text{W/(m}^2\ \text{K)}$) on the boundary of the third category is calculated using the following formula:

$$\left.\begin{aligned} &T \le 600\,^\circ\text{C}, \beta_T = 0.07 \\ &T > 600\,^\circ\text{C}, \beta_T = 0.07 + 8.75\times10^{-6}(T-600) \end{aligned}\right\} \quad (7.1)$$

4. The surfaces exposed to fire and the maximum size of a (rectangular) section of the structural members are shown in Table 7-1.

Experiment and Calculation of Reinforced Concrete at Elevated Temperatures

TABLE 7-1 Surfaces Exposed to Fire and the Maximum Size of a (Rectangular) Section of the Structural Members

	Surfaces exposed to fire	Maximum section
Planar member, e.g., slab, wall	One surface only	Thickness 200 mm
One-dimensional member, e.g., beam, column	Three surfaces	500 mm × 500 mm
	All four surfaces	600 mm × 600 mm
	Two adjacent surfaces	500 mm × 500 mm

If the conditions of the structural member analyzed do not fit with that listed above, the temperature field on the section can be determined approximately by interpolation or equivalent estimation method based on the similar conditions. Some examples are: the actual temperature–time curve is different from the ISO standard, the shape or size of the section is different, the values of the thermal parameters of the material or the coefficient of heat transfer on the boundary are different. The practical methods are explained below.

7.2 SLABS WITH ONE SURFACE EXPOSED TO FIRE

7.2.1 Calculation Tables

When one surface of a planar member, e.g., floor slab or wall, is exposed to fire, which lasts t = 30–180 min, the temperature on the section can be referred to and obtained directly from Table 7-2(a)–(f). These tables can be used separately for slabs of thickness h = 80, 100, 120, 150, 180, and 200 mm, and z is the depth away from the fire, i.e., the distance from the surface exposed to fire to the point at which the temperature is given (see also Fig. 7-1).

Considering the variation in the temperature gradient on a section, the intervals between the temperature points increase successively and are 5 mm for $z \le 20$ mm, 10 mm for z = 20–100 mm, and 20 mm for z = 100–200 mm, in order to improve the accuracy of temperature field on the section.

If the thickness of a slab is less than 200 mm and the duration of the fire is less than 180 min, the values of the thickness (h), time (t), and depth away from the fire (z) cannot be found in Table 7-2. The temperature values of approximate conditions can be found in these tables in advance and then interpolated to determine the required temperature value. For example, if the thickness of a slab is h = 100 mm, the temperature value at position z = 55 mm and at time t = 140 min is determined as follows: four values of temperature at z = 50 mm and 60 mm and at t = 120 min and 150 min are found in Table 7-2(b), and the required temperature value is obtained after linear interpolation twice.

If the thickness of a slab is $h > 200$ mm, the temperature value with $z \le 200$ mm can be taken approximately as that listed in Table 7-2(f), and the temperature value at $z > 200$ mm can be taken as a constant, i.e., the temperature at z = 200 mm.

7.2.2 Temperature Distribution Curves

All the data listed in Table 7-2 are drawn in the figures and connected smoothly; the curves of the temperature distribution on sections of the slabs of various thickness (h = 80–200 mm) and with one surface exposed to a fire, which lasts t = 30–180 min, are obtained (Fig. 7-2(a)–(f)). These figures show clearly the variation regularity of a one-dimensional temperature field in a slab.

Using these curves, the temperature field on the section of the slab with one surface exposed to fire can be found directly or obtained after the interpolation calculation.

TABLE 7-2 Temperature Value on a Section of Slab with One Surface Exposed to Fire (°C)

z (mm)	*t* (min)						*z* (mm)	*t* (min)					
	30	60	90	120	150	180		30	60	90	120	150	180
(a) Thickness of slab $h = 80$ mm							(b) Thickness of slab $h = 100$ mm						
							100	53	42	218	278	324	361
							90	58	154	236	301	352	392
80	95	218	307	371	417	453	80	69	173	261	330	384	429
70	105	238	334	404	456	496	70	87	200	293	366	424	471
60	126	267	370	446	502	545	60	114	237	335	411	472	521
50	159	310	418	497	556	602	50	151	286	387	466	528	579
40	208	368	479	560	621	669	40	203	349	454	533	596	648
30	276	444	556	638	699	746	30	274	430	536	615	678	728
20	372	544	654	732	791	837	20	370	534	638	715	775	822
15	433	605	712	787	843	887	15	432	597	699	773	829	875
10	505	674	776	847	899	940	10	504	668	766	836	889	931
5	591	753	847	912	960	997	5	591	749	840	904	953	991
0	693	842	926	982	1024	1057	0	693	840	922	978	1021	1054
(c) Thickness of slab $h = 120$ mm							(d) Thickness of slab $h = 150$ mm						
							150	23	51	92	132	169	201
							140	23	54	98	141	181	215
120	33	93	154	207	251	286	120	28	69	120	170	214	253
100	41	112	183	243	294	336	100	39	97	159	214	263	306
90	50	129	205	269	323	327	90	49	119	185	244	295	340
80	64	154	234	301	358	405	80	63	146	218	280	333	380
70	84	185	271	341	399	448	70	84	180	258	323	378	426
60	112	226	316	389	449	500	60	112	222	306	374	431	480
50	150	278	372	447	509	560	50	150	275	364	435	493	543
40	202	343	441	518	580	631	40	202	341	436	508	566	616
30	273	426	527	603	664	714	30	273	425	522	595	653	702
20	370	531	632	706	764	812	20	370	531	629	700	756	803
15	432	595	693	765	821	866	15	430	594	691	760	814	859
10	504	666	762	830	882	924	10	504	666	760	826	877	919
5	590	748	837	900	948	986	5	590	747	836	898	945	983
0	693	839	920	976	1018	1052	0	693	839	920	975	1017	1050

(Continued)

TABLE 7-2 Temperature Value on a Section of Slab with One Surface Exposed to Fire (°C)—cont'd

z (mm)	t (min)						z (mm)	t (min)					
	30	60	90	120	150	180		30	60	90	120	150	180
(e) Thickness of slab $h = 180$ mm							(f) Thickness of slab $h = 200$ mm						
							200	20	26	42	63	87	110
180	20	32	56	85	113	140	180	20	28	47	72	99	125
160	21	36	64	97	130	160	160	21	34	58	88	119	148
140	23	46	81	119	156	190	140	23	45	78	113	148	181
120	28	64	109	154	195	233	120	28	64	107	150	189	225
100	39	95	152	203	249	290	100	39	95	151	200	245	285
90	49	117	180	235	283	326	90	49	117	179	233	280	321
80	63	145	214	272	323	367	80	63	145	214	271	320	363
70	84	179	255	317	369	415	70	84	179	254	316	367	412
60	112	222	304	369	424	471	60	112	222	304	368	422	468
50	150	275	363	431	487	535	50	150	275	363	430	486	533
40	202	341	434	505	562	610	40	202	341	434	504	561	608
30	273	425	522	593	650	697	30	273	425	521	593	649	696
20	370	530	628	699	754	799	20	370	530	628	698	753	798
15	432	594	691	759	812	856	15	432	594	691	759	812	855
10	504	666	760	825	876	916	10	504	666	760	825	875	916
5	590	747	836	897	944	981	5	590	747	836	897	943	981
0	693	839	920	975	1016	1050	0	693	839	920	975	1016	1049

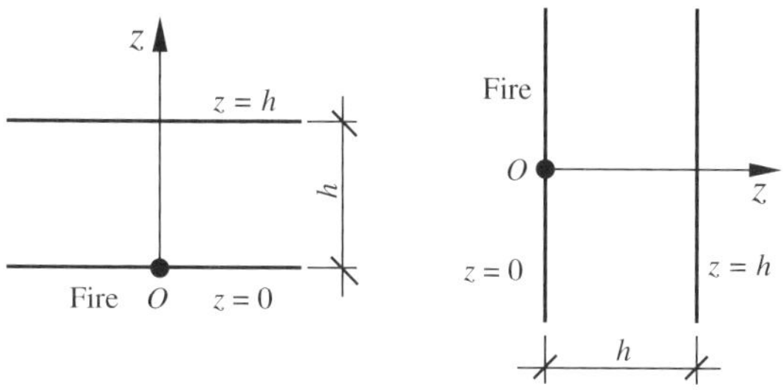

FIGURE 7-1 Calculation parameters for a slab with one surface exposed to fire.

7.3 BEAMS AND COLUMNS WITH THREE SURFACES EXPOSED TO FIRE

7.3.1 Calculation Tables

Usually, three surfaces, i.e., both sides and the bottom surface, of the sections of beams, columns, and similar members in a building are exposed to fire during an accident. Because of the symmetry of the section and the boundary conditions, the temperature value (T, °C) at certain points on the left half of the section only are given in Tables 7-3 to 7-9 for a fire lasting $t = 30$–180 min.

There are seven sizes of section (mm × mm) of structural members included in these tables: 200 × 300, 200 × 500, 300 × 300, 300 × 500, 400 × 400, 400 × 500, 500 × 500. The origin of the coordinates is located on the left bottom corner of the section (Fig. 7-3), and the coordinate of a point is (x,y), where the temperature value is given. Therefore, x and y are the distances from the point to the left side and the bottom surface exposed to fire, respectively.

When the section size of the structural member and the duration of the fire are identical to the conditions in the tables, the temperature field can be obtained directly. If they are not identical, various approximate methods can be used to determine the temperature values. Some examples are presented below.

The temperature values for larger sections (i.e., 200 mm × 500 mm, 300 mm × 500 mm, 400 mm × 500 mm, and 500 mm × 500 mm) of the member are given in the tables for $t = 120$ min and 180 min, but not for $t = 30$ min and 60 min. However, the temperature values of the latter can be determined approximately by referring to the values for the corresponding smaller section (i.e., 200 mm × 300 mm, 300 mm × 300 mm, or 400 mm × 400 mm).

If the section size of the structural member and the duration of the fire do not exceed the maximum values (500 mm × 500 mm and 180 min, respectively) listed in these tables, but cannot be found directly, the temperature value can be determined by the interpolation method, referring to the tables with similar conditions.

If the section size of the structural member exceeds the maximum value listed in these tables, e.g., 400 mm × 700 mm > 400 mm × 500 mm, the temperature value can still be determined using Table 7-8. The temperature values within the range of the lower part of the section ($y = 0$–200 mm) have the same values as in the same range listed in Table 7-8; the temperature values within the range of the upper part of the section ($y = 600$–700 mm) have the values in the range of $y = 400$–500 mm in Table 7-8; and the temperature values in the range of the middle part of the section ($y = 200$–600 mm) have the values at $y = 200$ mm and 400 mm in Table 7-8 with linear interpolation between them.

7.3.2 Graphs of Temperature Contours

Using all the data listed in Tables 7-3–7-9, the temperature contour lines on the section can be drawn and are shown in Figs. 7-4–7-10 for structural members of various section sizes and duration of fire (t). The shapes and sizes of these contours actually show the scope of different damage levels of concrete on the section of a structural member under fire.

The temperature field on the section of a structural member with three surfaces exposed to fire can be found directly from these graphs or obtained by the interpolation calculation. If the section size of the structural member and the duration of the fire are not identical to the conditions listed in these graphs, the methods suggested in Section 7.3.1 can be used.

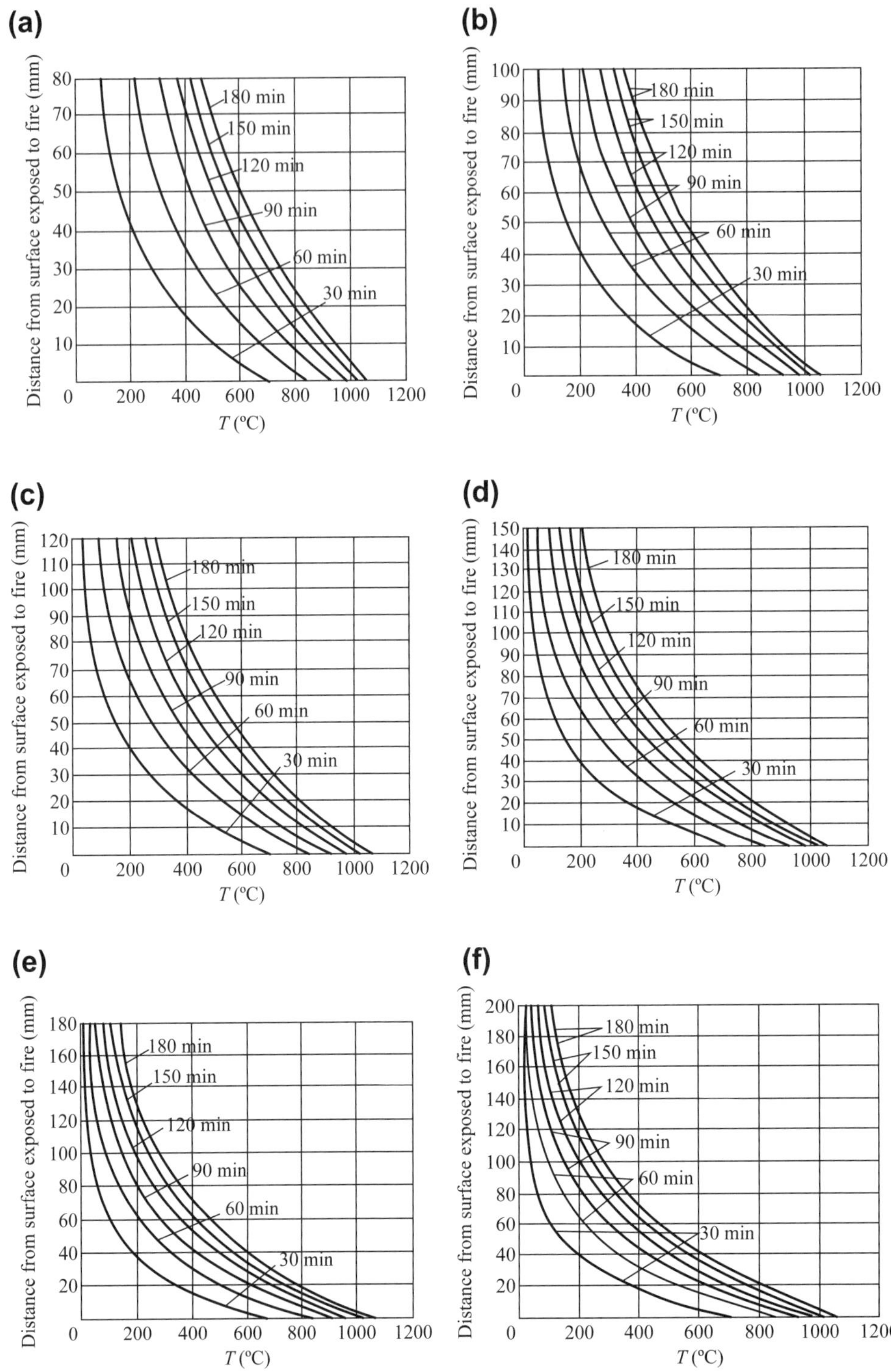

FIGURE 7-2 Curves of the temperature distribution on a section of a slab with one surface exposed to fire: (a) h = 80 mm; (b) h = 100 mm; (c) h = 120 mm; (d) h = 150 mm; (e) h = 180 mm; (f) h = 200 mm.

TABLE 7-3 Temperature Value on a Section (200 mm × 300 mm) of a Member with Three Surfaces Exposed to Fire (°C)

	x (mm)					*y* (mm)	*x* (mm)				
y (mm)	0	20	40	70	100		0	20	40	70	100
(a) $t = 30$ min						(b) $t = 60$ min					
300	648	322	174	76	50	300	779	445	285	167	134
200	693	370	202	87	57	200	840	535	352	209	167
150	693	371	203	88	58	150	841	539	357	217	176
100	695	376	212	100	72	100	846	556	383	250	212
70	701	395	239	136	110	70	855	588	429	308	274
40	722	456	324	240	220	40	874	661	532	435	408
20	754	555	456	396	382	20	897	750	661	593	574
0	816	754	722	701	696	0	933	897	874	856	851
(c) $t = 120$ min						(d) $t = 180$ min					
300	903	571	417	301	267	300	972	644	496	386	354
200	980	725	553	409	366	200	1058	845	690	555	514
150	983	737	570	432	390	150	1062	862	717	589	550
100	989	764	611	483	444	100	1068	891	761	647	611
70	997	800	664	551	517	70	1074	924	813	713	682
40	1010	864	762	675	649	40	1084	976	896	824	801
20	1024	930	865	809	791	20	1093	1027	978	933	919
0	1043	1024	1010	999	995	0	1106	1093	1084	1076	1073

TABLE 7-4 Temperature Value on a Section (200 mm × 500 mm) of a Member with Three Surfaces Exposed to Fire (°C)

y (mm)	*x* (mm)					*y* (mm)	*x* (mm)				
	0	20	40	70	100		0	20	40	70	100
(a) $t = 120$ min						(b) $t = 180$ min					
500	903	571	416	300	267	500	972	643	495	385	353
450	975	702	523	378	334	450	1051	810	641	501	458
400	979	722	548	404	360	400	1057	840	683	546	504
200	981	729	559	417	375	200	1060	854	704	574	533
150	983	737	571	433	392	150	1062	864	720	594	555
100	989	764	611	483	445	100	1068	892	762	648	612
70	997	800	664	551	517	70	1074	924	813	714	683
40	1010	864	762	675	649	40	1084	976	896	824	801
20	1024	930	865	809	791	20	1093	1027	978	933	919
0	1043	1024	1011	999	995	0	1106	1093	1084	1076	1073

TABLE 7-5 Temperature Value on a Section (300 mm × 300 mm) of a Member with Three Surfaces Exposed to Fire (°C)

y (mm)	*x* (mm)						*y* (mm)	*x* (mm)					
	0	20	40	70	100	150		0	20	40	70	100	150
(a) $t = 30$ min							(b) $t = 60$ min						
300	648	322	173	72	35	22	300	779	441	278	145	80	47
250	692	368	199	81	38	23	250	837	525	332	172	93	53
200	693	370	202	83	39	23	200	839	531	341	180	99	57
150	693	370	202	84	40	24	150	840	535	347	189	110	70
100	694	376	211	97	55	41	100	845	553	375	227	156	120
70	701	395	239	133	97	85	70	854	586	423	291	228	198
40	722	456	324	239	211	203	40	874	659	528	423	376	353
20	754	555	456	395	376	371	20	897	750	659	586	553	538
0	816	754	722	701	694	693	0	933	897	874	854	845	841

TABLE 7-5 Temperature Value on a Section (300 mm × 300 mm) of a Member with Three Surfaces Exposed to Fire (°C)—cont'd

y (mm)	*x* (mm)						*y* (mm)	*x* (mm)					
	0	20	40	70	100	150		0	20	40	70	100	150
(c) $t = 120$ min							(d) $t = 180$ min						
300	899	555	388	244	167	123	300	967	620	454	312	235	191
250	970	680	484	304	206	151	250	1044	775	582	400	300	243
200	975	701	511	331	230	174	200	1051	807	625	446	343	283
150	978	715	532	359	262	208	150	1055	828	658	488	390	333
100	985	748	580	423	337	289	100	1063	865	714	564	477	427
70	994	787	641	503	428	386	70	1070	903	776	646	571	528
40	1009	856	746	642	585	553	40	1082	963	872	778	722	690
20	1023	926	856	788	750	730	20	1092	1019	964	905	870	849
0	1043	1023	1009	994	986	981	0	1106	1092	1082	1071	1064	1060

TABLE 7-6 Temperature Value on a Section (300 mm × 500 mm) of a Member with Three Surfaces Exposed to Fire (°C)

y (mm)	*x* (mm)						*y* (mm)	*x* (mm)					
	0	20	40	70	100	150		0	20	40	70	100	150
(a) $t = 120$ min							(b) $t = 180$ min						
500	899	555	387	243	165	122	500	996	618	452	309	231	187
450	970	679	482	301	202	147	450	1043	772	577	394	291	234
400	974	697	504	321	218	160	400	1049	799	612	427	320	258
200	976	705	516	337	237	180	200	1053	815	638	461	358	299
150	978	716	533	360	264	210	150	1056	830	661	492	394	338
100	985	748	581	424	338	290	100	1063	865	715	565	478	428
70	994	787	641	503	428	387	70	1070	904	776	647	571	528
40	1009	856	746	642	585	553	40	1082	963	872	778	722	690
20	1023	926	856	788	750	730	20	1092	1019	964	905	870	849
0	1043	1023	1009	994	986	981	0	1106	1092	1082	1071	1064	1060

TABLE 7-7 Temperature Value on a Section (400 mm × 400 mm) of a Member with Three Surfaces Exposed to Fire (°C)

y (mm)	*x* (mm)							*y* (mm)	*x* (mm)						
	0	20	40	70	100	150	200		0	20	40	70	100	150	200
(a) $t = 30$ min								(b) $t = 60$ min							
400	648	322	173	72	35	21	20	400	779	441	277	145	78	34	25
350	692	368	199	81	38	21	20	350	837	522	332	171	90	37	26
200	693	370	202	83	39	21	20	200	839	531	342	180	97	41	29
150	693	370	202	84	40	23	22	150	840	535	347	188	108	54	43
100	694	376	211	97	55	40	39	100	845	553	375	227	154	108	99
70	701	395	239	133	97	84	83	70	854	586	423	290	227	189	182
40	722	456	324	239	211	202	202	40	874	659	528	423	275	347	343
20	754	555	456	395	376	370	370	20	897	750	659	586	553	535	532
0	816	754	722	701	694	693	693	0	933	897	874	854	845	840	839
(c) $t = 120$ min								(d) $t = 180$ min							
400	899	553	384	236	150	78	58		965	614	444	293	203	123	100
350	969	677	478	291	183	92	67		1042	765	565	371	254	151	121
200	976	703	512	327	215	121	95		1051	808	625	436	317	210	179
150	978	714	529	352	246	158	134		1055	825	650	472	360	261	232
100	985	746	578	418	324	249	229		1062	861	708	550	452	366	342
70	994	787	639	499	418	354	337		1070	901	771	636	551	478	458
40	1009	856	746	639	578	531	519		1081	962	869	771	709	655	640
20	1023	926	856	787	747	715	708		1092	1018	962	901	862	827	818
0	1043	1023	1009	994	985	978	977		1106	1092	1081	1070	1062	1055	1053

TABLE 7-8 Temperature Value on a Section (400 mm × 500 mm) of a Member with Three Surfaces Exposed to Fire (°C)

y (mm)	x (mm)							y (mm)	x (mm)						
	0	20	40	70	100	150	200		0	20	40	70	100	150	200
(a) $t = 120$ min								(b) $t = 180$ min							
500	899	553	384	236	150	78	57	500	965	614	444	293	202	122	99
450	969	677	478	291	182	92	67	450	1042	765	565	371	252	149	120
400	974	695	499	310	196	99	72	400	1048	791	598	401	276	165	132
200	976	703	512	327	216	121	95	200	1051	809	625	437	318	212	181
150	978	714	529	352	246	158	134	150	1055	825	651	472	360	261	233
100	985	746	578	418	324	249	229	100	1062	861	708	550	452	366	342
70	994	787	639	499	418	354	337	70	1070	901	771	636	551	478	458
40	1009	856	746	639	578	531	519	40	1081	962	869	771	709	655	640
20	1023	926	856	787	747	715	708	20	1092	1018	962	901	862	827	828
0	1043	1023	1009	994	985	978	977	0	1106	1092	1081	1070	1062	1055	1053

TABLE 7-9 Temperature Value on a Section (500 mm × 500 mm) of a Member with Three Surfaces Exposed to Fire (°C)

y (mm)	x (mm)							y (mm)	x (mm)						
	0	20	40	70	100	150	200		0	20	40	70	100	150	200
(a) $t = 120$ min								(b) $t = 180$ min							
500	899	553	384	235	148	72	32	500	965	613	443	290	197	108	54
450	969	677	478	291	181	85	35	450	1042	764	563	367	246	132	63
400	974	695	499	309	194	91	37	400	1048	790	596	397	269	145	70
200	976	703	512	326	214	115	62	200	1051	808	624	434	312	194	123
150	978	714	529	351	245	153	105	150	1055	824	649	470	355	247	183
100	985	746	578	417	323	245	207	100	1062	861	707	549	448	356	303
70	994	787	639	499	417	351	321	70	1070	901	770	645	549	470	427
40	1009	856	745	639	578	529	508	40	1081	962	869	770	707	650	619
20	1023	926	856	787	746	714	701	20	1092	1018	962	901	861	824	805
0	1043	1023	1009	994	985	978	975	0	1106	1092	1081	1070	1062	1055	1051

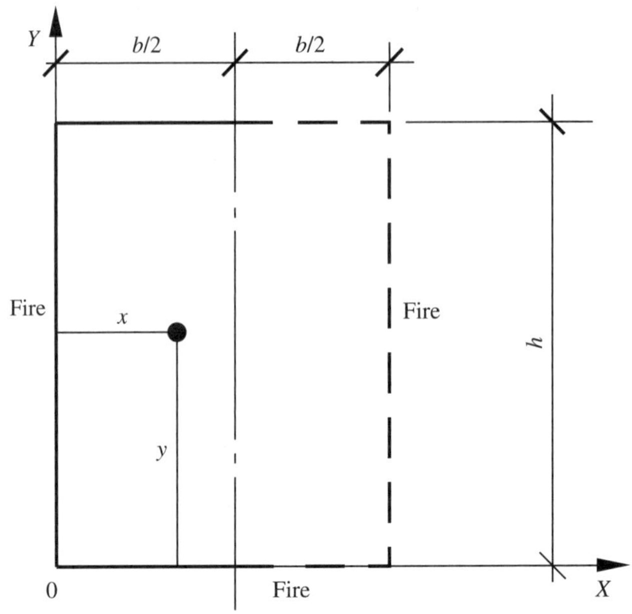

FIGURE 7-3 Calculation parameters for a member with three surfaces exposed to fire.

7.4 SQUARE COLUMNS WITH FOUR SURFACES EXPOSED TO FIRE

7.4.1 Calculation Tables

Sometimes all four surfaces of an arbitrary column of a square section in a building are exposed to fire during a fire accident. Because the section is symmetrical about both perpendicular axes, the temperature values (T, °C) are given in Tables 7-10–7-13 only for certain points on the left lower one-fourth of the section (Fig. 7-11) and the duration of the fire $t = 30$–180 mm. There are four section sizes (mm × mm) of the column included in these tables: 300 × 300, 400 × 400, 500 × 500, and 600 × 600. The origin of the coordinates is located on the left bottom corner of the section and the coordinates of the point are (x, y) where the temperature value is given.

The method for using these tables to determine the temperature field in a section of a square column with four surfaces exposed to fire is similar to that described in Section 7.3.1 for a structural member with three surfaces exposed to fire.

7.4.2 Graphs of Temperature Contours

Using all the data listed in Tables 7-10–7-13, the temperature contour lines on the section can be drawn and are shown in Figs. 7-12–7-15 for a square column of various section sizes and duration of fire (t). The method for using these graphs to determine the temperature distribution in a section of a square column with four surfaces exposed to fire is similar to that described above.

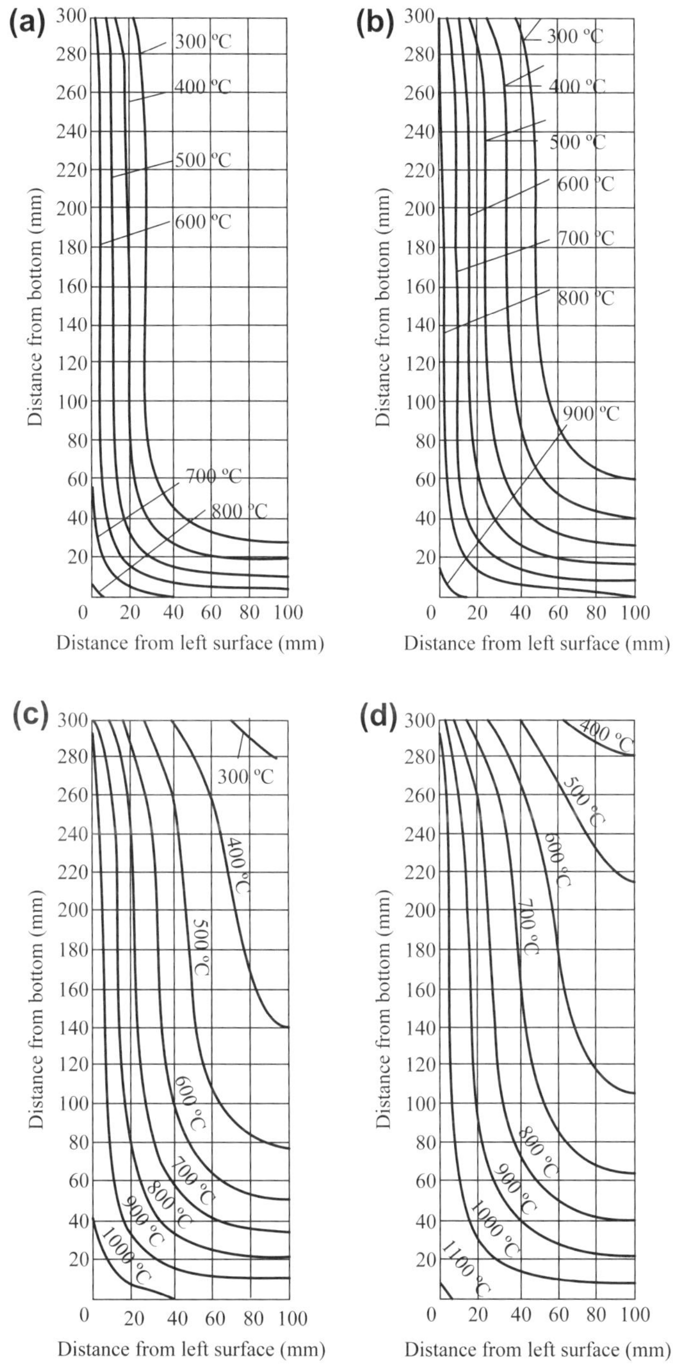

FIGURE 7-4 Temperature contour on a section (200 mm × 300 mm) of a structural member with three surfaces exposed to fire: (a) t = 30 min; (b) t = 60 min; (c) t = 120 min; (d) t = 180 min.

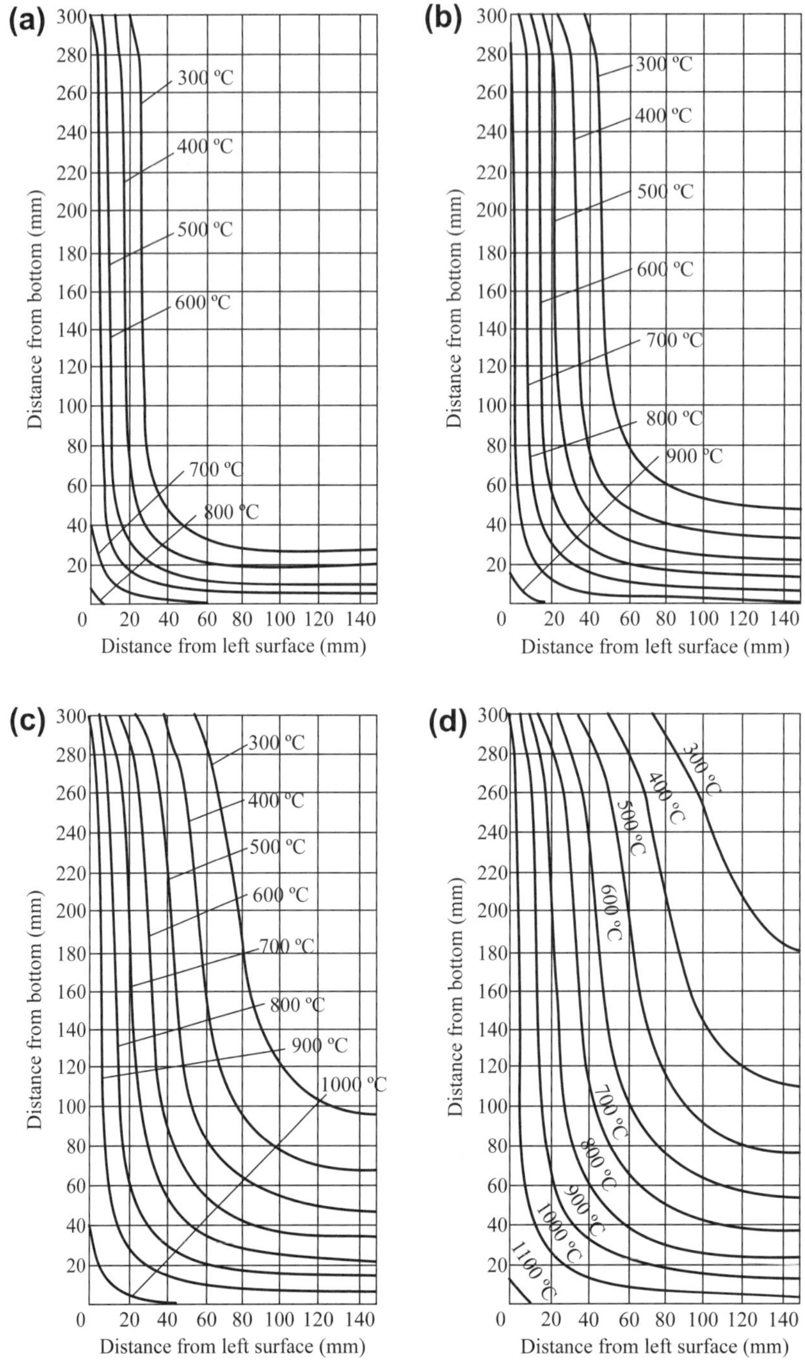

FIGURE 7-5 Temperature contour on a section (300 mm × 300 mm) of a structural member with three surfaces exposed to fire: (a) t = 30 min; (b) t = 60 min; (c) t = 120 min; (d) t = 180 min.

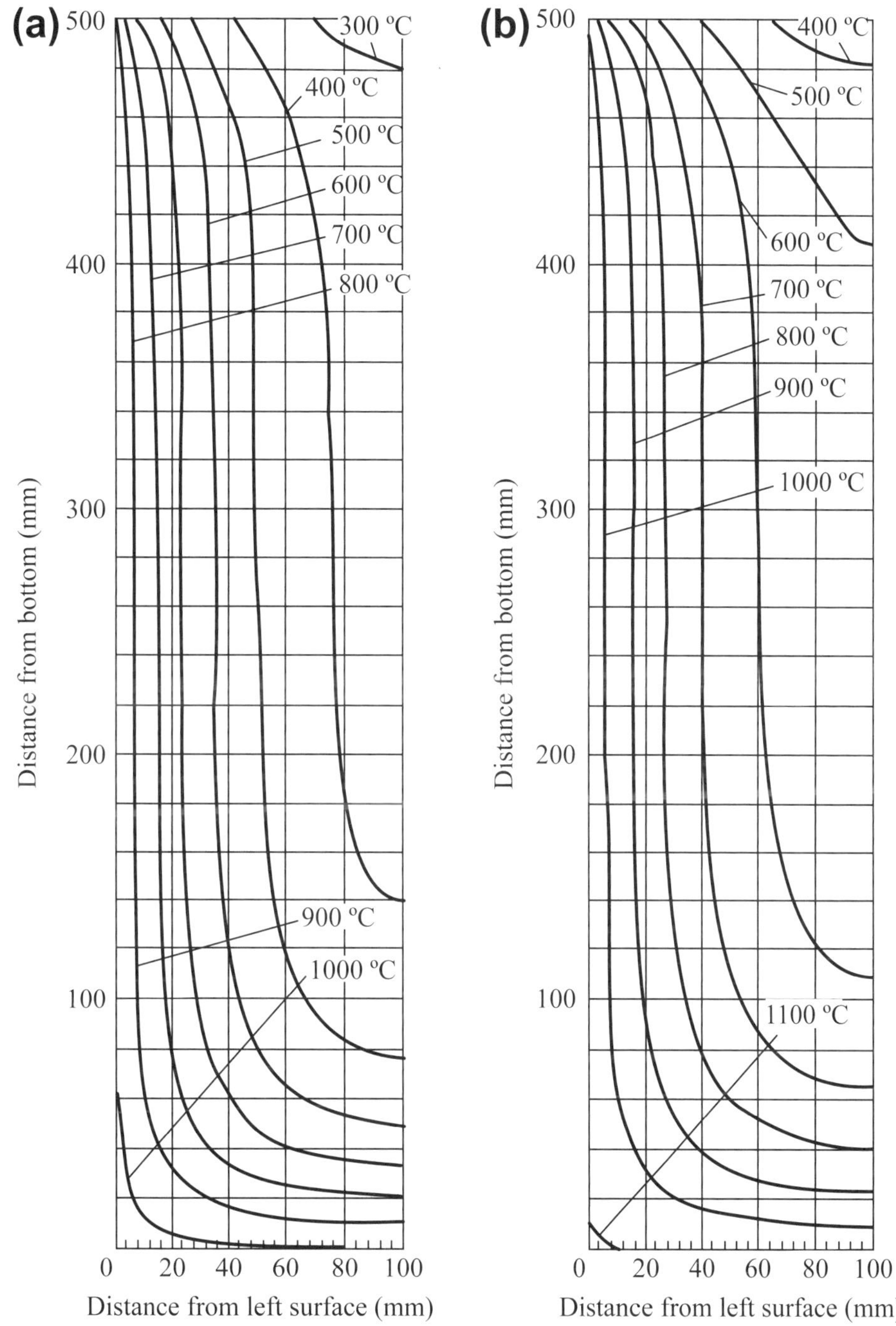

FIGURE 7-6 Temperature contour on a section (200 mm × 500 mm) of a structural member with three surfaces exposed to fire: (a) t = 120 min; (b) t = 180 min.

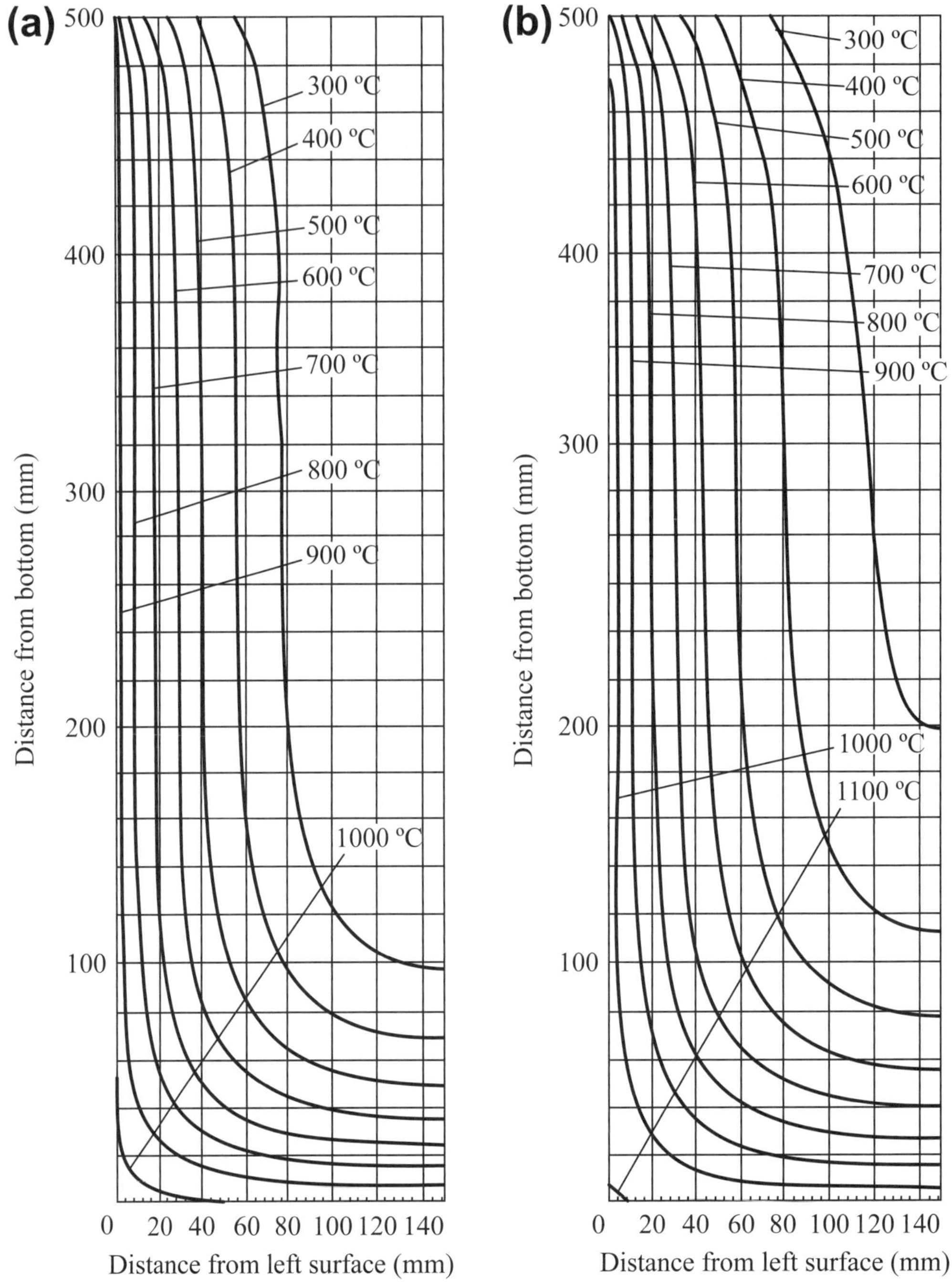

FIGURE 7-7 Temperature contour on a section (300 mm × 500 mm) of a structural member with three surfaces exposed to fire: (a) t = 120 min; (b) t = 180 min.

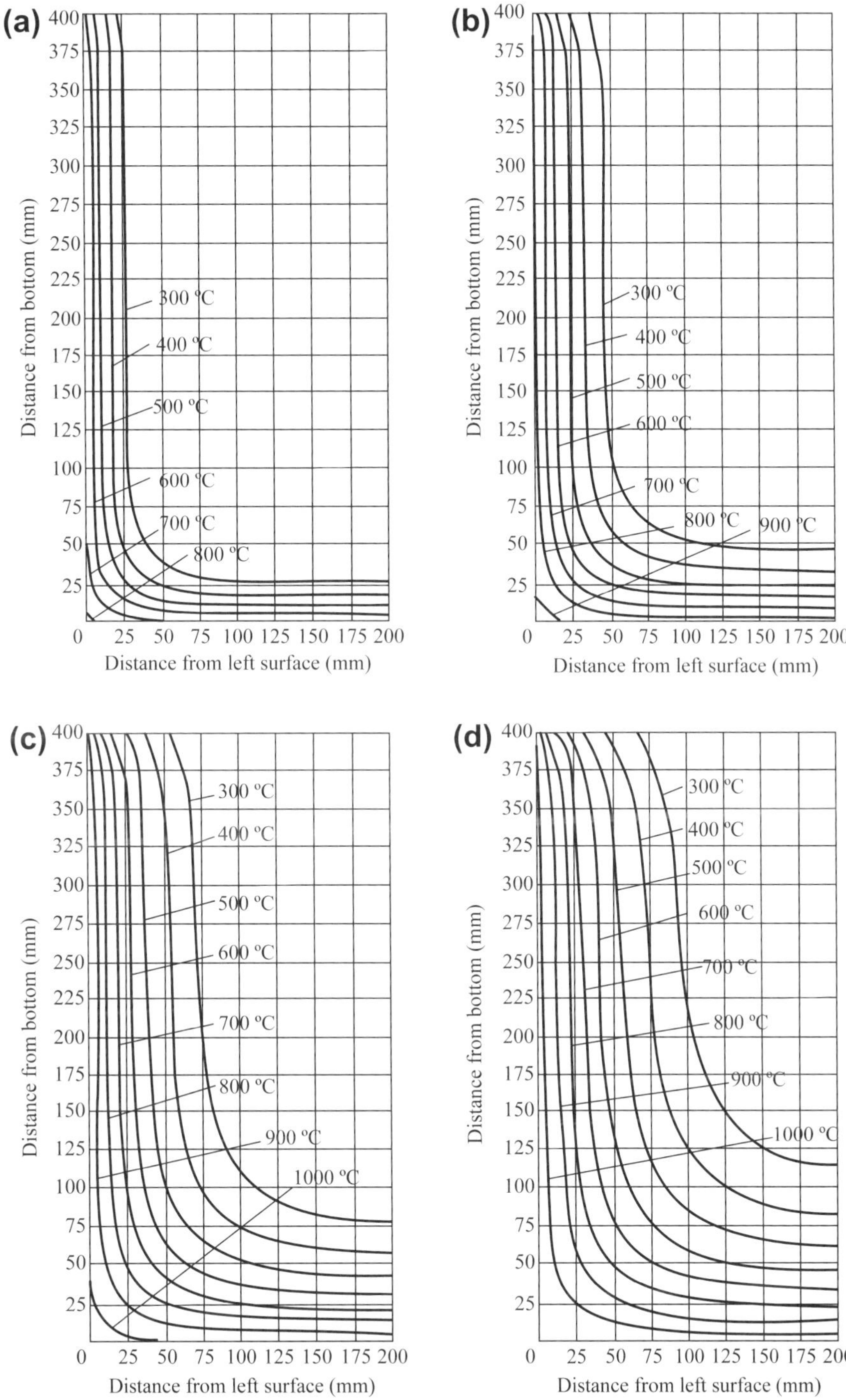

FIGURE 7-8 Temperature contour on a section (400 mm × 400 mm) of a structural member with three surfaces exposed to fire: (a) t = 30 min; (b) t = 60 min; (c) t = 120 min; (d) t = 180 min.

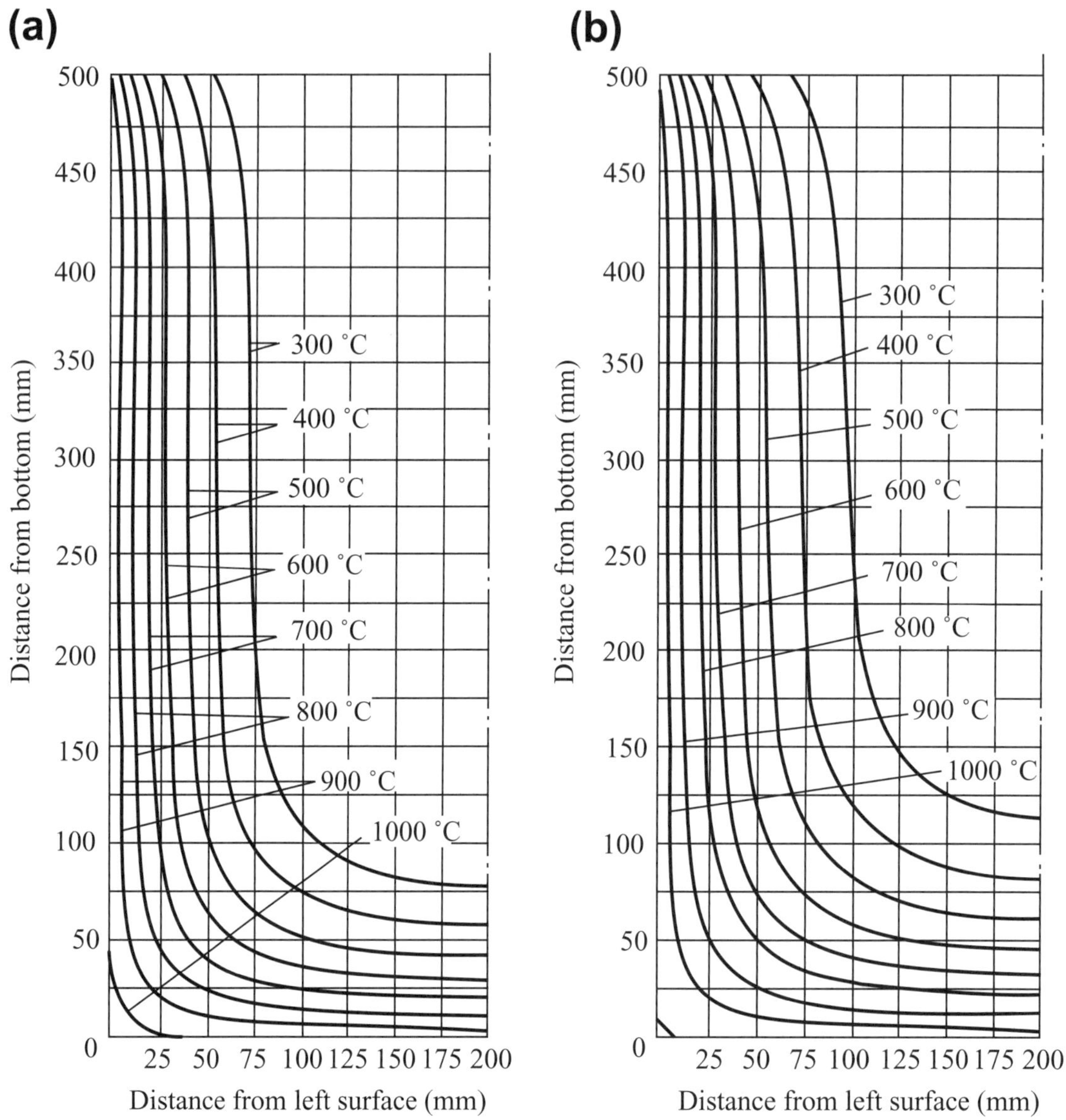

FIGURE 7-9 Temperature contour on a section (400 mm × 500 mm) of a structural member with three surfaces exposed to fire: (a) t = 120 min; (b) t = 180 min.

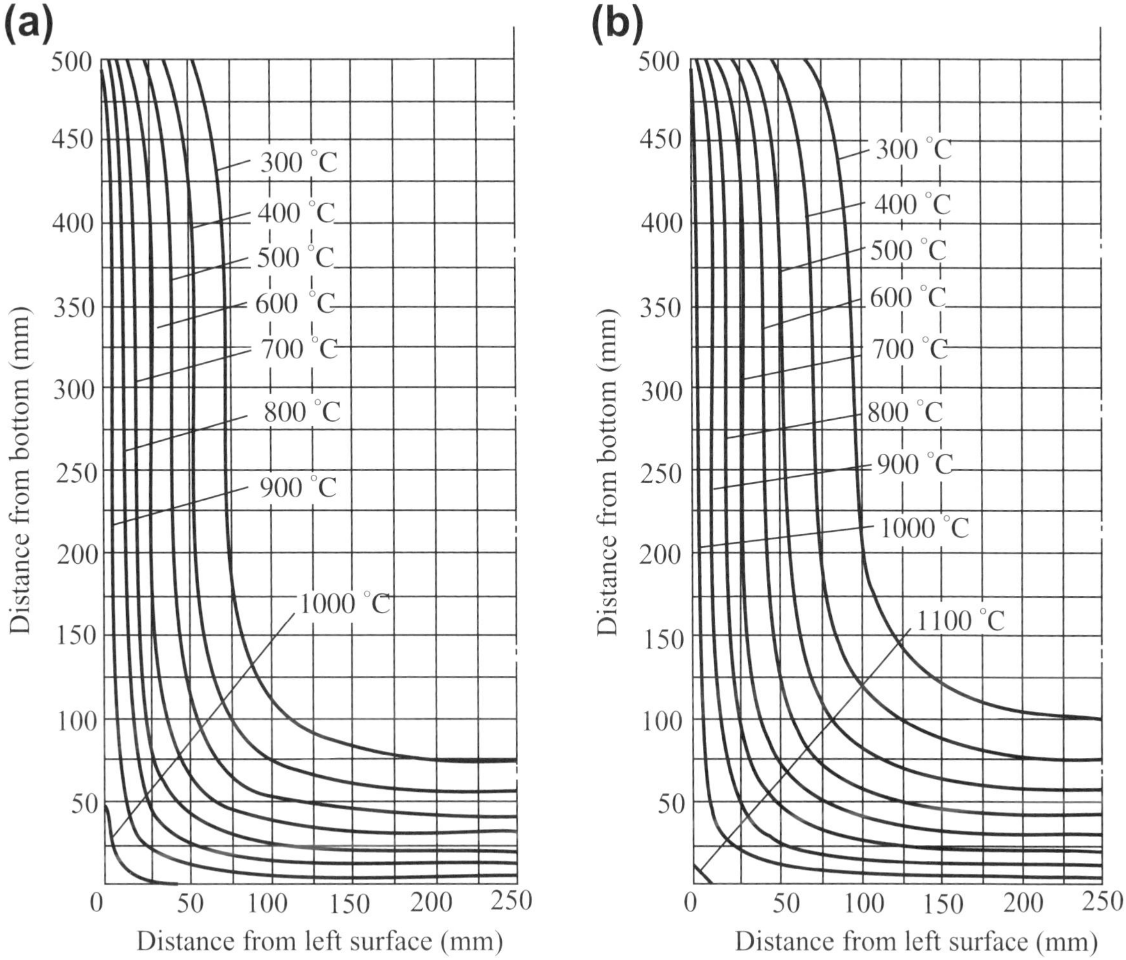

FIGURE 7-10 Temperature contour on a section (500 mm × 500 mm) of a structural member with three surfaces exposed to fire: (a) t = 120 min; (b) t = 180 min.

TABLE 7-10 Temperature Value on a Section (300 mm × 300 mm) of a Square Column with Four Surfaces Exposed to Fire (°C)

y (mm)	*x* (mm)						*y* (mm)	*x* (mm)					
	0	20	40	70	100	150		0	20	40	70	100	150
(a) $t = 30$ min							(b) $t = 60$ min						
150	693	371	203	85	41	26	150	841	538	354	199	123	84
100	694	376	211	97	55	41	100	845	553	376	229	158	123
70	701	395	239	133	97	85	70	854	586	423	291	229	199
40	722	456	324	239	211	203	40	874	659	529	423	376	354
20	754	555	456	395	376	371	20	897	750	659	586	553	538
0	816	754	722	701	694	693	0	933	897	874	854	845	841
(c) $t = 120$ min							(d) $t = 180$ min						
150	982	731	556	393	304	254	150	1060	853	697	542	454	403
100	986	751	587	433	350	304	100	1064	873	728	583	501	454
70	995	789	644	507	433	393	70	1071	907	782	656	583	542
40	1009	857	747	644	587	556	40	1082	965	875	782	728	697
20	1023	926	857	789	751	731	20	1092	1020	965	907	873	853
0	1043	1023	1009	995	986	982	0	1106	1092	1082	1071	1064	1060

TABLE 7-11 Temperature Value on a Section (400 mm × 400 mm) of a Square Column with Four Surfaces Exposed to Fire (°C)

y (mm)	*x* (mm)							*y* (mm)	*x* (mm)						
	0	20	40	70	100	150	200		0	20	40	70	100	150	200
(a) $t = 120$ min								(b) $t = 180$ min							
200	977	708	519	338	230	140	116	200	1053	818	641	460	347	248	220
150	978	715	531	354	250	164	140	150	1055	828	656	481	371	275	248
100	985	747	578	418	325	250	230	100	1062	862	710	553	455	371	347
70	994	787	639	499	418	354	338	70	1070	901	771	637	553	481	460
40	1009	856	746	639	578	531	519	40	1081	962	869	771	709	656	641
20	1023	926	856	787	747	715	708	20	1092	1018	962	901	862	828	818
0	1043	1023	1009	994	985	978	977	0	1106	1092	1081	1070	1062	1055	1053

TABLE 7-12 Temperature Value on a Section (500 mm × 500 mm) of a Square Column with Four Surfaces Exposed to Fire (°C)

y (mm)	x (mm)							y (mm)	x (mm)						
	0	20	40	70	100	150	200		0	20	40	70	100	150	200
(a) $t = 120$ min								(b) $t = 180$ min							
250	975	701	508	321	207	107	53	250	1051	805	619	427	304	186	115
150	978	714	529	351	245	152	107	150	1055	824	650	471	356	249	186
100	985	746	578	417	323	245	207	100	1062	861	707	549	449	356	304
70	994	787	639	499	417	351	321	70	1070	901	770	635	549	471	427
40	1009	856	745	639	578	529	508	40	1081	962	869	770	707	650	619
20	1023	926	856	787	746	714	701	20	1092	1018	962	901	861	824	805
0	1043	1023	1009	994	985	978	975	0	1106	1092	1081	1070	1062	1055	1051

TABLE 7-13 Temperature Value on a Section (600 mm × 600 mm) of a Square Column with Four Surfaces Exposed to Fire (°C)

y (mm)	x (mm)								y (mm)	x (mm)							
	0	20	40	70	100	150	200	300		0	20	40	70	100	150	200	300
(a) $t = 120$ min									(b) $t = 180$ min								
300	975	699	505	316	201	97	52	29	300	1050	800	611	415	288	162	98	60
200	976	703	512	326	214	114	72	51	200	1051	808	624	434	311	192	133	98
150	978	714	529	351	245	152	114	97	150	1055	824	649	470	355	245	192	162
100	985	746	578	417	323	245	214	201	100	1062	861	707	549	448	355	311	288
70	994	787	639	499	417	351	326	316	70	1070	901	770	634	540	470	434	415
40	1009	856	745	639	578	529	512	505	40	1081	962	869	770	707	649	624	611
20	1023	926	856	787	746	714	703	699	20	1092	1018	962	901	861	824	808	800
0	1043	1023	1009	994	985	978	976	975	0	1106	1092	1081	1070	1062	1055	1051	1050

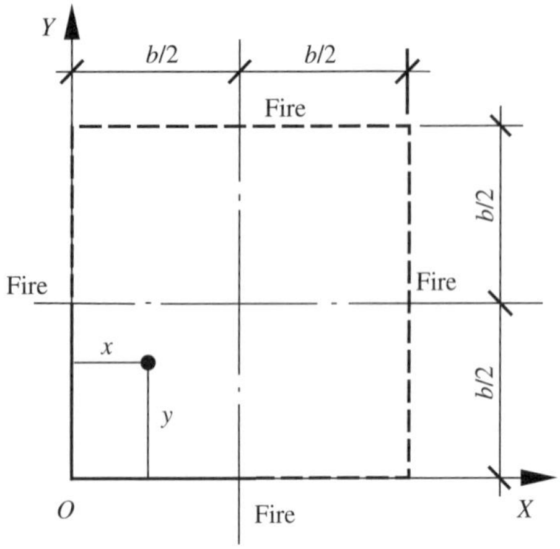

FIGURE 7-11 Calculation parameters for a square column with four surfaces exposed to fire.

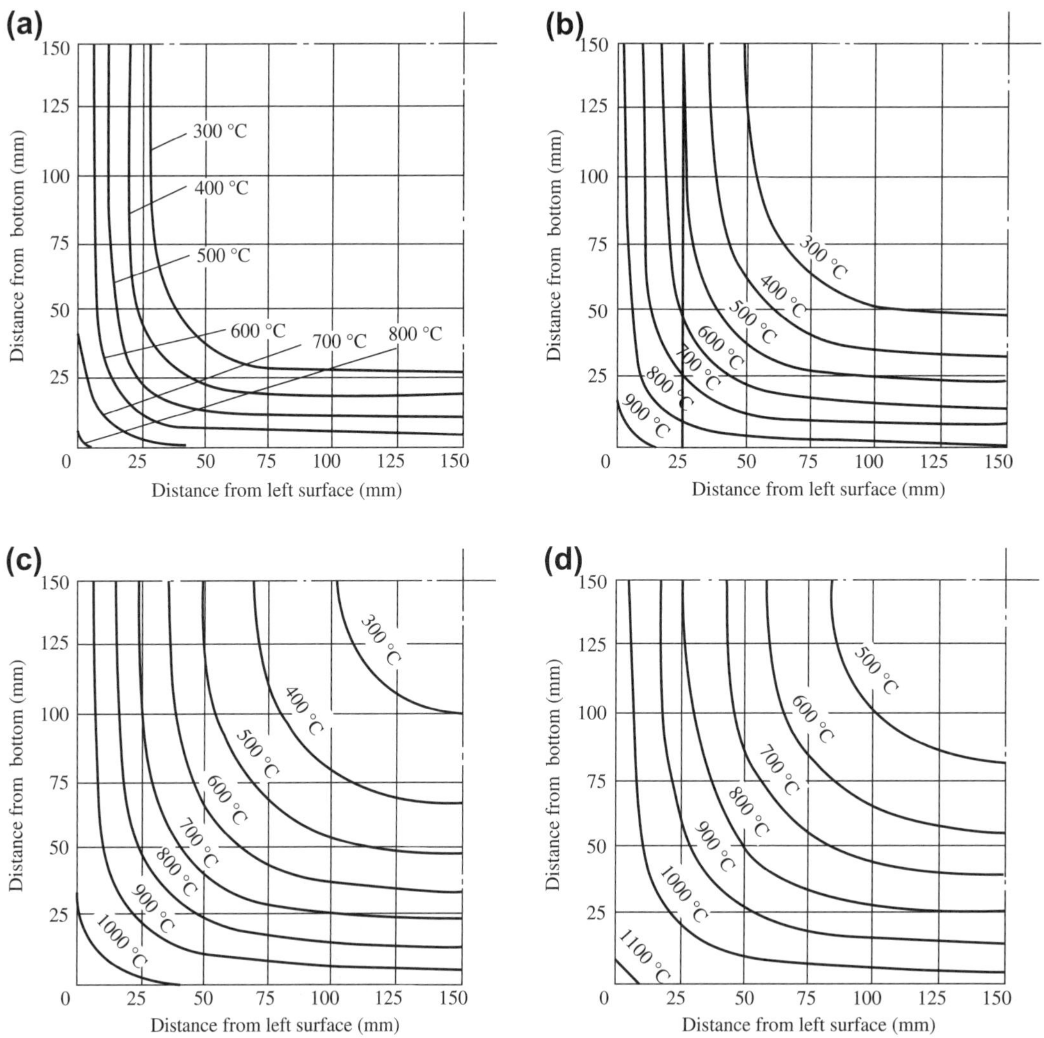

FIGURE 7-12 Temperature contour on a section (300 mm × 300 mm) of a square column with four surfaces exposed to fire: (a) $t = 30$ min; (b) $t = 60$ min; (c) $t = 120$ min; (d) $t = 180$ min.

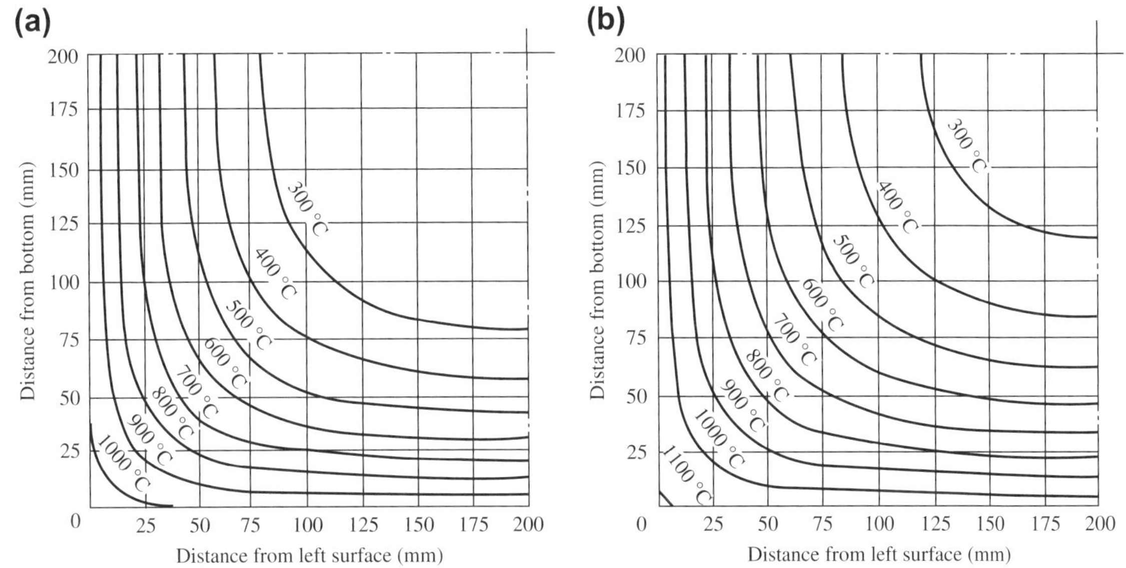

FIGURE 7-13 Temperature contour on a section (400 mm × 400 mm) of a square column with four surfaces exposed to fire: (a) t = 120 min; (b) t = 180 min.

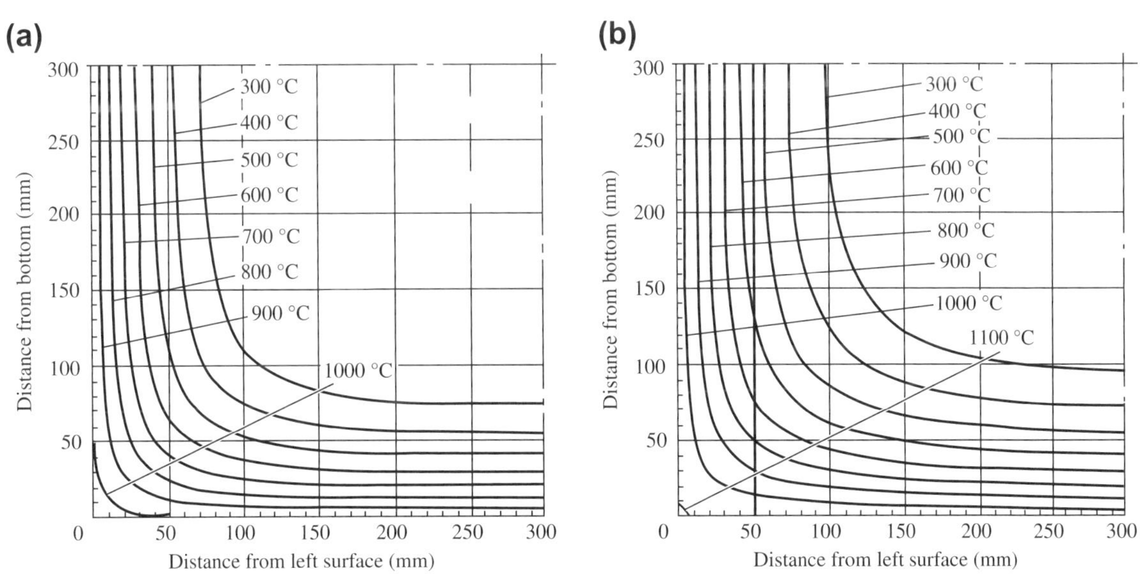

FIGURE 7-14 Temperature contour on a section (500 mm × 500 mm) of a square column with four surfaces exposed to fire: (a) t = 120 min; (b) t = 180 min.

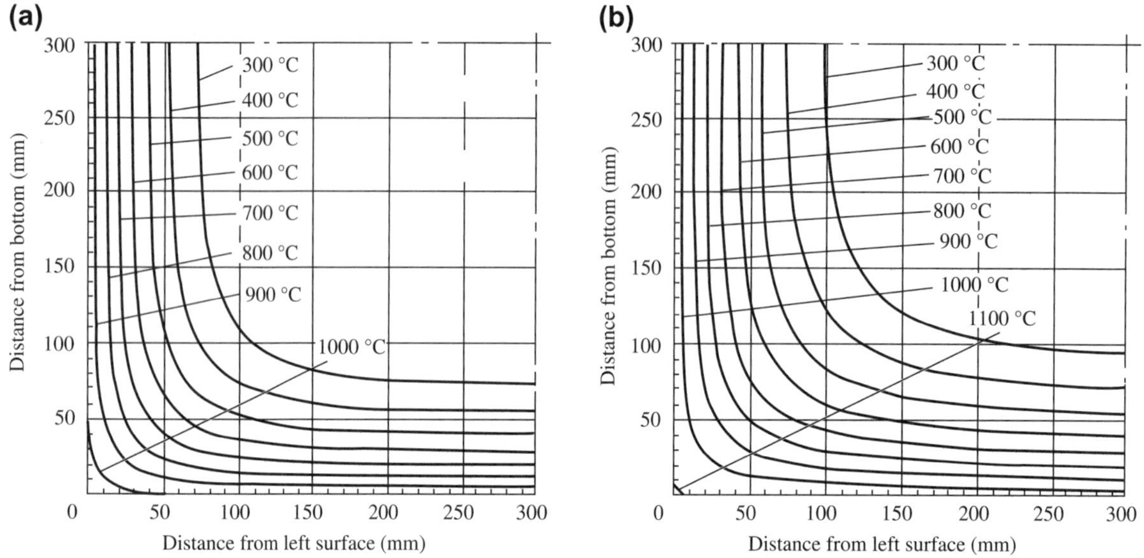

FIGURE 7-15 Temperature contour on a section (600 mm × 600 mm) of a square column with four surfaces exposed to fire: (a) t = 120 min; (b) t = 180 min.

7.5 SQUARE COLUMNS WITH TWO ADJACENT SURFACES EXPOSED TO FIRE

7.5.1 Calculation Tables

When a corner column of a building sustains a fire accident, its two adjacent surfaces may be exposed to fire. The temperature values (T, °C) at certain points on its square section are given in Tables 7-14–7-16 for the t = 30–180 min. These tables include three section sizes (mm × mm): 300 × 300, 400 × 400, and 500 × 500. The origin of the coordinates is located on the left bottom corner of the section (Fig. 7-16) and the coordinates of the point are (x,y) where the temperature value is given.

The method for using these tables to determine the temperature distribution on a section of a square column is similar to that described in Section 7.3.1.

7.5.2 Graphs of Temperature Contours

Using all the data listed in Tables 7-14–7-16, the temperature contour lines on the section can be drawn and are shown in Figs. 7-17–7-19 for a square column of various section sizes and duration of fire (t). The method for using these graphs to determine the temperature distribution on a section of a square column with two adjacent surfaces exposed to fire is similar to that described above.

CONCLUSIONS

When the fire-resistant behavior, including the bearing capacity at elevated temperature, fire endurance limit, and bearing capacity after fire, of a concrete structural member are analyzed, the transient temperature field, corresponding to the temperature–time curve of the fire, of a member has to be determined in advance. Based on the results of numerous calculations using the HTARC computer program (Chapter 6), the temperature fields on the sections are provided in this chapter for various structural members commonly used in engineering practice, and they can be used as an auxiliary method in fire resistance analysis of structural members.

TABLE 7-14 Temperature Value on a Section (300 mm × 300 mm) of a Square Column with Two Adjacent Surfaces Exposed to Fire (°C)

y (mm)	x (mm)								y (mm)	x (mm)							
	0	20	40	70	100	150	200	300		0	20	40	70	100	150	200	300
(a) $t = 30$ min									(b) $t = 60$ min								
300	648	322	173	72	35	21	20	20	300	779	441	277	145	78	34	22	20
200	693	370	202	83	38	21	20	20	200	839	531	341	179	96	40	26	22
150	693	370	202	84	40	23	21	21	150	840	535	347	188	107	54	40	34
100	694	376	211	97	55	40	38	35	100	845	553	375	227	154	107	96	78
70	701	395	239	133	97	84	83	72	70	854	586	423	290	227	188	179	145
40	722	456	324	239	211	202	202	173	40	874	659	528	423	375	347	341	277
20	754	555	456	396	376	370	370	322	20	897	750	659	586	553	534	530	441
0	816	754	722	701	694	693	693	648	0	933	897	874	854	845	840	839	779
(c) $t = 120$ min									(d) $t = 180$ min								
300	899	554	385	236	150	74	42	25	300	966	615	445	294	203	115	71	40
200	975	699	506	320	209	111	69	42	200	1050	800	612	421	299	182	122	71
150	978	714	528	350	243	151	111	74	150	1054	822	646	466	350	239	182	115
100	985	746	578	417	323	243	209	150	100	1062	861	706	547	446	350	298	202
70	994	787	639	498	417	350	320	236	70	1070	901	770	634	547	466	421	294
40	1009	856	745	639	578	528	506	384	40	1081	962	868	770	706	646	612	445
20	1023	926	856	787	746	714	699	554	20	1092	1018	962	901	860	822	800	615
0	1043	1023	1009	994	985	978	975	899	0	1106	1092	1081	1070	1062	1054	1050	965

TABLE 7-15 Temperature Value on a Section (400 mm × 400 mm) of a Square Column with Two Adjacent Surfaces Exposed to Fire (°C)

y (mm)	x (mm)									y (mm)	x (mm)								
	0	20	40	70	100	150	200	300	400		0	20	40	70	100	150	200	300	400
(a) $t = 120$ min										(b) $t = 180$ min									
400	936	587	417	266	175	90	50	24	21	400	965	613	443	290	197	107	61	28	22
300	1015	748	553	358	236	120	66	31	24	300	1048	791	598	399	272	147	84	39	28
200	1018	760	572	383	264	153	100	66	50	200	1051	807	623	433	310	190	130	84	61
150	1021	775	594	414	303	200	153	120	90	150	1055	824	649	469	354	244	190	147	107
100	1028	810	649	488	390	303	264	235	175	100	1062	861	707	548	448	354	310	271	297
70	1036	850	711	573	488	414	383	358	266	70	1070	901	770	634	548	469	433	399	290
40	1049	915	814	711	649	594	572	553	417	40	1081	962	869	770	707	649	623	598	443
20	1061	978	915	850	810	775	760	748	587	20	1092	1018	962	901	861	824	807	791	613
0	1078	1061	1049	1036	1028	1021	1018	1015	936	0	1106	1092	1081	1070	1062	1055	1051	1048	965

TABLE 7-16 Temperature Value on a Section (500 mm × 500 mm) of a Square Column with Two Adjacent Surfaces Exposed to Fire (°C)

y (mm)	x (mm)									y (mm)	x (mm)								
	0	20	40	70	100	150	200	400	500		0	20	40	70	100	150	200	400	500
(a) $t = 120$ min										(b) $t = 180$ min									
500	899	553	384	235	148	72	39	20	20	500	965	613	443	290	196	106	60	21	20
400	974	695	499	309	194	91	46	20	20	400	1048	790	596	397	268	142	77	22	21
200	976	703	512	326	214	114	71	46	39	200	1051	808	624	434	311	191	131	77	60
150	978	714	529	351	245	152	114	91	72	150	1055	824	649	470	355	245	191	142	106
100	985	746	578	417	323	245	214	194	148	100	1062	861	707	549	448	355	311	268	196
70	994	787	639	499	417	351	326	309	235	70	1070	901	770	634	548	470	434	396	290
40	1009	856	745	639	578	529	512	499	384	40	1081	962	869	770	707	649	624	596	443
20	1023	926	856	787	746	714	703	695	553	20	1092	1018	962	901	861	824	808	790	613
0	1043	1023	1009	994	985	978	976	974	899	0	1106	1092	1081	1070	1062	1055	1051	1048	965

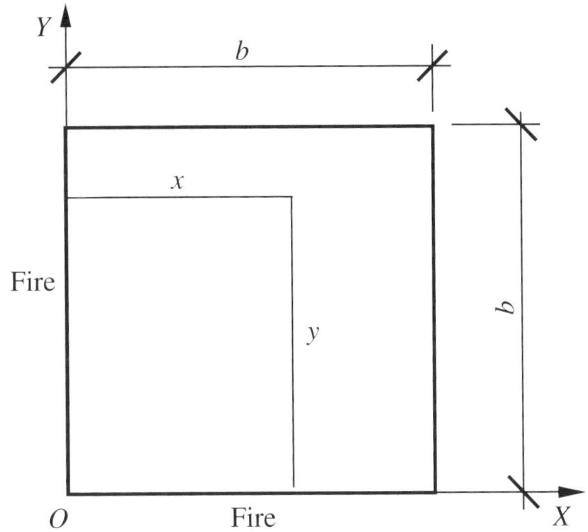

FIGURE 7-16 Calculation parameters for a square column with two adjacent surfaces exposed to fire.

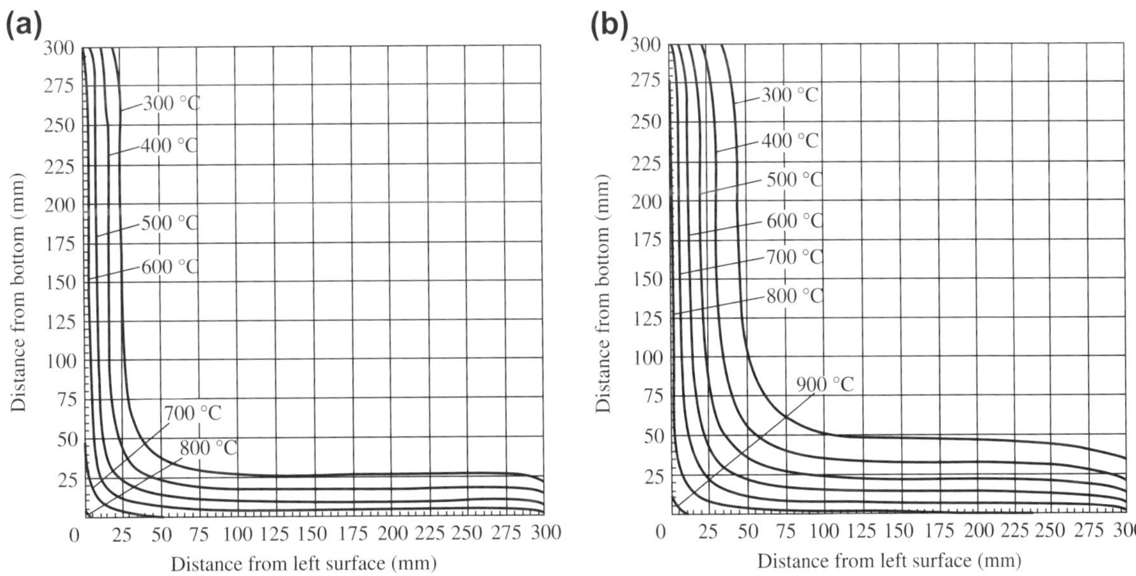

FIGURE 7-17 Temperature contour on a section (300 mm × 300 mm) of a square column with two adjacent surfaces exposed to fire: (a) $t = 30$ min; (b) $t = 60$ min; (c) $t = 120$ min; (d) $t = 180$ min.

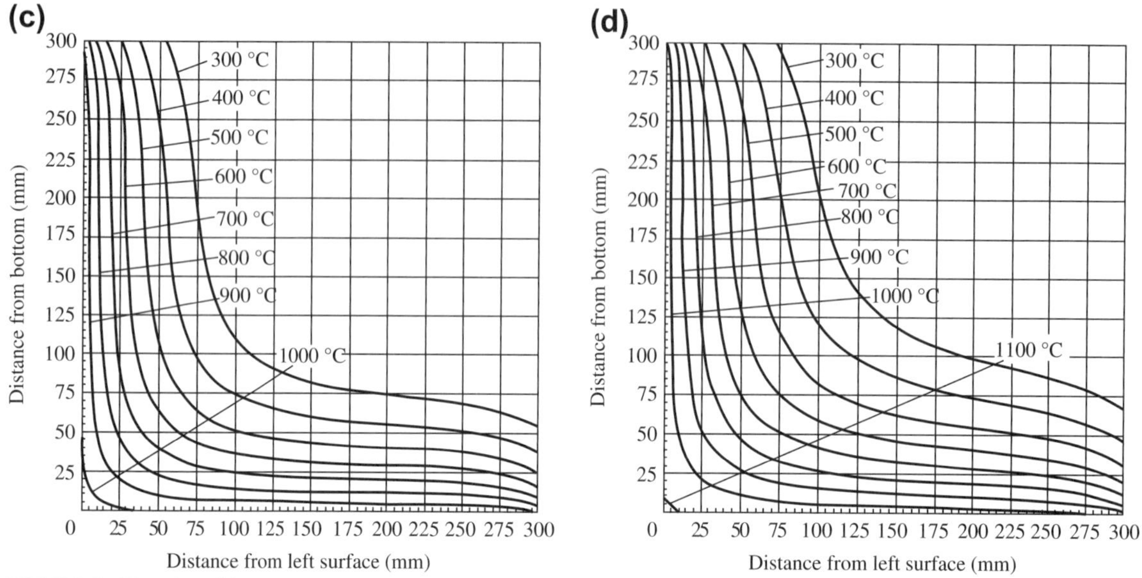

FIGURE 7-17—Cont'd

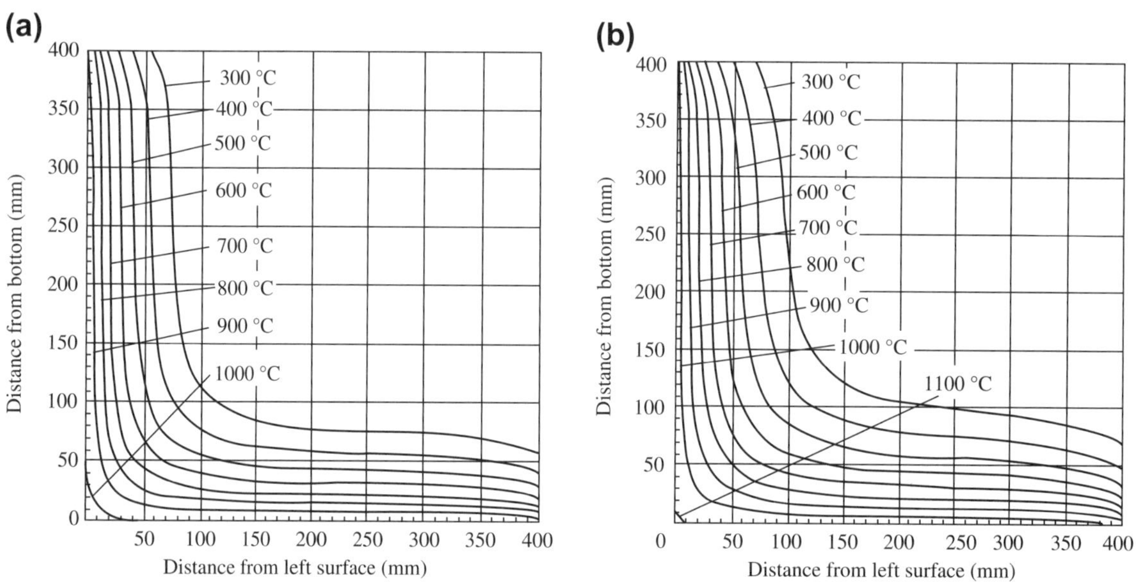

FIGURE 7-18 Temperature contour on a section (400 mm × 400 mm) of a square column with two adjacent surfaces exposed to fire: (a) t = 120 min; (b) t = 180 min.

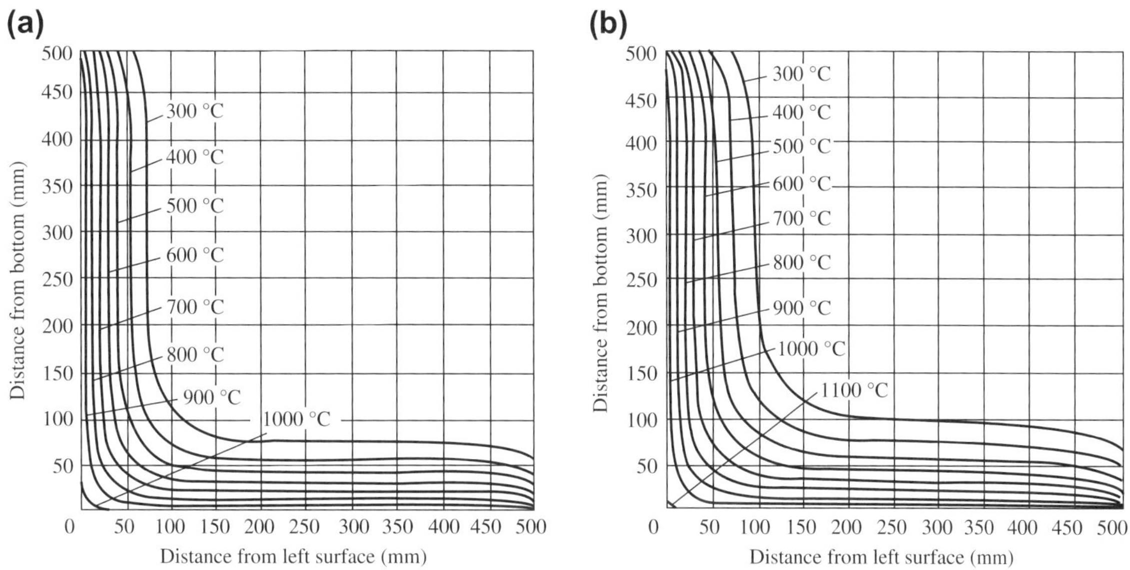

FIGURE 7-19 Temperature contour on a section (500 mm × 500 mm) of a square column with two adjacent surfaces exposed to fire (a) t = 120 min; (b) t = 180 min.

The main scope and conditions of the tables and graphs in this chapter are:

- temperature–time curve of fire: the ISO standard curve
  - duration of fire: 30–180 min
- type and section size of structural members exposed to fire:
  - slab with one surface exposed to fire: thickness 80–120 mm
  - beam or column with three surfaces exposed to fire: rectangular section 200 mm × 300 mm to 500 mm × 500 mm
  - square column with four surfaces exposed to fire: section 300 mm × 300 mm to 600 mm × 600 mm
  - square column with two adjacent surfaces exposed to fire: section 300 mm × 300 mm to 500 mm × 500 mm
- form of expression: table showing the temperature values at certain points on the section, temperature contour lines on the section

When the temperature field of a structural member needs to be determined, the temperature value and its distribution on the section of the member can be found directly from these charts, if the section size of the member and the duration of the fire are identical to the conditions in the chart. Otherwise, the temperature field is determined by interpolation calculations after the temperature values with approximate conditions are found in the tables.

PART III

Mechanical Behavior of Members and Structures at Elevated Temperatures

CHAPTER 8

Behavior of Flexural Members at Elevated Temperatures

8.1 TESTING METHOD AND DEVICE FOR STRUCTURAL MEMBERS

8.1.1 General Testing Program

The basic members of a reinforced concrete structure include mainly flexural members, such as the beam and the slab, and axial and eccentric compressive members. The mechanical behavior at elevated temperatures has to be measured from experiments using special heating and loading devices. These devices cannot be bought directly and no typical product exists. Generally, these devices are designed and manufactured by researchers, according to the research objective and program.

The existing devices used for testing a structural member at elevated temperatures are divided into two categories. One is a large furnace for fire endurance experiments on the structural member, and is built individually for experiments on different members, e.g., beam, slab, column, or wall plate.[8-1,8-2] A full-scale specimen is put into the chamber of the furnace and loaded to a certain level, the gas or fuel is sprayed into the chamber and set alight, and fire occurs immediately; the temperature is controlled and follows the predetermined temperature–time curve (e.g., the ISO standard, Eqn (5.1)), and the test is continued until failure of the specimen. The fire endurance is measured in hours. The experimental conditions of a device in this category approach those of a real fire accident. However, the scale of the device is large, the experimental technique is complicated, the initial investment is expensive, and the continuing expenses for the experiment are high. Therefore, devices of this category are normally installed in special research organizations and are suitable for the fire endurance experiments on structural and architectural members in practical engineering.

The second category is a small-scale furnace for testing structural members at elevated temperatures and it is usually heated by electrical power. Although it is not heated following the standard temperature–time curve and the fire endurance of a structural member cannot be measured directly, it still has some advantages:

- The device is easily manufactured by researchers, the investment is less, and the effect is obtained quickly.
- The temperature is easily controlled with high accuracy.
- The experimental technique and measurements are easier to solve and more items may be measured during testing.
- The size of the specimen is small and the experimental expense is lower.
- Experiments on various heating–loading conditions can be conducted with high efficiency. Therefore, a device of this category is suitable for systematic experimental and theoretical research on the mechanical behavior of structural members at elevated temperatures.

An experimental device has been designed and manufactured by the Structural Engineering Laboratory of Tsinghua University for the testing of

Experiment and Calculation of Reinforced Concrete at Elevated Temperatures

structural members on a small scale at elevated temperatures. The general scheme is shown in Fig. 8-1, and it is composed of three systems as follows.

1. Heating and temperature control system
This system includes an experimental furnace for the structural member and a temperature control box. The construction and specifications of the furnace are given in Section 8.1.2. A specimen is put into the chamber of the experimental furnace with one surface open during testing, then the electrical power is switched on and the specimen is heated. One end of the armored nickel–chromium and nickel–silicon thermocouple (WRNK-541) is inserted into the chamber of the furnace and the other end is connected to the temperature control box to control heating and to maintain the temperature of the furnace.

2. Support and loading system for the specimen
The experimental furnace and the specimen are supported separately on the testing machine or the loading floor of the laboratory, and a specimen with a positive bending moment (i.e., tension zone of the section exposed to high temperature) is loaded by the hydraulic testing machine (capacity of 2000 kN) or a hydraulic jack. The experimental loads are two symmetrical concentrated forces, which act on top of the specimen through a small beam. So an experimental section of pure bending (without shear force) is formed in the middle of the beam.

The specimen with a negative bending moment (i.e., compression zone of the section exposed to high temperature) is tested in the opposite position, and the middle section is put on the two supports and both ends are cantilevered (Fig. 8-2). There is a pair of loading frames at each end of the specimen and a hydraulic jack used for loading is hung below the beam of the frame. The experimental furnace is hung on top of the loading frames with the help of two adjustable bolts, which are used to properly adjust the position of the furnace. The loading method described in Zhang[8-3] is similar, but the supports in Fig. 8-2 are replaced by two hydraulic jacks and both ends of the beam are supported on the rollers, which are located between the upper and lower beams of the loading frame.

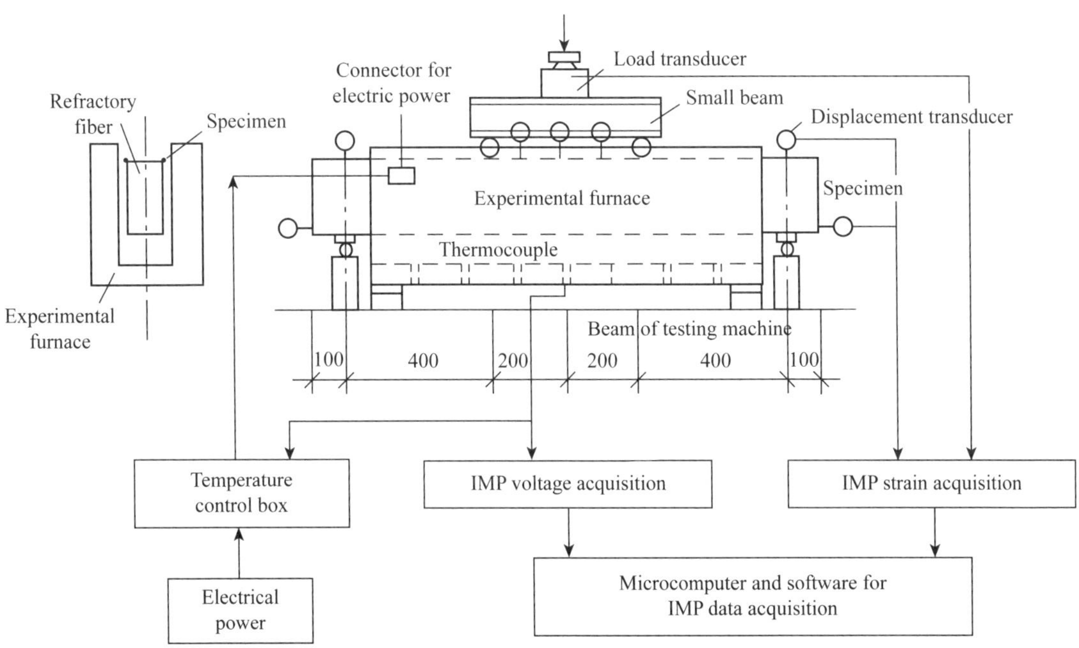

FIGURE 8-1 General scheme of testing devices for structural members at elevated temperatures.[1-12]

3. Data measurement and acquisition system
Three categories of experimental data need to be measured during the testing of structural members at elevated temperatures and different transducers are used to measure these data:

- The armored or ordinary (simplified) thermocouple is used to measure the temperatures in the furnace chamber and the interior of the specimen.
- A load transducer is used to measure the load value.
- Displacement transducers of different travel distance and accuracy can be selected and positioned properly to measure various deformations of the specimen, e.g., longitudinal and transverse deformation, deflection, curvature, rotating angle, even twisting angle. Various transducers are connected individually with the IMP voltage (35951A) or strain (35951B) data acquisition; the corresponding data and their variations (curves) are collected and displayed in real time on a microcomputer during the testing process.

The load and displacement transducers have to be placed some distance (>100 mm)[1-12,8-3] away from the surface of the specimen to avoid the influence of the high temperature of the specimen and steam evaporating from the water in the interior of concrete on the measuring accuracy of the transducer.

8.1.2 Design and Manufacture of a Furnace for Member Testing

When a building sustains a fire accident, generally both the sides and the bottom surface (three surfaces in total for a rectangular section) of a beam or column and all four surfaces of an arbitrary column are exposed to high temperature; only one surface of a floor slab or wall is exposed to high temperature. The experimental furnace is shaped like a channel (Fig. 8-3), which provides three sides for heating with one side open. The furnace is versatile and can be used not only for testing a structural member with three surfaces or one surface exposed to high temperature but also for members with four surfaces exposed to high temperature, when two furnaces are joined together. The furnace with one surface open is convenient for installing the specimen before testing, observing the response of the specimen during testing, and removing the specimen after testing.

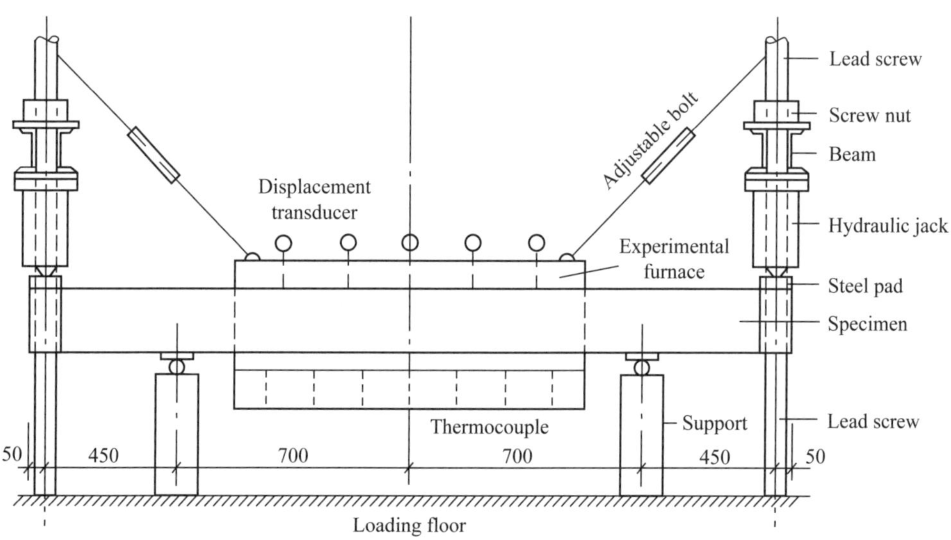

FIGURE 8-2 Method of loading a specimen with negative bending moment (compression zone exposed to high temperature).[9-12]

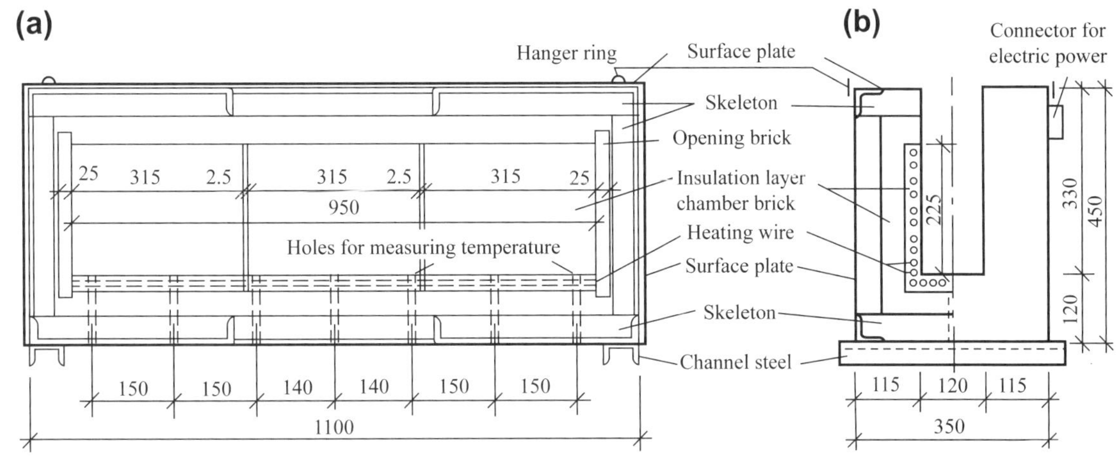

FIGURE 8-3 Construction of a furnace for testing members[1-12]: (a) longitudinal section; (b) transverse section.

The net width and height of the chamber in the experimental furnace are 120 mm and 330 mm, respectively, and the maximum section of the specimen that may be contained in it is 100 mm × 200 mm. The total length of the furnace is 1100 mm and the length of the heating section in the middle is 950 mm, which is four times greater than the depth of the specimen. Therefore, the experimental part of the specimen has the same temperature field on its section.

The experimental furnace is composed of the case structure, a heating element, an insulation layer, and accessories.

The case structure is welded and forms a skeleton using angle and belt steels. It is covered with a thin steel plate, therefore it is stiff and is used for supporting and fixing the furnace bricks and insulation material.

The chamber brick is one of the main parts of the heating element and is manufactured from refractory material. The original brick is a box section, and a channel section of net space 120 mm × 225 mm is formed after one side is cut off. Three pieces of bricks are connected one by one and a chamber 950 mm long is obtained. Both the sides and the bottom walls of the brick are 25 mm thick, and there are 26 longitudinal circle holes 12 mm in diameter in the walls. Five heating wires of 2 kW each pass reciprocally through these holes and the ends of the wires are linked with the connector outside the furnace case. One opening brick is put on each end of the chamber bricks for protection.

Siliceous aluminum refractory fiber is used as the thermal insulation layer and fills up the area between the chamber brick and the case structure.

In addition, two short steel channels are welded beneath both ends of the case and four rings are welded on top of the case for installing and hanging the case easily. The connector of the heating wires is located on the upper part on the side surface of the case, and electrical power is also connected to it. Seven holes for measuring temperature are located on the bottom of the case along its central line, and the thermocouples can be inserted into these holes.

A simple electrical circuit with relay is used in the temperature control box, and electrical power is connected to the experimental furnace through the control box (Fig. 8-1). When electrical power is switched on, the furnace is electrified and heated continuously if the feedback temperature signal of the thermocouple is lower than the predetermined value. The electricity supply stops immediately if the predetermined value is reached, or the furnace is electrified and heated again if the temperature in the chamber is lower than the predetermined value by the permissible tolerance.

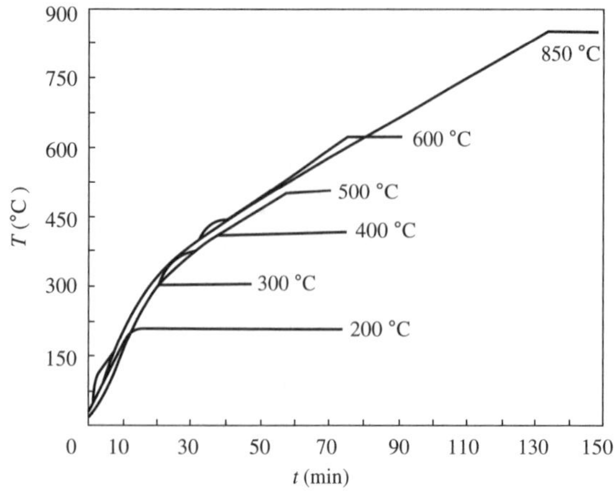

FIGURE 8-4 Various curves at elevating temperatures and constant temperatures for the experimental furnace.[1-12]

Although the experimental furnace and the temperature control box are manufactured separately, they are used together to measure the curves of elevation and for keeping the temperature at every thermocouple in the furnace constant. It has been demonstrated that the temperature is uniformly distributed within the middle section of the furnace, and the temperature–time curves measured from various experiments are stable and reproduce well (Fig. 8-4).

The specifications of the thermal behavior of the complete experimental device is as follows: the maximum cross section of the specimen is 100 mm × 200 mm, the length of the experimental section with high temperature is 900 mm, the maximum temperature may reach 1000 °C, the maximum heating velocity is 20 °C/mm, the control accuracy of constant temperature is within ±20 °C, and the temperature on the outer surface of the furnace case is less than 60 °C after heating for 5 h.

8.2 MECHANICAL BEHAVIOR AT ELEVATED TEMPERATURE

8.2.1 Testing Method and Specimens

When a reinforced concrete continuous beam or frame beam is being considered, the positive bending moment occurs in the middle part of the span and the negative bending moment occurs near the support. If three surfaces of the beam are exposed to a high temperature during a fire accident, the tension zone of the section within the middle part of the span and the compression zone of the section near the support are exposed to high temperature. Even if the section and the reinforcement along the beam are the same, the thermal behavior of these two sections differ considerably.

Beams with the tension and the compression zones exposed to high temperatures have different experimental and loading methods (Figs. 8-1 and 8-2). The specimens for both conditions have to be designed and manufactured separately, and the sizes and construction are shown in Fig. 8-5. The strengths of concrete and longitudinal reinforcements of the specimen at room temperature are listed in Table 8-1.

The mechanical behavior of a beam specimen at different temperatures is measured under the path of loading under constant temperature, and the method and procedure are as follows: the specimen is manufactured and cured for 28 days, kept in the laboratory and tested after aging for 60 days, the specimen is installed, and the longitudinal, transverse, and vertical positions of the experimental furnace are adjusted to place the experimental part of the specimen in the middle of the furnace. Various transducers are set up and connected to the measuring instrument, the experimental furnace is electrified and the specimen is heated; the supports of the specimen permit free expansion deformation. When the temperature in the chamber reaches the predetermined temperature (20–950 °C) and is held for 10 min, the specimen is loaded continuously until failure, i.e., loss of bearing capacity. In the meantime, the data on the temperature in the chamber, and the temperature, load, and deformation of the specimen are measured and recorded.

The thermocouples are fixed into the chamber of the furnace and the interior of the specimen, and can measure the temperatures and their variation at the corresponding places during the heating and loading process. Using these data, the vertical and transverse temperature distributions

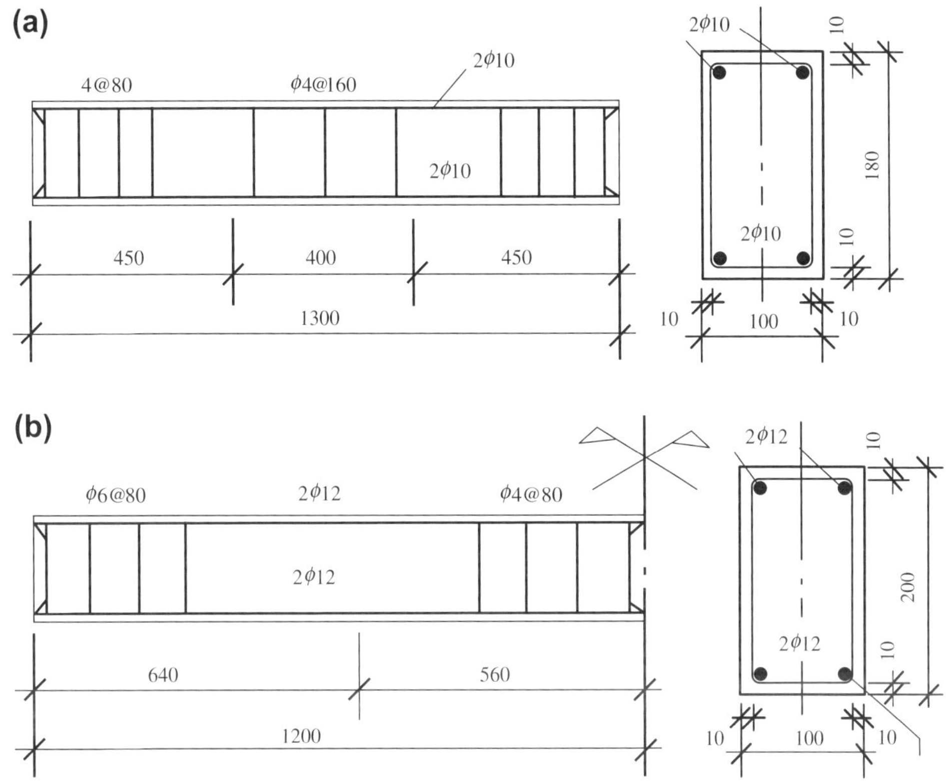

FIGURE 8-5 Specimens for measuring the behavior of beams at elevated temperatures: (a) tension zone exposed to a high temperature[8-4]; (b) compression zone exposed to a high temperature.[8-3]

TABLE 8-1 Strength of Specimens at Room Temperature

		Longitudinal reinforcement	
Specimen	**Concrete f_{cu} (MPa)**	**Diameter (mm)**	**f_y (MPa)**
Tension zone exposed to a high temperature	29.45	10	270
Compression zone exposed to a high temperature	39.20	12	310

on the middle section of the specimen at different times can be obtained (Fig. 8-6).

The maximum temperature occurs on the bottom and both side surfaces for a beam specimen with three surfaces exposed to high temperature, but it is slightly lower than that of the chamber, and the temperature on the side surface decreases gradually from bottom to top. Therefore, the temperature on the section is not uniformly distributed along both vertical and transverse directions. The temperature gradient within the range 20–30 mm of the outer layer of the section is high, but the temperature variation in the interior of the section is small. During the initial stages of heating, the temperature is not very high (e.g., <400 °C), but the temperature gradient in the outer layer is much greater, although it decreases gradually later. The top surface of the specimen does not

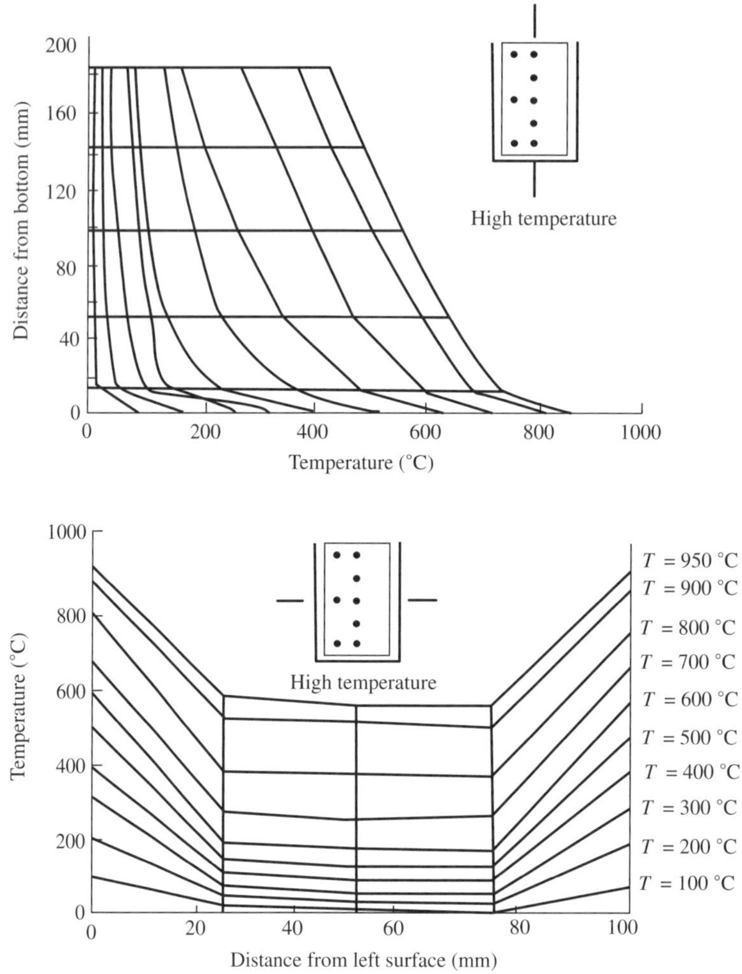

FIGURE 8-6 Temperature distribution on a section of the specimen.[8-5]

come in direct contact with the high temperature and has the lowest temperature. However, as the heating time continues and the temperature in the interior of the specimen increases, the temperature on the top surface also elevates gradually, because the heat being transferred to the top surface exceeds that transferred to the air. When the temperature (T) in the chamber of the experimental furnace is <600 °C, the temperature on the top surface is <80 °C. When the experimental temperature reaches 950 °C, the temperature on the top surface may reach 200–300 °C.

Under the same heating conditions, beams with tension and the compression zones exposed to high temperature have the same temperature field on their sections, but the mechanical behavior differs considerably. The existing experimental investigations[8-6–8-10] were conducted for the beams with compression zone exposed to high temperatures.

8.2.2 Beam with Tension Zone Exposed to High Temperature

Various macroscopic physical phenomena appear successively on the reinforced concrete beam when the temperature in the chamber of the furnace increases during the heating process. When the temperature in the chamber is T = 300 °C, water vapor escapes visibly from the gap between the specimen and the inner wall of the furnace. When $T = 400–500$ °C, the quantity of escaping vapor reaches a maximum and then decreases gradually. When $T = 700$ °C, no vapor is visible. The specimen is weighed after the test is finished and the weight loss is 3.1–6.7%.

The color of the surface of a specimen exposed to high temperatures varies as the temperature in the chamber elevates, and is similar to that of the cubic concrete specimen (Table 1-2), i.e., it is gray at room temperature, and turns gray-white, dark red, and red successively. After the test is finished and the specimen cools down, various damage phenomena, e.g., cracks, loosening, surface loss, and corners and edges spalling off and broken, are obvious on the surface of the concrete, and the level of this damage depends on the maximum temperature the specimen has reached. In addition, the changes to the surface and diameter of the reinforcement can be visible from the wide crack near the bottom of the beam.

A beam with the tension zone exposed to high temperatures experiences two testing stages: freely heating and loading under constant temperature. The measured bending moment–curvature curve of the section within the pure bending part in the middle of the beam is shown in Fig. 8-7. The heating process before loading the specimen causes curvature of the section and the deformation of the beam is convex toward the direction of high temperature, which is taken as a positive value, because of the nonuniformly distributed temperature and different thermal strain on the section. This is an initial curvature and its value is rather small when $T \leq 300$ °C, but increases quickly as the temperature elevates ($T \geq 400$ °C).

When the beam is loaded under constant temperature, the curvature caused by the bending moment and the initial curvature caused by the temperature are in the same direction (positive value), so the deformation of the beam increases continuously and is convex toward the direction of high temperature. Before yielding tensile reinforcement ($< f_y^T$), the curvature of the beam accelerates as the bending moment increases, and the slope of the bending moment–curvature curve and the stiffness of the beam decrease gradually. The higher the experimental temperature of the beam, the flatter the bending moment–curvature curve and the less stiff it is. When the tensile reinforcement in the beam is yielding (f_y^T), the bending moment–curvature curve turns obviously. The curvature then increases sharply as the bending moment increases slightly, and the bending process and the sudden failing of the beam are apparent to the naked eye. The higher the experimental

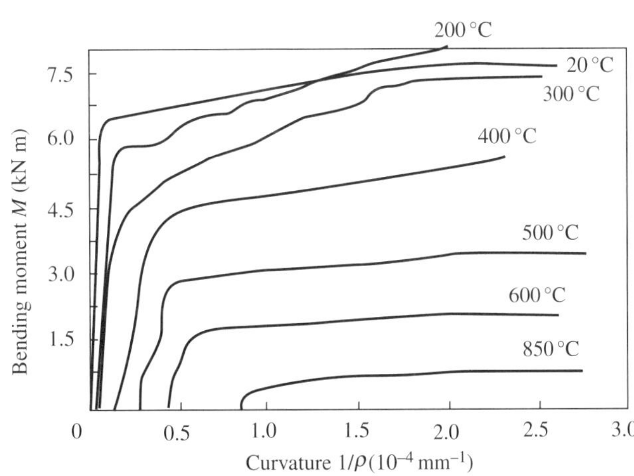

FIGURE 8-7 Bending moment–curvature relationship for a beam with the tension zone exposed to a high temperature.[8-5]

temperature of the beam, the smaller the yielding moment (M_y^T) and the shorter the failing process.

The beam is unloaded and cooled down after the test is completed, and it shows a noticeable flexible failure pattern. The experimental part in the middle of the beam is obviously bent and the residual deflection is far greater than that of the beam at room temperature. The tensile crack on the side surface extends up near the top from the bottom, and the number of cracks is less but the width of the cracks is greater. The height and the length of the compressive failure zone of concrete are small. The higher the experimental temperature of the beam, the wider and longer the tensile crack and the smaller the height of the compressive failure zone.

The ultimate bending moment of the beam with the tension zone exposed to high temperature (M_u^T) is calculated by the maximum load measured in the test. The ratio between it and the ultimate bending moment of the beam at room temperature (M_u) varies with the experimental temperature (in the chamber of the furnace) and is shown in Fig. 8-8. When the temperature is $T \leq 300$ °C, the ultimate bending moment of the beam is slightly lower than that at room temperature, but it is a little higher at $T = 200$ °C. The ultimate bending moment decreases sharply when $T \geq 400$ °C, and reaches only about 5% of that at room temperature when $T = 850$ °C. The variation regularity of the ultimate bending moment of the beam coincides with that of the yield strength of the reinforcement at elevated temperatures (f_y^T, Fig. 4-5).

The variation in the ultimate bending moment of the beam with the tension zone exposed to high temperature with the experimental temperature (T, °C) can be simulated by the following formula:

$$\frac{M_u^T}{M_u} = \frac{1}{1 + 55\,(T/1000)^{5.6}} \tag{8.1}$$

However, the formula can only be used to represent the variation regularity of the beam described above. When the beam is under other conditions, such as different shape and size of the section, or type and quantity of reinforcement, or temperature–time curve, the values of the parameters in the equation have to be calibrated.

The temperature in the beam specimen with three surfaces exposed to high temperature increases gradually during the heating stage, and increases continuously but slightly during the loading stage under constant chamber

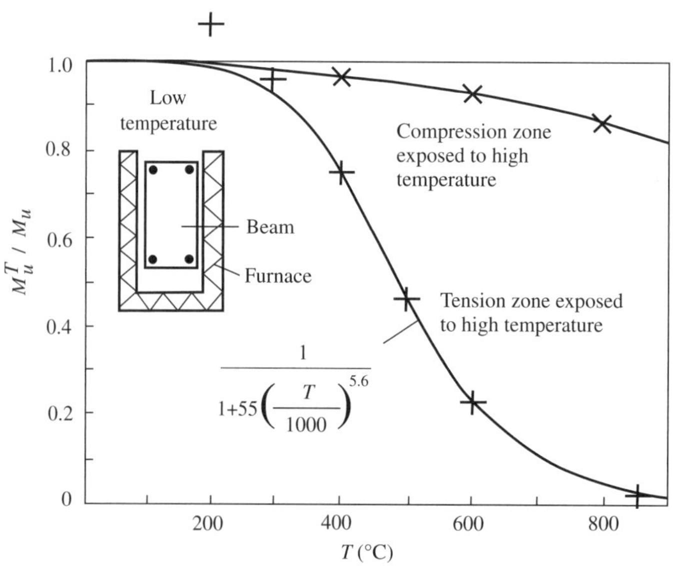

FIGURE 8-8 Ultimate bending capacity of a beam with three surfaces exposed to a high temperature.[8-5]

temperature. However, the temperature always distributes nonuniformly on its section during testing. The tensile reinforcement is located in the high temperature zone of the beam, and the temperature value is taken as that of the concrete adjacent to it, which can be obtained from the measured temperature distribution on the section (Fig. 8-6). The yield strength of the reinforcement decreases ($f_y^T < f_y$) before loading due to the high temperature. Because the tensile strength of concrete at elevated temperatures (f_t^T) is lower, the tensile force on the section of the beam after loading is carried mainly by the reinforcement; the stress of the reinforcement (σ) increases with the load or the bending moment. When the stress reaches the yield strength at elevated temperatures ($\sigma = f_y^T$), the yield bending moment (M_y^T) of the beam is reached and the beam deflects sharply, and soon afterward the ultimate bending moment of the beam (M_u^T) is reached. Therefore, the ultimate bending moment of the beam depends mainly on the yield strength of the tensile reinforcement at elevated temperatures (f_y^T).

At the ultimate stage of the beam at elevated temperatures, the tensile force of the reinforcement (A_s, f_y^T) is small and the height of the compressive concrete zone, which is needed to maintain the equilibrium conditions of the forces, is rather small, because the compressive zone is located on the low temperature side and the strength of concrete (f_c^T) loses less. However, the strain of the reinforcement at elevated temperatures is great (Fig. 4-9), so the curvature at elevated temperatures and the residual deformation after unloading and cooling of the beam are far greater than that of the corresponding beam at room temperature.

When a reinforced concrete beam fails due to a diagonal shear crack occurring near its end portion, the shear-bearing capacity (V_u) depends mainly on the shear span ratio of the load, the strength of the concrete ($f_c f_t$), and the contents and strength of the stirrup and longitudinal reinforcement of the beam.[0-2] The shear-bearing capacity of a beam (V_u^T) decreases under the action of high temperature. This has been investigated experimentally by Shi.[1-12] Less stirrup (Φ4@160 mm) is set up in the shear span of the specimen and the diagonal failure pattern occurs after loading at room temperature, but the failure pattern will change when the specimen is at elevated temperatures as follows: bending failure in the mid-span and shear failure on the end of the specimen occur together when $T = 200$ °C, bending failure occurs in the mid-span only when $T = 400$ °C, diagonal fine cracks appear on the end part but without failure signs, and pure bending failure occurs but no diagonal cracks appear on the end part when $T \geq 600$ °C.

The shear-bearing capacity of a beam depends mainly on the strength of the concrete located on the top part and the mid-height of the section, and the strength of the concrete there loses less because the temperature there is lower. Therefore, the decreasing amplitude of the shear-bearing capacity of a beam at elevated temperatures is smaller than that of the bending-bearing capacity. This causes a change in the failure pattern of the beam at elevated temperatures. Generally, the relative shear-bearing capacity of a beam is greater than the relative bending-bearing capacity ($V_u^T / V_u \geq M_u^T / M_u$), so its failure pattern is controlled by the latter.

8.2.3 Beam with the Compression Zone Exposed to High Temperature

The specimen and the loading form in a test of a beam with the compression zone exposed to high temperatures are different from that with the tension zone exposed to high temperatures, but the heating–loading process and the measuring method of both tests are the same. In addition, the physical phenomena occurring during the heating of the specimen, e.g., escaping water vapor, changing color of the concrete surface, and development of damage, are similar for both tests.[8-3]

A beam with the compression zone exposed to high temperatures also experiences two experimental stages: freely heating and loading under constant temperature. The measured bending moment–curvature relationship of the section within the mid-span of the beam is shown in Fig. 8-9.

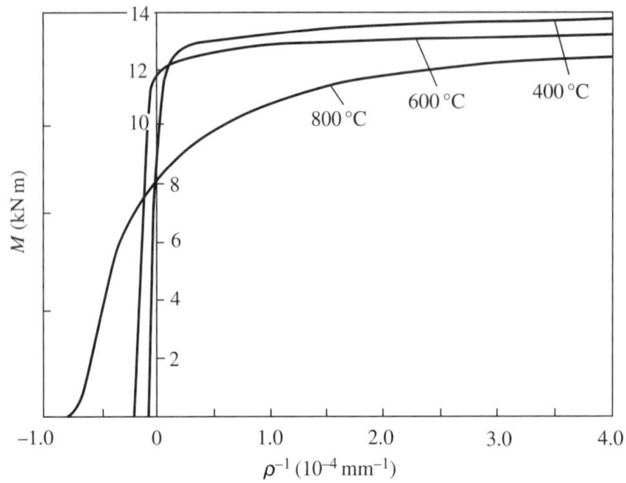

FIGURE 8-9 Bending moment–curvature relationship for a beam with the compression zone exposed to a high temperature.[8-3]

When the specimen is heating, the bending deformation is convex toward the high temperature zone (i.e., compression zone) because the thermal strain there is greater. The curvature of the section increases as the temperature elevates and the value is identical to that of the specimen with the tension zone exposed to high temperature, but the sign is taken as a negative one. When the specimen is loaded under constant temperature, the curvature caused by the action of the bending moment is convex toward the tension (i.e., lower temperature) zone (positive sign) and is opposite to the initial curvature during heating.

When the experimental temperature $T \leq 600$ °C, the temperature of the concrete in the interior of the specimen section is not yet high and the elastic modulus of the concrete reduces slightly; the curvature of the specimen increases almost linearly with the bending moment. When the tensile reinforcement, which is located in the low temperature zone, reaches its yield strength ($\sigma = f_y^T \approx f_y$), the curvature of the specimen increases suddenly, the bending moment curvature turns obviously, the deflection of the specimen increases quickly, and then the specimen fails. When the experimental temperature $T \geq 800$ °C, the concrete on the outer layer of the specimen is seriously damaged, the temperature of the concrete in the interior is high, and the elastic modulus (E_c^T) is lower. When the bending moment is relatively small ($M/M_u^T \leq 0.5$), the curvature of the specimen increases almost linearly. However, the strain of concrete at elevated temperatures increases considerably and the curvature accelerates under the action of a larger bending moment, so the slope of the bending moment–curvature curve decreases gradually and an arc is formed. After yielding the tensile reinforcement, the specimen deforms continuously and the tensile crack in the concrete expands and extends upward, the compression zone reduces gradually, and the ultimate bending moment (M_u^T) is reached when the concrete in the compression zone fails. However, there is no obvious turning (yielding) point on the bending moment–curvature curve. The ultimate curvature of the specimen is rather large, because the peak strain (ε_p^T) of the concrete at elevated temperatures is large.

The deformation of the beam with the compression zone exposed to high temperature is convex toward the compression zone (negative sign) during the initial heating stage, and it becomes convex toward the tension zone of low temperature (positive sign) at failure. Therefore, the bending moment–curvature curve has to pass a point with zero curvature, and the value of bending moment at this point varies considerably and depends on the experimental temperature (Fig. 8-9). However, the deformation at failure and the residual deformation after the unloading and cooling of the specimen are both convex toward the tension zone of low temperature, and the values of the deformations are noticeable and may reach 100 mm.

The final failure pattern of the specimen with the compression zone exposed to high temperature is also a flexural one. Compared with the specimen with the tension zone exposed to high temperature, more tensile cracks appear and distribute uniformly on the surface of the tension zone of low temperature, and larger height and longitudinal length are shown on the failure part of the compression zone for the specimen with the compression zone exposed to high temperature. The higher the experimental temperature of the specimen, the larger the area of failure on the compression zone.

The ratio between the ultimate bending moment of the specimen with the compression zone exposed to high temperature and that at room temperature (M_u^T / M_u) varies with the experimental temperature and is shown in Fig. 8-8. When the temperatures are $T \leq 600$ °C and $T = 800$ °C, the ratios are $M_u^T / M_u > 0.9$ and $M_u^T/M_u = 0.88$, respectively. The reduction in the ultimate bending moment of the specimen at elevated temperatures is limited. The relative ultimate bending moment (or M_u^T/M_u) of the specimen with the compression zone exposed to high temperature (when $T > 400$ °C) is several times higher than that with the tension zone exposed to high temperature and is also far higher than the relative strengths of the reinforcement and concrete materials at the same temperature ($f_y^T/f_y, f_c^T / f_c$, see earlier chapters).

The flexural failure pattern of a reinforced concrete beam is controlled by yielding of the tensile reinforcement, and its bearing capacity at elevated temperatures (M_u^T) depends on the tensile force ($f_y^T A_s$) of the reinforcement and the length of the internal force-arm of the section at the failure state. The bearing capacity of the beam with the tension zone exposed to high temperature reduces seriously as the experimental temperature increases; the main reason for this is that the yield strength of the tensile reinforcement at elevated temperatures (f_y^T) reduces sharply. However, in a beam with the compression zone exposed to high temperature, its tensile reinforcement is located on the low temperature zone, and it loses less yield strength ($f_y^T \approx f_y$). As the strength of the compression zone of the concrete at elevated temperatures decreases seriously ($f_c^T = f_c$), the height of the compression zone has to increase greatly to satisfy the equilibrium condition, and the length of the internal force-arm reduces correspondingly. Therefore, the decreasing amplitude of the ultimate bending moment (M_u^T) is limited. It is predicted that the ultimate bending moment will not reduce much if the beam is at an even higher temperature ($T > 800$ °C).

8.3 MECHANICAL BEHAVIOR UNDER DIFFERENT HEATING–LOADING CONDITIONS

When a real building sustains a fire accident or other source of high temperature, the reinforced concrete beams and slabs in it may experience a complicated variable heating–loading path (similar to Fig. 3-1). The behavior of the concrete beam at elevated temperatures introduced above occurs only when the beam experiences an extreme path, i.e., loading under constant temperature. The behavior of the beam in different thermal conditions or other heating–loading paths has to be investigated specially. The behavior of a beam under another extreme path, i.e., heating under constant load, and after the heating–cooling cycle is introduced in the following section.

8.3.1 Path of Heating Under Constant Load

The specimen, experimental devices, and measuring method for testing a concrete beam with three surfaces exposed to high temperature under the path of heating under constant load[8-11,8-12] are the same as those used for the specimen under the path of loading under constant temperature. The specimen is installed and its tension zone is exposed to a high temperature (Fig. 8-1). It is loaded at room temperature and the level of the load or bending moment is $P_0/P_u = M_u^T/M_u = 0\text{–}0.72$ where P_u and M_u are the ultimate load and bending moment, respectively, of the specimen at room temperature. Then, the load is maintained constantly and the experimental furnace is electrified continuously. The temperature of the specimen is elevated and the behavior of the specimen deteriorates gradually. Finally, the specimen fails because its deformation develops out of control. Various data, including the maximum (ultimate) temperature (T_u, °C) at the failure of the specimen, are measured and recorded during the testing process.

The macroscopic physical phenomena of the specimen occurring in the heating process under constant load, such as escape of internal water vapor, weight loss, change in the color of the concrete on the surface, damage developing on the surface, and their variations with the experimental temperature, are similar to that of the specimen under the path of loading under constant temperature. No obvious difference is observed even for specimens with different initial levels of bending moment (M_u^T/M_u).

The beam specimen experiences two testing stages, i.e., loading at room temperature and heating under constant load. The curvature of the beam caused in the loading stage at room temperature is negligible, compared with the thermal effect. The curvature of the section within the pure bending part in the mid-span of the specimen varies within the stage of heating under constant load and is shown in Fig. 8-10.

At the beginning of heating the specimen and when the temperature $T \le 300$ °C, the curvature of the specimen increases slowly and the differences between various specimens at different loading levels are not noticeable, because the thermal strain of concrete (ε_{th}) is not great and is confined by the linear deformation on the section. When the temperature $T > 300$ °C, the curvature of the specimen accelerates and the temperature–curvature curves of various specimens are deviated because of the different levels of the initial bending moment (M_u^T/M_u). When $P_0 = 0$ (i.e., $M_u^T/M_u = 0$) the specimen carries only its own weight and its curvature increases quickly with the temperature, as the thermal expansion strain of concrete accelerates. When the temperature elevates and reaches T = 600–800 °C, the curvature of the specimen does not increase rapidly again, because the thermal strain of concrete ceases (Fig. 2-2) and the temperature gradient on the section decreases (Fig. 8-6). For a specimen

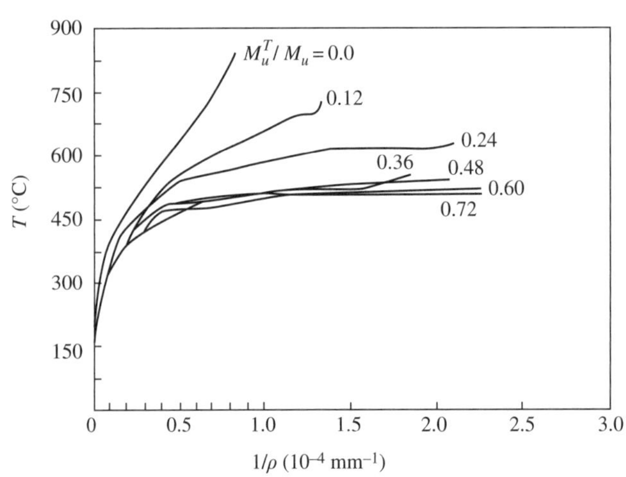

FIGURE 8-10 Temperature–curvature relationship for a specimen heated under constant load.[8-11]

carrying a load ($P_0 \neq 0$), the curvature increases very quickly with the experimental temperature, because not only the thermal expansion strains of concrete and reinforcement increase quickly but also the mechanical behavior of the materials deteriorates, i.e., the elastic modulus reduces, short time creep at elevated temperatures appears, and larger additional deformations are caused. The higher the level of the initial bending moment of the specimen, the larger the increasing rate of the curvature, and the lower the temperature value the ultimate state reaches when the deformation of the specimen is out of control.

The specimen is unloaded and cooled down after the testing is completed and the flexural failure pattern is observed, i.e., the tensile reinforcement yielded in advance and the concrete in the compression zone is crushed. The mid-span part under the high temperature of the specimen is obviously bent up and the residual deflection is great. The cracks in the tension zone extend up and near the top of the specimen, and their number is less but their width is greater. The height and length of the failure part of the concrete in the compression zone are both small. These signs of failure are similar to those of a specimen under the path of loading under constant temperature.

When the specimen under the testing path of heating under constant load is loaded initially to load (P_0), the stress state on the section is established correspondingly. Although the bending moment of the section of the specimen maintains a constant value during the heating process (P_0 = constant), the tensile stress (σ_s) of the longitudinal reinforcement increases slightly because the stress on the section is redistributed due to the thermal strain, concrete in the tension zone is out of work gradually, and the neutral axis on the section moves slightly. In the meantime, the temperature of the reinforcement increases continuously and its yield strength (f_y^T) reduces gradually. When the yield strength of the reinforcement at elevated temperatures is reduced to the stress value on the section at that time, the crack on the critical section of the specimen develops quickly, the deflection (curvature) increases sharply, and the specimen fails soon after. Therefore, the ultimate bending moment (M_u^T) of the specimen under the path of heating under constant load is also the initial bending moment, which corresponds to the initial load (P_0).

The experimental results for the ultimate temperature and the bending moment of specimens under the path of heating under constant load are plotted in Fig. 8-11. The ultimate temperatures of the specimens with the higher level of initial load (bending moment) ($P_0/P_u = M_u^T/M_u \geq 0.4$) are less different (Fig. 8-10), i.e., T = 555–513 °C. However, the lower the level of the initial load ($P_0/P_u \leq 0.36$), i.e., the lower the stress of the tensile reinforcement, the higher the temperature endured by the specimen.

The comparison of the bearing capacities between the beams under the path of heating under constant load and under the path of loading under constant temperature shows (Fig. 8-11) that the former is always higher than the latter. If the initial bending moment of the specimen under the path of heating under constant load is at a higher level ($M_u^T/M_u > 0.4$), and even the stress of tensile reinforcement of the specimen is at a higher level, the specimen fails only when the experimental temperature in the chamber of the furnace exceeds 500 °C and the temperature of the tensile reinforcement in the specimen is higher than 300–400 °C, at which the yield strength begins to reduce. Therefore, the ultimate temperature of the specimen is certainly higher than 500 °C. However, when the experimental temperature T = 300–500 °C, the ultimate bending moment of the specimen under the path of loading under constant temperature has already reduced $M_u^T < M_u$. If the initial bending moment of the specimen is at a lower level $M_u^T/M_u < 0.4$ or the experimental temperature is higher (T > 550 °C), the ultimate bending moments and temperatures of the specimens under both heating–loading paths are approaching each other. However, the specimen under the path of heating under constant load has slightly higher values of ultimate bending moment and temperature, because the strength of concrete

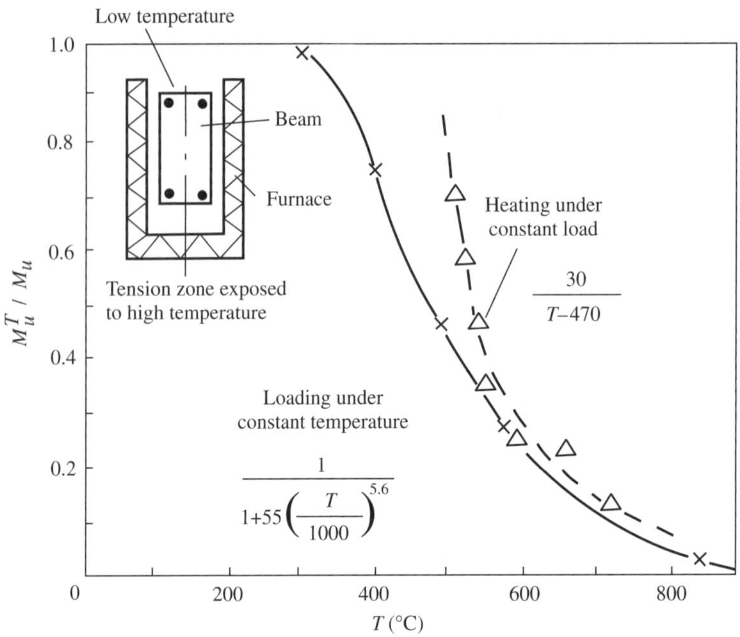

FIGURE 8-11 Comparison of the ultimate temperature–bending moment for beams with different heating–loading paths.[1-12]

under this path is higher (Fig. 3-5) and the tensile reinforcement sustains a shorter time at elevated temperatures.

The ultimate temperature (T_u, °C) of a beam with the tension zone exposed to high temperature and under the path of heating under constant load can be simulated by the formula:

$$T_u = 470 + \frac{30}{M_u^T / M_u} \tag{8.2a}$$

or

$$\frac{M_u^T}{M_u} = \frac{30}{T - 470} \quad (T > 500\ °\mathrm{C}) \tag{8.2b}$$

Similarly, the values of the parameters in these formulas have to be calibrated for beams of different materials and sections, or different temperature–time curves of heating.

8.3.2 Influence of the Cover Thickness of Reinforcement

The ultimate temperature and bearing capacity of a beam with the tension zone exposed to high temperature depends mainly on the yield strength of the tensile reinforcement at elevated temperatures. Because the temperature is distributed nonuniformly on the section of the beam and the temperature gradient of the outer layer is high (see Fig. 8-6), the concrete cover thickness of the reinforcement determines its temperature and has a great influence on the behavior of the beam at elevated temperatures.

Beams with different thicknesses of concrete cover are tested under the path of heating under constant load for comparison and three surfaces are exposed to high temperature. All the specimens are identical in length, width, effective depth of the section, and the materials and reinforcement construction. However, the thicknesses of the concrete covers below and beside the reinforcement and the total depth of the section are different for various specimens (Table 8-2). These specimens are loaded at room temperature in advance and the level of the initial bending moment is the same, i.e., $M_u^T/M_u = P_0/P_u = 0.411$. Then, the specimens are heated until the ultimate state is reached, when the development of deformation gets out of control.

TABLE 8-2 Thickness of the Concrete Cover and Ultimate Temperature of Specimens[8-13]

	Specimen 1-1	Specimen 1-2	Specimen 2-2	Specimen 3-2
Total depth of section (mm)	180	180	190	200
Thickness of bottom cover (mm)	10	10	20	30
Thickness of side cover (mm)	10	20	20	20
Ultimate temperature (°C)	540	544	644	663

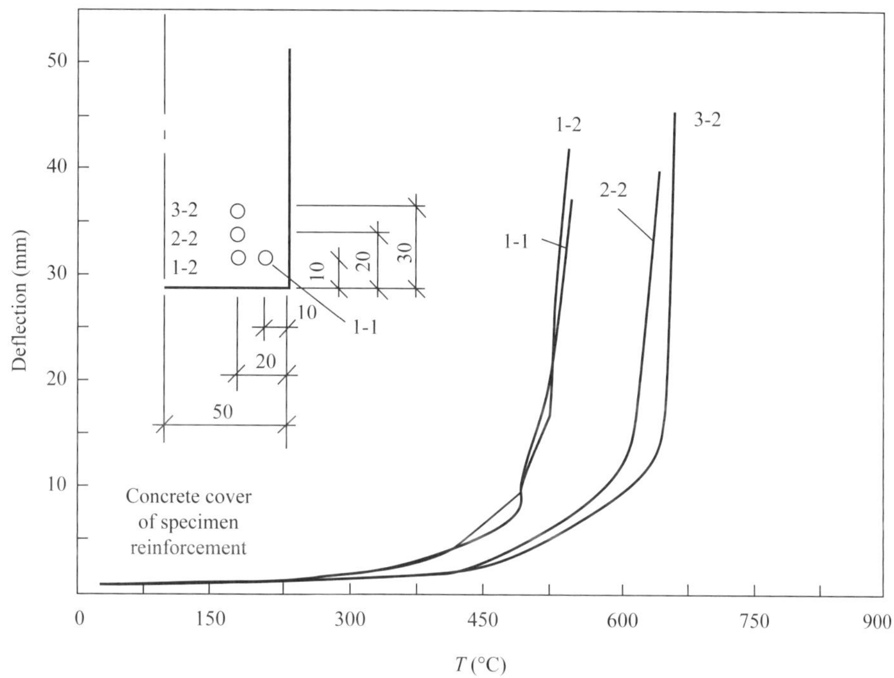

FIGURE 8-12 Deflection–temperature curves of specimens with concrete covers of different thickness.[8-13]

These specimens present the same failure pattern, i.e., the specimen fails when the tensile reinforcement is yielded at elevated temperatures, the crack extends up and near the top, and the concrete area of the compression zone is reduced and crushed. The specimen, after being unloaded and cooled, shows obvious residual flexural deformation and the small number of tensile cracks expand widely.

The deflection at the mid-span of the specimen varies during testing under the path of heating under constant load and is shown in Fig. 8-12. Although specimens 1-1 and 1-2 have the same thickness of bottom cover (10 mm) but different thickness of side cover (i.e., 10 mm and 20 mm, respectively), the measured deflection–temperature curves and the values of the ultimate temperature of both specimens are approaching each other. Because the level of the initial bending moment of the specimen is quite high ($0.411M_u$) and exceeds that of the cracking bending moment (e.g., $(0.2–0.3)M_u$ normally) at room temperature, the cracks existed on the tension zone of the specimen before heating. The thickness of the bottom cover of the reinforcement is small and has greater influence on the temperature value of the reinforcement. If only the thickness of the side cover of the reinforcement is increased, its temperature cannot be reduced effectively and the behavior of the beam at elevated temperatures cannot be improved effectively.

The thicknesses of both the bottom and side covers of the reinforcement in specimen 2-2 are increased to 20 mm; they are 10 mm for specimen 1-1. The deflection at the mid-span of specimen 1-1 accelerates when the temperature $T \geq 300$ °C and reaches one-fiftieth of the span length at about $T = 520$ °C. The specimen fails at $T = 540$ °C. Correspondingly, the temperatures are $T \geq 400$ °C, ~630 °C, and 644 °C, respectively, for specimen 2-2 when the above phenomena appear successively. Obviously, the behavior of specimen 2-2 at elevated temperatures is improved.

Specimen 3-2 has an even thicker concrete cover of reinforcement, and the bottom and side covers are 30 mm and 20 mm, respectively. Thermal deformation develops more slowly, the ultimate temperature is slightly higher, and the behavior of resisting the high temperature improves less but not obviously for specimen 3-2, compared with that of specimen 2-2. The main reason for this is that the width of the tensile crack is approximately proportional to the thickness of the concrete cover of the reinforcement,[0-2] and the temperature of the reinforcement on the cracking section does not decrease effectively if the thickness of the concrete cover is increased too much.

Some other experiments[2-6] show that concrete may spall off under high temperature. If the concrete cover is too thick, the concrete cover of the reinforcement in the specimen falls and the reinforcement is exposed directly to the high temperature. Therefore, this causes major deterioration in the behavior of resisting high temperature of the structural member. Therefore, the concrete cover has to be thickened properly in order to improve the thermal behavior of the structural member. A thicker concrete cover (e.g., 40–50 mm) is effective only when the protecting wire is put into it.[0-7,2-6]

8.3.3 After a Heating and Cooling Cycle

A structure sustaining a fire accident experiences at least one cycle of heating and cooling. The mechanical behavior of the structure at elevated temperatures determines the safety during the fire accident, and the residual behavior of the structure after it is cooled is the main basis to evaluate whether the structure can be used again and to work out the strengthening measures. The temperature, behavior of the materials, deformation, and bearing capacity of the structure under these working conditions are different than at room temperature, and are also different from each other. Therefore, these conditions should not be confused.

The method and device used to test a concrete beam under the two working conditions,[8-14] i.e., at elevated temperatures and after it is cooled, are basically the same (see Figs. 8-1 and 8-2). The length and cross section of the specimens are 2400 mm and 100 mm × 200 mm, respectively, the compressive strength of the concrete at room temperature is $f_e = 33.0$ MPa, the tensile and compressive reinforcements are each 2Φ12 mm, and the yield strength is $f_y = 234$ MPa. Four specimens are tested and three surfaces are exposed to high temperature. Two specimens (HT and HC) are heated to $T = 800$ °C and the temperature is kept constant, then positive and negative bending moments are applied, respectively, on the two specimens until their failure, corresponding to the conditions of the tension zone and the compression zone exposed to high temperature, respectively. Two other specimens (LT and LC) are also heated to $T = 800$ °C, but the temperature is maintained for 10 min only, then the specimens are taken out of the furnace and cooled down naturally to room temperature. After 16–20 h, positive and negative bending moments are applied on the two specimens until their failure. The numbers and the measured values of the yield and ultimate bending moments of the four specimens are listed in Table 8-3.

1. Beam with the tension zone exposed to high temperature

Specimen HT loaded at elevated temperatures experiences two stages, i.e., freely heating and loading at elevated temperatures; specimen LT loaded after it is cooled experiences three stages, i.e., freely heating, naturally cooling, and loading

TABLE 8-3 Comparison of the Characteristic Values of the Bending Moments of Specimens at Elevated Temperatures and After Cooling[8-14]

	Zone exposed to high temperature				Theoretical value of specimen at room temperature
	Tension zone		Compression zone		
	At elevated temperature	After cooling	At elevated temperature	After cooling	
	Specimen HT	Specimen LT	Specimen HC	Specimen LC	
Yield bending moment (kN m)	1.512	7.232	7.295	7.902	9.418
Ultimate bending moment (kN m)	3.826	8.514	7.832	8.583	9.835

at room temperature. The initial deformations (curvatures) before loading the specimens under both working conditions are approaching each other (Fig. 8-13); this shows that the recoverable deformation of the specimen during the cooling process is small.

When specimen HT is loaded at elevated temperatures, the deformation caused by the bending moment has the same direction (positive value) as the thermal deformation, the curvature accelerates, and the slope of the bending moment–curvature curve (also, the stiffness of the specimen) reduces gradually as the bending moment increases. When the crack in the tension zone widens, the tensile reinforcement is exposed locally to high temperature (800 °C) and its yield strength reduces seriously, and the value of the yield bending moment is low. The stress of the tensile reinforcement then increases continuously (Figs. 4-9 and 4-10), and the internal force-arm of the section also increases, the tensile crack expands, and the neutral axis moves up, so the ultimate bending moment of the specimen increases considerably and the bending moment–curvature curve varies correspondingly.

Specimen LT is loaded after it is cooled, the concrete is damaged during both the heating and the cooling processes, and internal cracks exist and are already widespread.[1-14,8-14] When the specimen is loaded at room temperature, from the beginning until the yielding of the reinforcement, the stiffness of the cross section is nearly

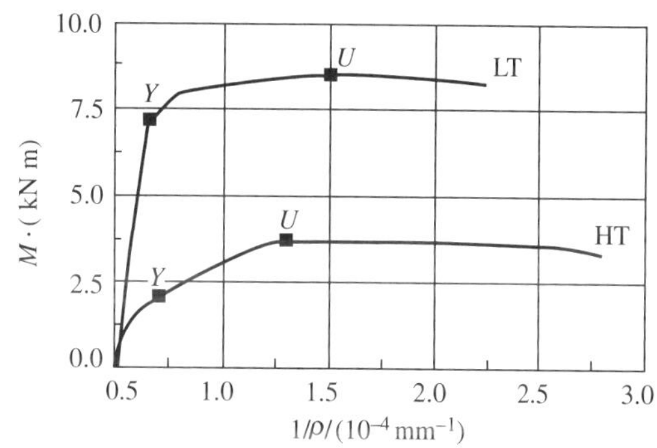

FIGURE 8-13 Bending moment–curvature relationship for beams with the tension zone exposed to a high temperature.[8-14]

constant and the bending moment–curvature curve is approximately a straight line. After the reinforcement reaches its yield strength, the deformation of the specimen increases quickly and the ultimate bending moment is soon reached as the load increases less (into the descending branch of the bending moment–curvature curve). This behavior of the specimen is similar to that of the beam at room temperature.

Comparing specimen HT loaded at elevated temperatures with specimen LT loaded after it is cooled, the yield and ultimate bending moments are considerably different. The reason is apparent. The reinforcement of the HT specimen is at elevated temperatures, so its yield strength (f_y^T) and the bearing capacity of the specimen decrease significantly. However, the yield strength of the reinforcement of the LT specimen recovers after cooling, and the bearing capacity is reduced slightly compared with the specimen at room temperature, because of the reductions in the concrete strength of the compression zone and the internal force-arm of the cross section. Therefore, the bearing capacity of the beam with the tension zone exposed to high temperature is reduced considerably at elevated temperatures, and most of the bearing capacity of the beam at room temperature can be recovered after it is cooled down to normal temperature.

Although the bending moments at yielding of reinforcement and the ultimate state of specimens HT and LT are considerably different, the values of the corresponding curvatures of both specimens are less different (Fig. 8-13). When specimen HT is loaded, its deformation is much greater than that of the beam at room temperature (Fig. 8-7), because the mechanical behavior of the concrete and the reinforcement deteriorate, i.e., their elastic moduli reduce, thermal deformation is greater, and the crack expands. Although specimen LT was cooled down to normal temperature before loading and the mechanical behavior of the reinforcement has already recovered, the mechanical behavior of concrete does not recover and is even worse (see Section 1.2.3). In addition, wider cracks already exist in specimen LT before loading and the characteristic value of its deformation under load approaches that of specimen HT at elevated temperatures. Therefore, the deformation behavior of a beam with the tension zone exposed to high temperature deteriorates seriously and does not improve even after it is cooled down.

2. Beam with the compression zone exposed to high temperature

After specimen HC is heated, the behavior of the concrete in the compression zone exposed to high temperature deteriorates. The specimen is loaded at elevated temperatures and its deformation is much greater than that of the beam at room temperature. However, the temperature of the concrete in the tension zone is lower and the compressive strain of concrete is greater at elevated temperatures, so the depth of

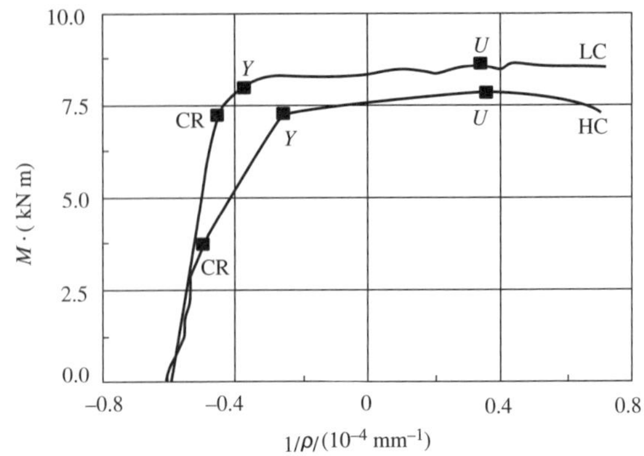

FIGURE 8-14 Bending moment–curvature relationship for beams with the compression zone exposed to a high temperature.[8-14]

the compression zone on the cross section is great and the tensile strain increases slowly. Therefore, the cracking bending moment of the specimen is quite high (Fig. 8-14). After the specimen cracks, its deformation accelerates with the load, the deformation increases sharply, and the bending moment–curvature curve clearly turns after the reinforcement yields. When the bending moment increases further but slightly, the ultimate bending moment of the specimen is reached, and the curve then turns into the descending branch.

After specimen LC experiences the heating–cooling cycle, no crack is found on the top surface of tensile concrete, which is located in the lower temperature zone. The strength and elastic modulus of the reinforcement have recovered to normal values, but the concrete on the compression zone is damaged and its behavior deteriorates. When the specimen is loaded at room temperature, its deformation increases approximately proportionally to the bending moment, and the cracking bending moment is higher than that of specimen HC. However, its deformation develops quickly after the concrete cracks, and the tensile reinforcement reaches the yield strength and the bending moment–curvature curve turns obviously. Then, the curve passes a flat yielding step, and turns into a flatter descending branch after the ultimate bending moment of the specimen is reached.

The tensile reinforcement of specimen HC is far away from the zone of high temperature and it loses less strength; that of specimen LC is at normal temperature and its strength is slightly higher. The concrete strength of the compression zone of specimen HC decreases significantly at elevated temperatures; that of specimen LC decrease even more significantly because it experienced a cooling process after heating. Therefore, the yield and ultimate bending moments of both specimens are approaching and the difference between them is less than 9%, but the bearing capacity of specimen LC after it is cooled is slightly higher.

The flexural deformation of the specimen with the compression zone exposed to high temperature is convex toward the compression zone (the curvature is taken as a negative value) during the heating process and recovers slightly during the cooling process, so the thermal deformations of specimens HC and LC are approaching each other before loading. When the specimen is loaded, its curvature and bending moment have the same direction and take positive values. When the tensile reinforcement yields, the thermal deformation of the specimen is not yet canceled out and the total curvature is still a negative value. It is seen that the thermal deformation of a structural member at elevated temperatures is very high. The deformation of the specimen then increases sharply and the total curvature turns into a positive value, i.e., the same direction as the bending moment, when the ultimate bending moment is reached. The values of the characteristic curvatures of specimen HC loaded at elevated temperatures and specimen LC loaded after it is cooled are also approaching each other.

3. Comparison between beams with the tension zone and the compression zone exposed to high temperature

The mechanical behavior of beams with the tension zone (HT) and the compression zone (HC) exposed to high temperature is compared in Section 8.2.3. When specimens LT and LC, which are loaded after cooling, are compared, the strengths of the tensile reinforcements of both specimens are approaching, but the strength of the compressive concrete of the LT specimen reduces less because it is located in the zone of low temperature. However, concrete strength has less influence on the yield and ultimate bending moments of the specimen, so the characteristic bending moments of both specimens are approaching (Table 8-3). This means that the residual bearing capacities of the parts with the positive and negative bending moments within the structural member are approaching after it is cooled.

Comparing Fig. 8-13 with Fig. 8-14, the total deformation at the ultimate bending moment of the beams (HT and LT) with the tension zone exposed to high temperature is several times greater than that of the beams (HC and LC) with the compression zone exposed to high temperature. The main reason for this is that the thermal and bending moment deformations of the former have the same sign and add together, and both deformations of the latter have opposite signs and are subtractive. If the deformation caused only by the bending moment of the specimen after loading is compared, the curvature values of the four specimens are about 0.1/m with small differences.

CONCLUSIONS

When a building structure sustains a fire accident, a beam with three surfaces and a slab with one surface exposed to fire are the most common working conditions. Under these conditions, the temperature distributes nonuniformly on the section of the structural member, and the difference between the temperatures on the top and the bottom of the section is considerable. The member or the section of a member bearing a positive bending moment is the condition of the tension zone exposed to high temperature; that bearing a negative bending moment is the condition of the compression zone exposed to high temperature. The bearing capacity and deformation of both conditions are considerably different.

Under the path of loading under constant temperature, the ultimate bearing capacity of a beam with the tension zone exposed to high temperature deteriorates significantly, because the yield strength of the reinforcement decreases sharply at elevated temperatures. The reducing amplitude of the bearing capacity is similar to that of the strength (f_y^T/f_y) of the reinforcement. The strength of the reinforcement recovers after the beam is cooled down and most of the bearing capacity of the beam can also be recovered. The bearing capacity of the beam with the compression zone exposed to high temperature decreases less, because the compressive strength of concrete reduces and the internal force-arm of the section decreases at elevated temperatures. However, the bearing capacity of the beam undergoes limited recovery after it is cooled down, because the compressive strength of concrete cannot be recovered.

Under the path of heating under constant load, the specimen bearing a high level of initial bending moment ($M_u^T/M_u > 0.4$) will not fail when the temperature is lower than 500 °C, and the ultimate temperature-bearing capacity exceeds that of the specimen under the path of loading under constant temperature. When the initial bending moment of the specimen is at a lower level ($M_u^T/M_u < 0.4$), the ultimate temperature increases as the initial bending moment decreases, and the variation regularity of this is similar to, but slightly higher than, that of the specimen under the path of loading under constant temperature.

Generally, a concrete beam and slab under the action of high temperature fail in a flexural pattern, which is controlled by the yielding of the reinforcement. The deformation (curvature, deflection) values of the structural member during heating and loading at elevated temperatures are far greater than at room temperature. Deformation develops very quickly after the yielding of the reinforcement and then it fails suddenly. The residual deformation after it is unloaded and cooled down is great and it bends up obviously. Flexural cracks appear on the surface of the structural member, and the number of the cracks is less but they are wider.

The shear resistance capacity of a concrete beam also reduces as the temperature increases, but the reducing amplitude is smaller than that of the bending moment resistance capacity. Therefore, if a structural member fails in a shear pattern at room temperature, it will fail in a flexural pattern at elevated temperature. The concrete cover of the reinforcement in a beam can effectively reduce the temperature of the reinforcement during a fire accident. When the thickness of the concrete is increased properly, the thermal resistance capacity of the structural member is improved. If the concrete cover is too thick, it may spall off and fall when a fire accident occurs. The reinforcement is then exposed directly to high temperature, which is unfavorable.

REFERENCES

[8-1] T.T. Lie, New facility to determine fire resistance of columns, Canadian Journal of Civil Engineering 7 (3) (1980) 551–558.

[8-2] A.H. Gustaferro, S.L. Selvaggio, Fire endurance of simply-supported prestressed concrete slabs, PCI Journal 12 (1) (1967) 37–53.

[8-3] J. Zhang, Experimental study of reinforced concrete columns and beams under loading and elevated temperature. Masters Dissertation, Tsinghua University, Beijing (1997).

[8-4] X. Shi, Z. Guo, Experimental investigation of mechanical behavior of simply supported reinforced concrete beam under high temperature, Journal of Structural Engineering (Supplementary Issue) 2 (3,4) (1991) 368–373.

[8-5] X. Shi, Z. Guo, Investigation on the behavior of reinforced concrete at elevated temperature, China Civil Engineering Journal 33 (6) (2000) 6–16.

[8-6] J.G. Church, L.A. Clark, The effect of combined thermal and force loads on the behavior of reinforced concrete beams, Structural Engineering 66 (16) (1988) 262–267.

[8-7] B. Ellingwood, T.D. Lin, Flexure and shear behaviors of concrete beams during fire, ASCE Journal of Structural Engineering 117 (2) (1991) 440–457.

[8-8] Z. Lu, B. Zhu, Y. Zhou, Experimental study on fire response of simply supported reinforced concrete beams, China Civil Engineering Journal 26 (3) (1993) 47–54.

[8-9] M.R. Kham, R. Royles, Post heat exposure behavior of reinforced concrete members, Magazine of Concrete Research 38 (135) (1986) 59–66.

[8-10] C. Wang, et al., Investigation on strength damage of simply supported reinforced concrete beam after heating and cooling, Journal of Southwest Jiaotong University 27 (2) (1992) 65–74.

[8-11] X. Shi, Z. Guo, Behavior of reinforced concrete beams under different load-temperature paths, Engineering Mechanics (Supplementary Issue) 2 (1995) 705–711.

[8-12] X.D. Shi, T.H. Tom, K.H. Tom, Z.H. Guo, Effect of force-temperature paths on behaviors of reinforced concrete flexural members, ASCE Journal of Structural Engineering 108 (3) (2002) 365–373.

[8-13] X. Shi, Z. Guo, Fire-resistant behaviors of reinforced concrete beams with different thickness of concrete cover, Industrial Construction 26 (9) (1996) 12–14.

[8-14] J. Sun, X. Shi, Z. Guo, Behaviors at elevating temperature and after cooling for reinforced concrete beams heated on their three surfaces, Building Structure 32 (1) (2002) 34–36.

CHAPTER 9

Behavior of Compressive Members at Elevated Temperatures

Usually, reinforced concrete columns and walls simultaneously bear the actions of the bending moment and axial compression. Normally, reinforced concrete beams and slabs bear only the bending moment under the action of load, but they also bear the additional action of axial force, when their longitudinal thermal deformation is constrained by a support member or an adjacent structure during a fire accident. These structural members are called eccentric compressive members or compressive–flexural members.

When a building sustains a fire accident, four, three, two, or one surfaces of the rectangular cross section of an eccentric compressive member may be exposed to high temperature, depending on its position. As for the central and eccentric compressive concrete members with peripheral surfaces exposed to fire, many experimental investigations have been reported, mostly involving fire endurance tests on prototype members using a large furnace.[9-1,9-2]

During a fire accident, three or one surfaces of a column are connected to the wall and a beam with a flange and wall exposed to fire; the two adjacent surfaces of a corner column in a room and a beam connected to the wall underneath are exposed to fire. These structural members are used commonly in buildings. Their thermal behavior is more complicated and is the main topic discussed in this chapter. A structural member with peripheral surfaces exposed to fire may be regarded as a special example.

9.1 CENTRAL COMPRESSIVE COLUMN WITH THREE SURFACES EXPOSED TO HIGH TEMPERATURE

9.1.1 Testing Method and Contents

The experimental devices and methods for testing a compressive concrete member at elevated temperatures[9-3,9-4] are basically the same as for flexural members (see Fig. 8-1). The heating and temperature control system and the data measurement and acquisition system can be used directly; only the support and loading system for the specimen have to be modified.

The vertical testing method (Fig. 9-1) is used for a central compressive column in order to avoid the influence of its weight on lateral displacement. The specimen is supported on the working table (or steel beam) of the 2000-kN hydraulic testing machine. Both ends of the specimen are in contact with knife hinges, making the acting point of the axial compression clearer. The experimental furnace (Fig. 8-3) is placed vertically on the work table of the test machine, and kept in the correct position relative to the specimen. The elevating temperature curve of the furnace is the same as in Fig. 8-4.

The size and the construction of the central compressive specimen are shown in Fig. 9-2. The specimens are made, cured, and stored in the laboratory for 60 days before the start of the test. The prismatic compressive strength of the concrete is f_c = 27.2 MPa when testing the specimen, and the yield strength of the longitudinal reinforcement is f_y = 340 MPa.

Experiment and Calculation of Reinforced Concrete at Elevated Temperatures

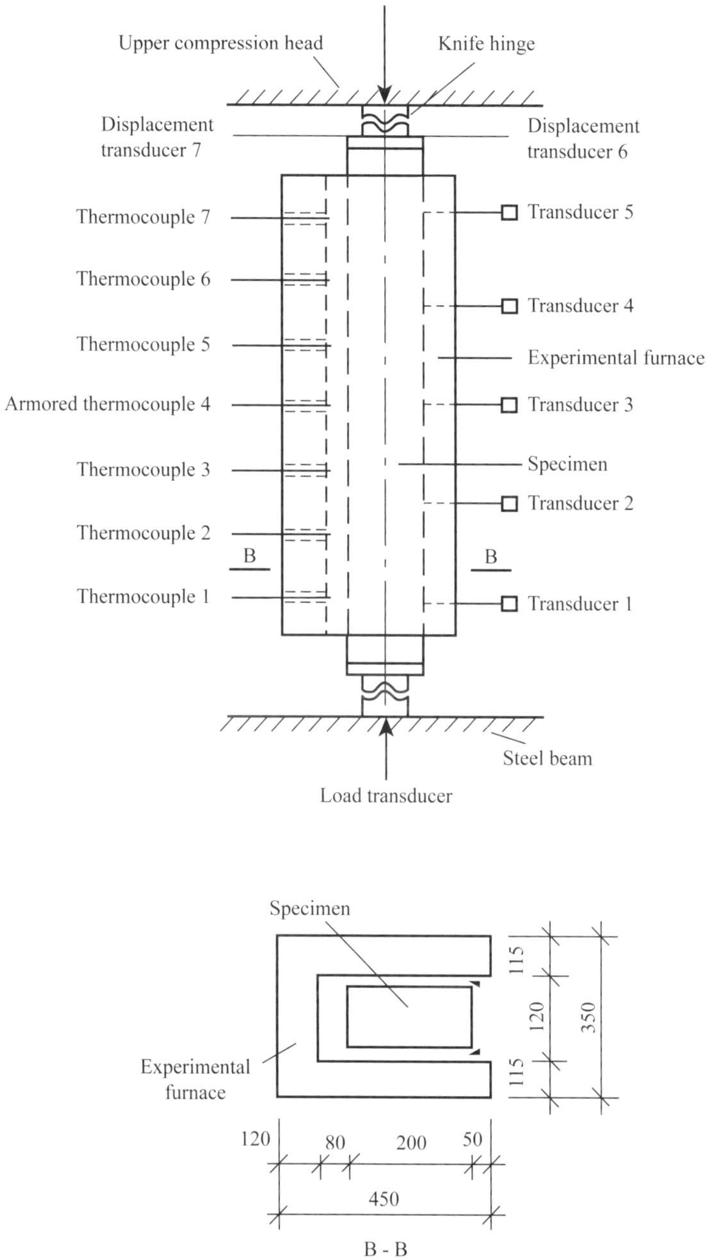

Figure 9-1 Scheme for central compression testing.

The testing procedure for the central compressive specimen is the same as that for the beam specimen (see Section 8.2.1). The temperature–load conditions tested include loading at constant temperature, heating under constant load, and a heating–cooling cycle. The temperature distributions in the furnace chamber and in the interior of the specimen, the ultimate temperature and load-bearing capacity, the lateral deflection at mid-span, and the axial deformation of

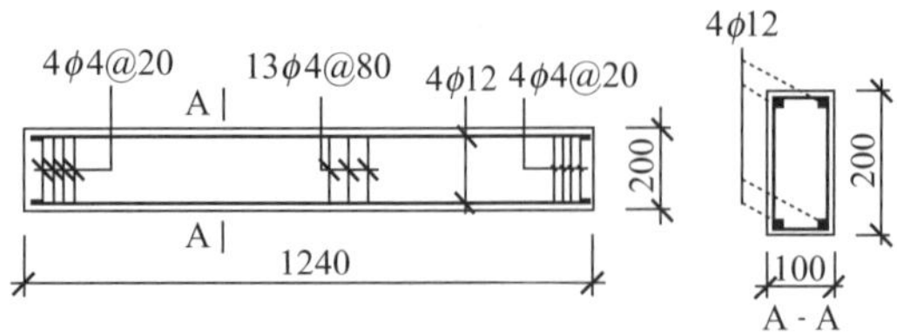

Figure 9-2 Specimen for central compression testing.[9-3]

the specimen are measured during the test. The measurement points are shown in Fig. 9-1.

Various macroscopic physical phenomena observed during the process of elevating the temperature of the compressive specimen are the same as those for the beam specimen (see Section 8.2.2), such as escape of water vapor, weight loss, change in the color of the surface, and damage to the surface layer (cracking, loosening, and spalling off). As the experimental furnace is set up vertically, the heating wire softens and its coil droops after the wire is electrified and its temperature is elevated. Therefore, the temperature in the lower part of the furnace chamber is slightly higher than in other parts,[9-3] but the temperature in the main testing part in the middle of the specimen changes less. The mechanical behavior of the specimens tested under different temperature–load paths is discussed in the following sections.

9.1.2 Mechanical Behavior at Elevated Temperatures

The mechanical behavior of the central compressive specimens at different temperatures is tested under the path of loading at constant temperature. After the specimen is set up, the experimental furnace is electrified and heated to a predetermined temperature (the maximum value is 950 °C) and the temperature is kept constant for 10 min; then the testing machine is started and the specimen is loaded until failure.

The central compressive specimen experiences two successive stages: freely heating and loading at constant temperature. The specimens are tested at different temperatures (in the furnace chamber); the lateral deflection at mid-span and the axial deformation vary with the axial force and are measured during the testing as shown in Fig. 9-3. The measured curves of the rotation angle at the end of the specimen are similar.[9-3]

During the freely heating stage, three surfaces of the specimen section are heated and the temperature distributes nonuniformly in the interior of the specimen. The flexural deflection is convex toward the surface of high temperature (which is taken as a positive value), because the expanding strain of concrete (ε_{th}) on this surface is larger. In the meantime, the average expanding strain of the specimen section causes axial elongation (which is taken as a negative value). Both the lateral deflection and axial deformation increase with the testing temperature. When the experimental temperature $T \leq 200$ °C, the temperature in the interior of the specimen is even lower and the temperature gradient is small, so the deformation of the specimen is small. When the temperature $T > 400$ °C, the thermal strain of concrete increases quickly and the temperature gradient on the section also increases, so the deflection and axial deformation of the specimen develop rapidly. After the temperature reaches $T > 700$ °C, the thermal strain of concrete stagnates (Fig. 2-2), although the temperature inside and outside the specimen increases continuously, so the temperature gradient on the section reduces (Fig. 8-6) and the deformation rate of the specimen decreases. When the experimental temperature is 900 °C, the lateral deformation and axial elongation of the specimen are measured to be about 7.7 mm and 9.0 mm, respectively.

During the stage of loading at constant temperature, the temperature on the section of the specimen distributes nonuniformly, and the loss of elastic modulus and strength of concrete are correspondingly different. Even when the specimen is acted on by a central force and the compressive stress on the section distributes uniformly, the strain still develops nonuniformly on the section, i.e., the compressive strain on the high temperature side is larger and that on the low temperature side is smaller. So, the deflection caused is convex toward the surface of low temperature and is opposite to that caused during

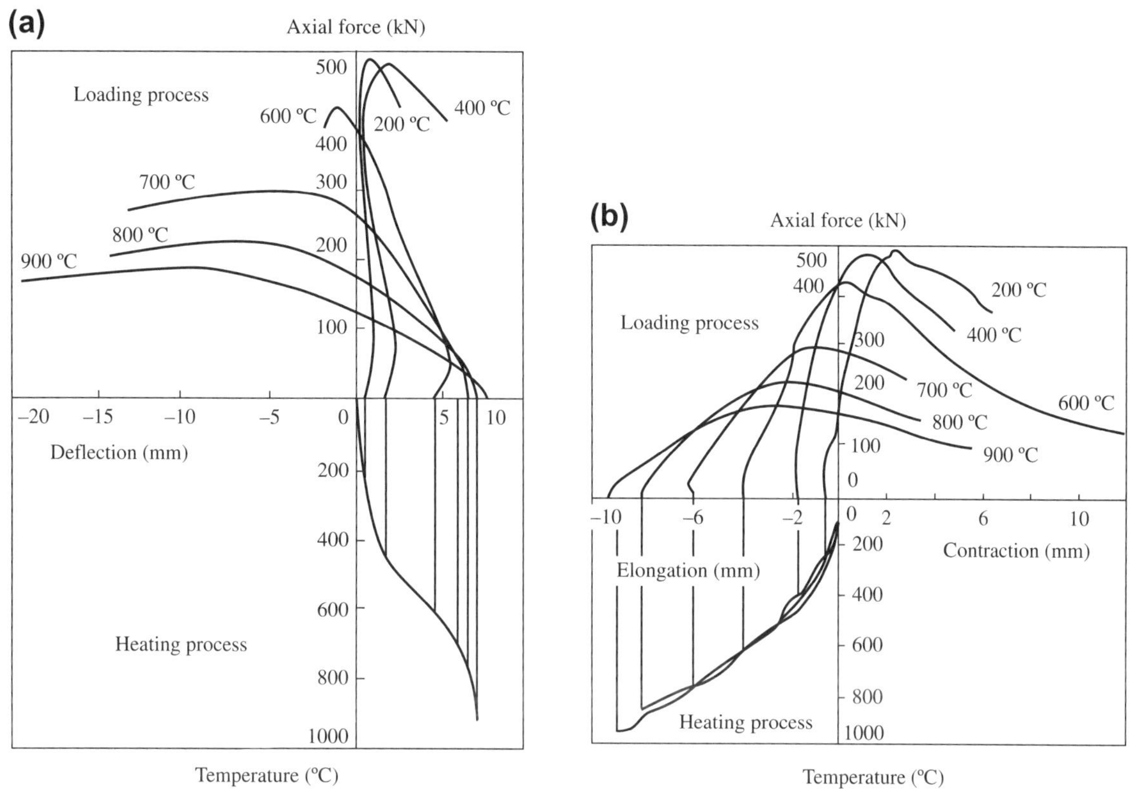

Figure 9-3 Deformation of a central compressive column with three surfaces exposed to high temperature[9-5]: (a) lateral deflection at mid-span; (b) axial deformation.

heating. For a specimen tested at a temperature $T \leq 400$ °C, the lateral deflection caused by the axial force is small because the internal temperature of the concrete is lower and the loss of the material behavior is limited. The deflection of the specimen at the ultimate axial force (N_u^T) is still convex toward the high temperature side (positive value) and the thermal deflection caused during heating has not yet been canceled out. The concrete on the low temperature side fails first because of lower peak strain (ε_p^T, Fig. 2-5) and the axial force enters the descending branch. Then the specimen is flexed continuously toward the high temperature side. As for the specimen tested at a temperature $T \geq 600$ °C, the lateral deflection caused by the axial force is large because the internal temperature of the concrete is high and the material behavior deteriorates enormously. The deflection of the specimen at the ultimate axial force (N_u^T) becomes convex toward the low temperature side (negative value) and the thermal deflection caused during heating has been canceled out. The concrete on the high temperature side fails first and its strain increases considerably, the load carrying capacity of the specimen reduces, and the lateral deflection is continuously convex toward the low temperature side and develops considerably.

Therefore, all the specimens with three surfaces exposed to high temperature fail with the pattern of smaller eccentricity, but the final deflection may be convex toward the high temperature or low temperature side, and this depends on the value of the experimental temperature (see Fig. 9-5). After the failed specimen cools down, the residual flexural deformation is apparent. Greater height and longitudinal length of the compressive failure zone are seen on the concave side, but some fine and short transversal tensile cracks are seen on the convex side.

The axial deformation of the specimen is elongation (negative value) during the stage of freely heating and contraction (positive value) caused by the axial force. When the specimen is tested at the experimental temperature $T \leq 400$ °C, the initial thermal expanding deformation is small and the stiffness reduces slightly; the compressive deformation caused at ultimate strength N_u^T cancels out the expanding deformation and the total deformation is contraction. The compressive deformation develops more quickly as the axial force decreases. When the specimen is tested at the experimental temperature $T \geq 600$ °C, the stiffness reduces obviously as the temperature increases, so larger compressive deformation occurs after loading. However, the total deformation at ultimate strength N_u^T is still elongation, because the initial value of the thermal expanding deformation also increases with the experimental temperature. The concrete on the high temperature side fails first and the axial force enters the descending branch; the axial compressive deformation develops rapidly and turns, finally, into contraction.

The ultimate strength N_u^T of the central compressive specimen with three surfaces exposed to high temperature reduces monotonically as the experimental temperature increases (Fig. 9-4), but the variation regularity is obviously different from that of the compressive strength of concrete material (see Fig. 2-7). When the specimen is tested at an experimental temperature $T \leq 400$ °C, the ultimate axial force is obviously lower than that of the central compressive member at normal temperature. It is actually an eccentric compressive condition and the initial eccentricity of the axial force is the thermal deflection caused during the heating stage, although the internal temperature of the concrete is not high and strength loss of the concrete is limited. When the specimen is tested at a higher temperature (≥600 °C), the ultimate axial force reduces continuously, as the mechanical behavior of concrete and reinforcement is damaged significantly and the initial eccentricity

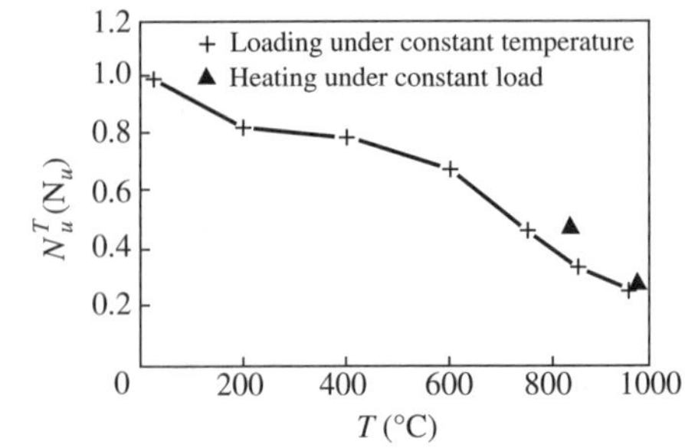

Figure 9-4 Ultimate bearing capacity of a central compressive specimen with three surfaces exposed to high temperature.[9-5]

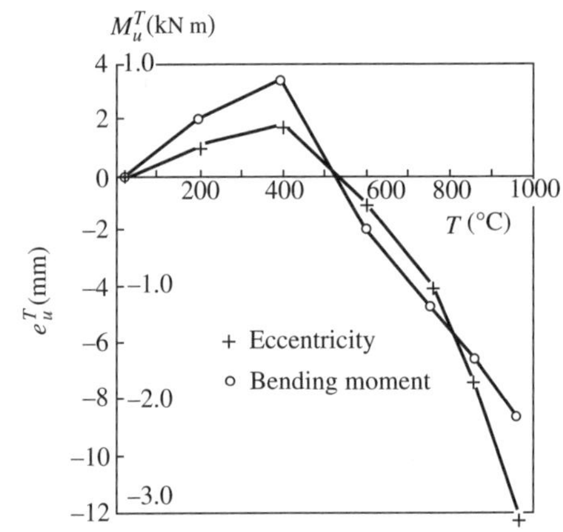

Figure 9-5 Eccentricity and bending moment of a central compressive member at ultimate conditions.

caused during the heating stage increases. However, the ultimate axial force of the specimen tested at the highest experimental temperature (950 °C) is about 30% of that at normal temperature, and this ratio is apparently much higher than the residual strengths (<10%) of concrete and reinforcement materials at the same experimental temperature. The main reason is that the specimen experiences only a short time (about 2.5 h) at high temperature, so the internal temperature is lower and the average strength of the material is higher.

When a specimen with three surfaces exposed to high temperature is tested, the lateral deflection, i.e., the initial eccentricity of the axial force, is formed after the heating stage and additional eccentricity is caused during the stage of loading at constant temperature, and the total eccentricity and the ultimate bending moment at the ultimate axial force (N_u^T) are e_u^T and $M_u^T = N_u^T \cdot e_u^T$, respectively. The values change from positive (deflection convex toward the high temperature side) to negative (deflection convex toward the low temperature side) as the experimental temperature increases; the variable processes are shown in Fig. 9-5. Therefore, although the axial force of the specimen acts on the geometric center of the initial section and acts on the geometric center of both end sections after testing, additional eccentricity and larger bending moments are caused on the mid-span section and the asymmetrical failure pattern of smaller eccentricity is finally formed because the temperature distributes nonuniformly on the section and the thermal strain and mechanical behavior of the materials are inhomogeneous.

9.1.3 Mechanical Behavior Under the Path of Heating Under Constant Load

When a central compressive specimen with three surfaces exposed to high temperature is tested under the path of heating under constant load, the experimental device and the method used are the same as previously (Fig. 9-1) but the testing procedure is different. The initial central force ($N_0 / N_u = N_u^T / N_u = 0.3, 0.5$) is applied first to the specimen at normal temperature and kept constant, then the experimental furnace is electrified and heated continuously until deformation of the specimen is out of control and the bearing capacity is exhausted. The value of the ultimate temperature (T_u) is measured and the test is complete.

The lateral deflection and axial deformation of the central compressive specimen with three surfaces exposed to high temperature and under the path of heating under constant load are shown in Fig. 9-6. A central load is applied to the specimen at normal temperature and the lateral deflection is zero. When the experimental temperature $T < 500$ °C, the temperature distributes nonuniformly on the specimen section during the stage of heating under constant load. Larger expanding strain occurs on the high temperature side and the deflection of the specimen is slightly convex toward the same side (positive value). But the temperature in the interior of the concrete is not high and the deterioration level of the material is low, so the deflection of the specimen is small. When the experimental temperature $T > 600$ °C, the behavior of the concrete and the reinforcement deteriorates rapidly and the deformation of the specimen is considerable. The compressive strain on the high temperature side is obviously greater than that on the low temperature side; the total deflection of the specimen converts gradually to become convex toward the low temperature side (negative value) and increases quickly. When the ultimate temperature is approached, the reinforcement on the high temperature side

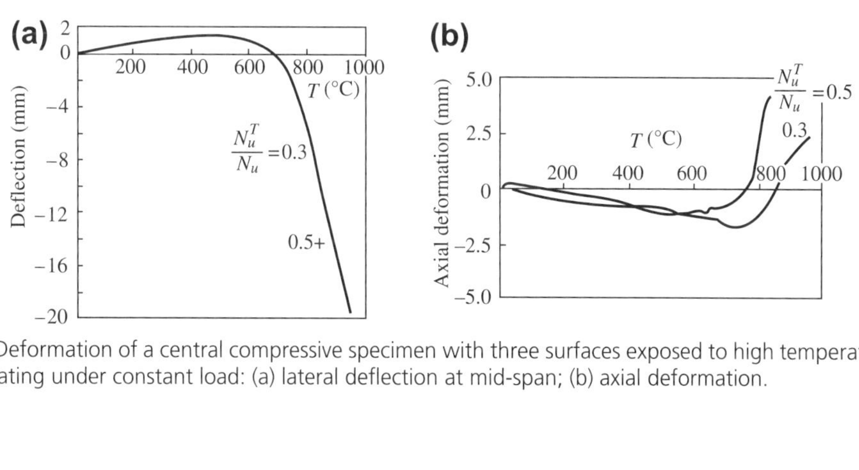

Figure 9-6 Deformation of a central compressive specimen with three surfaces exposed to high temperature and under the path of heating under constant load: (a) lateral deflection at mid-span; (b) axial deformation.

yields under compression and the strain of the concrete and deflection of the specimen increase quickly, and the specimen soon fails. The final failure pattern of the specimen is the same as that for the specimen tested under the path of loading at constant temperature. The residual flexural deflection is apparent and is convex toward the low temperature side. Larger compressive failure areas are shown on the concave side and a few fine and short transverse cracks appear on the convex side.

The central compressive force is applied on the specimen at normal temperature; the contraction deformation (positive value) occurs with a small value in the axial direction. The deformation converts to axial elongation (negative value) during the process of heating under constant load, as the concrete expands with the elevating temperature. But the elongation deformation increases slowly because of the restraint of compressive stress in concrete (see Fig. 3-9). When the experimental temperature is higher than 600–700 °C, depending on the level of axial compression (N_0/N_u), the behavior of the concrete and reinforcement deteriorates dramatically and the compressive strain increases considerably. Then, the axial deformation converts again to contraction (positive value) and develops rapidly until the ultimate temperature is reached and the specimen fails.

Central compressive force is applied to the specimen at normal temperature and the strain and stress distributed on its section are uniform. During the process of heating under constant load, the temperature distributes nonuniformly on the section and the thermal strain and mechanical behavior of the material are no longer homogeneous, so a flexural deformation is formed and an additional eccentricity is caused on the middle part of the specimen. Although the central force applied is kept unchanged, the uniformly distributed stress on the section can no longer be maintained and stress redistribution has to occur continuously, but the resultant stress on the section has to maintain balance with the axial force throughout. When the ultimate temperature of the specimen is reached, the compressive strength of the concrete at elevated temperatures (f_c^T) on part of the section reduces to the value of the stress actually acted; the deformation of the specimen increases sharply and the specimen fails (see Fig. 11-9).

The measured values for the ultimate temperature-bearing capacity of the central compressive specimen with three surfaces exposed to high temperature and under the path of heating under constant load are plotted in Fig. 9-4, and they are above the connecting line for the measured values of the specimens tested under the path of loading at constant temperature. It shows that the former has the advantage of the behavior of resisting high temperature. The indices for the main behavior at the ultimate conditions of the specimens tested under both paths are compared in Table 9-1.

When the specimen tested under the path of heating under constant load reaches the ultimate temperature (T_u), the ultimate central force is greater than that of the specimen tested under the path of loading at constant temperature. If the additional eccentricity that exists practically is considered, the ultimate bending moments make a bigger difference. Furthermore, the higher the level of the initial axial force (N_u^T / N_u), the more advantage the previous one has. This conclusion agrees with the comparison of the behavior of beams with three surfaces exposed to high temperature under both testing paths (see Fig. 8-11). In addition, the signs of axial deformation of the specimens under both paths are just the opposite. Obviously, these comparisons reflect the behavior difference of concrete material under both testing paths (see Fig. 3-5 and Fig. 3-9).

The specimen is unloaded and cooled down to room temperature after the end of testing under the path of heating under constant load, and then the axial compressive force is applied again on the center of both end sections. As the larger eccentricity, i.e., residual lateral deflection already exists at the mid-span section, the residual ultimate bearing (capacity) of the specimen is far lower than that at high temperature (Table 9-1).

TABLE 9-1 Comparison of the Mechanical Behavior at Ultimate Conditions of Central Compressive Specimens Tested Under Different Heating–Loading Paths

Testing path	Specimen number	Temperature T_u (°C)	Central force N_u^T (kN)	Lateral deflection e_u^T (mm)	Ultimate bending moment M_u^T (kN m)	Axial deformation (mm)	Ultimate load after cooling (kN)
Loading under constant temperature	T8N	850	222.9	−7.30	−1.627	−2.4	—
Heating under constant load	N5T	840	300.0	−14.33	−4.299	+4.2	133
Loading under constant temperature	T9N	950	178.1	−12.07	−2.150	−3.1	—
Heating under constant load	N3T	965	180.0	−19.62	−3.532	+2.4	126

Lateral deflection, i.e., eccentricity at the mid-span section, is taken as a negative value when it is convex toward the low temperature side. Axial deformation of elongation is taken as a negative value and that of contraction is taken as a positive value.

9.2 ECCENTRIC COMPRESSIVE COLUMNS WITH THREE SURFACES EXPOSED TO HIGH TEMPERATURE

9.2.1 Testing Method and Contents

The experimental device and the method used to test the eccentric compressive specimen with three surfaces exposed to high temperature are the same as for the flexural member at elevated temperatures (see Fig. 8-1), but the support and loading systems are different (Fig. 9-7). The experimental furnace (see Fig. 8-3) is placed horizontally and suspended using four adjustable bolts located at the top corners, in order to conveniently adjust its vertical and horizontal positions. The specimen is also placed horizontally and two roller supports are used temporarily at both ends, which do not obstruct free thermal deformation of the specimen. The hydraulic jack is fixed on the block support and exerts a horizontal load on the specimen through a spherical hinge on the front and a knife hinge at other end of the specimen. When the load reaches about 5 kN, the roller supports are moved away and the load acts continuously and equally until the specimen fails.

The size and construction of the specimen are shown in Fig. 9-8. The cross section in the middle of the specimen is 100 mm × 200 mm and is reinforced symmetrically. There are two corbels at both ends of the specimen, which are used to apply the eccentric load and prevent local failure of the concrete under compression. The specimens are produced in batches[8-3,9-3,9-6] and the concrete is aged for 60 days before testing. At normal temperature, the cubic and prismatic compressive strengths of the concrete in the testing area are f_{cu} = 32.7–38.8 MPa and f_c = 26.0–28.8 MPa, respectively, and the yield strength of the longitudinal reinforcement is f_y = 340 MPa.

The initial eccentricity e_0 of the load acting on the specimens with three surfaces exposed to high temperature is from +0.6h to −0.6h, and the positive and negative values represent, respectively, the loading point at the low temperature side (or positive bending moment and the tension zone exposed to high temperature) and at the high temperature side (or negative bending moment

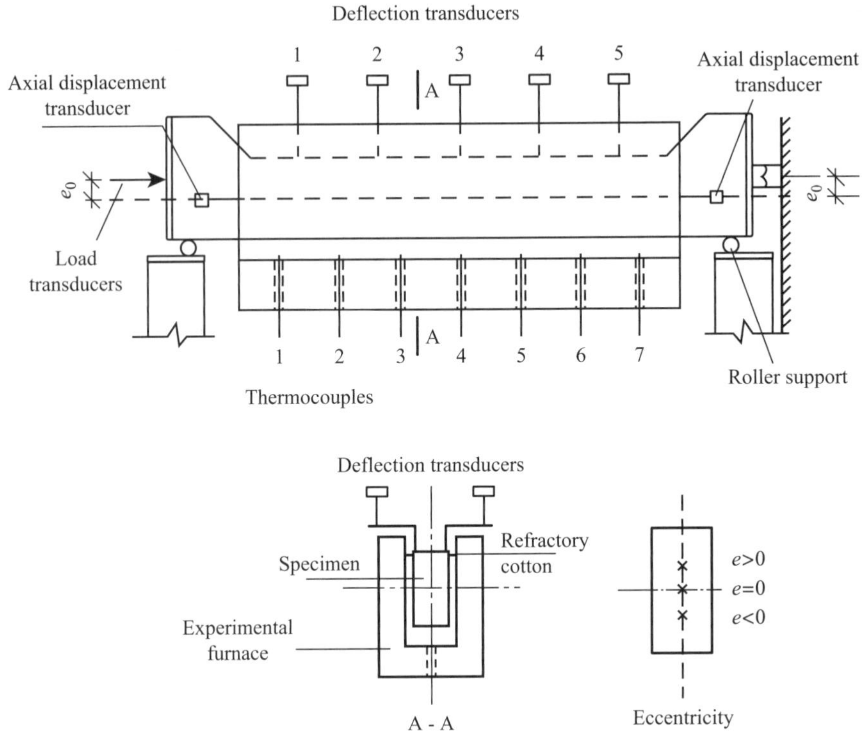

Figure 9-7 Scheme for the experimental device for an eccentric compressive specimen with three surfaces exposed to high temperature.[9-6]

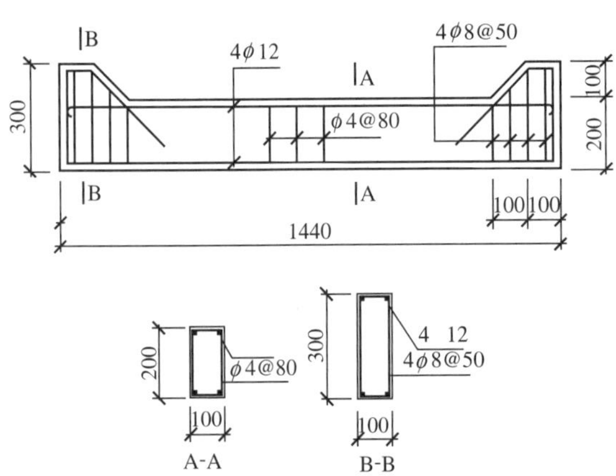

Figure 9-8 Eccentric compressive specimen.[9-6]

and the compression zone exposed to high temperature). In addition, the flexural members with the tension and the compression zones exposed to high temperature (Chapter 8) correspond to the eccentricities $e_0 = +\infty$ and $-\infty$, respectively.

The testing procedure for the eccentric compressive specimen at elevated temperatures is the same as for the flexural and central compressive specimens. The heating–loading conditions tested also include the path of loading at constant temperature, the path of heating under constant load, and a heating–cooling cycle. The macroscopic physical phenomena observed and the contents (including the temperature distribution on the section) measured during the testing process are also the same as or approximate to that of the flexural and central compressive specimens.

9.2.2 Deformation at Elevated Temperatures

The eccentric compressive specimen with three surfaces exposed to high temperature and under the testing path of loading at constant temperature also experience two stages: freely heating and loading at constant temperature. The variation curves for lateral deflection at mid-span and axial deformation of the specimens are shown in Fig. 9-9.

During the freely heating stage, the curvature and lateral deflection (positive values) of the specimen are convex toward the high temperature side and are caused by the elevated temperatures acting on three surfaces, and axial elongation deformation (negative value) is also caused. The deformation values of the specimen (length 1440 mm) at different testing temperatures are listed in Table 9-2; they develop rapidly at elevated temperatures and their values are considerable.

During the stage of loading at constant temperature, the variation in deflection at the mid-span of the specimen depends mainly on the eccentricity of the load. When the eccentricity $e_0 > 0.2h$, the high and the low temperature zones of the section are stretched and shortened, respectively. Thus, the curvature of the section has the same sign as that during heating and the deflection is convex toward the high temperature side and develops continuously under loading. As the load on the specimen increases, the cracking of the concrete in the tension zone tends to be more serious, the strain of the reinforcement increases, the neutral axis moves up, the compression zone on the section reduces, the compressive strain of the concrete also increases, and the deflection at the mid-span of the specimen develops quickly. When the load approaches and reaches the ultimate strength (N_u^T), the reinforcement yields and the crack expands in the tension zone with the high temperature, the strain of the concrete in the compression zone with the low temperature increases considerably, the deflection of the specimen accelerates, and it leads to failure of the specimen. The total deflection (Δ_u^T) of the failed specimen is large and after the specimen cools down, it is obviously bent and convex toward the high temperature side.

When the eccentricity $e_0 < 0.1h$, or the load is acted on the high temperature side, the curvature of the specimen caused during loading is opposite to that caused during heating. Correspondingly, as the load increases, the deflection of the specimen caused during heating counteracts gradually and develops in reverse, i.e., convex toward the low temperature side (negative value). When the load approaches and reaches the ultimate strength (N_u^T), the reinforcement in the tension zone at low temperature yields and the strains of the concrete and the reinforcement in the compression zone at high temperature increase quickly; the deflection of the specimen increases suddenly and the specimen fails with the pattern of larger eccentricity. Alternatively, the strength of the concrete is reached, the reinforcement yields in the compression zone at high temperature, and the deformation of the specimen increases quickly, but only a few transverse cracks occur in the tension zone at low temperature; the specimen fails with a failure pattern of smaller eccentricity.

The axial deformation of the specimen with three surfaces exposed to high temperature is elongation (negative value) during the freely heating stage and increases with the test temperature (Table 9-2). During the stage of loading at constant temperature, the axial deformations of the specimens tested at 400 °C and 600 °C are contraction (positive value). Thus, the elongation of the specimen caused during heating is counteracted gradually and the total axial deformation has converted to contraction when the ultimate strength (N_u^T) is reached. When the specimen is tested at 800 °C and the load eccentricity $e_0 \leq 0.4h$, the axial compressive deformation caused by the load is not large enough to counteract the considerable expanding deformation caused during heating, and the total deformation is still elongation when the ultimate strength is reached. Furthermore, when the load eccentricity $e_0 = 0.6h$, the bending moment, the area of the tension zone, and the strain at elevated temperatures of the specimen are greater; the average axial strain of the section is elongation during loading and has the same sign as the strain during heating, so the axial elongation of the specimen develops continuously. Therefore, the value of axial deformation of the specimen at ultimate strength (N_u^T) varies considerably with the test temperature and the load eccentricity.

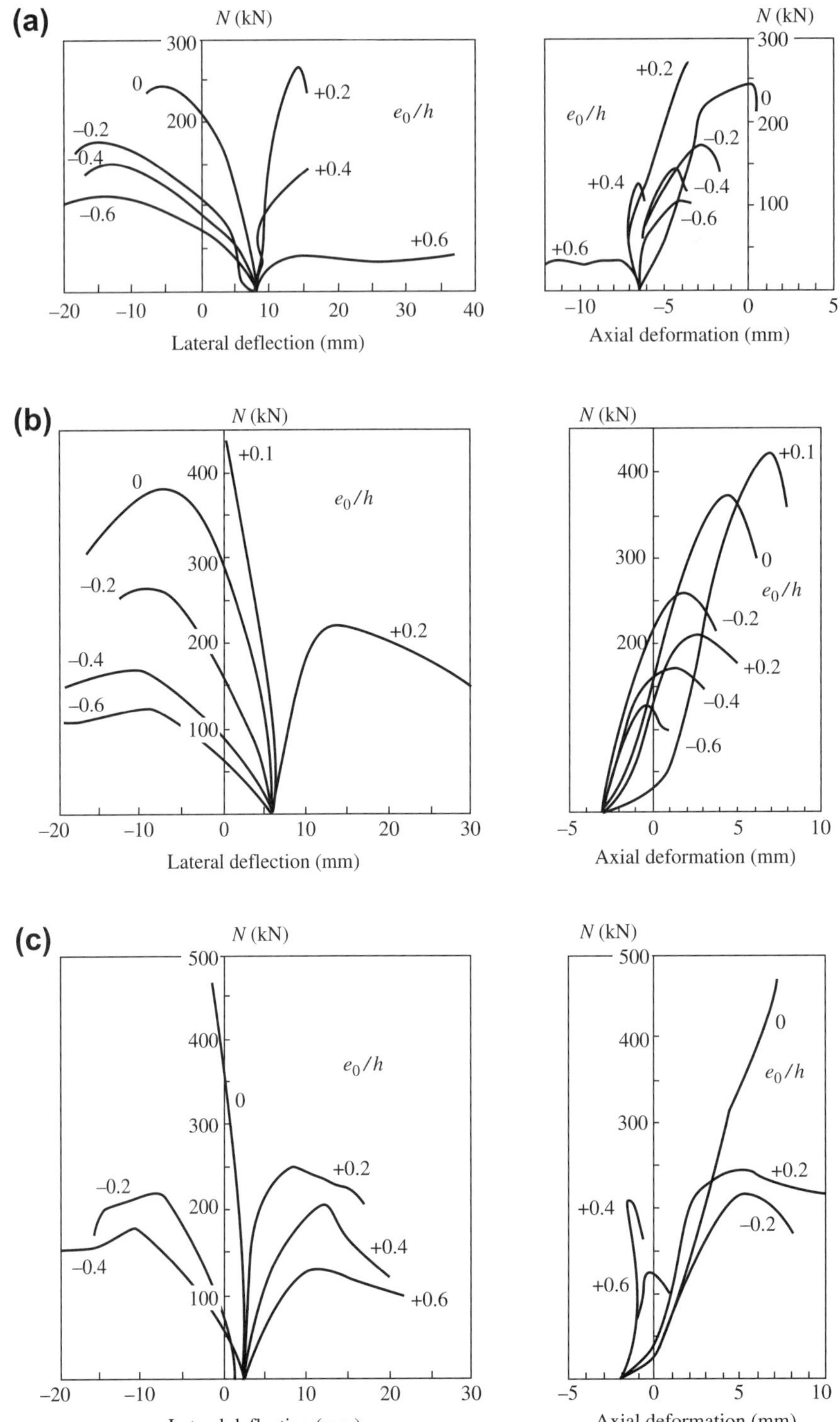

Figure 9-9 Lateral deflection and axial deformation of an eccentric compressive specimen with three surfaces exposed to high temperature[9-7–9-9]: (a) $T = 800$ °C; (b) $T = 600$ °C; (c) $T = 400$ °C.

TABLE 9-2 Deformation of a Specimen After Free Heating[9-7–9-9]

Testing temperature (°C)	400	600	800
Lateral deflection at mid-span (mm)	2.2 to 3.0	5.5 to 6.8	7.2 to 8.5
Axial deformation (mm)	−1.4 to −2.1	−2.0 to −3.0	−6.0 to −7.0

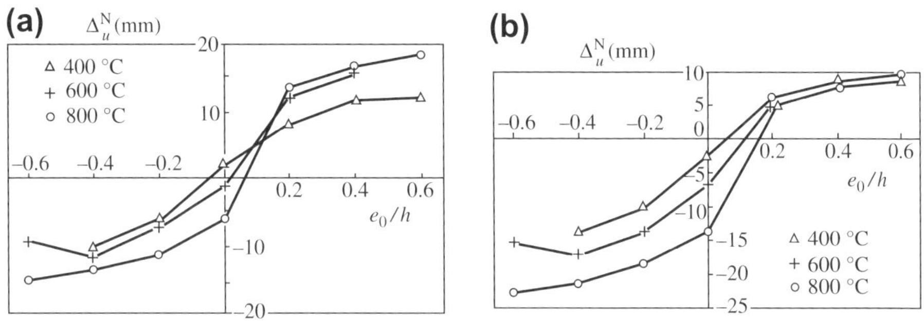

Figure 9-10 Deflection at the mid-span of a specimen at ultimate strength: (a) total deflection at mid-span; (b) deflection caused during loading.

The total deflection of the specimen at ultimate strength (N_u^T) is also an additional eccentricity (Δ_u^T) of the axial load, and it varies with the test temperature and the initial load eccentricity as shown in Fig. 9-10(a). When the initial load eccentricity moves from +0.6h to −0.6h, the ultimate deflection of the specimen (Δ_u^T) reduces gradually and converts from being convex toward the high temperature side to convex toward the low temperature side. The ultimate deflection varies greatly when $|e_0|<0.2h$, but tends to converge gradually when $|e_0|>0.4h$. The variation regularity of the ultimate deflection of the specimen is similar when the testing temperature T = 400–800 °C, and the absolute value of the deflection increases with the test temperature.

The ultimate deflection (Δ^T) of the specimen minus the deflection occurring during the heating stage (Δ_u^T) is the deflection caused by the load (N_u^T) applied during the stage of loading at constant temperature:

$$\Delta_u^N = \Delta_u^T - \Delta^T \tag{9.1}$$

Because the deflection during heating (Δ^T) depends only on the test temperature and is not related to the load eccentricity, the curve $\Delta_u^T - (e_0/h)$ is moved parallel and downward by Δ^T and the curve $\Delta_u^N - (e_0/h)$ is obtained (Fig. 10-9(b)). The deflection (Δ_u^N) caused by the ultimate load varies slightly for the specimen with eccentricity $e_0 > 0.2h$, when the test temperature is within the range of T = 400–800 °C.

9.2.3 Ultimate Strength and Optimum Eccentricity

The ultimate strength (N_u^T) of the eccentric compressive specimen with three surfaces exposed to high temperature varies with the test temperature and the initial load eccentricity (e_0) (Fig. 9-11). The maximum strength (N_0) of a column with a symmetrical section at normal temperature occurs when central compression ($e_0 = 0$) is applied. When the initial load eccentricity increases and whether the load acts on the left or the right side, the ultimate strength of the column decreases monotonically, and the curve $N_u - (e_0/h)$ is symmetrical about the ordinate.

When three surfaces of a specimen are exposed to high temperature, the temperature distributes nonuniformly on the section, the deterioration in the mechanical behavior of concrete and reinforcement at elevated temperatures is variable,

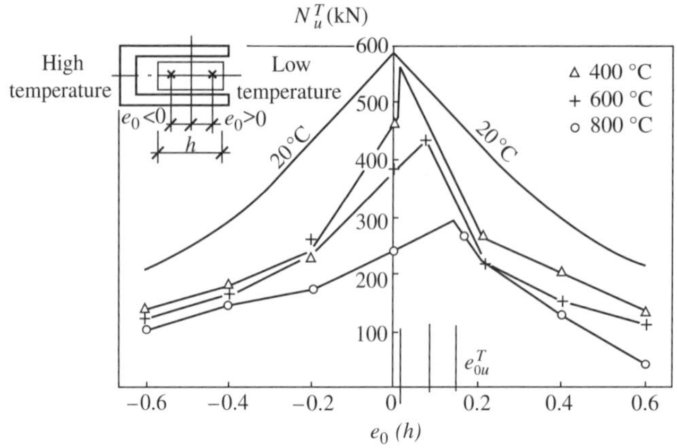

Figure 9-11 Relation between the ultimate strength of a specimen and the initial eccentricity of the load.[8-3,9-3]

inhomogeneous and asymmetrical fields of strength and strain of the material are formed on the section, and the resultant center of the material strength moves toward the low temperature side. Thus, the ultimate strength of the specimen depends on not only the value of load eccentricity but also its direction. When the load acts on the low temperature side ($e_0 > 0$) and the high temperature side ($e_0 < 0$), the ultimate strength of the specimens at elevated temperatures are different and the curve $N_u^T - (e_0/h)$ is asymmetrical about the ordinate.

When central compression ($e_0 = 0$) is applied to the specimen with three surfaces exposed to high temperature, the strength of the specimen is not the maximum value because the strength of the material on the section cannot be used fully. If the load moves properly toward the low temperature side ($e_0 > 0$), the higher strength of the concrete in the zone of low temperature can be utilized and the strength of the specimen increases. However, the strength of the specimen reduces again if the eccentricity is too much. The maximum strength of a specimen under the same test temperature is called the optimum ultimate strength, the corresponding eccentricity is called the optimum eccentricity ($e_{0u}^T > 0$), and the point of the load applied is called the optimum center, which is always on the low temperature side. When the specimen is tested at normal temperature, the behavior of the material on the section is homogeneous and the optimum eccentricity should be $e_{0u}^T = 0$. As the test temperature of the specimen is increased, the difference in behavior in the material on the section zones of high and low temperature is enlarged and the optimum eccentricity has to be increased gradually. The data in Fig. 9-11 show that the value of the optimum eccentricity e_{0u}^T/h is slightly greater than zero when the test temperature is T = 400 °C, and the values are $e_{0u}^T/h \approx 0.1$ and 0.15 when T = 600 °C and 800 °C, respectively. (The abscissa e_0 in Fig. 9-11 is the initial eccentricity of the specimen or the eccentricity at both end sections after testing. When the ultimate strength of the specimen at elevated temperatures is reached, additional eccentricity, i.e., the total deflection (Δ_u^T) occurs at the critical section in the middle of the specimen and the actual ultimate eccentricity is ($e_u = e_0 + \Delta_u^T$). So, only if this is used as the abscissa, accurate optimum center and eccentricity are achieved.)

In addition, the optimum eccentricity (e_{0u}^T) is an important boundary of the behavior of the specimen in an asymmetrical temperature field. When the initial eccentricity of an eccentric compressive specimen is greater than the optimum eccentricity ($e_0 > e_{0u}^T$), it means that the load is located on the low temperature side and the specimen will fail in the pattern of smaller eccentricity, which is controlled by the compressive strengths of the concrete and the reinforcement on the low temperature side, or in the pattern of greater eccentricity, which is controlled by the yield strength of tensile reinforcement on the high temperature side, and the ultimate deflection is convex toward the high temperature side. The ultimate strength of the specimen decreases quickly as the eccentricity increases, so the slope of the curve $N_u^T - (e_0 / h)$ is greater. On the other hand, when the initial eccentricity is smaller than the optimum eccentricity ($e_0 < e_{0u}^T$, algebraic value), this means that the load is located on the high temperature side and the specimen will fail in the pattern of smaller eccentricity, which is controlled by the compressive strengths of the concrete and reinforcement on the high temperature side, or in the pattern of greater eccentricity, which is controlled by the yield strength of the tensile reinforcement on the low temperature side, and the ultimate deflection is convex toward the low temperature side. The ultimate strength of the specimen decreases slowly as the eccentricity decreases, so the slope of curve $N_u^T - (e_0 / h)$ is smaller. When the load eccentricity is near the optimum eccentricity ($e_0 \approx e_{0u}^T$), the flexural deformation of the specimen is very small and is similar to that of the specimen subjected to central compression at room temperature.

If the initial load eccentricity of the specimen is the same, its strength (N_u^T) reduces monotonically as the test temperature increases (Fig. 9-12). The reducing slope of the curve $N_u^T - T$ of the central compressive specimen (e_0 = 0) is at the maximum. When the load acts on the high temperature side (e_0 < 0), the curve $N_u^T - T$ of the specimen is always below that of the central compressive specimen and the curves never intersect. However, when the load acts on the low temperature side (e_0 > 0) and the test temperature is within a certain range, the ultimate strength of the specimen is greater than that of the central compressive specimen (Fig. 9-11), and the curve $N_u^T - T$ also reduces monotonically but intersects with that of the central compressive specimen (Fig. 9-12).

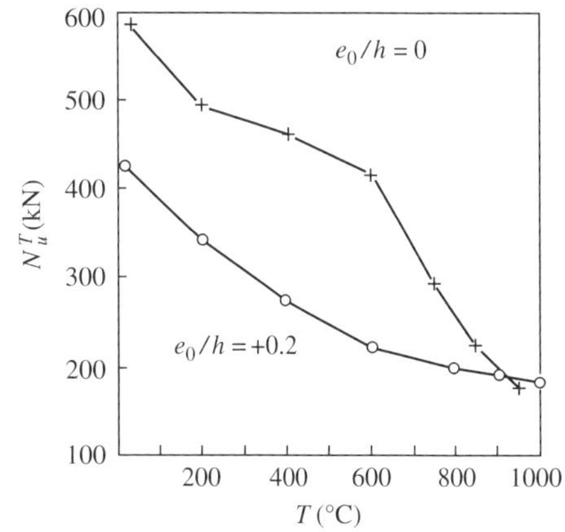

Figure 9-12 Relation between the ultimate strength and the test temperature.[9-10]

The values of axial compression and bending moment carried by a structural member with specific sections and materials can be presented by a corresponding ultimate envelope. The test results for the beams and the central and eccentric compressive columns with three surfaces exposed to high temperature, introduced earlier, are plotted in Fig. 9-13, in which the dimensionless coordinates N_u^T / N_0 and $M_u^T / (N_0 h_0)$, but $M_u^T / (f_c b h_0^2)$ for the beam, are used. N_0 is the ultimate central compression of the specimen at room temperature, and b and h_0 are the width and effective depth, respectively, of the specimen section. Considering the additional eccentricity (i.e., deflection $\Delta_u^T z$) at the ultimate condition, the ultimate bending moment of the specimen at elevated temperatures can be calculated from

$$M_u^T = N_u^T \left(e_0 + \Delta_u^T \right) \tag{9.2}$$

The test data for the beams is located on the abscissa, and the beams with the tension and the compression zones at high temperature are located in the positive and the negative directions, respectively.

The ultimate envelope of the axial compression–bending moment of a reinforced concrete column of rectangular section and symmetrical reinforcement is symmetrical about the ordinate when it is at room temperature, and a tip point (discontinuous point) of the envelope is located

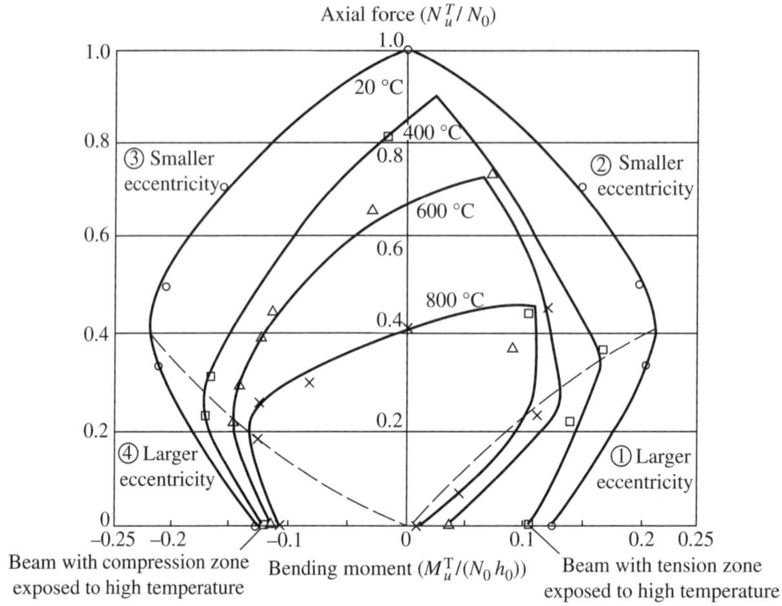

Figure 9-13 Ultimate envelope of the axial compression–bending moment of a specimen with three surfaces exposed to high temperature.[9-9]

on the ordinate, corresponding to the condition of central compression (N_0, $e_0 = 0$). Two boundary points, which distinguish between the failure patterns of larger and smaller eccentricities, are located symmetrically on both sides of the ordinate. When the test temperature of the specimen increases, the ultimate envelope of the axial compression–bending moment reduces gradually toward the inside, the tip point corresponding to the optimum eccentricity e_{0u}^T and the ultimate strength drifts down toward the right-hand side, and the asymmetrical envelope gradually becomes more obvious and significant. The boundary points of the failure patterns on both sides move inside and toward the origin, but they are not located symmetrically again.

The connecting lines of the tip points and the boundary points divide the envelopes into four areas with different failure patterns and they are from right to left (or $e_0 = +\infty$ to $-\infty$):

1. failure pattern of larger eccentricity, which is controlled by the tensile reinforcement yielding on the high temperature side
2. failure pattern of smaller eccentricity, which is controlled by the compressive strength of the concrete on the low temperature side
3. failure pattern of smaller eccentricity, which is controlled by the compressive strength of the concrete on the high temperature side, and
4. failure pattern of larger eccentricity, which is controlled by the tensile reinforcement yielding on the low temperature side.

As the test temperature increases, area 3 expands and area 2 contracts obviously, but area 1 expands and area 4 contracts slightly.

The intervals between the ultimate envelopes of the axial compression–bending moment at different test temperatures vary in different ways. When the test temperature $T \geq 400$ °C, the intervals of area 1 (larger eccentricity) and area 3 (smaller eccentricity) expand rapidly, which shows that the strength of the specimen decreases significantly at higher temperature. The intervals of area 2 (smaller eccentricity) and area 4 (larger eccentricity) are smaller, which shows that the decrease in the strength of the specimen is limited.

9.3 ECCENTRIC COMPRESSIVE COLUMNS WITH THREE SURFACES EXPOSED TO HIGH TEMPERATURE UNDER DIFFERENT CONDITIONS

9.3.1 Behavior Under the Path of Heating Under Constant Load

The experimental device and method used to test an eccentric compressive specimen with three surfaces exposed to high temperature and under the path of heating under constant load are the same as the pervious ones (see Fig. 9-7 and Fig. 8-1), and the size and materials of the specimen are the same as shown in Fig. 9-8. In order to demonstrate the difference between this specimen and a specimen under the path of loading at constant temperature, paired specimens are used and compared directly after testing.[9-6,9-11]

For the path of loading at constant temperature, the specimen is heated to the predetermined temperature, then the eccentric load is applied until the specimen fails and the ultimate load (N_u^T) and the corresponding deformation are measured. The specimen numbers are T8N4, T6N2, etc.

For the path of heating under constant load, the specimen is loaded eccentrically at normal temperature and the value of the load is taken as the ultimate load (N_u^T) of another specimen in pairs. Then, the specimen is heated until it fails and the ultimate temperature is measured. The specimen numbers are N4T8, N2T6, etc.

The symbols T and N in the specimen numbers represent heating and loading, respectively, and the order shows the sequence in the test path. The number after the symbol T multiplied by 100 °C is the experimental temperature, and the number after the symbol N multiplied by 0.1h is the eccentricity of the load.

The lateral deflections at the mid-span of the specimens vary with temperature and are shown in Fig. 9-14(a). The specimens (T–N series) tested under the path of loading at constant temperature are heated freely, and the deflections at the mid-span increase and accelerate gradually and are not related to the eccentricity of the load, so the deflection–temperature curves of specimens T8N4, T8N2, and T8N0 are similar. The specimens (N–T series) tested under the path of heating under constant load are loaded first at room

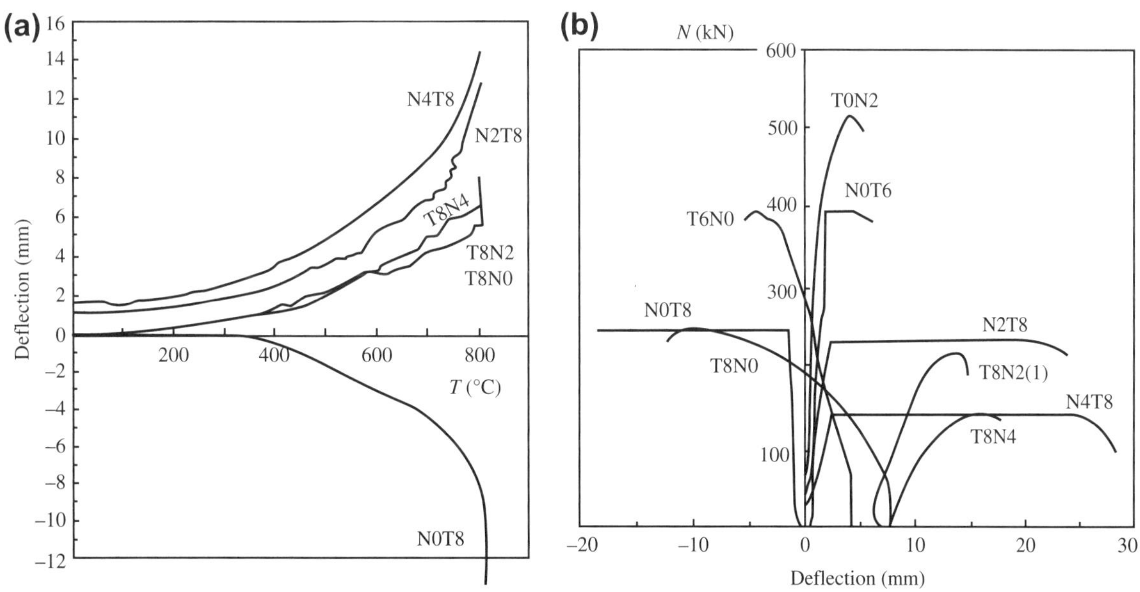

Figure 9-14 Comparison of deflections at the mid-span of specimens under different heating and loading paths[9-6]: (a) deflection–temperature; (b) deflection–load.

temperature, and initial deflections of different values, although not very different, are caused because of the different values and eccentricities of the load. During the subsequent heating process, the deflections of the specimens vary considerably and depend on the load eccentricity. For the specimens with eccentricity $e_0 > 0.2h$, the tensile zone is exposed to high temperature and its deflection during heating and that during loading are in the same direction. Thus, the total deflections develop gradually during heating, and the deflection–temperature curves are nearly parallel with that of the specimens in the T–N series. The specimen with eccentricity $e_0 \leq 0$ deflects after heating but is convex toward the direction of the low temperature zone, which is the opposite direction to the specimens in the T–N series, because the compressive strain on the high temperature zone is large. This is shown in Fig. 9-14(a) by comparing specimens N0T8 and T8N0.

The deflections at the mid-span of the specimens vary with the load and are shown in Fig. 9-14(b). The deflections of the specimens (T–N series) tested under the path of loading at constant temperature are similar to those shown in Fig. 9-9. However, the comparative specimens (N–T series) are loaded first at room temperature, so their deflections are small with nearly linear variation during loading, and they increase along the horizontal lines during the heating stage under constant load. The direction of the deflection depends on the eccentricity of the load. After the ultimate temperature is reached, the deflection of the specimen increases sharply and its strength reduces, so the specimen fails suddenly.

When the specimens are compared in pairs, the failure pattern of both heating–loading paths is identical, the direction of residual deflection is the same, and the failure depth of the compressive zones on the section and the development of tensile cracks are similar. The ultimate temperature and axial load of the paired specimens are compared in Table 9-3.

The specimens tested at temperature $T \geq$ 800 °C are discussed as an example. When the load eccentricity $e_0 \geq 0.2h$, the specimens fail on the pattern of larger eccentricity, which is controlled by the tensile reinforcement yielding in the high temperature zone, so the ultimate temperatures and axial loads measured from the paths of loading at constant temperature and of heating under constant load are approaching

TABLE 9-3 Main Experimental Results for Eccentric Compressive Specimens with Three Surfaces Exposed to High Temperature and Under the Path of Loading at Constant Temperature Compared with the Path of Heating Under Constant Load[9-6]

Initial eccentricity e_0/h	Experimental path and specimen number	Ultimate temperature T_u (°C)	Ultimate strength N_u^T (kN)	Ultimate deflection Δ_u^T (mm)	Ultimate bending moment M_u^T (kN m)
0	T6N0	600	396.4		
	N0T6	584			
	T8N0	800	247.9	−10.5	−2.603
	N0T8	852		−21.5	−5.33
0.2	T4N2	400	331.3		
	N2T4	395			
	T6N2	600	296.8		
	N2T6	580			
	T8N2	800	239	+12.0	12.428
	N2T8	800	233	+22.3	14.561
0.4	T8N4	800	140.9	+14.5	13.315
	N4T8	800		+24.5	14.724

each other. But the latter shows slightly larger additional eccentricity (also ultimate deflection) at failure, and the ultimate bending moment is also slightly larger. The central compressive specimens ($e_0 = 0$) fail on the pattern of smaller eccentricity, which is controlled by the compressive concrete in the high temperature zone, and the specimen tested under the path of heating under constant load has higher values for the ultimate temperature and bending moment. There is a lack of experimental data for an eccentric compressive column of load eccentricity $e_0 < 0$. According to nonlinear analysis and considering the coupling constitutive relation of concrete, the theoretical values[9-6] for the ultimate temperature and strength of eccentric compressive columns under both heating–loading paths are shown in Table 9-4. This demonstrates that the column under the path of heating under constant load has higher values for the ultimate temperature and bending moment when the axial load is equal.

The envelopes of ultimate axial load–bending moment of the eccentric compressive specimens with three surfaces exposed to high temperature ($T \geq 800$ °C) and under two extreme heating–loading paths are shown in Fig. 9-15. Both the experimental data and the theoretical values (broken lines in the figure) demonstrate that the envelope corresponding to the path of heating under constant load (N–T) is located on the outside of the envelope corresponding to the path of loading at constant temperature (T–N). Therefore, the fire resistant behavior of the former is better than that of the latter.

The behavior of the eccentric compressive specimens with three surfaces exposed to high temperature is different under various heating–loading paths. The main reason is that the compressive strength of the concrete material under the path of heating under constant load is higher

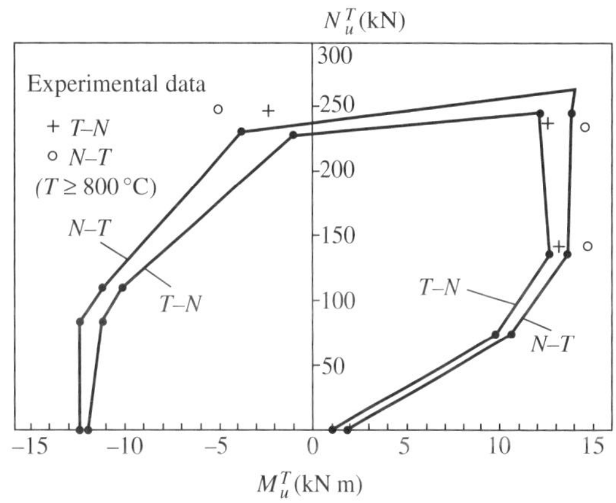

Figure 9-15 Envelopes of the ultimate axial compression–bending moment of eccentric compressive specimens with three surfaces exposed to high temperature and under two extreme heating–loading paths.

TABLE 9-4 Comparison of Ultimate Temperature–Strength (Theoretical Values) of Eccentric Compressive Columns with Three Surfaces Exposed to High Temperature but Under Different Heating–Loading Paths

Initial eccentricity e_0/h	Experimental path and specimen number	Ultimate temperature T_u (°C)	Ultimate strength N_u^T (kN)	Ultimate deflection Δ_u^T (mm)	Ultimate bending moment M_u^T (kN m)
−0.6	T8N6A	800	83	−19.6	−11.59
	N6T8A	982		−29.7	−12.43
−0.4	T8N4A	800	106	−16.5	−10.22
	N4T8A	1043		−24.6	−11.09
−0.2	T8N2A	800	150	−8.6	−7.29
	N2T8A	983		−17.5	−8.63
0	T8N0	800	227	−4.68	−1.06
	N0T8	855		−18.20	−4.13

than that under the path of loading at constant temperature (see Fig. 3-5). When the experimental temperature $T \leq 600$ °C, the temperature distributed on most parts of the specimen section is still low, differences in the strength of the concrete under different heating–loading paths are less, and the behavior of the specimens at elevated temperatures is similar (Table 9-3). When the experimental temperature $T \geq 800$ °C, especially for the specimen with eccentricity $e_0 \leq 0$ and a failure pattern of smaller eccentricity, the ultimate strength of the specimen depends mainly on the compressive strength of the concrete in the high temperature zone, so the specimen under the path of heating under constant load shows better behavior at elevated temperatures. If the specimen has load eccentricity $e_0 \leq -0.4h$ and a failure pattern of larger eccentricity, its ultimate strength depends mainly on the yield strength of the tensile reinforcement in the low temperature zone, and the differences in the behavior of the specimen at elevated temperatures is smaller for both paths.

9.3.2 Behavior After Temperature Sustained and After Cooling

When a fire accident occurs in a building, the structural members in the building may experience many temperature variations. As well as the two basic heating–loading paths described above, the temperature typically has to reduce eventually to a normal value after the high temperatures are sustained for a certain period of time.

Three types of temperature conditions[9-12,9-13] are selected to compare eccentric compressive specimens with three surfaces exposed to high temperature:

Condition H. After the experimental temperature reaches 800 °C and is maintained for 10 min, the specimen is loaded until failure. This is the same as testing under the path of loading at constant temperature, described in Section 9.2.

Condition P. After the temperature reaches 800 °C and is maintained (±25 °C) for 2 h, the specimen is loaded until failure.

Condition L. After the temperature reaches 800 °C and is maintained for 10 min, the specimen cools down naturally to room temperature and is then loaded 16–20 h later until failure.

The experimental devices and procedures needed for these tests are the same as the previous one (see Figs. 9-7 and 8-1), and the size and materials of the specimens are the same as shown in Fig. 9-8; the initial eccentricity of the load is $e_0 = +0.6h$ to $-0.6h$.

The curves for axial load–deflection at the mid-span of the eccentric compressive specimens under the temperature conditions P and L are plotted in Fig. 9-16. The corresponding curves for the specimens under the temperature condition H, which experience two stages, i.e., freely heating and loading at constant temperature, are shown in Fig. 9-9(a). The specimens under the temperature condition P experience three stages, i.e., freely heating, maintaining a high temperature, and loading at constant temperature. The deformation of the specimen during heating in the first stage is the same as that under the temperature condition H, and the deflection at the mid-span is about 8 mm. The deflection of the specimen reaches about 13 mm and increases by 62% after maintaining the high temperature for 2 h. In addition, the stiffness of the specimen reduces and its deformation increases quickly after loading, and the axial load–deflection curve tends to be flatter. The deflection value of specimen P at the ultimate strength (N_u^T) is much greater than that of the specimen under the temperature condition H (see Fig. 9-17).

The specimen under the temperature condition L experiences three stages: freely heating, naturally cooling, and loading at room temperature. The deformation of the specimen during heating in the first stage is the same as that under the temperature conditions H and P. The deflection of the specimen during cooling recovers slightly but only 1 mm or even less is recovered after it is cooled and after 6 h. Then, the specimen is loaded at room temperature and its axial load–deflection curve is similar to that of the

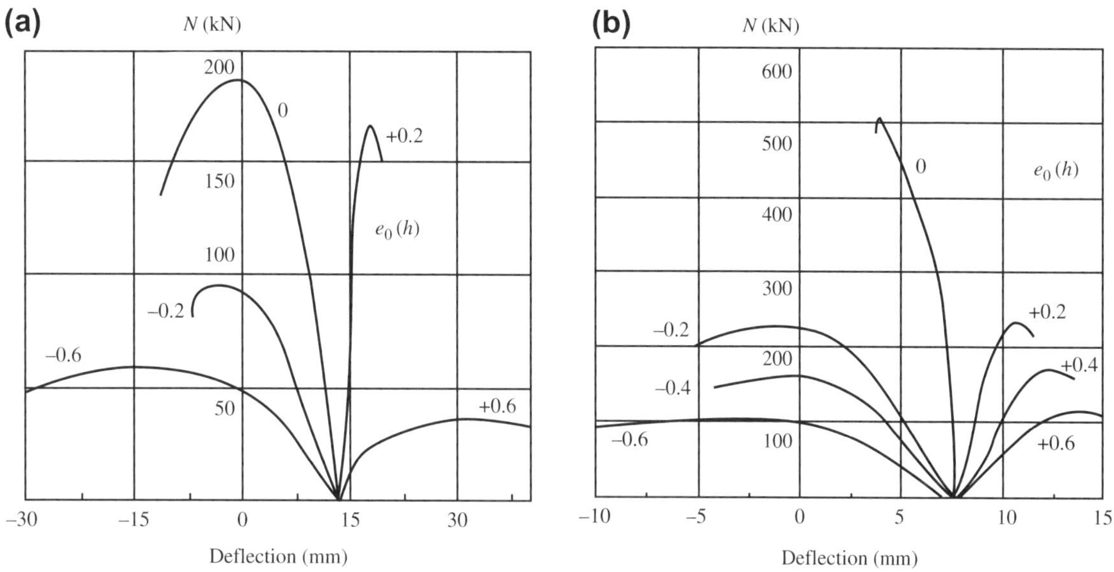

Figure 9-16 Curves of axial load–deflection at the mid-span of eccentric compressive specimens under different temperature conditions[9-12]: (a) condition P; (b) condition L.

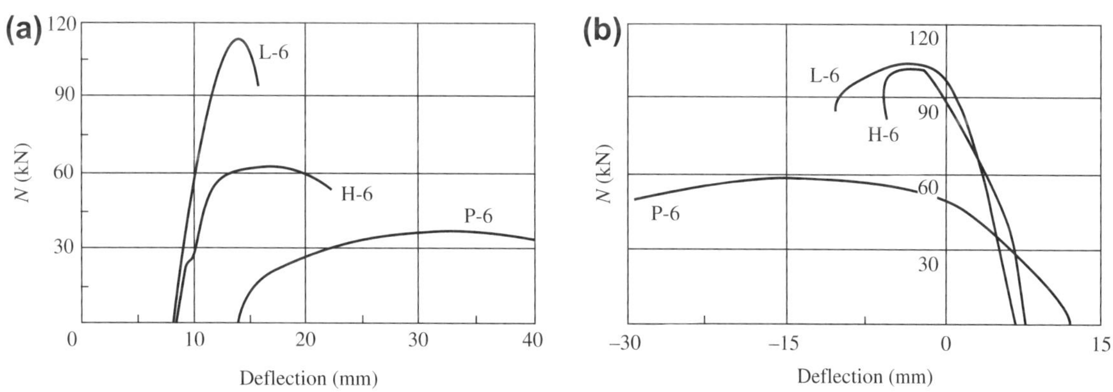

Figure 9-17 Comparison of load–deflection curves of eccentric compressive specimens under different temperature conditions[9-12]: (a) $e_0 = 0.6h$; (b) $e_0 = -0.6h$.

H series specimens, because the mechanical behavior of concrete does not recover after it is cooled.

The direction of deflections during the loading of all the specimens under these temperature conditions is identical. When the load eccentricity $e_0 \geq 0.2h$, the direction of deflection of the specimen during loading is the same as that during heating, so the deflection of the specimen increases continuously and is convex toward the high temperature zone at failure. When the load eccentricity $e_0 \leq 0$, the direction of deflection of the specimen during loading is opposite to that during heating, so the deflection of the specimen is counteracted gradually. The deflection of the specimen at failure turns and is convex toward the low temperature zone, if the load eccentricity $e_0 \leq 0.2h$.

The axial load–deflection curves of the specimens under the three temperature conditions, but with the same initial eccentricity (e_0 = $0.6h$

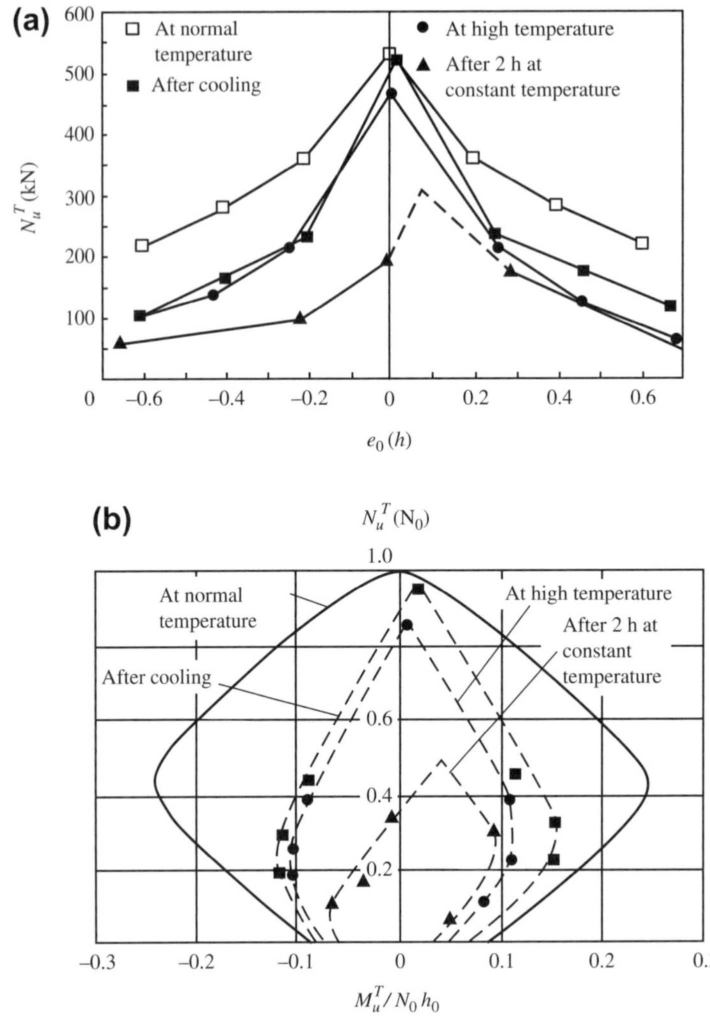

Figure 9-18 Comparison of the ultimate strength of eccentric compressive specimens under different temperature conditions[9-12]: (a) $N_u^T - e_0/h$; (b) $N_u^T - M_u^T$.

or −0.6*h*), are compared in Fig. 9-17. However, the ultimate loads of the specimens loaded at high temperature (H series) and after cooling (L series) vary considerably, i.e., they differ greatly when $e_0 = 0.6h$, whereas they are approaching when $e_0 = -0.6h$; but the slopes of the axial load–deflection curves (or stiffness of the specimen) before the ultimate load and the deflection values at the ultimate load are approaching each other. However, the stiffness of the specimen (P series) loaded after maintaining a high temperature for 2 h reduces obviously, and the deflection caused by loading is much greater than that of the H and L series specimens.

The ultimate strength–eccentricity $(N_u^T - e_0)$ curves and the ultimate envelopes of the axial compression–bending moment $(N_u^T - M_u^T)$ of the eccentric compressive specimens under the three temperature conditions and at a high temperature of 800 °C are shown in Fig. 9-18. Obviously, these curves are asymmetrical and incline to the right-hand side of the ordinate, and they are notably lower than the corresponding curves for the specimen at room temperature.

When the specimen loaded at high temperature (H series) is compared with that loaded after cooling (L series), the failure pattern of both specimens with $e_0 \leq 0.2h$ is the same, i.e., the pattern of smaller eccentricity or the pattern of larger eccentricity is controlled by tensile reinforcement in the low temperature zone. The material strength and the ultimate strength of both specimens are approaching. When the load eccentricity $e_0 \geq 0.4h$, the failure pattern of the specimens is of larger eccentricity and is controlled by tensile reinforcement in the high temperature zone, so the ultimate load of the L series specimens is obviously higher than that of the H series specimens, as the strength of the reinforcement in the L series specimens has recovered to the value at room temperature.

For the specimens loaded after maintaining a high temperature for 2 h (P series), the internal temperature on the section is elevated, the material behavior of most areas in the section deteriorates, the deformation of the specimen increases considerably, and significant ultimate strength is lost, and the area of the envelope reduces sharply and is very inclined.

9.4 ECCENTRIC COMPRESSIVE COLUMN WITH TWO ADJACENT SURFACES EXPOSED TO HIGH TEMPERATURE

9.4.1 Testing Method and Contents

When eccentric compressive specimens, including central compressive and flexural specimens, with two adjacent surfaces exposed to high temperature are tested, a special furnace needs to be developed; the other experimental devices, including the temperature control system, support and loading device for the specimen, and the instruments for measuring and recording data, can be the same as those used for specimens with three surfaces exposed to high temperature (see Fig. 8-1 and Fig. 9-7).

The construction of the special furnace used to test specimens with two adjacent surfaces exposed to high temperature, developed by the Structural Engineering Laboratory of Tsinghua University, is shown in Fig. 9-19. The length of the furnace body is 1100 mm, its transverse outline is 500 mm × 500 mm; the net size of the chamber is 250 mm × 250 mm, and the openings are each 200 mm on two adjacent sides.

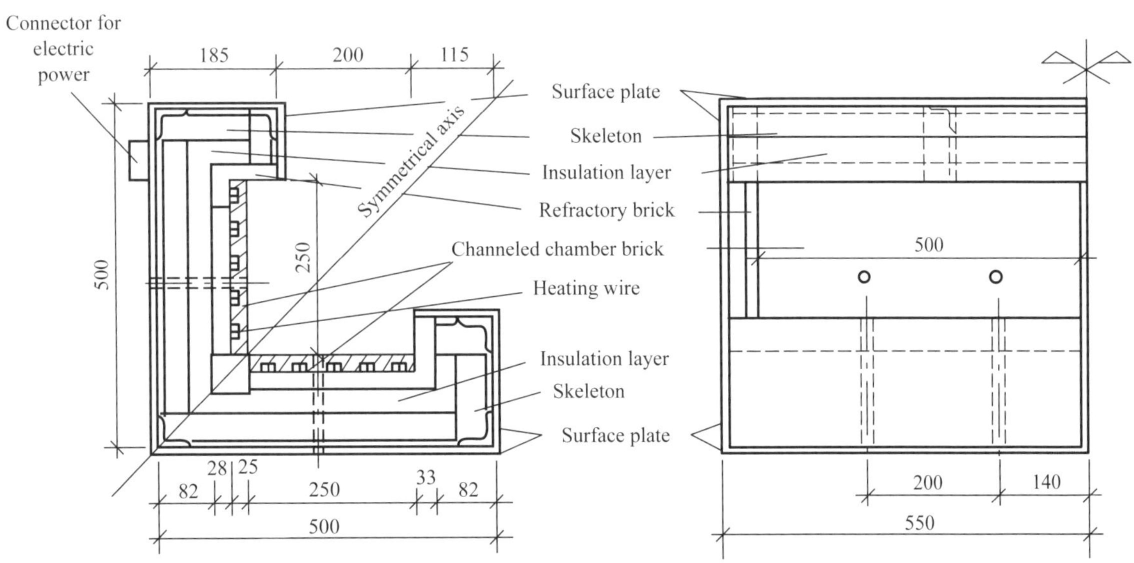

Figure 9-19 Construction of the furnace used for specimens with two adjacent surfaces exposed to high temperature.[9-6]

The construction of the furnace body is symmetrical about a 45° diagonal line. The furnace case is welded by angle and belt steels to form a skeleton and is covered with thin steel plates. Two pieces of channeled chamber brick plate (250 mm × 500 mm), which is made of refractory material, are connected on each side of the furnace. Four 2 kW heating wires pass through those channels and the ends of the wires are linked to the connector outside the furnace case. The refractory bricks are placed outside the channeled brick and the opening bricks are placed at both ends of the channeled brick for protection. Siliceous aluminum refractory fiber is used as the thermal insulation layer and is placed between the chamber brick and the furnace case.

After the specimen (cross section 100 mm × 200 mm) is put into the chamber of the furnace, the electric power is switched on and the furnace is heated. The temperature in the chamber increases with the heating time as shown in Fig. 9-20. The rate of increase in the temperature in the furnace used for the specimen with two adjacent surfaces exposed to high temperature is slightly lower than that used for the specimen with three surfaces exposed to high temperature because the total electrical power of the heating wires (8 kW < 10 kW) is less and the specimen has one more surface dispersing heat. Four thermocouples are placed on each side of the furnace and the temperatures measured in the chamber are similar.[9-6] When the experimental temperature reaches 800 °C and is maintained for 1 h, the temperature on the outer surface of the case is lower than 60 °C. This shows that the temperature in the chamber is distributed uniformly and the heat insulation behavior of the furnace is good.

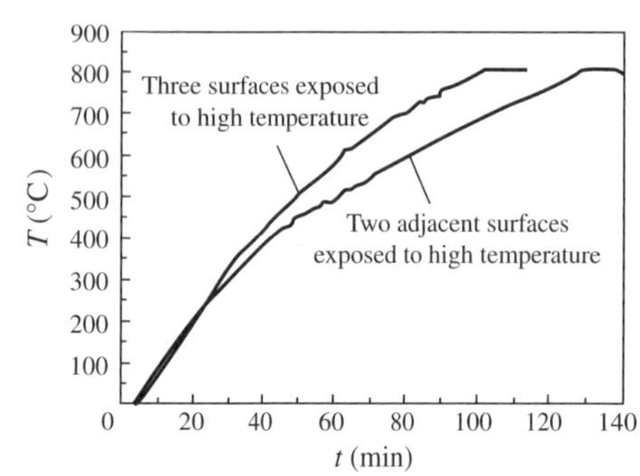

Figure 9-20 Elevating temperature curves for the experimental furnaces.[9-14]

The support and loading method of the specimen are the same as shown in Fig. 9-7. Two spherical hinges are placed at both ends of the specimen and the specimen can bend freely in two directions after it is loaded using a hydraulic jack.

The size and construction of the eccentric compressive specimen are the same as shown in Fig. 9-8, but the specimens are produced in a separate batch. The concrete age of the specimens during testing is 60–90 days, the cubic strength at room temperature is f_{cu} = 24.7 MPa, and the yield strength of the reinforcement at room temperature is f_y = 340 MPa.

All the specimens with two adjacent surfaces exposed to high temperature are tested under the path of loading at constant temperature (T = 800 °C). The number and initial eccentricities (e_{0x}, e_{0y}) of the applied loads are shown in Fig. 9-21 for all specimens. The first and second (if any) digits in the specimen number represent the vertical and horizontal relative eccentricities ($e_{0y}/0.1h$, $e_{0x}/0.1b$). The letter A after the digit shows the load applied in the high temperature zone (negative eccentricity), otherwise it shows the load applied in the low temperature zone (positive eccentricity).

9.4.2 Temperature Distribution and Deformation

During the free heating stage, water vapor starts to escape from the gap between the specimen and the furnace body when the temperature in the chamber is about T = 350 °C. The quantity of vapor escaping reaches a maximum and water drops seep onto the nonheating surfaces of the specimen when T = 500–600 °C; the amount of water vapor reduces gradually and the water on the surfaces evaporates, but some water vapor still escapes at T = 800 °C. This phenomenon is

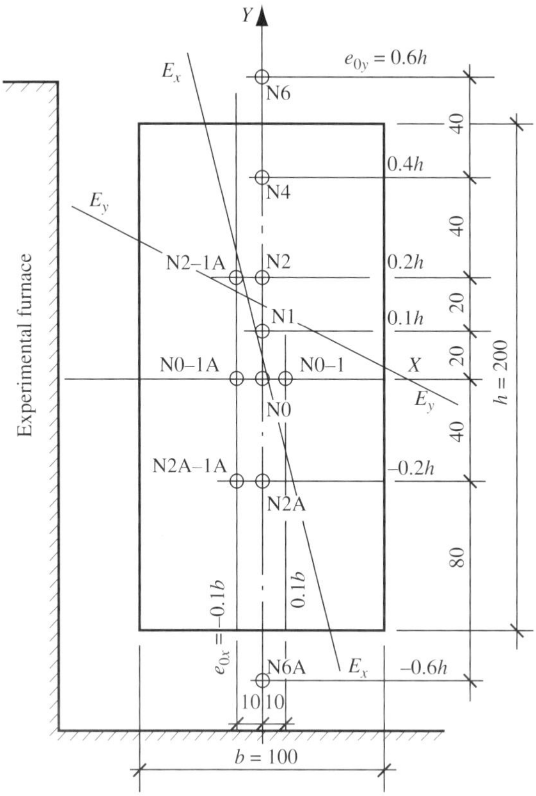

Figure 9-21 Initial positions of the load for all specimens.[9-6]

similar to that of the specimen with three surfaces exposed to high temperature. The color change on the heated surfaces is also the same.

When the experimental temperature of the specimen with two adjacent surfaces exposed to high temperature is T = 800 °C, the measured temperature in its middle section (Fig. 9-22) is not distributed symmetrically in both the horizontal and vertical directions. The temperature at the heated corner (left bottom) of the specimen shows the maximum value, the temperature on both heated surfaces is nearly 800 °C, and the temperature near the unheated surfaces reduces gradually. The temperature on most of the unheated surfaces is rather low (<100 °C). Therefore, a considerable temperature difference (gradient) is formed on the section and along the horizontal and vertical directions.

Comparing the specimen with two adjacent surfaces exposed to high temperature with the specimen with three surfaces exposed to high temperature, the temperatures within 20 mm thickness from the heated surfaces are similar for both specimens, but the temperature on the other parts and the average temperature of the section of the former is much lower than that of the latter; the temperature gradient of the former is greater than that of the latter.

During the free heating process, the deformation of the specimen with two adjacent surfaces exposed to high temperature develops rapidly as the temperature increases (Fig. 9-23). The vertical deflection (Δ_y^T) and the axial deformation Δ_z^T of this specimen are similar to the specimen with three surfaces exposed to high temperature. However, because the total temperature differences (ΔT) along the horizontal and vertical directions of the specimen section are similar but the width of the section is less than its depth ($b < h$), the horizontal curvature (in inverse proportion to the width b) and deflection of the specimen are greater than the vertical curvature (in inverse proportion to depth h) and deflection $\Delta_x^T > \Delta_y^T$, respectively, of the same specimen. Bidirectional flexural deformations appear and the horizontal deflection is greater than the vertical deflection after heating the specimen with two adjacent surfaces exposed to high temperature. This is the main characteristic of the specimen and is different from that of the specimen with three surfaces exposed to high temperature.

During the stage of loading at constant temperature, the vertical deflection (Δ_y) of the specimen with two surfaces exposed to high temperature depends on the eccentricity of the axial load (Fig. 9-24). When the specimen is loaded with vertical eccentricity only (e_{0y} = 0.6h to –0.6h and e_{0x} = 0), the flexural deflection (Δ_y) caused by the load is convex toward the high temperature zone and is in the same direction as the heating deflection if $e_{0y} \geq 0.2h$; the deflection caused by the load develops in the opposite direction if $e_{0y} \leq 0.1h$. When the eccentricity of the specimen is $e_{0y} \leq 0$, the deflection at the ultimate strength is convex toward the low temperature zone. The regularity

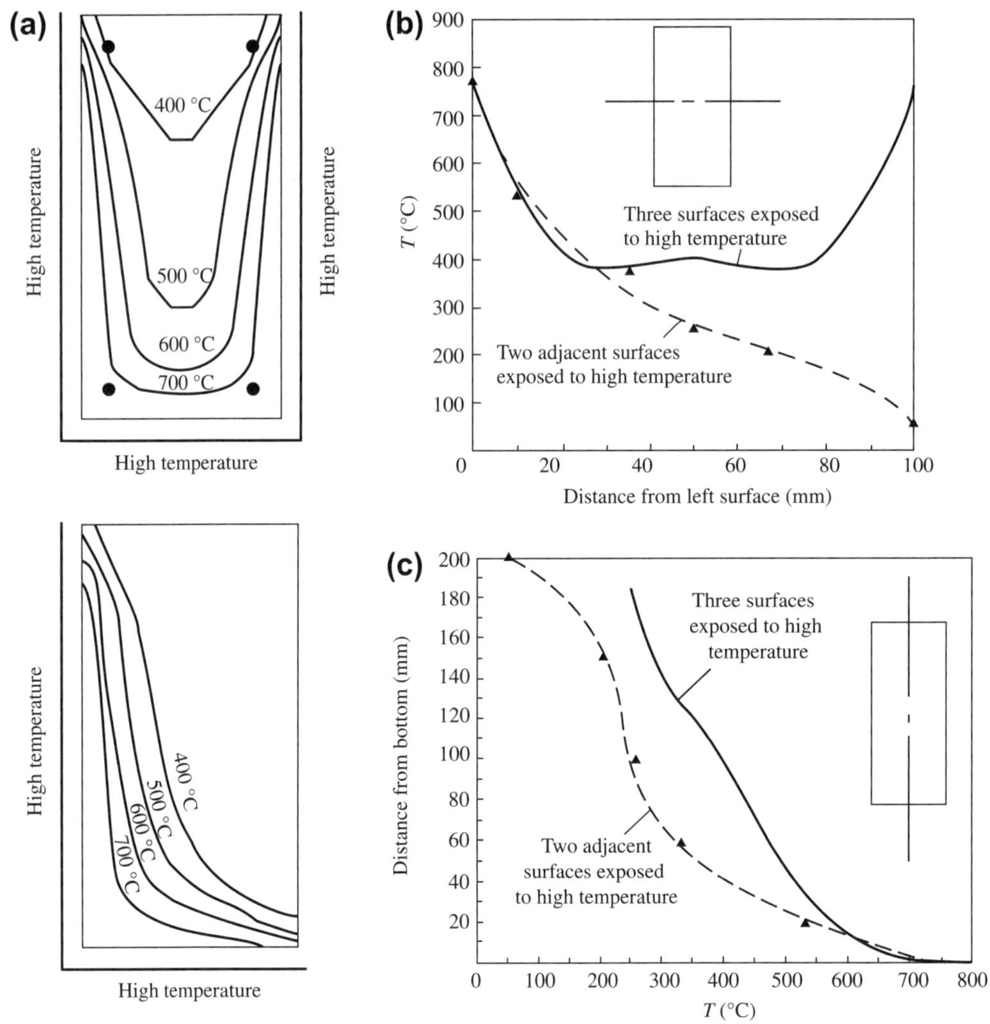

Figure 9-22 Temperature distribution on the section of the specimen[9-6]: (a) temperature contours; (b) along the width of the section; (c) along the depth of the section.

of deformation and the failure pattern of the specimens are similar to those of the specimen with three surfaces exposed to high temperature (see Fig. 9-9).

When the specimen is loaded with horizontal eccentricity ($e_{0x} \neq 0$), the vertical deflection (Δ_y) caused during loading may be convex toward the high or the low temperature zones (Fig. 9-24(b)). The section can be divided into two areas (see Fig. 9-21) according to the experimental results[9-6]: when the initial position of the load is located on the upper right part (low temperature zone) of the line E_y–E_y, the deflection caused during loading is convex toward the high temperature zone and is in the same direction as the deflection during heating; when the initial position of the load is located on the lower left part (high temperature zone) of the line E_y–E_y, the deflection caused during loading is opposite to that caused during heating. Furthermore, the further the load position from the line E_y–E_y, the flatter the curve N–Δ_y, and the larger the deflection caused by the load.

The horizontal deflection (Δ_x) of the specimen with two adjacent surfaces exposed to high temperature also depends on the eccentricity of the axial load (Fig. 9-25) during the stage of loading at constant temperature. When the load position is located on the right side (low temperature zone) of the line E_x–E_x (see Fig. 9-21) on the section, the deflection caused during loading is convex toward the high temperature zone and is in the same direction as the deflection during heating. When the load position is located on the left side (high temperature zone) of the line E_x–E_x, the deflection caused during loading is opposite to that during heating. Also, the further the load position from line E_x–E_x, the flatter the curve N–Δ_x and the larger the deflection caused by the load.

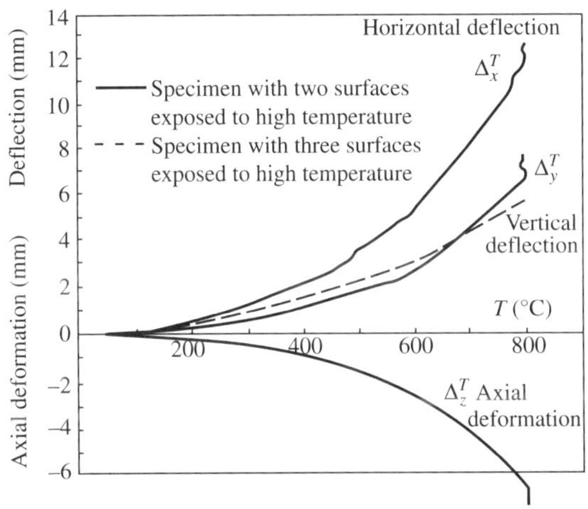

Figure 9-23 Deformation of the specimen with two adjacent surfaces exposed to high temperature during heating.[9-6]

The total horizontal and vertical deflections of the eccentric compressive specimens with two surfaces exposed to high temperature are Δ_{ux}^T and Δ_{uy}^T, respectively, at the ultimate strength (N_u^T), and the deflections caused by the load are Δ_{ux}^N and Δ_{uy}^N, respectively, after deducting the corresponding deflections Δ_x^T and Δ_y^T during heating (Eqn (9.1)). They vary also with the eccentricity of the load (Fig. 9-26). When the specimen is loaded with vertical eccentricity only ($e_{0x} = 0$), the vertical deflections (Δ_{uy}^T and Δ_{uy}^N) turn from being convex toward the high temperature zone into being convex toward the low temperature zone, when the eccentricity e_{0y} changes from $+0.6h$ to $-0.6h$. The variation regularity is similar to that of the specimen with three surfaces exposed to high temperature (see Fig. 9-10). The horizontal deflection (Δ_{ux}^N) caused by the load also changes from positive to negative, but the total deflection (Δ_{ux}^T) of the specimen is always convex toward the high temperature zone (positive value) because the deflection caused during heating is considerable ($\Delta_x^T \approx 12$ mm).

When the specimens are loaded with horizontal eccentricity only ($e_{0y} = 0$), the horizontal deflections (Δ_{ux}^T) are convex toward the high

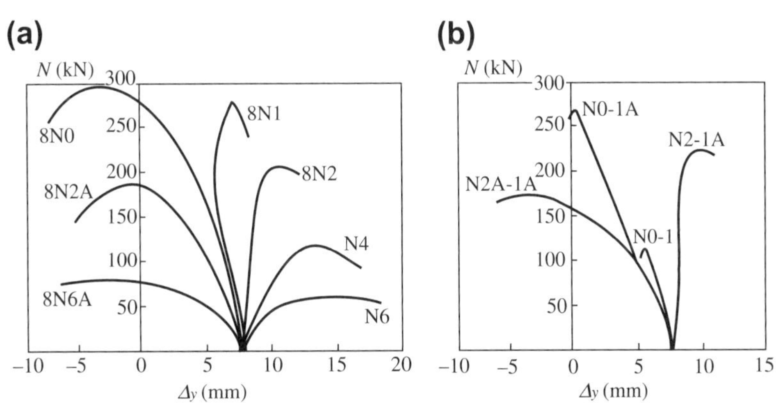

Figure 9-24 Vertical deflection of eccentric compressive specimens with two adjacent surfaces exposed to high temperature[9-6]: (a) $e_{0x} = 0$, $e_{0y} \neq 0$; (b) $e_{0x} \neq 0$.

temperature zone and the vertical deflections (Δ_{uy}^{T}) change from being convex toward the high temperature zone to being convex toward the low temperature zone (Fig. 9-26(b)) when the eccentricity changes from +0.1*b* to −0.1*b*.

When the axial compression is applied, the axial deformation (Δ_z) of the specimen with two adjacent surfaces exposed to high temperature is a contraction (Fig. 9-27) regardless of the direction and value of the load eccentricity, and this causes a reduction in the expansion deformation during heating. When parts of the specimens reach their ultimate strengths, the expansion deformations during heating are totally counteracted and the total deformations are contractions.

When the eccentric compressive specimen with two adjacent surfaces exposed to high temperature is tested, three angle transducers are set up at both ends and in the mid-span of the specimen in order to measure the torsional angle. The results show that the bidirectional flexural deflections of the specimen are dominant during the heating and loading process, and the torsional deformation is small and negligible.

9.4.3 Ultimate Strength

When the specimen with two adjacent surfaces exposed to high temperature fails under the path of loading at constant temperature, the failure pattern bidirectional eccentric compression and its residual flexural deflection is obvious in both directions. The compressive failure and tensile crack zones on its section are not symmetrical and its neutral axis should be inclined (Fig. 9-28); this applies even with the load of the specimen with the eccentricity in one direction only or without eccentricity ($e_{0x} = e_{0y} = 0$). When the

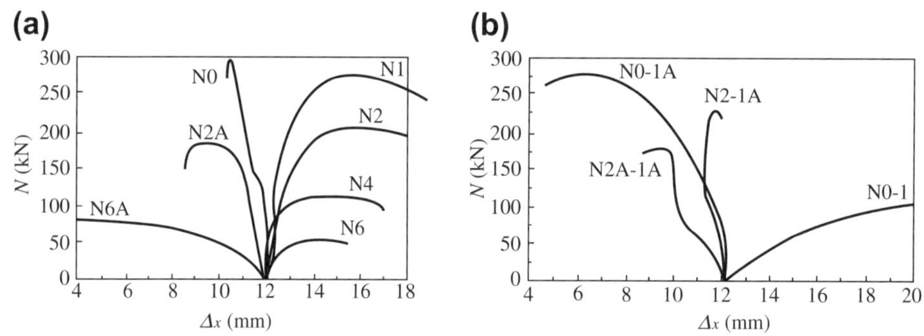

Figure 9-25 Horizontal deflection of eccentric compressive specimens with two adjacent surfaces exposed to high temperature[9-6]: (a) $e_{0x} = 0$, $e_{0y} \neq 0$; (b) $e_{0x} \neq 0$.

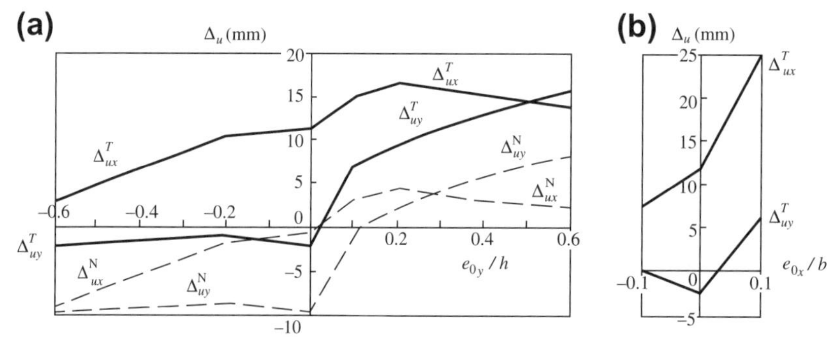

Figure 9-26 Horizontal and vertical deflections at the ultimate strength of eccentric compressive specimens with two adjacent surfaces exposed to high temperature[9-6]: (a) $e_{0x} = 0$; (b) $e_{0y} = 0$.

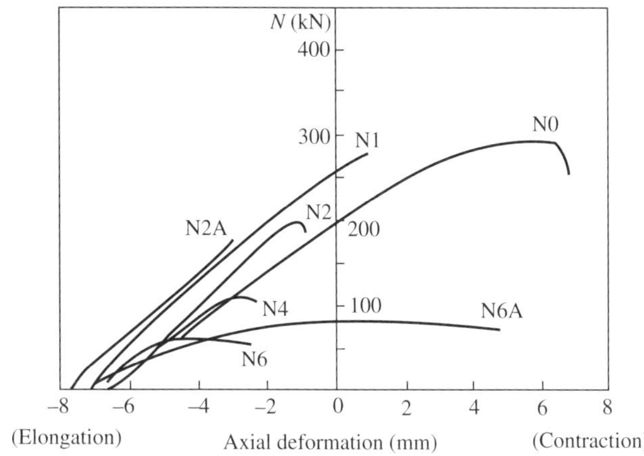

Figure 9-27 Axial deformation of eccentric compressive specimens with two adjacent surfaces exposed to high temperature.[9-6]

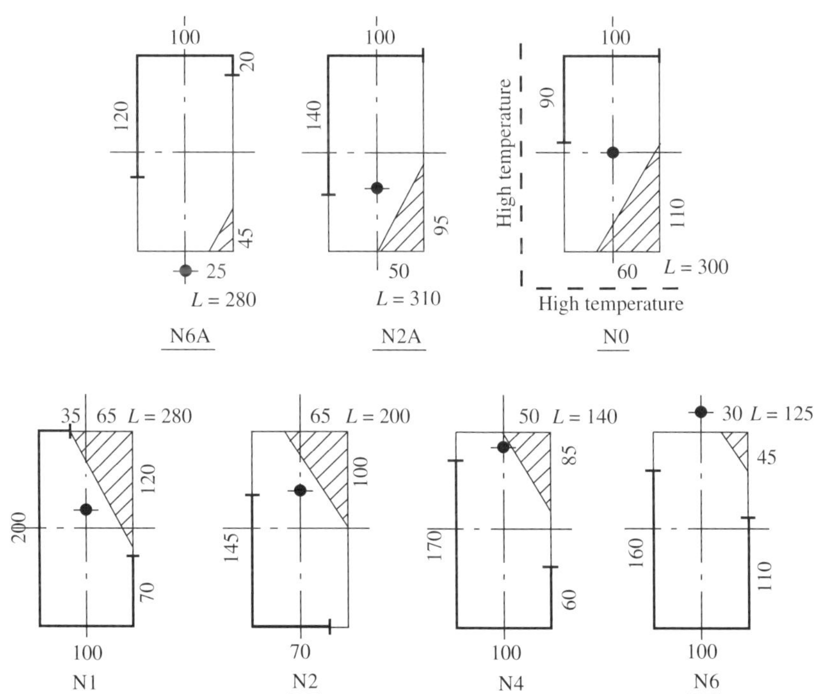

Figure 9-28 Failure zone on the section of a compressive specimen with two adjacent surfaces exposed to high temperature and with one-directional eccentricity ($e_{0x} = 0$) only.[9-14] The black dot in this figure shows the initial position of the load, the shaded area represents the compression zone, L is the longitudinal length of the compression failure zone, and the thick lines along the periphery denote the scope of the tensile crack and the figure noted shows the length along the periphery of the section.

specimen is loaded with vertical eccentricity only ($e_{0x} = 0$), the compressive failure zone is located on the upper right part (low temperature corner) of the section and the tensile crack zone is located on the lower left part (high temperature corner) of the section if the eccentricity $e_{0y} \geq 0.1h$; the compressive failure zone turns to the lower right part and the tensile crack zone is located in the upper left part if the eccentricity $e_{0y} \leq 0$. As the eccentricity of the load increases, the area of the compressive failure zone on the section reduces, but the area of the tensile crack zone increases and the width of the crack also increases.

The ratios between the measured ultimate strengths (N_u^T) of the eccentric compressive specimens with two adjacent surfaces exposed to high temperature and the strength (N_0) of the central compressive specimen at room temperature are written at the initial positions of the load applied, and the ratios vary with the eccentricity in one direction of the load (e_{0y}/h and e_{0x}/b, separately) (Fig. 9-29). The ultimate strength of the specimens distributes asymmetrically along both X and Y axes, and the initial position of the maximum ultimate load is located in the low temperature zone (the first quadrant of the section), i.e., the optimum eccentricities on both directions are greater than zero but are not equal.

The variation regularity of the ultimate strength of the eccentric compressive specimen with two adjacent surfaces exposed to high temperature is similar to that with three surfaces exposed to high temperature (see Fig. 9-11) as the eccentricity changes. When the load eccentricity is greater than the optimum one, i.e., the load applied is located in the low temperature zone, the deflection during loading is convex toward the high temperature zone and it is the same as the deflection during heating, so the additional eccentricity at the ultimate state is high. In addition, the yield strength of the tensile reinforcement located in the high temperature zone reduces significantly, so the ultimate strength of the specimen decreases quickly. On the other hand, when the load eccentricity is smaller than the optimum one, i.e., the load applied is located in the high temperature zone, the direction of the deflection during loading is opposite to that during heating and the additional eccentricity is small, and the yield strength of the tensile reinforcement located in the low temperature zone reduces less, so the ultimate strength of the specimen decreases slowly.

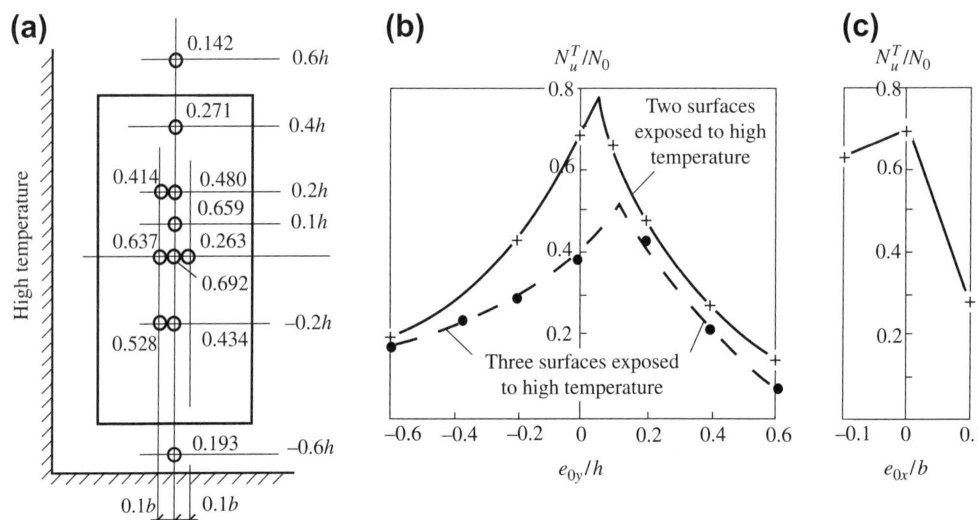

Figure 9-29 Ultimate strength of an eccentric compressive specimen with two adjacent surfaces exposed to high temperature[9-15]: (a) distribution of the ultimate strength; (b) $e_{0x} = 0$, $e_{0y} \neq 0$; (c) $e_{0y} = 0$, $e_{0x} \neq 0$.

When the specimen with two adjacent surfaces exposed to high temperature is loaded eccentrically in one direction only ($e_{0x} = 0$, e_{0y}), the vertical ultimate bending moment (M_{uy}^T) and the envelope of the axial compression–bending moment ($N_u^T - M_{uy}^T$) are asymmetrical about the ordinate (Fig. 9-30). In addition, because the temperature on the left side of the section is higher and the mechanical behavior of the materials there deteriorates significantly, the horizontal deflection arises during both heating and loading stages, and the additional horizontal eccentricity (Δ_{ux}^T, see Fig. 9-26) and corresponding horizontal ultimate bending moment ($M_{ux}^T = N_u^T \cdot \Delta_{ux}^T$) are not small at the ultimate state, even though the initial horizontal eccentricity of the load is zero ($e_{0x} = 0$). The horizontal ultimate bending moment has a positive value (deflection convex toward the high temperature zone) and is also asymmetrical about the ordinate (Fig. 9-30(a)). The maximum value occurs when the eccentricity is near the optimum one and it reduces gradually as the vertical eccentricity moves out to both sides.

The eccentric compressive specimens with two adjacent surfaces and with three surfaces exposed to high temperature have the same size and materials, and they also have similar test methods and heating–time curves, but the ultimate strengths are different because the temperature fields in their sections are different (see Fig. 9-22), and hence the behavior of the material, the deformations, and the stress states are also different. Comparing the specimens with vertical eccentricity only ($e_{0x} = 0$, e_{0y}), as an example, the ultimate strengths (N_u^T) of both specimens are similar when the eccentricity is greater (i.e., $e_{0y} \geq 0.2h$ or $e_{0y} \leq -0.4h$) according to Figs. 9-29 and 9-30. However, when the eccentricity is smaller ($e_{0y} = -0.4h$ to $0.2h$) and the specimens have a failure pattern of smaller eccentricity, the ultimate strength (N_u^T) of the specimen with two adjacent surfaces exposed to high temperature is much higher than that with three surfaces exposed to high temperature. Obviously, although the specimen with two adjacent surfaces exposed to high temperature presents an asymmetrical temperature distribution in both directions of the section

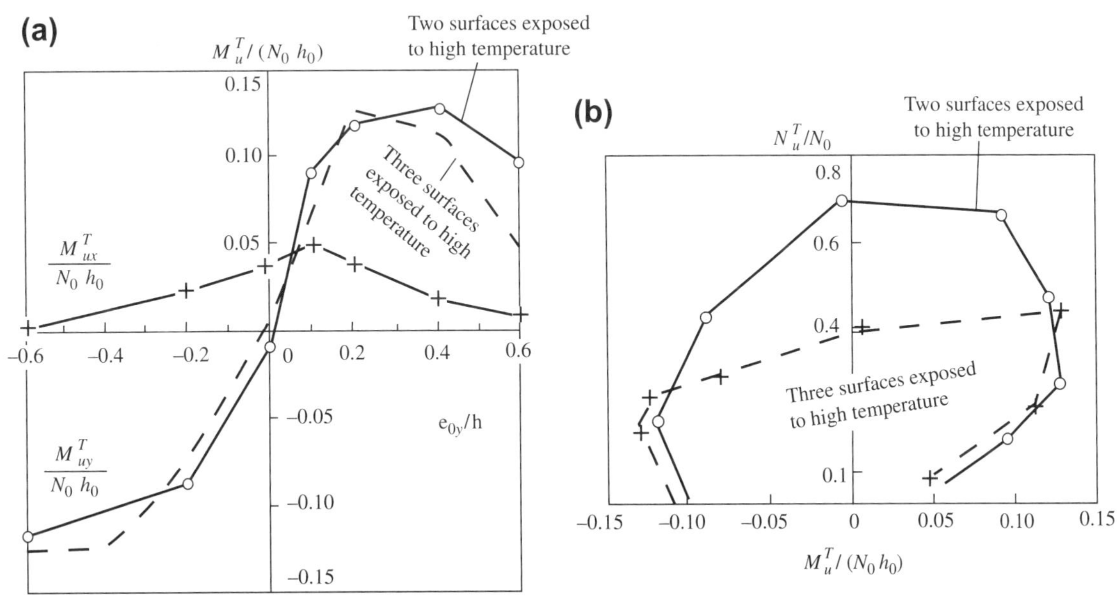

Figure 9-30 The ultimate bending moment and the envelope of the axial compression–bending moment of eccentric compressive specimens ($e_{0x} = 0$) with two adjacent surfaces exposed to high temperature[9-14]: (a) ultimate bending moment; (b) ultimate envelope of the axial compression–bending moment.

and appears to have a larger horizontal deflection, the ultimate strength is still higher because the failure pattern is controlled by the compressive resistance of concrete when the eccentricity is smaller, the area with high temperature in its section is smaller, and the total loss of material strength is less.

CONCLUSIONS

When a structural member sustains a fire, but its periphery is not wholly exposed to high temperatures, e.g., only three surfaces or two adjacent surfaces of a rectangular section are exposed to high temperature, a nonuniform and asymmetrical temperature field is formed in the section. Hence, the distributions of the material behavior in the section are also nonuniform and asymmetrical, and the mechanical behavior of the structural member is also asymmetrical.

There is a vertically symmetrical axis in a section of a compressive member with three surfaces exposed to high temperature, and the member is flexed in one direction only when the load acts in the plane of the symmetrical axis. However, when the axial load acts on the geometric center of the section, the ultimate strength of the compressive member is not the maximum and the deflection is obvious. Only when the axial load acts in the low temperature zone can the maximum ultimate strength of the compressive member be achieved and the deflection is very small. The point where the load is applied is called the optimum center, the distance from it to the geometric center is called the optimum eccentricity, and the corresponding strength is called the optimum ultimate strength of the compressive member. Obviously, the optimum center and the ultimate strength of the compressive member vary with the temperature field in the section, and the optimum eccentricity increases and the optimum ultimate strength reduces when the temperature increases.

When the axial load acts on the low temperature side and its eccentricity is greater than the optimum, the compressive member fails with the pattern of smaller eccentricity, which is controlled by the compressive concrete in the low temperature zone, or fail with the pattern of larger eccentricity, which is controlled by yielding of the tensile reinforcement in the high temperature zone. The ultimate deflection of the compressive member is convex toward the high temperature zone and the ultimate strength reduces quickly as the eccentricity increases. On the other hand, when the eccentricity of the axial load is smaller than the optimum and the load acts on the high temperature side, the compressive member fails with the pattern of smaller eccentricity, which is controlled by the compressive concrete in the high temperature zone, or fails with the pattern of larger eccentricity, which is controlled by yielding of the tensile reinforcement in the low temperature zone. The ultimate deflection of the compressive member is convex toward the low temperature zone and the ultimate strength reduces slowly as the eccentricity increases.

There is no symmetrical axis in both the X and Y directions of a section of a compressive member with two adjacent surfaces exposed to high temperature and biflexural deflections occur during both the heating and loading processes; even the horizontal deflection is greater than the vertical one. The optimum center of the section is located in the low temperature zone and is away from the geometric center; both horizontal and vertical eccentricities exist between both centers and they increase with the temperature. The variations in the ultimate strength and the deflection direction of the compressive member depend on the comparison between the load eccentricity and the optimum eccentricity, and the regularity is similar to that of the compressive member with three surfaces exposed to high temperature.

Although the actions of temperature and load are asymmetrical and greater deflections are formed in both directions of the section, the area of deteriorated concrete in the section is smaller and slighter, so the ultimate strength of the compressive member with two adjacent surfaces exposed to high temperature is greater than that with three surfaces exposed to high temperature under the same environmental temperature and eccentric load conditions.

The ultimate temperature–strength of a structural member with three surfaces exposed to high temperature and under the path of loading at constant temperature is the lower boundary among various temperature–load paths. The behavior of the structural member under the path of heating under constant load is better than that under the path of loading at constant temperature, and the ultimate temperature–strength of the former is obviously higher, especially when the temperature is much higher and the load eccentricity is much smaller.

The mechanical behavior of eccentric compressive (including central compressive and bending) structural members deteriorates at elevated temperatures; cracking and deformation increase and its strength decreases. The main reasons are the damage to the materials in the concrete and reinforcement at elevated temperatures and the additional deformation caused by nonuniform and asymmetrical behavior of the materials in the section. The deterioration in behavior of the structural member tends to be more serious as the temperature increases or is sustained. When the structural member at elevated temperatures is maintained for a period of hours, the temperature inside the section continues to increase gradually, the area of damaged material expands, and the ultimate strength reduces continuously and considerably.

When the structural member cools down to room temperature after experiencing a high temperature, the strength and modulus of elasticity of the reinforcement recover, but the mechanical behavior of the concrete does not recover. Therefore, most of the ultimate strength of a structural member at room temperature recovers if its failure is controlled by yielding of the tensile reinforcement, which includes failure patterns of larger eccentricity compression and bending. But the ultimate strength of the structural member cannot be recovered if the failure pattern is of smaller eccentricity compression and controlled by the compression of the concrete.

REFERENCES

[9-1] N. Su, T. Lin, T.T. Lie, Fire-resistant behavior of reinforced concrete column, China Civil Engineering Journal 25 (1) (1992) 25–36.

[9-2] A.H.B. Ng, M.S. Mirza, T.T. Lie, Response of direct models of reinforced concrete column subjected to fire, ACI Structural Journal 87 (3) (1990) 313–323.

[9-3] H. Li, Experimental investigation of reinforced concrete columns under loading and elevated temperature. Masters dissertation, Tsinghua University, Beijing (1994).

[9-4] H. Li, X. Shi, Z. Guo, Researches on failure characteristic and deformation regularity of central compression column of reinforced concrete at elevated temperature, Special Structures 14 (1) (1997) 34–37.

[9-5] X. Shi, H. Li, Z. Guo, Experimental investigation on mechanical behavior of reinforced concrete central compression column with three surfaces exposed to fire, Journal of Building Structures 18 (4) (1997) 13–22.

[9-6] J. Yang, Experimental research, theoretical analysis, and practical calculation of eccentrically compressed reinforced concrete members under loading and elevated temperature. Doctoral thesis, Tsinghua University, Beijing (2000).

[9-7] X. Shi, H. Li, Z. Guo, Experimental investigation of eccentric compression column of reinforced concrete with three surfaces exposed to high temperature, Journal of Fuzhou University (Science and Technology) (Supplementary Issue) 24 (1996) 138–144.

[9-8] X. Shi, H. Li, Z. Guo, Research on ultimate strength of reinforced concrete compression column with three surfaces exposed to high temperature, Engineering Mechanics (Supplementary Issue) (1996) 49–53.

[9-9] J. Zhang, X. Shi, Z. Guo, Experimental investigation of compressive-flexural reinforced concrete members with different eccentricities at elevated temperature. Symposium of Fifth National Conference on Basic Theory and Engineering Application of Reinforced Concrete Structures, Tianjin University Press, Tianjin (1998) 10–15.

[9-10] J. Yang, W. Ding, X. Shi, Z. Guo, Experimental investigation of compressive-flexural reinforced concrete members under different temperature. Symposium of Fifth National Conference on Basic Theory and Engineering Application of

Reinforced Concrete Structures, Tianjin University Press, Tianjin (1998) 212–216.

[9-11] J. Yang, X. Shi, Z. Guo, Experiment and analysis of mechanical behavior of eccentrically compressed reinforced concrete members under two heating-loading paths, Engineering Mechanics 18 (3) (2001) 81–90.

[9-12] J. Sun, Comparative research on behaviors of reinforced concrete members under several temperature situations. Masters dissertation, Tsinghua University, Beijing (2001).

[9-13] J. Sun, Y. Li, X. Shi, Z. Guo, Experimental investigation on behaviors of eccentrically compressed reinforced concrete members heated on three surfaces under different temperature conditions, Journal of Building Structures 22 (4) (2001) 84–89.

[9-14] J. Yang, X. Shi, Z. Guo, Comparison of mechanical behaviors between reinforced concrete eccentric compression members heated on two and three surfaces, Industrial Construction 30 (6) (2000) 34–37.

[9-15] J. Yang, X. Shi, Z. Guo, Experimental research on reinforced concrete members with flexural and axial load and two surfaces exposed to high temperature, Building Structure 30 (2) (2000) 23–27.

CHAPTER

10

Behavior of Statically Indeterminate Structures at Elevated Temperatures

10.1 INVESTIGATION OF CONTENT AND TESTING METHODS

One of the outstanding advantages of concrete is that it is made by molding, when it is used as a main material in structural engineering. An integral structure is formed easily after concrete is cast in a mold, regardless of the variations in shape and size. Therefore, most reinforced concrete structures in practice are statically indeterminate systems, and the typical form is a frame structure connected integrally by beam and column members.

The mechanical behavior of the basic structural members, including flexural and compressive members, with the emphasis on analyses of their cross section, at elevated temperatures was introduced in detail in the previous chapters. This chapter summarizes the experimental regularities of simple (statically determinate) specimens.

When a building sustains a fire accident or another temperature effect, the behavior and safety of the statically indeterminate concrete structure depend not only on the behavior of the structural members (cross sections) at elevated temperatures, but also, to a greater extent, on the distribution of internal forces and their variation within the structure. Because of the thermal inertia of concrete, a severe nonuniform temperature field is formed on the section of a structural member under the action of high temperature. Correspondingly, considerable flexural and axial deformations of the member occur and are restrained by other structural members; the stiffness of the member (section) is changed due to deterioration in the behavior of the materials at elevated temperatures. Therefore, the internal forces in a statically indeterminate structure redistribute significantly and thus the deformation, failure pattern, and ultimate strength change.

Experimental investigations on statically indeterminate concrete structures at elevated temperatures are limited so far. Some existing experiments[10-1,10-2] provide only the fire endurance (hours) level of the structural member, but lack measurements and analyses on the variation in the processes of deformation and internal forces of the structure, because of difficulty in measuring these at elevated temperatures.

An experimental measuring technique and two types of specimens of statically indeterminate structure (i.e., a continuous beam of two spans and a single-bay and single-story frame) have been designed, produced, and tested under heating and loading[1-12] in the Structural Engineering Laboratory of Tsinghua University. Complete data and experimental results[10-3–10-5] are available after observing and measuring during the testing process.

10.1.1 Specimen Design and Testing Content

A continuous beam of two spans is the simplest statically indeterminate structure containing only one indeterminacy, but it reflects the basic characteristic behavior of a statically indeterminate structure. The mechanical conditions are simple and clear; the deformations, redistribution of internal force, and failure process are

Experiment and Calculation of Reinforced Concrete at Elevated Temperatures

observed and measured easily during testing. The experimental devices needed are readily available and the experimental costs are low. Therefore, it is a suitable device for testing a statically indeterminate structure at elevated temperatures.

The continuous beam specimen (Fig. 10-1) is designed with two equal spans of 1.2 m each. The total length and section are 2.5 m and 100 mm × 180 mm, respectively, which fit within the existing furnace (see Fig. 8-3) for testing a structural member with three surfaces exposed to high temperature. The upper and lower reinforcements on the section of the specimen are symmetrical and its construction is shown in Fig. 10-1. Both ends of the specimen are supported on rollers, which permit longitudinal (horizontal) displacement, and one concentrated load acts symmetrically on each span.

Six specimens of a continuous beam were produced to investigate the load position (the distance (βl) between the concentrated load and the end support), the initial loading level (the value of constant load), and the heating conditions (heating simultaneously on both spans or heating on one span only while the other span is kept at room temperature). The numbers and experimental parameters of the specimens are listed in Table 10-1.

A single-bay and single-story frame was selected as the statically indeterminate structure. Although it is the simplest form of frame structure, it includes both basic structural members, i.e., beam and column, and has three degrees of indeterminacy. Therefore, it still reflects the main characteristics of the frame structure.

The construction of the frame specimen is shown in Fig. 10-2. Two concentrated loads act symmetrically on the third point of the beam. The span and height of the frame and the sizes of beam and column sections fit well with the existing experimental furnace (see Fig. 8-3). The bottoms of the two columns are connected with two stiff base beams.

Five frame specimens were designed and their numbers and experimental parameters are listed in Table 10-2. One specimen (TFC-1), used for comparison, was tested at room temperature, and the value of the ultimate load obtained is P_u (kN). Four other specimens were tested at elevated temperatures and the factors investigated are the initial load level (P_0/P_u) and the ratio between the linear elastic rigidities of the beam and column:

$$i_b = \frac{Eb_b h_b^3}{12L}, \quad i_c = \frac{Eb_c h_c^3}{12H}$$

$$\therefore \quad \frac{i_b}{i_c} = \frac{b_b h_b^3 H}{b_c h_c^3 L} \tag{10.1}$$

All the specimens were made in the laboratory using a steel mold. The continuous beams were cast in a normal position and the frames were cast horizontally. They were prepared for testing after casting, compacting, and curing of the concrete. The raw materials used for the concrete specimens were slag cement, river sand, and crushed limestone with maximum particle size of 15 mm. The mix of the concrete and its strength at room temperature while testing are shown in Table 10-3.

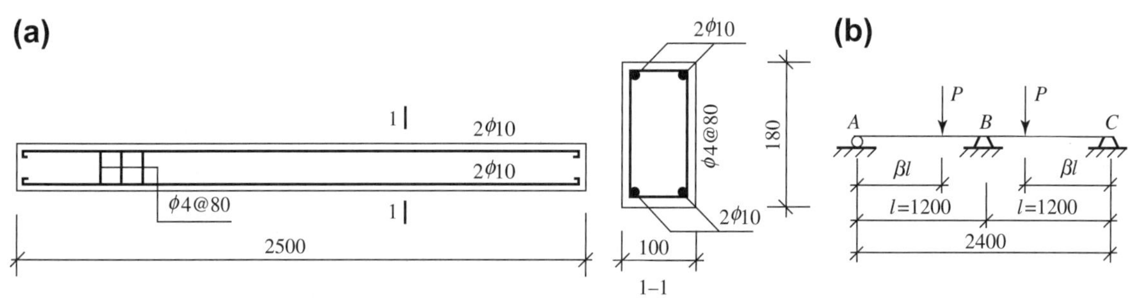

FIGURE 10-1 Construction and loading pattern of a continuous beam specimen[1-12]: (a) specimen; (b) loading pattern.

TABLE 10-1 Numbers and Experimental Parameters for Continuous Beam Specimens

Number of specimen	Load position (β)	Heating spans	Value of constant load (kN)
TCB1-1	$\frac{1}{3}$	2	10
TCB1-2		2	20
TCB1-3		1	20
TCB2-1	$\frac{2}{3}$	2	25
TCB2-2		2	35
TCB2-3		1	10

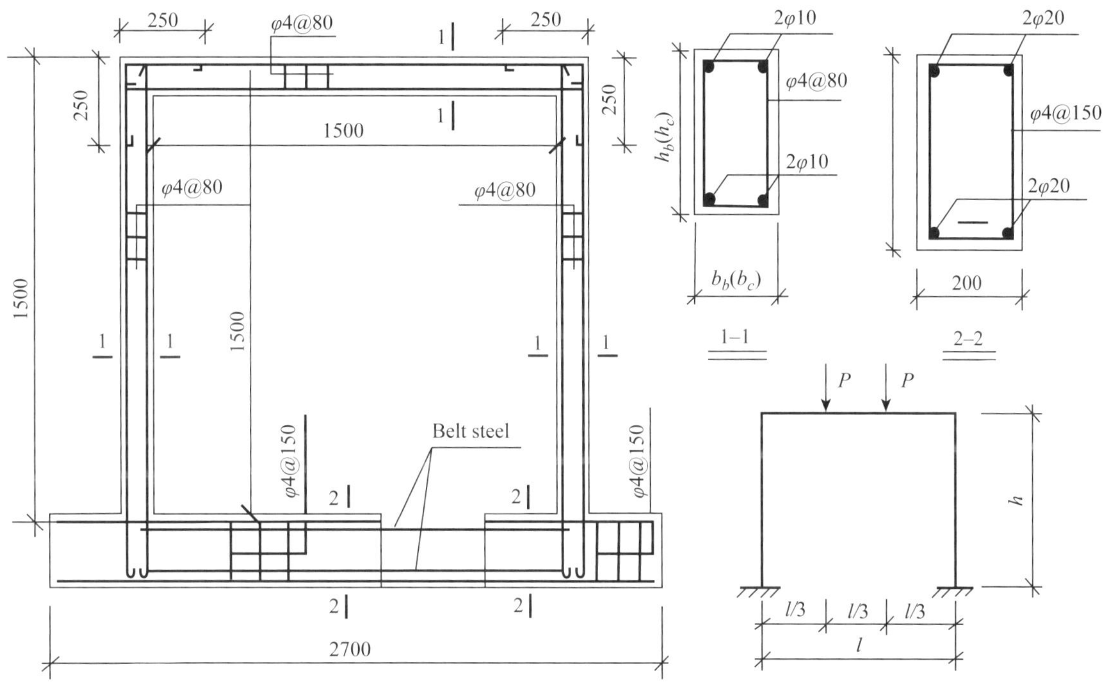

FIGURE 10-2 Construction and loading pattern of a frame specimen.[1-12]

TABLE 10-2 Numbers and Experimental Parameters for Frame Specimens

Number of specimen	Span of beam L (mm)	Height of column H (mm)	Depth of beam section h_b (mm)	Depth of column section h_c (mm)	Ratio between linear elastic rigidities of beam and column (i_b/i_c)	Load level P_0/P_u
TFC-1	1700	1425	150	200	0.354	1.0
TFC-2	1700	1425	150	200	0.354	0.49
TFC-3	1700	1425	150	200	0.354	0.30
TFC-4	1650	1425	150	150	0.877	0.29
TFC-5	1650	1400	200	150	2.011	0.31

The width of the beam and column sections is $b_b = b_c = 100$ mm. The beam span L and the column height H are varied because of the different depths of their sections, and are adjusted to fit the size of the experimental furnace.

TABLE 10-3 Mix and Strength of Concrete

Number of specimen		Grade of cement	Mix in weight ratio				Cubic strength at room temperature f_{cu} (MPa)
			Cement	Water	Sand	Crushed stone	
Continuous beam	TCB (All)	325	1	0.48	1.54	2.99	29.50
Frame	TFC-1	325					29.94
	TFC-2						
	TFC-3						
	TFC-4	425					47.81
	TFC-5						

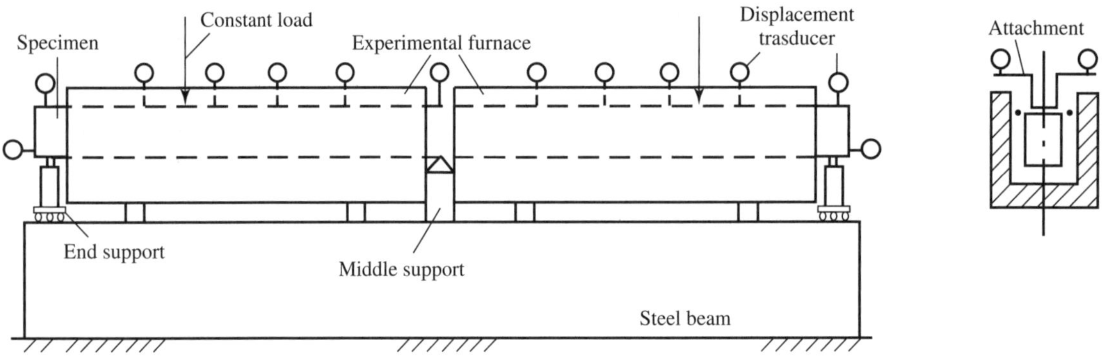

FIGURE 10-3 Experimental devices and measuring points for a continuous beam.[1-12]

The longitudinal reinforcement used in all the specimens was grade I (diameter is 10 mm), its yield strength at room temperature was $f_y = 270$ MPa, and the net thickness of the concrete cover was 10 mm. The stirrups were made of cold drawing steel wire (No. 8) after tempering.

10.1.2 Testing Method and Measuring Technique

The testing method of heating and loading of the continuous beam and frame specimens is similar to that for single beam or column specimens. The heating and temperature control system, and the transducers and recording instruments for measuring various items, which are introduced in Chapters 8 and 9, can also be used. Only the support system and loading equipment for the statically indeterminate specimens have to be modified, and a measuring method for the unknown redundant reactions has to be developed.

The continuous beam is supported on a long steel beam, each span is heated using an experimental furnace (see Fig. 8-3) with three surfaces of the specimen exposed to high temperature, and its three supports out of the furnace and in an area at room temperature (Fig. 10-3). The middle support is a hinge and free rotation of the specimen is permitted; two movable and rotary roller supports with reaction measurement capabilities are set up at both ends of the specimen. The displacement transducers are set up on top (low temperature area) of the specimen to measure the transverse deflection. To avoid the influences of high temperature and water vapor on the transducer, an attachment is fixed on top of the specimen, and then the deflection of the specimen is moved to the low temperature area and measured. In addition, two displacement transducers are set up horizontally at both ends of the specimen to measure the longitudinal deformation.

The continuous beam is a statically indeterminate system of one indeterminacy and the distribution of its internal force can be obtained if only one of the three reactions is measured. Two supports with reaction measurement capabilities are set up at both ends of the beam, then the reactions measured can be checked and the specimen can be erected and adjusted easily. The construction of the roller supports with reaction measurement capabilities is shown in Fig. 10-4. The main part includes three pairs of adjustable screw nuts and bolts of 50 mm diameter. The screw bolts are welded on the lower 30-mm-thick steel plate and are arranged in an isosceles triangle on the plane, and a row of steel rollers is put underneath the plate. Another steel plate is put on top of the screw nuts and two force transducers (15 kN each) are put on the plate. The support can measure the vertical reaction; it also satisfies all the requirements for rotation, horizontal displacement, and height and slope adjustments of the specimen. After many tests, the support is convenient to use, works successfully and has high measurement accuracy and a low friction coefficient of horizontal movement.

Three experimental furnaces (see Fig. 8-3) for a structural member with three surfaces exposed to high temperature are used to heat separately the beam and columns of the frame specimen. A steel cover plate with thermal insulation is set up on each side of the two joints between the beam and column of the frame, and is connected integrally with the furnace by an attached steel connector. Four adjustable bolts are linked with the connectors and are hung on the loading frame (Fig. 10-5) for convenient adjustment of the relative positions of the furnace and the specimen.

There is no thermal resource on the inner side of the steel cover plate at the joint, but the gap between the plate and the joint is linked up with the chambers of the experimental furnaces on both sides. Therefore, the joint of the specimen is heated indirectly by the high temperature airflow, and the temperature there is about 20–100 °C lower than that in the furnace chamber, according to experimental measurements.

The top surface of the beam and the outside surfaces of the two columns of the specimen are located in the low temperature area, and 13 displacement transducers are set up there for measuring separately the deflections and axial deformations of the beam and columns. Also, the deformations are moved out to the room temperature area through the attachments and are measured. Two base beams at the bottom of the two columns of the specimen are located in the low temperature area, and two displacement transducers are set up on each beam for measuring the deformations of the supports.

The frame specimen is a statically indeterminate structure of three degrees of indeterminacy and the lower ends of the two columns should be connected stiffly with the base beams. The base beams are specially designed and constructed

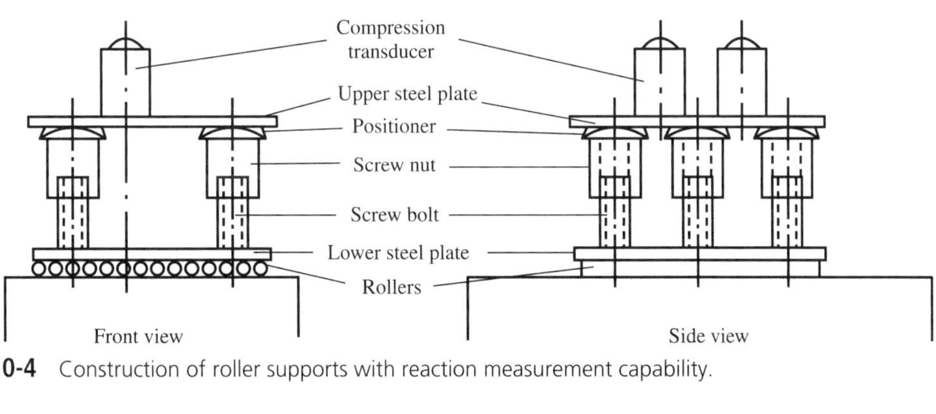

FIGURE 10-4 Construction of roller supports with reaction measurement capability.

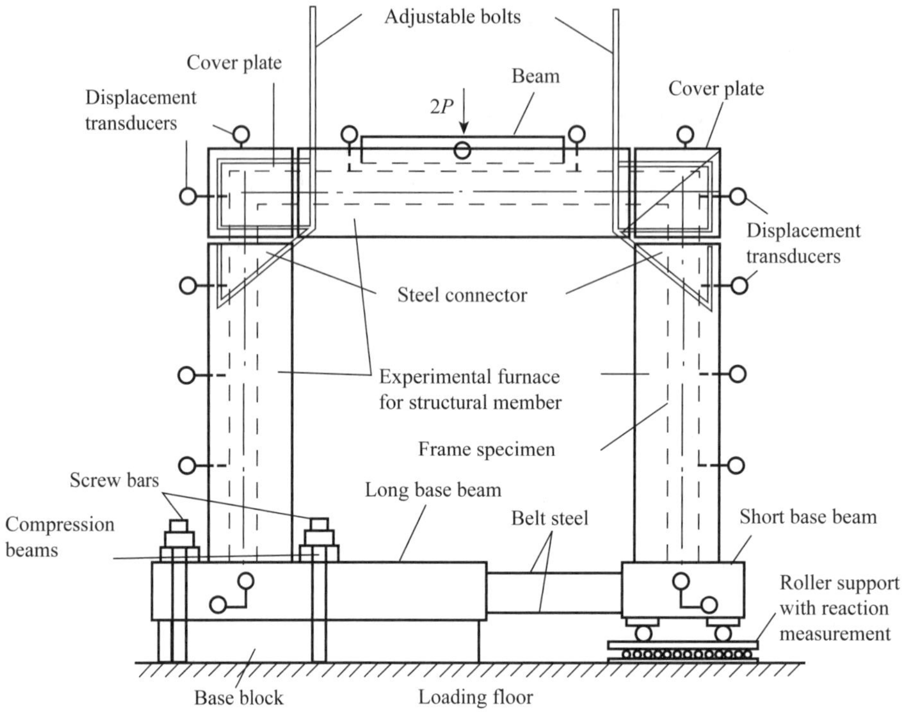

FIGURE 10-5 Experimental devices and measuring points of a frame.[1-12]

(Fig. 10-5) in order to measure the three unknown redundant reactions of the specimen. One base beam is set up at the bottom of each column and two base beams are linked with two steel belts. The base beam on the left column is longer and is compressed tightly on the loading floor by the compression beams and screw bars, and then an ideal fixed end is composed. The base beam on the right column is short and a roller support with reaction measurement capability is set up underneath; horizontal displacement of the right column is restrained by the two steel belts on its left side. A construction like this can measure unknown redundant forces and satisfy the basic requirement of a fixed end of the column.

The construction of the roller support with reaction measurement capability is shown in Fig. 10-6(a). It is composed of three thick-walled steel tubes, steel plate, and rollers. The thick-walled steel tubes are used as the compression transducers and welded and arranged in an isosceles triangle on a 30-mm steel plate, under which there is a row of steel rollers and the bottom steel plate. The short base beam of the right column of the specimen is set up on the roller support.

The measuring and calculation methods (Fig. 10-6(b)) for the unknown redundant reactions or the internal forces on the bottom section of the right column of the frame specimen are as follows: four strain gauges of electric resistance are pasted on each of the three compression transducers and the two steel belts between the base beams, and each transducer is calibrated separately and accurately before testing; the strain gauges of the two compression transducers on the left-hand side of the support are linked serially and measure the value of reaction R_1 during testing. The compression transducer on the right-hand side measures the value of reaction R_2, and the strain gauges on the steel belts measure the values of the horizontal constrained forces H_1 and H_2; then the internal forces on the bottom

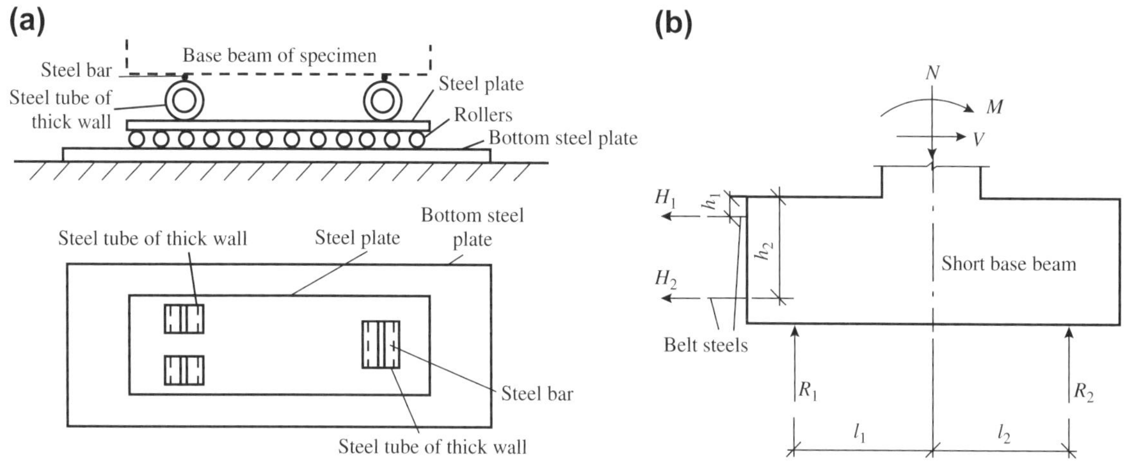

FIGURE 10-6 Roller support with reaction measurement capability and calculation of the internal forces on the bottom section of a column: (a) construction of the roller support; (b) calculation of the internal forces at the bottom of the column.

section of the right column can be calculated after determining the positions of all the transducers:

$$\text{Axial force: } N = R_1 + R_2$$
$$\text{Shear force: } V = H_1 + H_2 \qquad (10.2)$$
$$\text{Bending moment: } M = R_2 l_2 - R_1 l_1 - H_1 h_1 - H_2 h_2$$

According to these values, the distributions of all the internal forces of the statically indeterminate frame specimen can be obtained. Their redistributions during the testing process of heating and loading can also be obtained in the same way.

Experience with many heating and loading tests of the statically indeterminate continuous beam and frame specimens has shown that the integral testing program, measuring technique, and attached devices, which are introduced above, are reliable, convenient to use, and accurate.

10.2 BEHAVIOR OF A CONTINUOUS BEAM AT ELEVATED TEMPERATURES

10.2.1 Macroscopic Processes of Deformation and Failure

All the continuous beam specimens are tested under the path of heating under constant load. First, a concentrated load (P_0) is applied at the predetermined position on each span of the specimen and is kept constant. Then, the experimental furnaces are heated continuously and the temperatures on both side surfaces and the bottom of the specimen increase. When the deformation of the specimen is out of control, the value of its ultimate temperature (T_u, °C) is measured. However, the end of specimen TCB2-3 rises up from the support when the experimental temperature is higher than 800 °C (see Fig. 10-9) and it will not fail under the action of the initial load, so the specimen is loaded again until failure.

The macroscopic phenomena, including escape of water vapor, weight loss, color change, and cracking on the surface, and their variations in the continuous beam specimen during the process of heating under constant load, are similar to that of single beam or column specimens.

Four of the continuous beam specimens are heated simultaneously on both spans, so their behavior is symmetrical. The maximum deflection in the middle of the span increases monotonically during testing (Fig. 10-7). The specimen is loaded at room temperature and the deflection in the span increases slightly with the initial load level, but the value of the deflection is much less than that during heating. Flexural deflection of the specimen occurs during heating and is convex toward the high temperature zone, because a

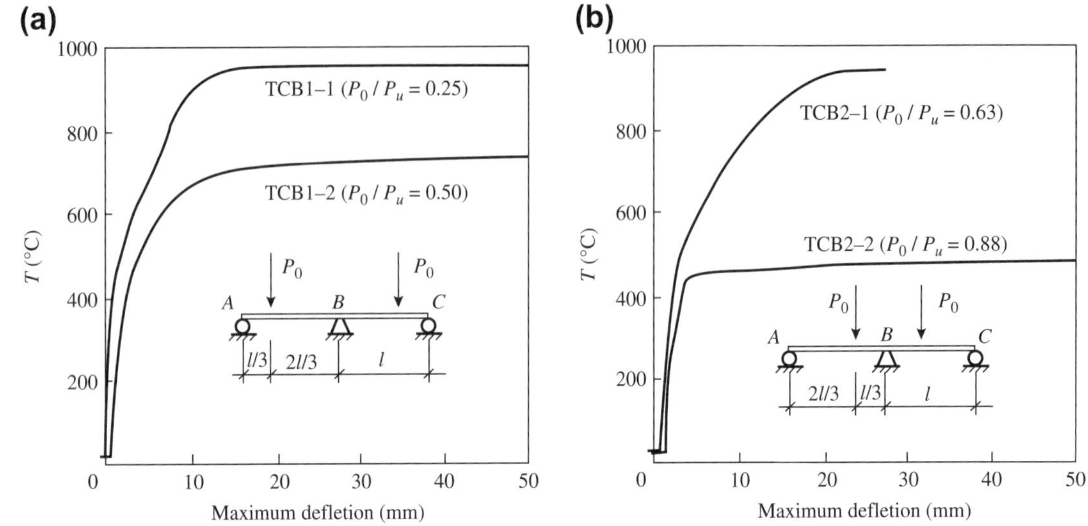

FIGURE 10-7 The maximum deflection in the span of a continuous beam specimen heated in both spans[10-4]: (a) load position $\beta = 1/3$; (b) load position $\beta = 2/3$.

nonuniform temperature field and thermal strain are formed on the section. The deflection in the span of the continuous beam is certainly smaller than that of a single member (e.g., Fig. 9-9) under the same conditions, because the continuous beam is restrained by the middle support.

The mechanical behavior of the concrete and the reinforcement of the specimen deteriorates after heating, their strain increases, and the deflection in the span increases rapidly. The deflection in the span increases less when the experimental temperature $T \leq 400$ °C and increases obviously when $T > 400$ °C, and the temperature–deflection curve gradually becomes horizontal and its slope reduces. When the plastic hinges occur successively at the mid-span and the middle support of the specimen and a mechanism of one degree of freedom is formed, the specimen soon fails because its deformation is out of control. Nevertheless, the failing process of a continuous beam is slower than that of a simply supported beam (Chapter 8).

Deformation during heating and the ultimate temperature of the specimen vary considerably with different positions of the concentrated load (or the distribution diagram of the bending moment). In addition, comparing specimens TCB1-2 with TCB1-1 or TCB2-2 with TCB2-1, the deflection in the span of the former is obviously greater than that of the latter at the same temperature, and the ultimate temperature reduces greatly as the initial load increases.

The failed specimen is removed from the furnace chamber after it cools down. It is seen that the residual deformation is great and the local bends appear clearly near its mid-spans and middle support. Tensile cracks in the span of the specimen are less. The critical tensile cracks at the positions of the concentrated loads expand widely and extend nearly to the top surface of the specimen. Horizontal cracks appear in the compression zone and are caused by compressive failure of the concrete, but the failure area is small. There are many tensile cracks on the top part near the middle support; a few of them expand widely and the extended length is about one-half to two-thirds of the section depth. The compressive depth is on the lower part of the section and no obvious symptoms of compressive failure are found. These phenomena in a continuous beam are the same as the failure pattern of a beam with the tension or the compression zone exposed to high temperature (Chapter 8).

Two other continuous beam specimens (TCB1-3 and TCB2-3) are also loaded symmetrically at room temperature, but only one span is heated later until failure. Therefore, the mechanical responses appear asymmetrical. The maximum deflections of the heated and unheated spans of the specimen are measured during testing and shown in Fig. 10-8.

When the initial loads are applied on the specimen at room temperature, the deflections in both spans are equal and small. When specimen TCB1-3 is heated, the maximum deflection in the heated span increases with the experimental temperature and the variation regularity is similar to that of the previous four specimens, but the maximum deflection in the unheated span is basically unchanged. When the specimen approaches the ultimate temperature, a mechanism of one degree of freedom is formed after two plastic hinges appear successively in the heated span and near the middle support; then the specimen fails and its failure pattern is the same as that of the previous four specimens. Although the specimen is heated on one span only and the temperature, deformation, and strength of its two spans are no longer equal (symmetrical), it demonstrates that the diagrams of the bending moment and shear force on both spans are still symmetrical. Therefore, the maximum value of the bending moment in the unheated span is equal to that of the ultimate bending moment (M_u^T) of the heated span and is far less than that of the ultimate bending moment (M_u) of the same section at room temperature. After the specimen fails, only a few fine tensile cracks appear in the unheated span, and the residual flexural deformation and other failure symptoms are not evident.

The deformation of specimen TCB2-3 is also small during loading at room temperature. The maximum deflection in the unheated span of the specimen during heating is also basically constant; the deflection in the heated span varies specifically. The maximum deflection in the heated span varies less when the experimental temperature is $T \leq 200$ °C, and increases slightly when $T = 200$–400 °C, but the end of the specimen rises up gradually from the support when $T > 400$ °C (Fig. 10-9). The end of the specimen leaves the support and rises up about 15 mm at $T = 823$ °C, so the deflection in the heated span reduces slightly instead. The specimen is then loaded again until failure at a constant temperature of 823 °C; its deflection turns to increase again and the risen end recovers gradually.

The end of specimen TCB2-3 rises up at elevated temperatures ($T \geq 400$ °C) and the reaction there is zero. Therefore, the bending moment on the section between the concentrated load and the end support (length βl) is also zero, and that between the concentrated load and the middle support has a negative value. Then, the

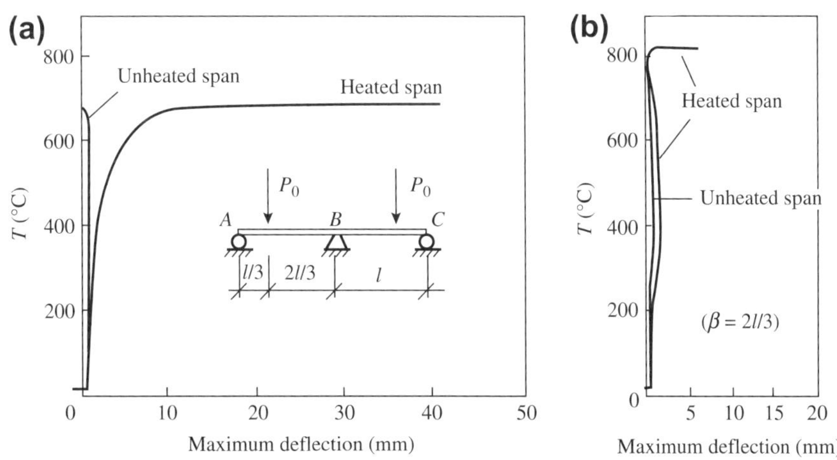

FIGURE 10-8 The maximum deflections in the spans of a continuous beam specimen heated in one span only[10-4]: (a) specimen TCB1-3; (b) specimen TCB2-3.

continuous beam is changed into a bicantilever in view of the mechanical condition. During the second loading stage, the first plastic hinge appears at the middle support,[10-6] then the deformation of the specimen increases considerably and its end falls down to the support, and a positive bending moment occurs again in the span. When the second plastic hinge is formed in the heated span, the specimen undergoes a mechanism of one degree of freedom and fails. The failure pattern is the same as that of specimen TCB1-3.

10.2.2 Ultimate Strength at Elevated Temperatures

When the six continuous beam specimens, introduced above, fail, plastic hinges are formed in the heated span(s) and near the middle support, and the ultimate bending moments at elevated temperatures on the corresponding sections are achieved. The measured ultimate temperatures (T_u) and strengths (P_u^T) of the specimens tested are summarized in Table 10-4.

When the continuous beam specimen TCB1-1 is compared with TCB1-2, the ultimate temperature reduces from 950 °C to 743.2 °C and the reducing amplitude is 21.8%, as the load level is increased from $P_0/P_u = 0.25$ to 0.5, although both specimens have the same load position (or calculation model) and heated spans. When specimen TCB2-1 is compared with TCB2-2, the ultimate temperature reduces from 937 °C to 490.9 °C and the reducing amplitude reaches 47.6%, as the load level (P_0/P_u) is increased from 0.63 to 0.88.

When the position of the concentrated load acting on the continuous beam is changed, the distribution diagram of the bending moment or the ratio between the bending moments in the span and at the middle support changes. For example, the load positions are $\beta = 1/3$ and 2/3 for the two series of specimens tested, and the corresponding ratios of the bending moments are $M_0/M_B = 1.18$ and 0.87. This means that the bending moment in the span is decreased, while that at middle support is increased, when the second series of specimens (TCB2) is compared with the first series (TCB1). On the other hand, the cross sections in the span and at the support of the continuous beam at room temperature are the same and the ultimate

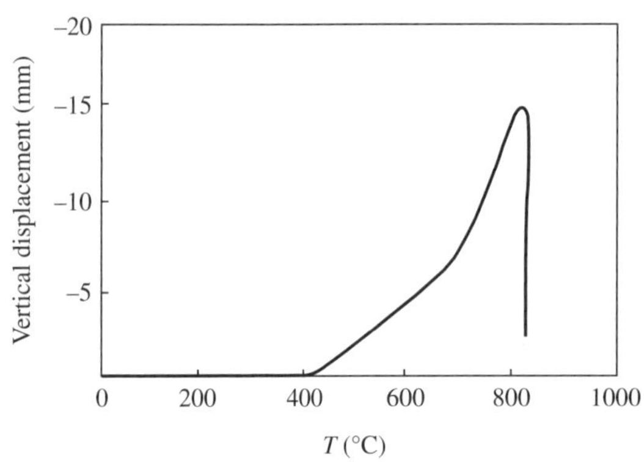

FIGURE 10-9 Vertical displacement of the end of the heated span of specimen TCB2-3.

TABLE 10-4 Ultimate Strength of Continuous Beam Specimens at Elevated Temperatures

Number of specimen	Constant load P_0 (kN)	$\frac{P_0}{P_u}$	$\frac{M_0}{M_u}$	$\frac{M_B}{M_u}$	T_u (°C)	P_u^T (kN)
TCB1-1	10.0	0.25	0.259	0.222	950.0	10.0
TCB1-2	20.0	0.50	0.519	0.444	743.2	20.0
TCB1-3	20.0	0.50	0.519	0.444	697.9	20.0
TCB2-1	25.0	0.63	0.602	0.692	937.0	25.0
TCB2-2	35.0	0.88	0.843	0.972	490.9	35.0
TCB2-3	10.0	0.25	0.241	0.278	823.0	24.7

P_u and M_u are the theoretical values of the ultimate load and bending moment, respectively, on the section of the continuous beam specimen at room temperature, and M_0 and M_B are the maximum bending moments of the sections in the span and at the middle support of the specimen, respectively, under the action of load P_0 and at room temperature.

values of their bending moments (M_u) are also the same. However, the ultimate bending moment in the span of the specimen at elevated temperatures reduces significantly because the tension zone is exposed to high temperature; the reduction in the ultimate bending moment near the middle support is limited (see Fig. 8-8) because the compression zone is exposed to high temperature. Therefore, the two ultimate bending moments differ considerably. The distribution of the bending moments of the second series of specimens coincides with the varying tendency of the ultimate strength of the continuous beam after heating, so that the plastic hinge in the span occurs later and the ultimate temperature and strength of the specimen are higher. For example, the load level P_0/P_u = 0.63, ultimate load $P_u^T = 25\ kN$, and ultimate temperature T_u = 937 °C of specimen TCB2-1 are much higher than the corresponding values for the specimen TCB1-2, i.e., P_0/P_u = 0.50, $P_u^T = 20\ kN$, and T_u = 743.2 °C.

When the specimen (TCB1-2) is heated in both spans and a concentrated load of the same position and value is acted on each span, the mechanical behavior is symmetrical about the middle support and the angular rotation of the section at the middle support is zero throughout the loading and heating process. However, as the specimen (TCB1-3) is heated in one span only, angular rotation of the section at the middle support is certain to occur, and then the bending moment there has to be reduced and the maximum bending moment in the span is increased, correspondingly, because the rigidity of the heated span is smaller than that of the unheated span, although the bending moment diagrams on both spans are still symmetrical. This leads the first plastic hinge to appear earlier in the heated span, so the ultimate temperature of the continuous beam heated in one span only is lower than that heated in both spans (697.9 °C < 743.2 °C).

10.2.3 Redistribution Process of Internal Force

A continuous beam of two spans is a statically indeterminate structure of one degree of indeterminacy. When the specimen is loaded at room temperature, the distribution of the bending moment calculated by the measured values of the load and reactions fits well with the results of the theoretical analysis based on homogeneous linear elastic material. Even if the specimen (TCB2-2) is loaded to a high level (P_0/P_u = 0.88), and the concrete cracks have already appeared in the tension zone and the nonelastic strain of concrete has occurred in the compression zone, the difference between the bending moments measured and analyzed elastically is still less (<10%).

When the continuous beam specimen is heated under a constant load, large thermal deformations are caused because a nonuniform temperature field is formed on the sections. The flexural deformation (curvature = $1/\varrho$, see Figs. 8-7 and 8-9) among them is constrained by the redundant support, so redistribution of the bending moment and shear force of the specimen result. However, the longitudinal expansive deformation (see Figs. 9-3 and 9-9) is not restrained and no axial force results, because the roller supports are set up on both ends of the specimen.

The specimens TCB1-1 and TCB1-2 are similar. The reactions at their supports measured during testing vary with the elevating temperature and are shown in Fig. 10-10. The maximum bending moment ($M_0 = R_A\beta l$) in the span of the specimen is directly proportional to the reaction at the end support ($R_A = R_C$). According to the variations in the reactions at the supports (i.e., also the bending moment in the span) shown in the figure, there are three stages in the redistribution process of internal force in a continuous beam during heating.

1. Initial development stage. When the specimen is heated and the experimental temperature increases from the normal value to 300–400 °C, the temperature in its interior is not yet high but the temperature gradient on the section is large, so the expansive strain of concrete is not uniform and flexural deformation, which is convex toward the high temperature zone, results and the reaction at the middle support (R_B) increases. At the

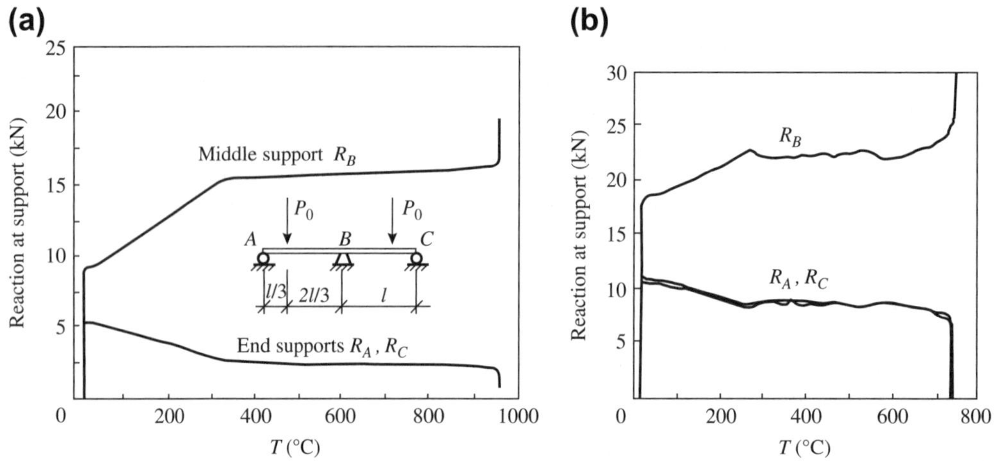

FIGURE 10-10 Variation of the reaction at the support of a continuous beam specimen at elevated temperatures[10-7]: (a) TCB1-1 ($P_0/P_u = 0.25$); (b) TCB1-2 ($P_0/P_u = 0.50$).

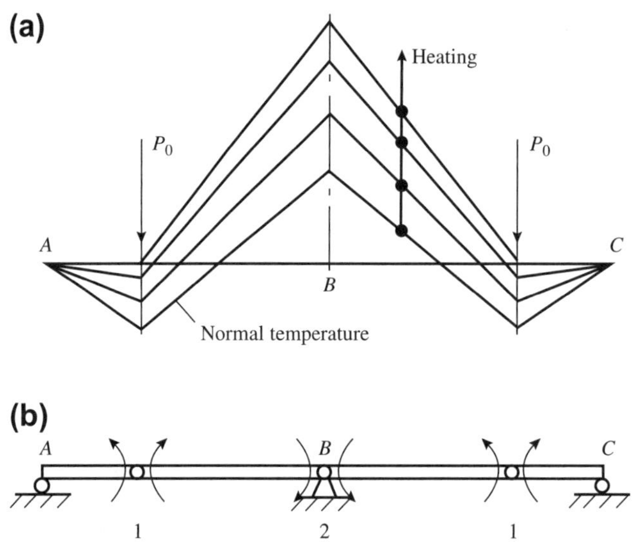

FIGURE 10-11 Redistribution of the internal force and failure mechanism of a continuous beam: (a) redistribution of the bending moment; (b) position and order of appearance of plastic hinges.

same time, the reactions at the end supports ($R_A = R_C$) have to reduce because the loads on the specimen are constant (P_0 = const.). Therefore, the bending moment in the span (M_0) reduces and the bending moment at the section in the middle support (M_B) increases gradually (Fig. 10-11).

2. Stable development stage. When the specimen is heated continuously, the temperature in the interior of its section increases gradually and the behavior of the concrete deteriorates more significantly. The reducing amplitude of section rigidity in the span area (of $+M$) with the tension zone exposed to high temperature on the continuous beam is far greater than that in the area near the middle support (of $-M$) with the compression zone exposed to high temperature, so nonuniform distribution of

section rigidity results. In the meantime, the thermal deflection is continuously convex toward the high temperature zone, although the temperature gradient on the section reduces slightly. Due to the actions of these two factors, the redistribution of internal forces in the continuous beam develops continuously and stably, and the bending moment in the span reduces while the bending moment at the middle support increases (Fig. 10-11).

3. Stage after the plastic hinge is formed. As the experimental temperature increases constantly, the bending moments in the span and at the middle support continuously decrease and increase. However, the ultimate bending moment (M_{0u}^T) in the span section with the tension zone exposed to high temperature reduces even more quickly. This leads the existing bending moment in the span to first reach the ultimate value of the bending moment, and then the first plastic hinge is formed there. As the experimental temperature increases continuously, the ultimate value of the bending moment at the plastic hinge under high temperature reduces further and the reaction at the end support decreases correspondingly. In the meantime, the reaction at the middle support increases and the bending moment of the section there increases quickly. When the bending moment there reaches the ultimate value of the bending moment of the specimen with the compression zone exposed to high temperature, the second plastic hinge appears near the section on the middle support (Fig. 10-11(b)) and the specimen composes a mechanism and fails.

The load and heating conditions of both specimens TCB1-1 and TCB1-2 are the same, and the redistribution processes of internal force are similar. However, when TCB1-1 is compared with TCB1-2, its load level is higher (P_0/P_u = 0.50 > 0.25), so the second stage is reached earlier, the first plastic hinge also appears earlier, and the ultimate temperature is lower. If the ratio between the reactions at the end support is reduced when the first plastic hinge is formed and that before heating is taken as an index of the redistribution amplitude of the internal force of the continuous beam after heating, the index of the specimen TCB1-2 is obviously lower than that of the specimen TCB1-1 due to the higher load level.

The measured reactions at the supports of specimen TCB2-1 vary during the heating process and are shown in Fig. 10-12(a). Because the

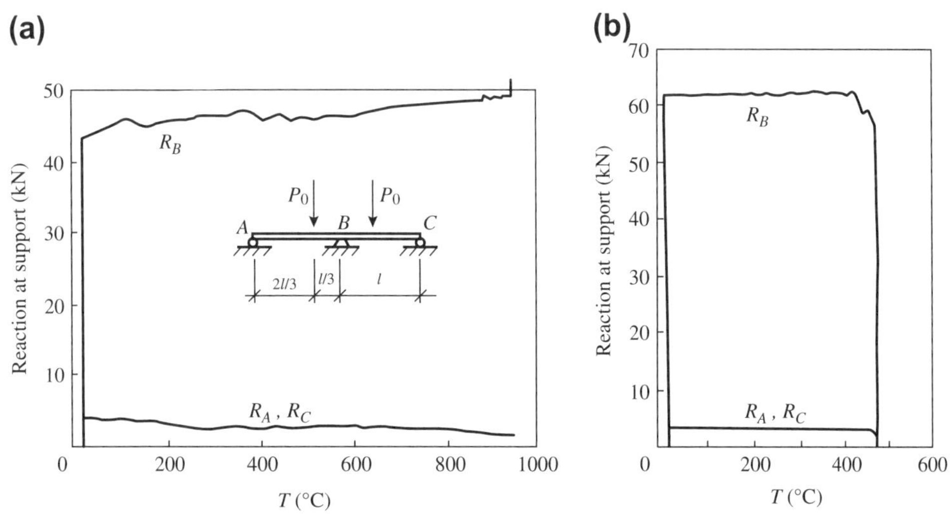

FIGURE 10-12 Variation of the reaction at the support of a continuous beam specimen at elevated temperatures[10-7]: (a) TCB2-1 (P_0/P_u = 0.63); (b) TCB2-2 (P_0/P_u = 0.88).

concentrated loads on the specimen are located near the middle support ($\beta = 2/3$), the reaction at the end support is far smaller than that at the middle support ($R_A = R_B$) and the bending moment in the span is relatively small (M_0/M_B). Therefore, the rigidities of the sections in the span and near the middle support at elevated temperatures are not so different, the internal force redistributes slowly, and the turn between the first and second stages is not clear. As the bending moment in the span under the action of load is reduced, it is helpful for delaying the appearance of the plastic hinge in the redistribution process of the internal force during heating and for increasing the ultimate temperature of the specimen. The appearance order of plastic hinges and the failure mechanism of specimen TCB2-1 are the same as for specimen TCB1-1 and others, but the load level of specimen TCB2-1 is higher ($P_0/P_u = 0.63$) and the redistribution amplitude of the internal force is smaller.

The load level of specimen TCB2-2 is quite high ($P_0/P_u = 0.88$) and the ratio between the initial bending moment and the ultimate bending moment on the section in the span reaches $M_0/M_u = 0.843$ (Table 10-4). After the specimen is heated, the bending moment in the span reduces slightly and the redistribution process of the internal force is short. However, the ultimate bending moment at elevated temperatures reduces quickly and the first plastic hinge appears in the span when the experimental temperature $T < 500$ °C (Fig. 10-12(b)), so the redistribution amplitude of the internal force is very small. Another plastic hinge then appears near the middle support, and the specimen composes a mechanism and then fails.

Two other continuous beam specimens are heated in one span only during testing and the measured reactions at the supports during heating vary as shown in Fig. 10-13. It is seen that the behavior of specimen TCB1-3 is similar to that of specimens TCB1-1 and TCB1-2. However, the section at the middle support of specimen TCB1-3 is rotated during heating in one span and the deformations in both spans are adjusted and are different, so the redistribution amplitude of the internal force is small and the turn from the first to the second stage is not clear.

The concentrated load acting on specimen TCB2-3 is located near the middle support and the load level is $P_0/P_u = 0.25$, so the initial

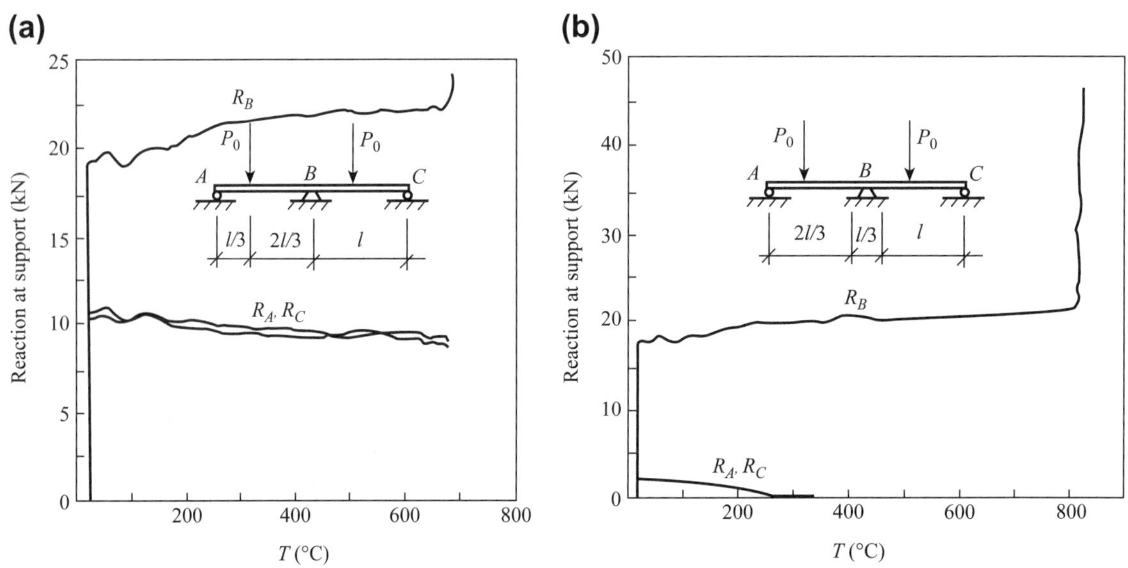

FIGURE 10-13 Variation of the reaction at the support of a continuous beam specimen heated in one span only[10-7]: (a) TCB1-3 ($P_0/P_u = 0.50$); (b) TCB2-3 ($P_0/P_u = 0.25$).

reaction at the end support ($R_A = R_C$) is quite low. The flexural deformation of the specimen after it is heated is convex toward the high temperature zone and redistribution of the internal force occurs, so the reaction at the end support reduces gradually and even reduces to zero at experimental temperature $T \approx 400$ °C. When the specimen is heated continuously, the thermal deformation makes the end support rise up constantly (Fig. 10-9); then the area between the end support and concentrated load is free of internal force, and the calculation model of the specimen is changed to the bicantilevers from the continuous beam of two spans (Fig. 10-11). At this time the reactions at the supports keep constant values, i.e., $R_A = R_C = 0$ and $R_B = 2P_0$. When the experimental temperature reaches 823 °C, the ultimate bending moment on the section of the middle support with the compression zone exposed to high temperature (M_u^T) reduces slightly (Fig. 8-8) but is still far greater than the value of the bending moment there caused by the load ($P_0 = 10$ kN), so the specimen does not fail. It is estimated that the specimen will not fail at all even if it is heated continuously, so it is loaded again and the experimental temperature is kept constant (823 °C). When the load reaches $P_0 = 24.7$ kN, the first plastic hinge is formed near the section on the middle support and considerable rotation occurs there, so both ends of the specimen fall down to the end supports and the mechanical behavior of the continuous beam of two spans is recovered. When the reaction at the end support increases again and the corresponding value of the bending moment in the span increases to that of the ultimate bending moment of the section with the tension zone exposed to high temperature, the second plastic hinge appears there and the specimen composes a mechanism and fails.

Generally, the redistribution process for the internal force of a reinforced concrete continuous beam with two spans at room temperature and monotonic loading is as follows: before the tensile concrete cracks, the distribution of the internal force agrees well with the result of linear elastic analysis; after the concrete cracks, the rigidity of the section is reduced and the value of the rigidity differs from section to section, so the redistribution of the internal force is slight; after the first plastic hinge is formed near the middle support, the internal force is redistributed sharply and the bending moment in the span increases quickly with the load and that at the middle support remains constant. When the second plastic hinge has appeared in the span, the continuous beam composes a mechanism and then fails.

The redistribution processes for the internal force of six reinforced concrete continuous beams with two spans under monotonic heating and different loading conditions are presented in detail above. The macroscopic regularity is similar to that at room temperature, e.g., before and after the appearance of the first plastic hinge, the redistribution of the internal force is caused by different factors, and the redistributed amplitudes are also different; when several plastic hinges are formed successively, the continuous beam is composed of a mechanism of one degree of freedom and then fails. However, the redistribution of the internal force of the continuous beam during the heating process is larger and more complex; the redistribution of the internal force may be larger before the appearance of the first plastic hinge; even the calculation model is changed; the ultimate value of the bending moment at the plastic hinge reduces continuously as the temperature increases; and the order of appearance of the plastic hinges may have more than one possibility.

10.3 BEHAVIOR OF FRAMES AT ELEVATED TEMPERATURES

10.3.1 Macroscopic Processes of Deformation and Failure

Four single-bay and single-story frame specimens are tested under the path of heating under constant load. A pair of concentrated loads act symmetrically on third points of the frame beam and maintained at a predetermined value (P_0). Then the experimental furnace is electrified and heated; both sides and the inner surfaces of the frame beam and columns are heated continuously. When the specimen fails because its deformation is out of control, the value of the ultimate temperature (T_u) is measured. If the specimen does not fail when the experimental temperature $T > 950$ °C is reached, the temperature is kept constant for 60 min and then the specimen is loaded again until failure.

During the heating process of the specimen, water vapor in the interior of the two columns escapes less from the outer surfaces; most of it rises up and escapes from the joints on the top. Other macroscopic phenomena, i.e., weight loss, color change, and cracking on the surface, and variations with elevating temperature, are similar to that of a single beam or column specimen.

Specimen TFC-1 is tested at room temperature for comparison. The deflection in the beam span, the lateral displacements of the columns, and the axial deformations of the beam and columns of the specimen are very small during the loading process, and, basically, they are directly proportional to the value of load. As the load increases, the tensile crack appears first on the top surface at the end of the frame beam because of the larger bending moment there; several tensile cracks then appear successively on the bottom of the frame beam between the two concentrated loads, i.e., in the span zone of the pure bending moment (shear force is zero). Each of the cracks extends and expands gradually but its width is small. As the ultimate load ($P_u = 27.9$ kN) is approached and reached later, tensile cracks appear successively on the outer surface near the top and on the inner surfaces near the bottom of the frame columns; then three plastic hinges are formed successively at both ends and the mid-span of the frame beam and the frame specimen composes a beam mechanism and fails. After the load is removed, most of the cracks are closed and the residual deformation is not clear on the failed specimen, except for the cracks and bend deformations near the plastic hinges. At this time, the maximum deflection at the mid-span of the specimen is only 7.5 mm and the axial elongation is less than 1 mm.[1-12] This is the general failure pattern of a statically indeterminate structure.

When the specimens TFC-2 to TFC-5 are loaded to the predetermined values (the maximum is $P_0/P_u = 0.49$) at room temperature, no obvious tensile cracks appear on the outer surfaces of the frame beam and columns, and the flexural and axial deformations are small. During the heating process under constant load of the specimen, the deflection at the mid-span of the beam (i.e., downward displacement related to the top of the column) and the axial deformations of the beam and column increase with the experimental temperature (Figs. 10-14 and 10-15), and the deflection curves of the frame beam and columns are shown in Fig. 10-16.

The initial load level $P_0/P_u = 0.49$ of specimen TFC-2 corresponds approximately to the service load of a practical structure used normally. When the experimental temperature reaches $T \approx 200$ °C, the cracks appear successively on the top surface at the end of the beam and on the outer surface at the top of the column, and they develop slowly. When $T \geq 400$ °C, the cracks in the concrete and the flexural and axial deformations of the beam and column increase gradually because of serious deterioration in the behavior of the concrete. Although the relative deformation of the beam is already great when $T = 500$ °C, the absolute deflection is still smaller than that at room temperature (Fig. 10-16(a)), because the axial elongations of both columns are even greater. The lateral displacement of the column is convex toward its outer surface during loading at room temperature, but it turns to the opposite direction and is convex toward the inner surface of the column when $T = 500$ °C, because the temperature in the inner side is higher and the expansive strain there is greater than that on the outer side; in the meantime the axial deformation of the beam is elongated. As the experimental temperature increases further, the behavior of the concrete deteriorates more significantly; the flexural deformation in the beam span develops quickly and is even greater than the expansive elongation of the column, so the absolute value of the deflection at the mid-span of the beam increases quickly. In addition, many transverse cracks appear at a uniform

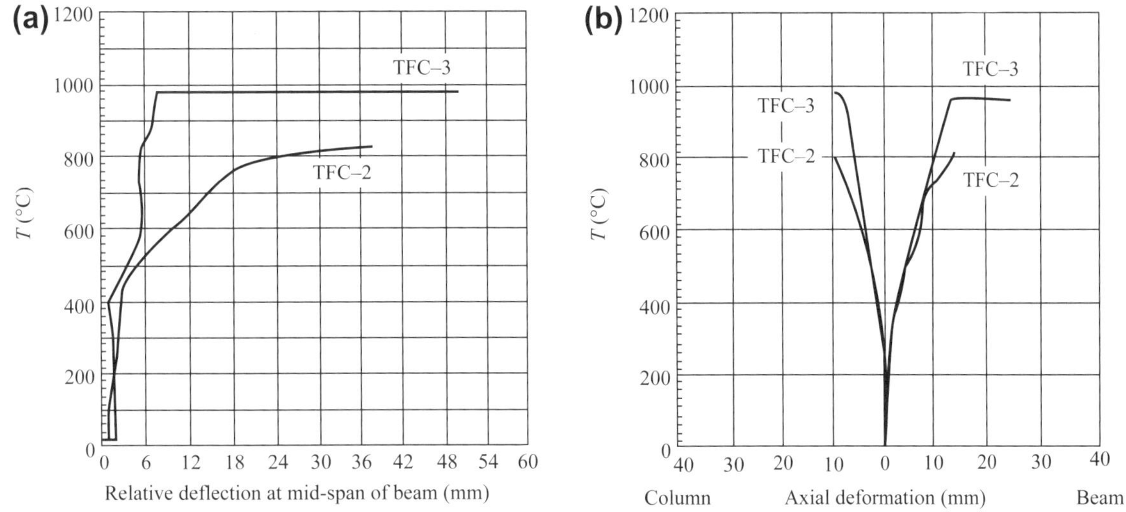

FIGURE 10-14 Deformation of frame specimens TFC-2 and TFC-3 during heating[10-8]: (a) relative deflection at the mid-span of the beam; (b) axial deformations of the beam and column.

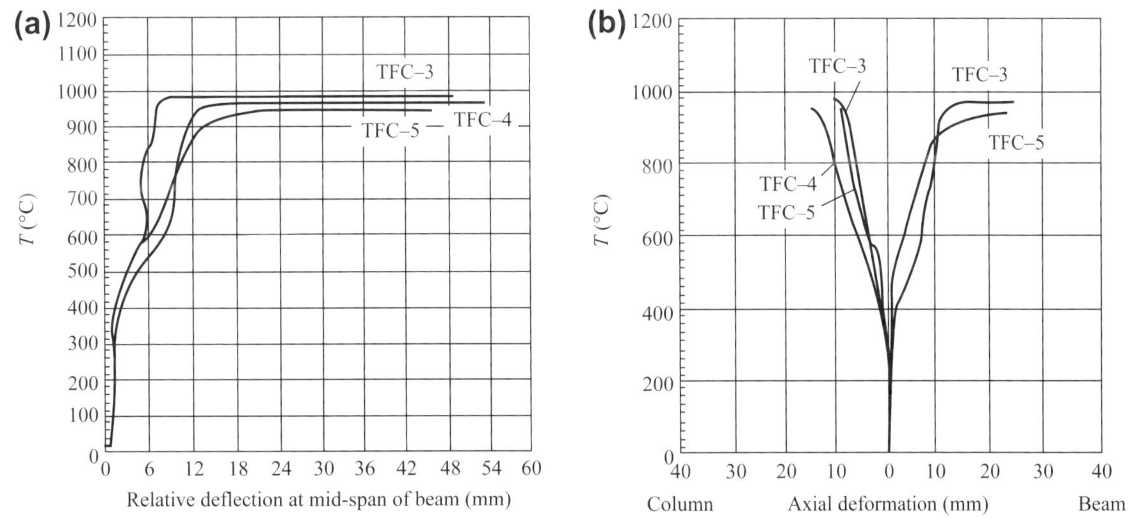

FIGURE 10-15 Deformation of frame specimens TFC-3 to TFC-5 during heating[10-8]: (a) relative deflection at the mid-span of the beam; (b) axial deformations of the beam and column.

distance on the outer surfaces of the column from top to bottom, because the compressive strain of concrete at high temperatures in the inside of the column increases considerably. Therefore, the flexural deformation of the column turns again to the opposite direction, i.e., is convex toward the outer surface, and develops rapidly.

When the experimental temperature approaches and reaches 820 °C, the ultimate bending moment in the span of the beam reduces significantly because the tension zone there is exposed to high temperatures, the tensile reinforcement yields, and the first plastic hinge is formed there. Then, the bending moment at the end of the beam

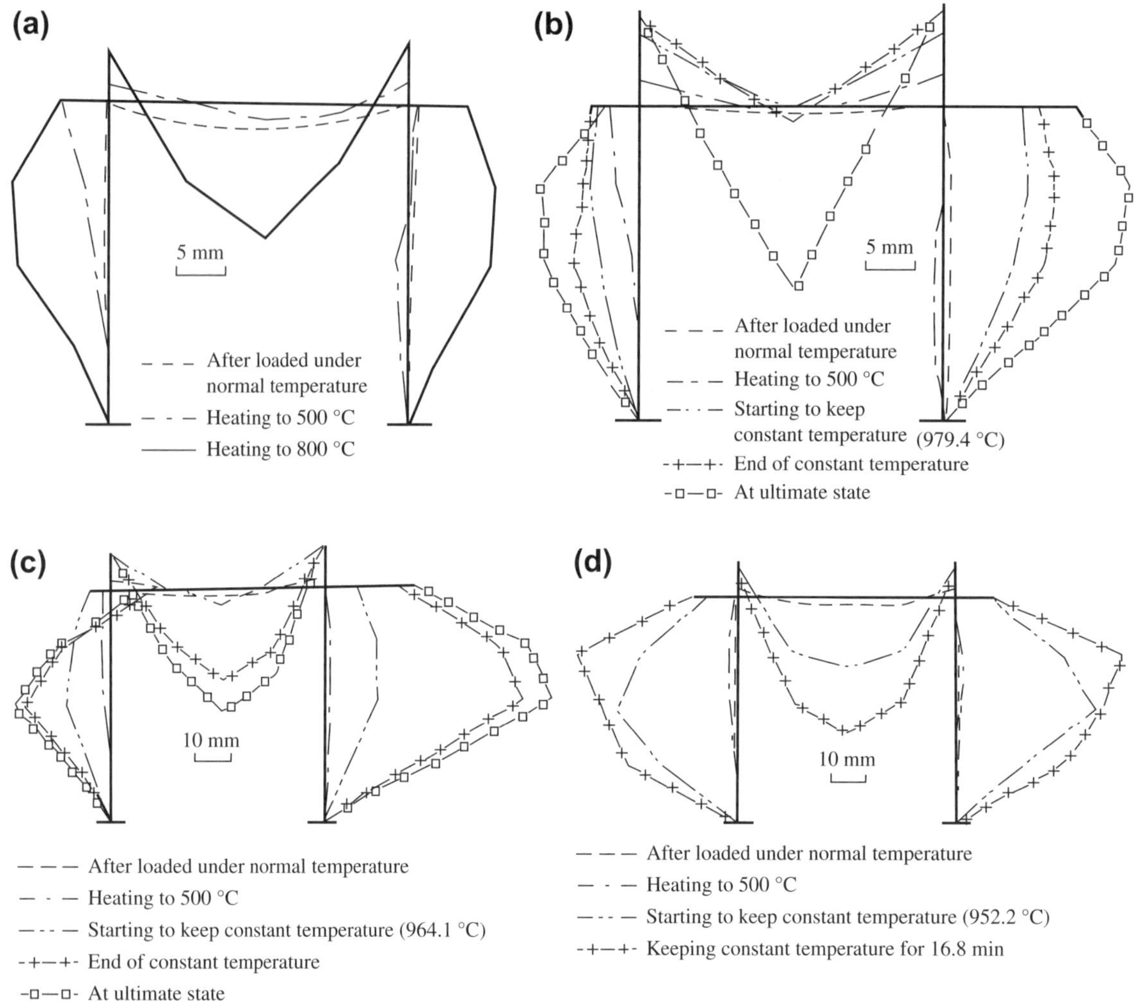

FIGURE 10-16 Deflection curves of the beam and column of the frame specimen: (a) TFC-2; (b) TFC-3; (c) TFC-4; (d) TFC-5.

increases gradually as the internal force is redistributed, but the ultimate value of the bending moment there reduces slowly because the compression zone is exposed to a high temperature. When the values of the existing bend moment and the ultimate bending moment are equal, two plastic hinges opposite in direction to the first one are formed at each end of the beam, and the frame specimen composes a beam mechanism (Fig. 10-17) and then fails. When specimen TFC-2 is compared with specimen TFC-1 tested at room temperature, the failure mechanism and the positions of the plastic hinges are identical, but the appearance order of the plastic hinges is just the opposite. The exterior of specimen TFC-2 after it fails is also similar to that of specimen TFC-1, but the residual deformation is much greater. The local bend deformations and wide cracks are obvious near the positions of the plastic hinges at mid-span and at both ends of the frame beam. The convex deflection of the column is clear but is smaller than that of the frame beam because the rigidity of the column is greater (i_b/i_c = 0.354), and the width of the cracks on

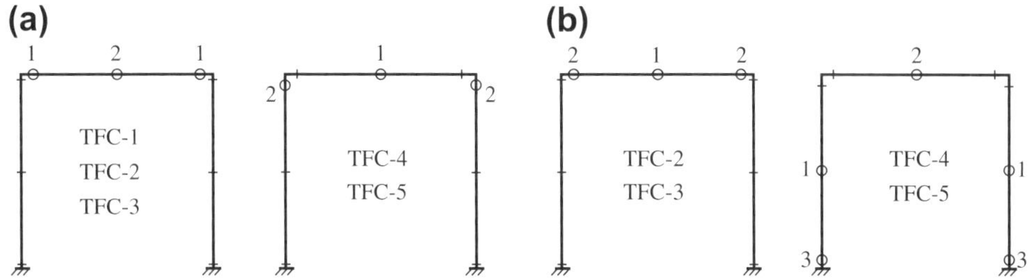

FIGURE 10-17 Position and order of appearance of plastic hinges of a frame specimen: (a) at normal temperature (theoretical analysis); (b) at elevated temperatures (tested).

the column is smaller. The axial deformations of the beam and column reach 10 mm and 6 mm, respectively.

The frame specimens TFC-3 and TFC-2 have the same cross section, beam length, and column height, and are also tested under the same conditions. The only difference between them is the initial load level ($P_0/P_u = 0.30 < 0.49$). Therefore, the internal force and deformation of specimen TFC-3 caused during loading at room temperature are even less and no cracking occurs. The deflection in the mid-span of the beam, the lateral displacement of the column, and the axial deformations of the beam and column of the frame specimen are slightly less throughout the heating process (Figs. 10-14 and 10-16). When the cracks appear first on the top surface near the end of the beam and on the outer surface near the top of the columns, the experimental temperature is about 450 °C, which is higher than that of specimen TFC-2. When the experimental temperature is increased further to T = 979.4 °C, the deformation and cracks in the specimen are large, but are still less than the corresponding values for specimen TFC-2 when T = 800 °C, and no failure symptom has yet shown up. The load and temperature of the specimen are then kept constantly for 60 min, deformation and cracks develop further, but still no failure symptom appears. When the specimen is loaded again and the load reaches P_0 = 17.3 kN, three plastic hinges are formed successively in the mid-span and at both ends of the frame beam, and the frame specimen composes a beam mechanism and then fails. The failure mechanism and pattern of the specimen are identical with that of specimen TFC-2, but the residual deformation of specimen TFC-3 is greater.

Frame specimens TFC-4 and TFC-5 have similar values for the initial load levels (P_0P_u = 0.29 and 0.31, respectively) as specimen TFC-3 (P_0/P_u = 0.30), and the testing conditions are identical. The differences between them are the depths of the sections and the values of the ratio between the linear rigidities of the beam and column (i_b/i_c, Table 10-2). Comparing specimens TFC-4 and TFC-5 with specimen TFC-3, the values of the ratio of linear rigidity are increased, because the depth of the beam is increased or the depth of the column is decreased. When the two specimens are loaded at room temperature, the difference in the internal force of the specimens causes different values of deformations, but all the deformations are small and no cracking occurs on the specimens. When the two specimens are heated and the experimental temperature reaches $T \geq$ 200 °C, a crack occurs first on the outer surface near the top of the column because of the weakness there. Then, transverse cracks appear successively at a uniform distance along the height of the column, and they extend and expand gradually; no cracks are found on the top surface of the beam. As the rigidity of the column is reduced, the flexural deformation caused during loading is greater than that caused during heating, so the lateral displacement of the column is continuously convex toward the outer surface (Fig. 10-16(c) and (d)). The relative deflection in the mid-span of the frame beam is greater than that of specimen TFC-3 (Fig. 10-15(a)), because the constraint action of both columns is weakened. The axial deformations of the beam and column are related to the size of their section, and they are decreased and increased, respectively, when the specimens are compared with specimen TFC-3, but the differences between them are limited (Fig. 10-15(b)).

When the experimental temperature is $T \geq 400$ °C, the behavior of the concrete material deteriorates more significantly and the cracks on the specimen expand gradually, and the deformations of the beam and column increase rapidly along the previous directions. When $T \geq 950$ °C, the flexural deformation of the column is convex toward its outer surface and can be seen by naked eye, but still no failure symptoms appear. Then, the experimental temperature is kept constant, and the deformations of the beam and column of the specimen increase greatly again. Specimen TFC-5 fails after the temperature is kep constant for 16.8 min, because its deformation is out of control. However, specimen TFC-4 is loaded again after the temperature is kept constant for 60 min, and it fails when the load (P_0) reaches 9.8 kN from 8.5 kN.

The failure conditions for specimens TFC-4 and TFC-5 are examined after testing[1-12]: fine tensile cracks are found only on the outer surface near the top of the column; other parts on and near the joint between the beam and column are basically undamaged; the plastic rotation zones are found at the mid-span of the beam and the bottom of both columns because the reinforcements in the inner side are subjected to elevated temperatures (when these three plastic hinges are formed, the frame specimen composes a statically determinate arch of three hinges, but this is not enough to compose a changeable mechanism); considerable thermal flexural deformation is caused within a long area in the middle part (longer than half the height) of the column due to less rigidity, and larger additional eccentricity appears there; the compressive deformation of the concrete on the inner side of the column is considerable as it is located in the high temperature zone; even the concrete cover there falls down and the compressive reinforcement yields, causing a large plastic rotation angle and a plastic hinge of compression type is formed there. Then the frame specimen composes a changeable mechanism of mixed type of beam and column with five plastic hinges (Fig. 10-17(b))[10-8] and fails.

Every frame specimen composes a changeable mechanism of one degree of freedom and then fails, after several plastic hinges are formed successively under the loading and heating process. A specimen with a different ratio between the rigidities of the beam and the column may show various failure mechanisms. If the five frame specimens of similar size and construction are loaded at room temperature, all of them form a mechanism of beam type (Fig. 10-17(a)) and then fail. However, three plastic hinges are formed successively at both ends and at the mid-span of the beam for specimens TFC-1, TFC-2, and TFC-3; they are formed successively in the mid-span of the beam and near the top of both columns for specimens TFC-4 and TFC-5 because of the smaller section of the column. The failure mechanisms of the frame specimens observed during testing at elevated temperatures (Fig. 10-17(b)) are different from that observed at room temperature. They are different in the position or the order of appearance of the plastic hinges, even in various types of mechanisms.

10.3.2 Ultimate Strength at Elevated Temperatures

The ultimate temperature (T_u) and strength (load P_u^T) of the five frame specimens measured in during testing are summarized in Table 10-5.

When a reinforced concrete frame specimen is tested at elevated temperatures, considerable deformation occurs because of serious deterioration in the behavior of the concrete and the reinforcement. The ultimate strength reduces considerably compared with that of the specimen at room temperature, e.g., the relative strengths of the four specimens are reduced by 38–70% when the experimental temperature is higher than 800 °C.

The higher the initial load level on the frame specimen, the worse the fire (high temperature) resistant behavior. Specimen TFC-3, with an initial load level $P_0/P_u = 0.30$, does not fail after a high temperature of 979.4 °C is maintained for 60 min, and the load that it can bear is doubled. However, specimen TFC-2, with an initial load level $P_0/P_u = 0.49$, fails when the experimental temperature increases to 820 °C.

If the ultimate strengths of the frames at room temperature are identical but the ratios between the rigidities of the beam and column are not equal, the resistant capacities at elevated temperatures may be obviously different. For example, the initial load levels of frame specimens TFC-3, TFC-4, and TFC-5 are nearly the same ($P_0/P_u \approx 0.30$), but the ratios between the linear rigidities of the beam and column range from $i_b/i_c = 0.354$ to 2.011, respectively. Although all experience high temperatures of more than 950 °C, the specimen with the highest ratio fails earlier, i.e., specimen TFC-4 fails when the value of the second loading is small after a high temperature is maintained for 60 min; even specimen TFC-5 fails during the process of constant high temperature. Therefore, the ultimate strength of the frame structure at elevated temperatures is reduced gradually, as the ratio between the rigidities of the beam and column increases, i.e., the section of the beam increases or the section of the column decreases.

10.3.3 Redistribution Process of Internal Forces

Frame specimen TFC-1 is loaded at room temperature until failure. The actual values of the bending moment of the specimen are calculated (see Eqn 10.2) from the measured reactions at the support (R_1 and R_2) and the horizontal constraint forces (H_1 and H_2), and they are compared in Table 10-6 with the results of linear elastic analysis. The measured bending moments at the bottom and the top of the column and at the end of the beam are slightly smaller; that at the mid-span of the beam is slightly greater. The reason is that a tiny rotation occurs on the base of the specimen (Fig. 10-18). The maximum error between the measured and calculated bending moments is less than ±8%, when the load level is

TABLE 10-5 Ultimate Strength of Frame Specimens at Elevated Temperatures

Number of specimen	Value of constant load P_0 (kN)	$\frac{P_0}{P_u}$	Ratio between rigidities of beam and column i_b/i_c	Ultimate temperature T_u (°C)	Time constant temperature maintained (min)	Ultimate Load P_u^T (kN)
TFC-1	27.9	1.00	0.354	20.0	0	27.9
TFC-2	13.7	0.49	0.354	820.0	0	13.7
TFC-3	8.5	0.30	0.354	979.4	60	17.3
TFC-4	8.5	0.29	0.877	964.1	60	9.8
TFC-5	9.3	0.31	2.011	952.2	16.8	9.3

TABLE 10-6 Comparison of Bending Moments on Sections of Frame Specimen TFC-1

Position of section	Bottom of column			Top of column and end of beam			Mid-span in beam		
Load P/P_u	0.25	0.50	0.75	0.25	0.50	0.75	0.25	0.50	0.75
Measured bending moment M_t (kN·m)	1.057	2.197	3.151	2.114	4.394	6.302	1.827	3.488	5.522
Calculated bending moment M_c (kN·m)	1.118	2.236	3.354	2.236	4.472	6.707	1.705	3.410	5.115
M_t/M_c	0.945	0.983	0.939	0.945	0.983	0.940	1.072	1.023	1.080

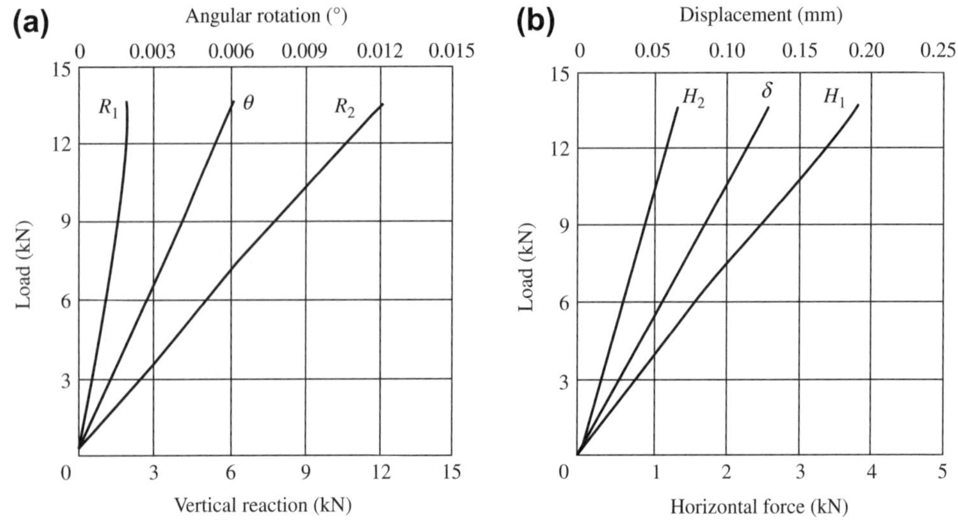

FIGURE 10-18 Reactions and deformations of the base of frame specimen TFC-2 under loading: (a) vertical reactions and angular rotation; (b) horizontal forces and displacement.

$P_0/P_u \leq 0.75$. This shows that the redistribution amplitude of the internal force of the frame, due to the appearance of tensile cracks and nonelastic compressive strain in the concrete, is small before the formation of the plastic hinge. In the meantime, the testing program and technique used to measure the unknown redundant reactions are shown to be reasonable and all the transducers work normally with proper accuracy.

Frame specimen TFC-2 is loaded ($P_0/P_u = 0.49$) first at room temperature, and the reactions measured at the support (R_1, R_2) and the horizontal constraint forces (H_1, H_2), the angular rotation and the horizontal displacement of the short base beam or the bottom of the right column (see Figs. 10-5 and 10-6), increase approximately linearly with the value of the load (Fig. 10-18). A positive bending moment (with tension in the inner side of the column) is caused on the bottom of the right column under loading, so the angular rotation of the short base beam is clockwise, and the reaction there is $R_2 > R_1$ and the horizontal constraint force is $H_1 > H_2$.

When frame specimen TFC-2 is heated under constant load, the inner and outer sides of the beam or column are at high and low temperatures, respectively, and nonuniform temperature fields are formed on their sections. Then, nonuniform thermal strain and deterioration in the behavior of the material result and the redistribution of the internal force of the frame specimen is complicated (Fig. 10-19). During the early heating stage, the thermal expansive strain of concrete (curvatures $1/\rho$ are convex toward the inner side with the high temperature of the beam and column) is predominant, and a negative bending moment is caused on the bottom section of the column due to the constraint of its support. Therefore, the positive bending moment (with tension in the inner side of column) caused during loading is reduced gradually and reaches zero when the experimental temperature $T \approx 200$ °C. The bending moment then turns to a negative value (with tension in the outer side of the column) and reaches a maximum at $T \approx 400$ °C. Correspondingly, the angular rotation at the bottom of the column (or base beam) turns counterclockwise from the clockwise direction, the reaction R_1 and horizontal force H_2 increase, and the reaction R_2 ($R_1 + R_2 = P_0 =$ const.) and the horizontal force H_1 decrease. At this time, the positive bending moment in the mid-span of the frame beam decreases and the negative bending moment at its end increases, and the

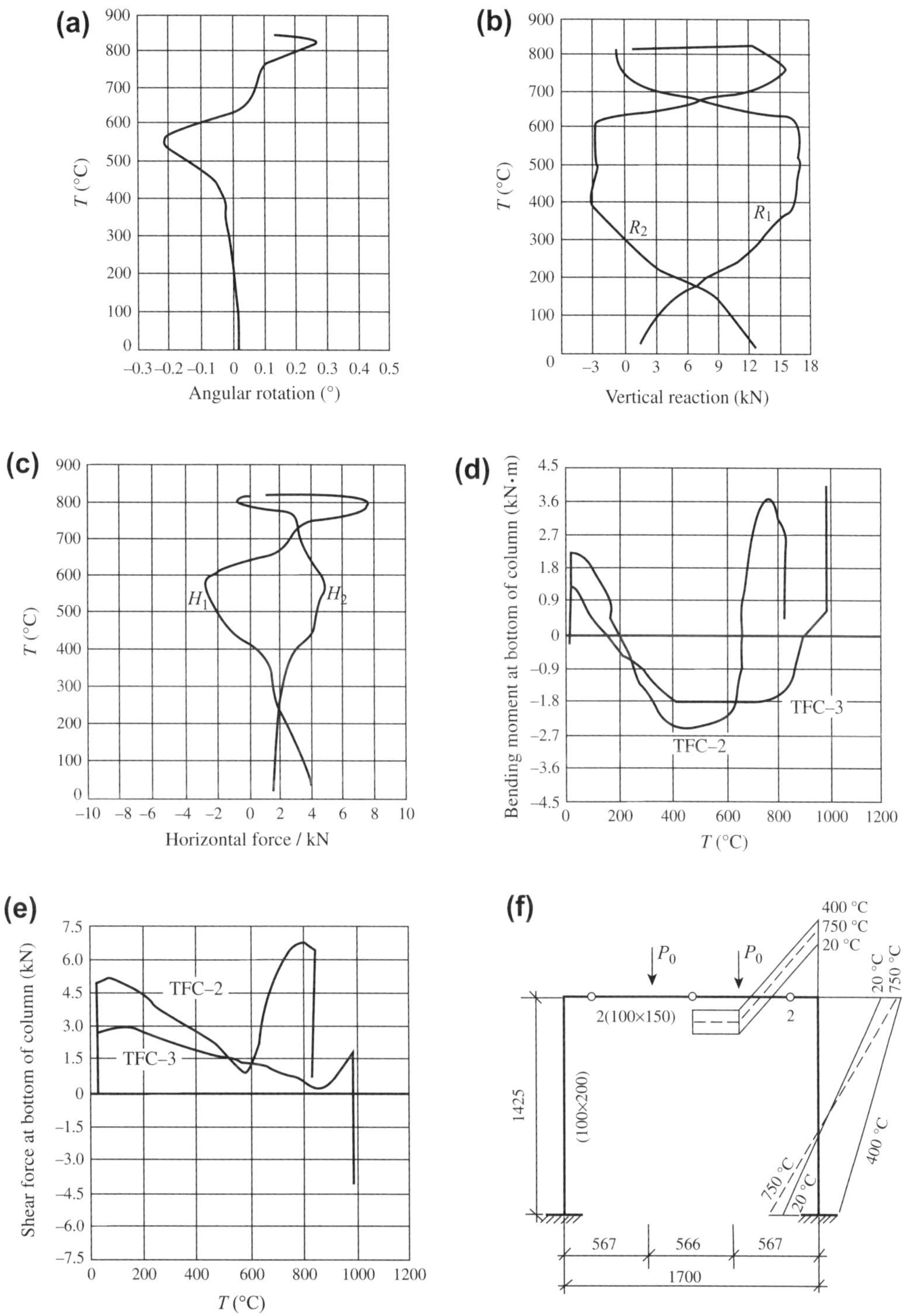

FIGURE 10-19 Variations of the reactions at the support and internal forces on the section of frame specimen TFC-2 at elevated temperatures: (a) angular rotation at the bottom of the column; (b) vertical reactions; (c) horizontal forces; (d) bending moment at the bottom of the column; (e) shear force at the bottom of the column; (f) variation of bending moment diagram.

negative bending moment is distributed over the entire height of the column.

When the experimental temperature is higher ($T > 400$ °C), compressive stress in the inner side of the column section increases under the action of the negative bending moment and the behavior of the concrete deteriorates significantly at high temperature. The compressive strain increases sharply and flexural deformation of the column is convex toward its outer surface and develops quickly (see Fig. 10-16). As the deformation is also constrained by the support, the negative bending moment at the bottom section of the column reduces gradually. When the experimental temperature reaches $T \approx 600$ °C, the redistribution of the internal force of the specimen develops rapidly, and the bending moment at the bottom section of the column is zero at $T \approx 650$ °C and turns to positive afterward. In the meantime, the angular rotation at the bottom of the column (or base beam) turns again to clockwise from the counterclockwise direction, and the reaction R_1 and horizontal force H_2 decrease while reaction R_2 and horizontal force H_1 increase. The bending moment, reactions, and deformation at the bottom of the column experience a process that is repeated twice, i.e., positive–negative–positive or negative–positive–negative. Correspondingly, the bending moments in the mid-span, at the end of the beam, and at the top of the column also experience a process that is repeated twice, i.e., decreasing and increasing or vice versa.

As frame specimen TFC-2 is heated continuously, the internal force and deformation develop constantly. When the experimental temperature approaches 820 °C, the bending moment in the mid-span of the beam reaches the ultimate value of the bending moment on the section with the tension zone exposed to high temperature, and the first plastic hinge appears there. Then, the bending moments at both ends of the beam soon reach the ultimate value of the bending moment on the section with the compression zone exposed to high temperature and two plastic hinges are formed on both ends of the beam; so the frame specimen composes a mechanism of beam type and fails (see Fig. 10-17(b)). As the section depth and strength of the column are greater than that of the beam, no plastic hinge is formed on the column. The frame specimen is unloaded after failure, and the internal force and deformation recover slightly.

Frame specimen TFC-3 is different from TFC-2 only in the lower level of the constant load ($P_0/P_u = 0.30 < 0.49$). The variations in the reaction and deformation, the redistribution regularity of the internal force (Fig. 10-19(d) and (e)), and the failure pattern of both specimens measured during the loading and heating processes are similar, but are quantitatively different. For example, the bending moment on the bottom section of the column of specimen TFC-3 turns from a positive to a negative value at about 150 °C and turns back to a positive value again at about 890 °C. The temperature interval (740 °C) at which the negative bending moment appears is obviously longer than that of specimen TFC-2 (450 °C) and the variable gradient or algebraic difference of the positive and negative bending moments of specimen TFC-3 is obviously smaller. In addition, the variable amplitude of the shear force on the bottom section of the column is smaller, and the variable amplitude of the bending moment at the end of the beam (also at the top of the column) is greater than that of specimen TFC-2 (Fig. 10-20). Of course, frame specimen TFC-3 has a higher ultimate temperature (T_u) and longer fire (high temperature) resistance endurance (hours).

Frame specimens TFC-3, TFC-4, and TFC-5 are a series of comparative specimens with approximately equal initial load levels and the same heating conditions, but with different ratios between the linear rigidities of their beam and column. The bending moment and shear force at the bottom section of the column are also measured during testing, and they vary with the experimental temperature (Fig. 10-21). The reactions on the support, deformation, etc. of the specimens vary correspondingly.

As the ratio between the linear rigidities of the beams and columns of the frame specimens increases (i.e., the beam section increases or the column section decreases), the positive bending moment and shear force at the bottom section of the column, caused under the action of load, reduce gradually. However, the value

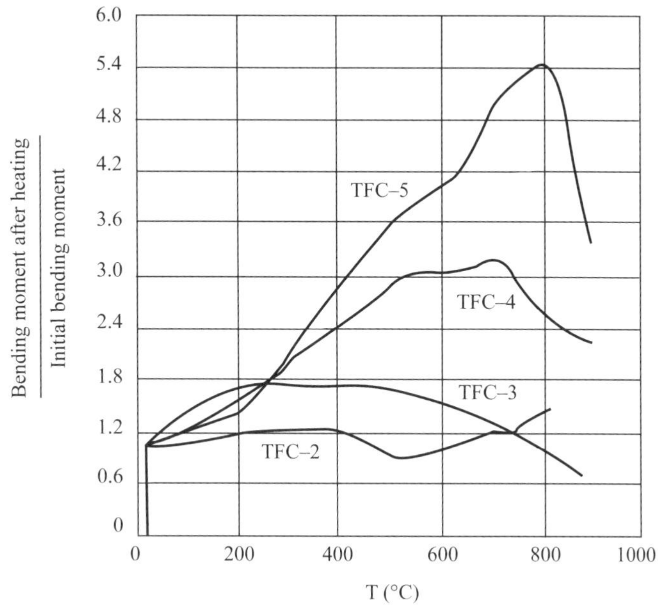

FIGURE 10-20 Variation of the bending moment at the end of the beam (top of column) in a frame specimen at elevated temperatures.

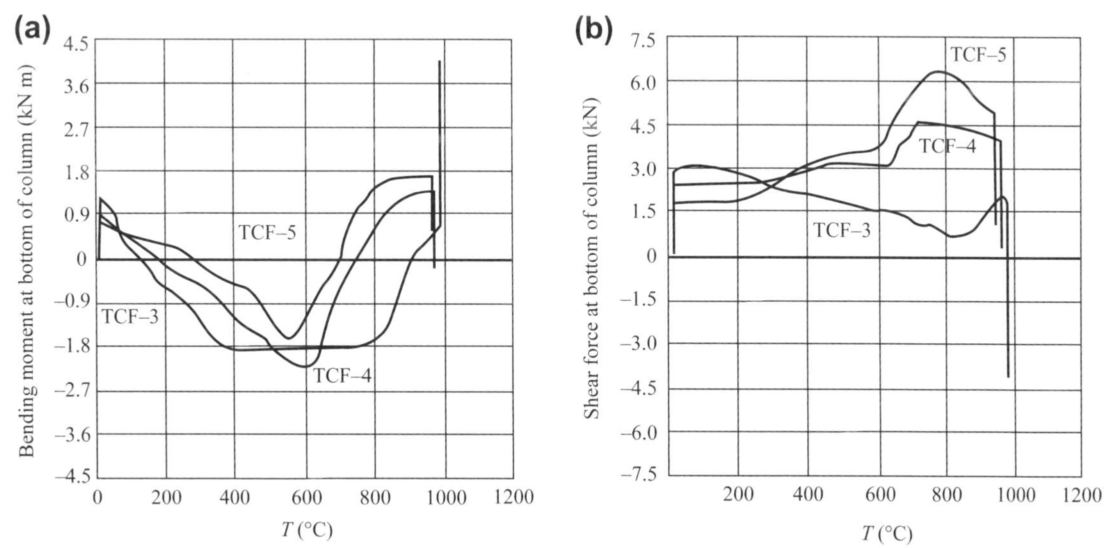

FIGURE 10-21 Variation of the internal force at the bottom section of a column in a series of frame specimens at elevated temperatures: (a) bending moment at the bottom section of the column; (b) shear force at the bottom section of the column.

of the temperature, when the positive bending moment turns into a negative one during heating, increases, but it decreases when the negative bending moment turns back to a positive one; so the temperature interval where the negative bending moment appears at the bottom section of the column is shortened obviously and the variable gradient of the positive and negative bending moments reduces. However, the variation regularity of the shear force at the bottom section of the column is different. The shear force of frame specimen TFC-5 with a smaller section of the column increases monotonically, because the time period when the positive bending moment is acting is longer. The bending moment at the end of the beam (also at the top of the column) also increases rapidly (Fig. 10-20) due to the larger section of the beam, and the value of the bending moment at T = 800 °C is five times greater than that before heating. The increment of the bending moment at the top of the column is disadvantageous for the column with smaller sections, so a plastic hinge of compression type is formed in the middle of the column (Fig. 10-17(b)) and the frame specimen fails eventually. The ultimate strength of frame specimen TFC-5 at elevated temperatures is much lower than that of frame specimen TFC-3 (see Table 10-5).

CONCLUSIONS

When a statically indeterminate reinforced concrete structure bears an action of high temperature, the mechanical behavior is more complicated and the ultimate strength is reduced obviously, compared with the same structure at room temperature. According to the tests on a simple continuous beam and frame specimens at elevated temperatures, the general characteristics of the thermal behavior of statically indeterminate structures is known.

When a statically indeterminate structure at elevated temperatures is approaching failure, its deformation develops quickly and the failing process is quite short. The residual deformation and cracking after failure are apparent. However, the failing process is slower than that of a statically determinate structural member, so it is more advantageous for fire resistance.

The redistribution process of the internal force of a statically indeterminate structure at elevated temperatures can be divided into three stages, and the main factor in each stage that influences the process is different and as follows:

1. The initial stage (T = 20–400 °C). The flexural deformation ($1/\varrho$) of the structural member is caused by nonuniform thermal expansive strain of concrete.
2. The middle stage of heating (T > 400–500 °C). The behavior of concrete and reinforcement materials deteriorates significantly and their strains increase sharply, so the rigidity of the section of the structural member reduces considerably.
3. The stage after formation of the first plastic hinge. The ultimate bending moment at the section of every plastic hinge formed successively reduces continuously and the angular rotation increases there.

The redistribution process of the internal force of a statically indeterminate structure at elevated temperatures varies sharply and is complex. The internal forces on the section of each structural member not only vary considerably in quantity, but also increase and decrease repeatedly; even their positive and negative values change reciprocally. Furthermore, the calculation model of the structure may be modified under certain conditions. The regularity and amplitude of the redistribution of the internal force of the structure depend on the value of the rigidity of every structural member, the relative ratio between rigidities of the structural members, the initial load level, and the heating conditions.

The statically indeterminate structure composes a variable mechanism and then fails after several plastic hinges are formed successively. The type and shape of the failure mechanism, and the positions and order of appearance of the plastic hinges in the structure at elevated temperatures are quite different from that at room temperature.

The plastic hinges formed in a concrete structural member at elevated temperatures can be divided into the hinges with the tension and the compression zones exposed to high temperature. In addition, the plastic hinge of the compression type is formed possibly in the structural member with the greatest axial compression, and the area of the hinge zone is long and the deformation (rotation) is considerable. The ultimate strength (M_u^T) at the section of the plastic hinge does not remain constant, but reduces continuously when high temperature is sustained. These are the important differences between the plastic hinges at elevated temperatures and at room temperature.

The ultimate temperature–strength of a statically indeterminate structure reduces considerably as the level of constant load increases, and also reduces as the ratio between the rigidities of the beam and column increases (or the rigidity of the column decreases).

Large flexural and axial deformations of the structural members, i.e., beam and column, are caused at elevated temperatures. When a fire accident occurs locally in a building, the structural member at elevated temperatures is constrained by the surrounding structural members at room temperature, and considerable additional internal forces are caused.

REFERENCES

[10-1] J.M. Becker, B. Bresler, Reinforced concrete frames in fire environments, Journal of the Structural Division 103 (1) (1977) 211–223.

[10-2] F.J. Vechio, J.A. Sato, Thermal gradient effect in reinforced concrete frame structures, Structural Journal 87 (3) (1990) 262–275.

[10-3] X. Shi, Z. Guo, Experiments of reinforced concrete continuous beam and frame at elevated temperature, The First National Conference on Fire Prevention of High-Rise Building, Beijing (1991).

[10-4] X. Shi, Z. Guo, Experimental investigation of behavior of reinforced concrete continuous beams at elevated temperature, China Civil Engineering Journal 30 (4) (1997) 26–34.

[10-5] X. Shi, Z. Guo, Experimental investigation of behavior of reinforced concrete frames at elevated temperature, China Civil Engineering Journal 33 (1) (2000) 36–45.

[10-6] X. Shi, Z. Guo, Researches on plastic hinge and failure mechanism of statically indeterminate reinforced concrete structure at elevated temperature. Symposium of Third National Conference on Basic Theory and Engineering Application of Reinforced Concrete Structures, Science and Technology Press of Hunan Province, Changsha (1993) 877–884.

[10-7] X. Shi, Z. Guo, Investigations of failure mechanism and internal force redistribution of reinforced concrete continuous beam at elevated temperature, Building Structure 7 (1996) 34–37.

[10-8] X. Shi, Z. Guo, Investigations of failure characteristic and mechanism of reinforced concrete frame at elevated temperature. Symposium of Fifth National Conference on Structural Engineering (Volume II). Haikou (1996) 14–18.

PART IV

Theoretical Analysis and Practical Calculation Method

CHAPTER

11

General Mechanical Characteristics of Inhomogeneous Sections

11.1 STRUCTURAL MEMBERS OF INHOMOGENEOUS SECTIONS IN ENGINEERING PRACTICE

Many experimental investigations have demonstrated that a seriously nonuniform temperature field is caused in the sections of a reinforced concrete structure and its members at elevated temperatures (after a fire accident) because of the thermal inertia of concrete material. Correspondingly, the strength value and stress–strain (constitutive) relationship at each point on the section are different depending on the deterioration level of the material at that point. Furthermore, they vary constantly with different temperature conditions surrounding the structural member and the time at high temperature. Therefore, an original section of homogeneous material is changed into a section of dynamic inhomogeneous material under the action of a high temperature. This is the essential difference between structural members at elevated temperatures and at room temperature.

Many important phenomena are revealed by testing a structural member at elevated temperatures. For example, when a structural member is heated under asymmetric conditions, considerable flexural deformation and additional eccentricity are caused, the ultimate strength–eccentricity curve and the ultimate envelope of axial compression–bending moment are asymmetrical, and the acting point of maximum strength is not located at the geometric center of its section, but moves to the optimum center located in the low temperature zone of the section. These are the mechanical responses of the structural member of an inhomogeneous section. The behavior of the deformation and the strength of the structural member depend on the nonuniform distribution and the variation regularity of the material in its section.

There are still many similar situations of inhomogeneous sections of structural members in structural engineering. For example, nonuniform damage is caused on the surface layer and the inner side of the section of a structural member due to bad durability of the concrete material, after the structure is used for a long period of time; damage is caused at different levels in a section of a structural member after a severe earthquake. The characteristic of these structural members is that the mechanical behavior of concrete varies continuously along one or two directions in the section. There are other examples. When a beam or column is strengthened with its section enlarged, the original and new concrete become an inhomogeneous section. Various types of sections of steel-reinforced concrete composite beams and columns are another kind of inhomogeneous section in structural members. An ordinary reinforced concrete beam or the column itself is an inhomogeneous section composed of concrete and reinforcement, whose behavior is totally different. A pure concrete beam can also be considered as an inhomogeneous section within a member, because the strengths and stress–strain relationships of concrete in the compression and tension zones are different. The characteristics of these structural members are that the materials and their behavior in the different parts of the section change abruptly.

Experiment and Calculation of Reinforced Concrete at Elevated Temperatures

In addition to the definitions of inhomogeneous sections of structural members described above, random inhomogeneous sections of structural members also occur in structural engineering. For example, random inhomogeneity, not only on the transverse section but also on the longitudinal direction, may arise in the interior of a large structural member because the concrete is located at different depths, inside and outside the reinforcement skeleton, and is cast at different times, compacted differently, and cured under different conditions, and as a result of occasional differences in the raw materials and the mix of each concrete mixture. Generally, the variable amplitude of random inhomogeneity in a concrete section is small and is within the permissible range of engineering accuracy, so special mechanical analysis is not necessary.

In addition, different parts on a section of structural members are not loaded simultaneously and a phase difference in the stress history exists between them. For example, a composite concrete beam or slab is composed of precast and cast-in-situ parts. When the precast part is loaded first during the construction stage, the initial stress state is established on its section. When the structural member is completed integrally after the concrete is cast in situ and the service load is applied later, the stress state on the precast part varies successively while the stress state on the cast-in-situ part has just begun. If a strengthened column is unloaded partially before strengthening, a residual stress state should exist on its original section; after it is strengthened with the additional section and loaded again later, the beginnings of stress increments of the original and additional sections are different although they work together. A section with a structural member with this kind of stress history can be considered a generalized inhomogeneous section.

Briefly, in engineering practice, the structural member of an inhomogeneous section is formed due to differences in materials or in material behavior, or the phase difference of the stress history, or both factors together.

All the structural members of inhomogeneous sections should have some common characteristics of mechanical behavior and the same or similar basic regularity. The analysis methods and conclusions can be used to refer one to the another. Analysis of a structural member in an inhomogeneous section of linear elastic materials is discussed in Section 11.3. The experimental investigations and theoretical analysis method of a reinforced concrete structural member at elevated temperatures are given in Chapters 8, 9, 10, 12, and 13. They provide references for analysis of other kinds of structural members of inhomogeneous sections.

11.2 MECHANICAL CHARACTER POINTS ON A SECTION

Resistance of materials is an important theory system, based on an ideal, homogeneous, linear elastic material, and it provides the fundamental formulas to calculate the stress and deformation of structural members that resist various internal forces, including axial tension and compression, bending moment, shear force, and torsion. In the past, the analyses of stress and deformation, checks on the bearing capacity, and design of the cross section of the structure and its members made of various engineering materials, including steel and concrete, were based theoretically on the resistance of materials. Even now, the fundamental formulas are still used in some provisions of the existing design code of reinforced concrete structures,[0-1] e.g., checking a structural member for fatigue and analysis of prestressed concrete members.

Several mechanical concepts, such as neutral axis, linear strain hypothesis, moment of area, moment of inertia, and core of cross section, are summarized in the resistance of materials, and the formulas for sectional stress and deformation of various structural members are also derived from it. All of these have universal significance.

Although the existing formulas derived in the resistance of materials cannot be used directly for the calculation of structural members of various inhomogeneous sections in engineering practice, mentioned above, the concepts and methods of the resistance of materials can be used as a reference for analyzing the members. Several important basic concepts of inhomogeneous sections are given below and some simple conditions are analyzed.

An inhomogeneous rectangular section of one dimension is taken as an example. The concept and definition of every mechanical character point on the section are given, and a section of two dimensions can be analyzed in the same way. It is assumed that the material and its behavior at every point of the rectangular section ($b \cdot h$) are identical along its width direction, but vary along its depth direction following a certain rule. If the coordinate origin is located at the center of the section, the stress–strain relationship and the ultimate strength of the material at a point being z apart from the origin are $\sigma_z - \varepsilon_z$ and f_z, respectively (Fig. 11-1), and both are functions of the ordinate z. The main mechanical character points needed for the sectional analysis based on the linear strain hypothesis are given below.

1. Geometric center (G)

The geometric center of a section is also the center of its weight, and depends only on its geometric shape. It has no relationship with the stress–strain relationship of the material at every point on it. When the conclusion of resistance of materials is used directly, the moment of area of the section is

$$S=\int_{-h/2}^{h/2} b\,dz \cdot z=0 \qquad (11.1a)$$

It is found that the geometric center of the section is located at its center and the eccentricity is

$$e_G=0 \qquad (11.1b)$$

and has no relationship with the inhomogeneity of the section.

2. Deformation center (D)

When the strain on the section of a structural member is distributed uniformly, i.e., ε_z = constant, or $d\varepsilon_z/dz=0$ and $1/\rho=0$, the corresponding nonuniform distribution of stress is formed on the section (Fig. 11-2) because of the difference in the stress–strain relationship of the material at different positions on the section.

The resultant (axial force N_D) and resultant couple (bending moment M_D) of distributed stress on the section can be obtained according to the equilibrium formulas:

$$N_D=\int_{-h/2}^{h/2}\sigma_z(\varepsilon_z)\,b\,dz,\quad M_D=\int_{-h/2}^{h/2}\sigma_z(\varepsilon_z)\,bz\,dz \qquad (11.2a)$$

or they are transformed into the axial force N_D and the corresponding eccentricity

$$e_D=\frac{M_D}{N_D} \qquad (11.2b)$$

The latter is the distance from the deformation center (D) to the geometric center (G) on the section.

Therefore, when the axial force acts on the deformation center of an inhomogeneous

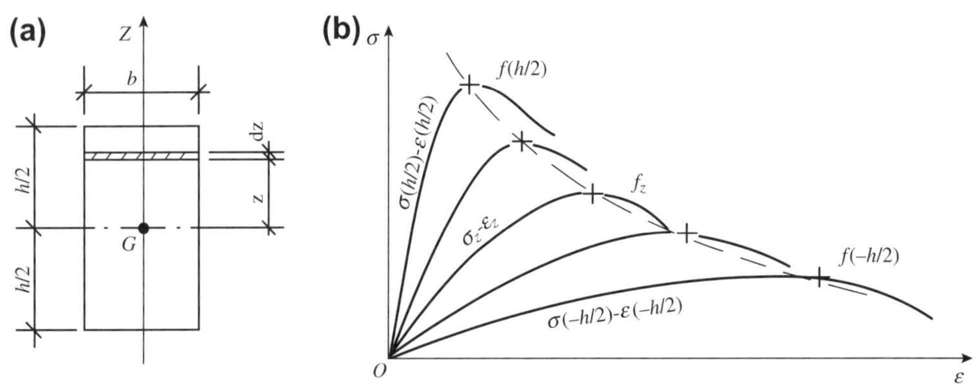

FIGURE 11-1 Inhomogeneous section and variation in material behavior. (a) Section; (b) stress–strain relationship at each point.

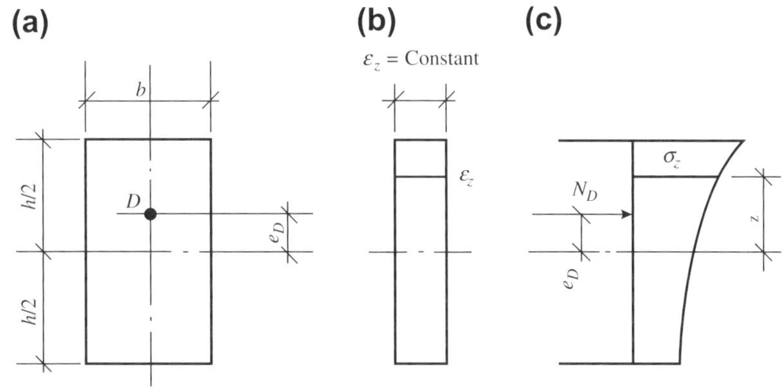

FIGURE 11-2 Deformation center: (a) section; (b) strain distribution; (c) stress distribution.

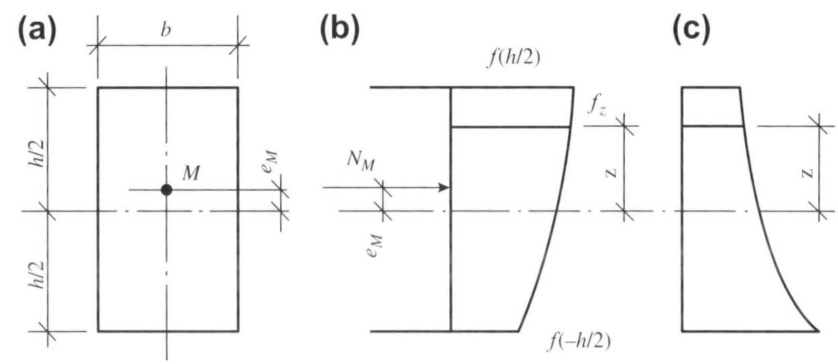

FIGURE 11-3 Absolute strength center: (a) section; (b) distribution of ultimate strength; (c) distribution of strain corresponding to ultimate strength.

section, the strain is equal anywhere on it and its curvature ($1/\varrho$) is zero. Obviously, there is one corresponding deformation center (D) for every particular value of strain in the section, so the position of the deformation center (e_D) and the corresponding axial force (N_D) are functions of the strain value (ε_z).

3. Absolute strength center (M)

If the stress at every point on the section reaches the ultimate strength of the material at the same point separately (Fig. 11-3), the corresponding axial force, bending moment, and eccentricity of the section can be calculated:

$$N_M = \int_{-h/2}^{h/2} f_z b \, dz, \quad M_M = \int_{-h/2}^{h/2} f_z b z \, dz, \; e_M = \frac{M_M}{N_M} \quad (11.3)$$

The absolute strength center (M) of the section is at a distance e_M from the geometric center (G).

Obviously, the absolute maximum strength (N_M) corresponding to the absolute strength center is the maximum possible strength of the section, only when the potential strengths of all the materials on the section are utilized simultaneously and fully. However, the distribution of the strain corresponding to the ultimate strength of the material in the section is not consistent, under general conditions, with the compatibility condition of deformation (linear strain hypothesis), so it is not realizable. The absolute maximum strength may be reached or approached only when under some special conditions, e.g., an ideal elastic (or nonelastic) plastic material of a homogeneous or inhomogeneous section.

4. Optimum center (P)

The ultimate strength of a structural member of an inhomogeneous section varies with the eccentricity of the axial load (force) (see Figs. 9-11 and 11-4). The ultimate strength of the structural member (N_G) is not the maximum when the load is acting on the geometric center of the section (G, e = 0); it increases gradually and reaches the maximum when the load is moved continuously toward the side of higher strength of material on the section (e > 0), but it reduces if the load eccentricity increases further. The maximum strength (axial force) that the structural member can reach practically is called the optimum ultimate strength (N_P). The corresponding loading point is called the optimum center (P) and its distance from the geometric center is called the optimum eccentricity:

$$e_P = \frac{M_P}{N_P} \tag{11.4}$$

As far as homogeneous linear elastic and nonlinear materials are concerned, the geometric center, deformation center, absolute strength center, and optimum center on the section of a structural member coincide, i.e., $e_D = e_M = e_P = 0$, and their corresponding ultimate strengths are identical. However, the four centers on the section of a structural member of inhomogeneous linear elastic and nonlinear materials are located separately. The eccentricities are not equal to one another and vary with the inhomogeneity of the material field on the section and the stress–strain relationship of the material. Moreover, the corresponding ultimate strengths are also not equal to one another; the optimum ultimate strength of the section is not greater than the absolute maximum strength, but is not less than the ultimate strength when the load is acted on the deformation center, so

$$N_M \geq N_P \geq N_D \tag{11.5}$$

Generally, these three ultimate strengths are greater than the ultimate strength (N_G) when the load is acted on the geometric center (e = 0) of the section.

In addition, the core of the section is also a mechanical character point on the section of a structural member. There are two cores on an asymmetrical inhomogeneous section and they are located asymmetrically on the upper and the lower parts of the section. When the load is acted on one of them, the stress (strain) at either the top or the bottom side of the section is zero. This is discussed in detail below.

11.3 ANALYTICAL SOLUTION OF LINEAR ELASTIC MATERIAL

The constitutive relation of linear elastic material is the simplest, and it is also a special example and the basis for analyzing various nonlinear materials. When the inhomogeneous section of a structural member is composed of linear elastic materials, the mechanical behavior can be calculated analytically. Two examples[11-1,11-2] of different inhomogeneous sections are discussed in the following sections. The analysis method and the calculated results are given to illustrate the

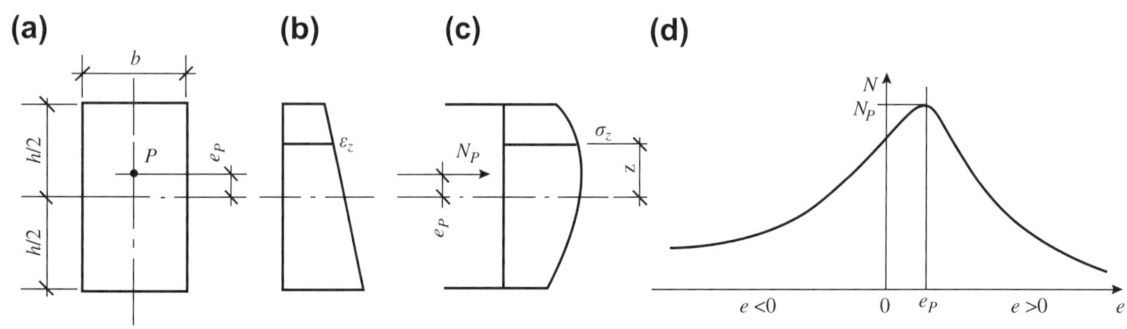

FIGURE 11-4 Optimum center: (a) section; (b) strain on section; (c) stress distribution; (d) ultimate strength–eccentricity relation.

general regularity of the mechanical behavior of a structural member of an inhomogeneous section.

11.3.1 Rectangular Section Composed of Two Materials

It is assumed that a structural member of rectangular section $(b \cdot h)$ is composed of two kinds of linear elastic materials (M_1 and M_2) and each occupies one half of the section (Fig. 11-5). If the strengths are equal ($f_1 = f_2 = f$), but the moduli of elasticity are not equal ($E_1 > E_2$), the stress–strain relations are

$$\sigma_1 = E_1\varepsilon \leq f \text{ and } \sigma_2 = E_2\varepsilon \leq f \qquad (11.6a)$$

respectively. So the ultimate strains are

$$\varepsilon_1 = \frac{f}{E_1} \text{ and } \varepsilon_2 = \frac{f}{E_2} \qquad (11.6b)$$

respectively, and the ratio between them is

$$\frac{\varepsilon_1}{\varepsilon_2} = \frac{E_2}{E_1} \qquad (11.6c)$$

When the axial load (N) is acted with an eccentricity e and the strains at the top and the bottom of the section (ε_t and ε_b) are taken as

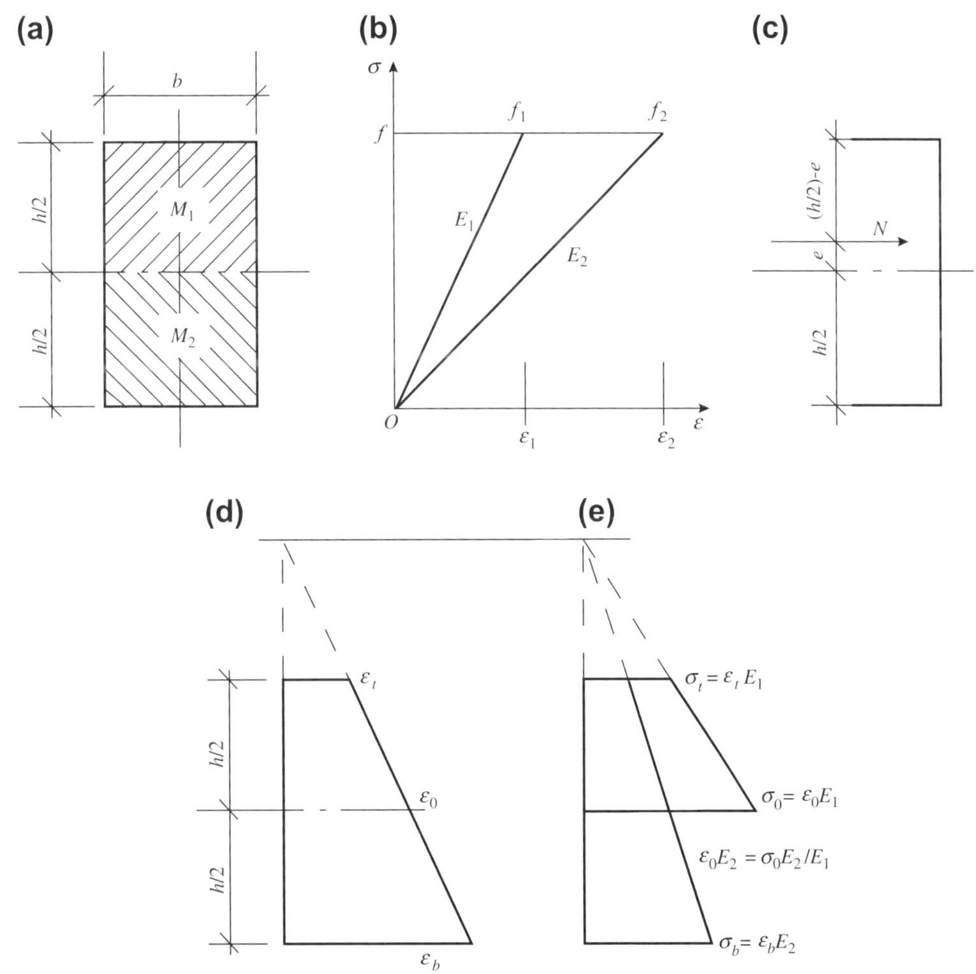

FIGURE 11-5 Analysis of a section composed of two materials: (a) section; (b) stress–strain relationship; (c) eccentric load; (d) distribution of strain; (e) distribution of stress.

basic unknowns, two equilibrium equations of forces can be established assuming the strain on the section is distributed linearly:

$$\left.\begin{aligned}\sum X=0,\ N&=\frac{bh}{8}\left[\varepsilon_t(3E_1+E_2)+\varepsilon_b(E_1+3E_2)\right]\\ \sum M=0,\ Ne&=\frac{bh^2}{48}\left[\varepsilon_t(5E_1-E_2)-\varepsilon_b(5E_2-E_1)\right]\end{aligned}\right\} \tag{11.7a}$$

The equations are solved and the ratio between both strains is obtained:

$$\frac{\varepsilon_t}{\varepsilon_b}=\frac{(5E_2-E_1)+(3E_2+E_1)(6e/h)}{(5E_1-E_2)-(3E_1+E_2)(6e/h)} \tag{11.7b}$$

The stress at the top, the bottom, and the geometric center (on the side of M_1) of the section can then be calculated:

$$\left.\begin{aligned}\sigma_t&=\varepsilon_t E_1=\frac{8E_1}{(3E_1+E_2)+(E_1+3E_2)(\varepsilon_b/\varepsilon_t)}\cdot\frac{N}{bh}\\ \sigma_b&=\varepsilon_b E_2=\frac{8E_2}{(E_1+3E_2)+(3E_1+E_2)(\varepsilon_t/\varepsilon_b)}\cdot\frac{N}{bh}\\ \sigma_0&=\frac{1}{2}(\varepsilon_t+\varepsilon_b)E_1=\frac{1}{2}\left(\sigma_t+\frac{E_1}{E_2}\sigma_b\right)\end{aligned}\right\} \tag{11.8}$$

and the distribution of stress on the section is shown in Fig. 11-5(e).

These basic formulas are used to calculate the strain, stress, and curvature of the inhomogeneous section under the combined actions of axial compression and bending moment, and the ultimate values of the axial compression and bending moment are obtained, correspondingly. If the strengths of both materials are equal, $f_1 = f_2 = f$, and the moduli of elasticity are $E_1 = 2E_2$, the ultimate strains should be $\varepsilon_2 = 2\varepsilon_1$. Several mechanical characteristic conditions of the section will appear successively when the eccentricity of the axial compression moves from $e = -\infty$ to $+\infty$ continuously. The calculations are listed in Table 11-1, and the relationships between the ultimate axial compression and eccentricity and between the ultimate envelope of axial compression and the bending moment of the inhomogeneous section are shown in Fig. 11-6.

From the analysis above, the important characteristics of the mechanical behavior of the structural member of an inhomogeneous rectangular section composed of two types of linear elastic materials can be found and compared with those of a section composed of a single homogeneous linear elastic material.

1. Although the strain on a section of a structural member agrees with the linear strain hypothesis when any load is acted, the stress diagram on the section varies suddenly at the place where the modulus of elasticity changes.
2. When the load is acted at the geometric center (G) of the section or the compression is acted centrically ($e = 0$), the strain is distributed nonuniformly on the section and the curvature of the section will not be zero ($1/p \neq 0$). The value of the strain is greater where the modulus of elasticity is smaller, and vice versa.
3. The curve of the ultimate axial compression–eccentricity is asymmetrical about the ordinate (N) axis, and the optimum center (P) and optimum (also maximum) ultimate strength appear on the right side of the ordinate or on the side of greater value of the modulus of elasticity on the section. The ultimate envelope of axial compression–bending moment is also asymmetrical about the ordinate (N) axis and is oblique up and toward the right side; the highest tip corresponds to the optimum ultimate strength.
4. When the optimum center coincides with the deformation center ($e_P = e_D$), the ultimate strengths of both conditions are equal ($N_P = N_D$).
5. The inhomogeneous section in this example is composed of two materials, so there are three points where failure may occur: top (material M_1), bottom (M_2), and mid-depth (M_1, but M_2 will not fail because of the ultimate strains $\varepsilon_2 > \varepsilon_1$). Correspondingly, when the eccentricity of the load moves from $e = -\infty$ to $+\infty$, four failure modes, i.e., tension at top, compression at bottom, compression at mid-depth, and compression at top, will transit successively. The control region and eccentricity range of each failure mode can be found in Fig. 11-6 and Table 11-1.

TABLE 11-1 Analysis of Inhomogeneous Linear Elastic Sections

Position of load	$\frac{e}{h}$	Distributions of strain and stress on section	Ultimate strength	Curvature $1/\varrho$	Failure controlled by
Negative bending moment $-M$	$-\infty$	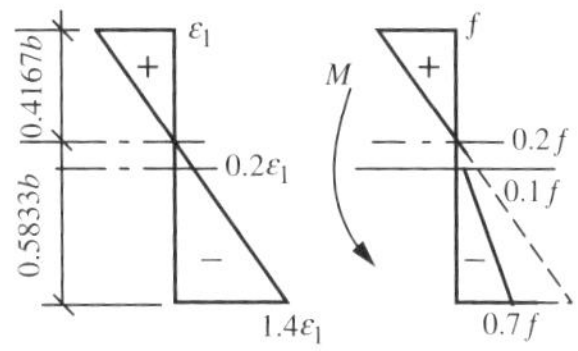	Nu = 0 $\frac{M_u}{fbh^2} = -0.1325$	$-2.4\frac{\varepsilon_1}{h}$	Tension at top
	−0.8333	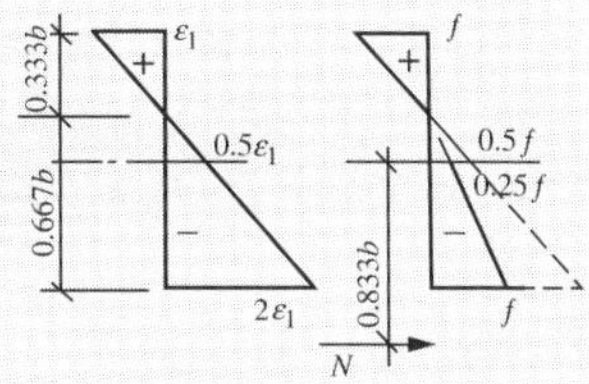	$\frac{N_u}{fbh} = 0.1875$ $\frac{M_u}{fbh^2} = -0.1562$	$-3\frac{\varepsilon_1}{h}$	Tension at top and compression at bottom
Lower core of section C_b	−0.1	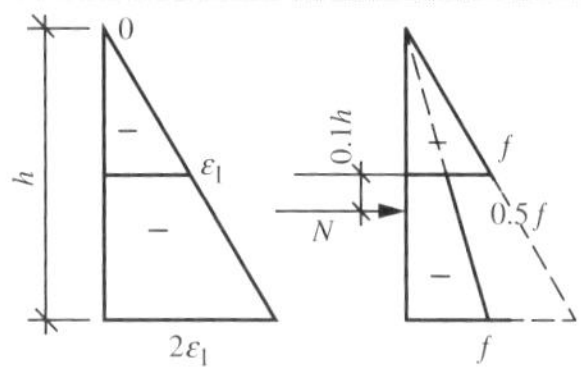	$\frac{N_u}{fbh} = 0.625$ $\frac{M_u}{fbh^2} = 0.0625$	$-2\frac{\varepsilon_1}{h}$	Compressions at bottom and mid-depth
Geometric center G	0	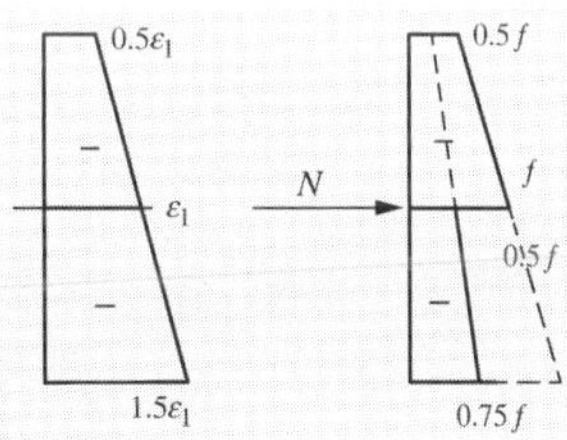	$\frac{N_u}{fbh} = 0.6875$ $M_u = 0$	$-\frac{\varepsilon_1}{h}$	Compressions at mid-depth
Deformation center D (optimum center P)	0.0833	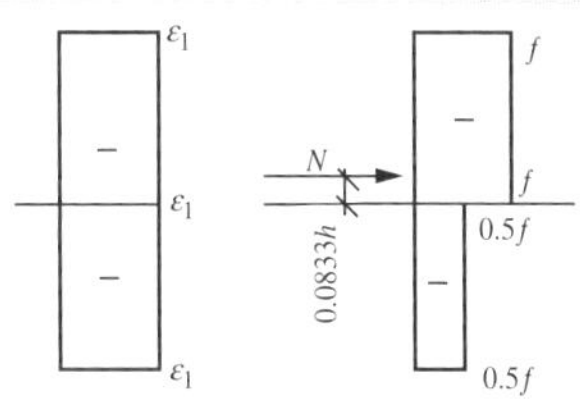	$\frac{N_u}{fbh} = 0.75$ $\frac{M_u}{fbh^2} = 0.0625$	0	Compressions at mid-depth and top
Upper core of section C_t	0.2143	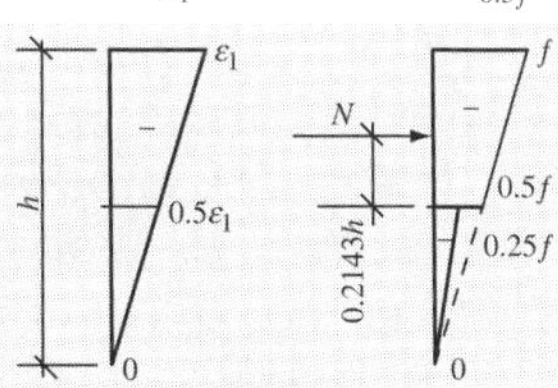	$\frac{N_u}{fbh} = 0.04375$ $\frac{M_u}{fbh^2} = 0.0938$	$\frac{\varepsilon_1}{h}$	Compression at top
Positive bending moment $+M$	$+\infty$	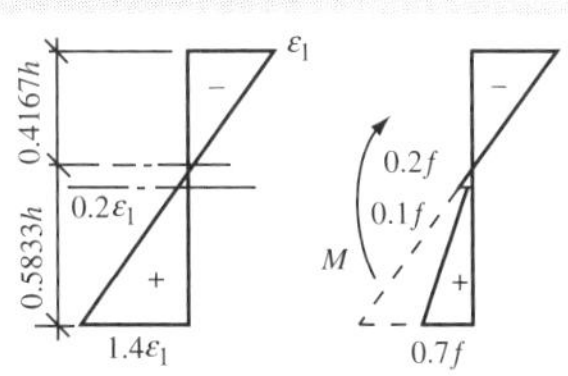	$N_u = 0$ $\frac{M_u}{fbh^2} = 0.1375$	$2.4\frac{\varepsilon_1}{h}$	Compression at top

TABLE 11-1 Analysis of Inhomogeneous Linear Elastic Sections—cont'd

Position of load	$\frac{e}{h}$	Distributions of strain and stress on section	Ultimate strength	Curvature $1/\varrho$	Failure controlled by
Absolute strength center M	0	(Compatibility condition of deformation (strain) on section is not satisfied) (see diagram below)	$\frac{N_u}{fbh}=1$ $M_u=0$	—	Every point on section

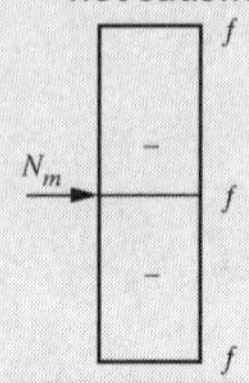

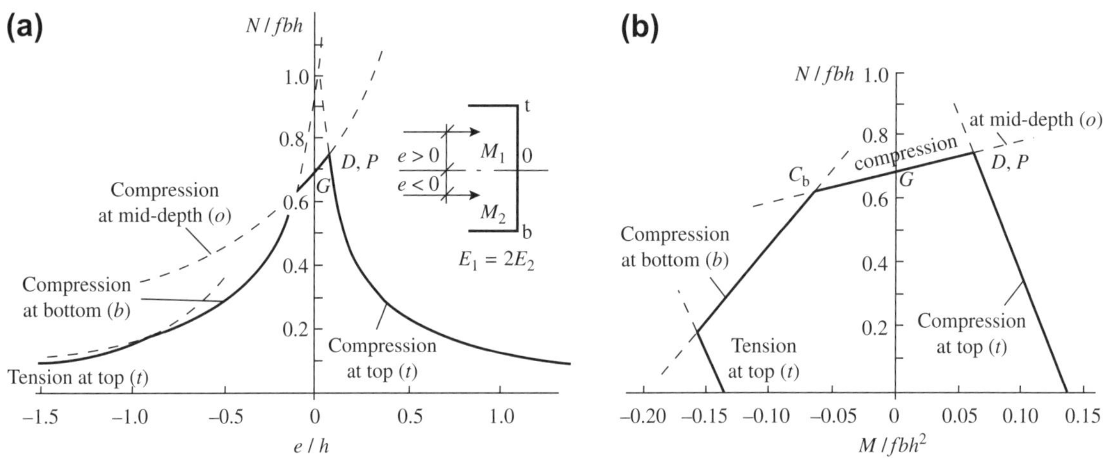

FIGURE 11-6 Ultimate strength of an inhomogeneous linear elastic section: (a) relationship between the ultimate axial compression and eccentricity; (b) ultimate envelope of the axial compression–bending moment.

Obviously, the main mechanical characteristics of the structural member loaded eccentrically with three surfaces exposed to high temperature (see Chapter 9) have much in common with this example.

11.3.2 Section Composed of Material with Continuously Variable Behavior Along Its Depth

It is assumed that a structural member of rectangular section ($b \cdot h$) is composed of an inhomogeneous linear elastic material, and the strength of the material at any point on the section is identical ($f_1 = f_2 = f$), but the modulus of elasticity of the material is E_1 at the top and decreases linearly to E_2 at the bottom of the section (Fig. 11-7). Correspondingly, the ultimate strain of the material is ε_1 at the top and increases nonlinearly to ε_2 at the bottom of the section. If the coordinate origin is located at the top of the section, the modulus of elasticity of the material at z from the origin is

$$E_z = E_1 - (E_1 - E_2)\frac{z}{h} \tag{11.9a}$$

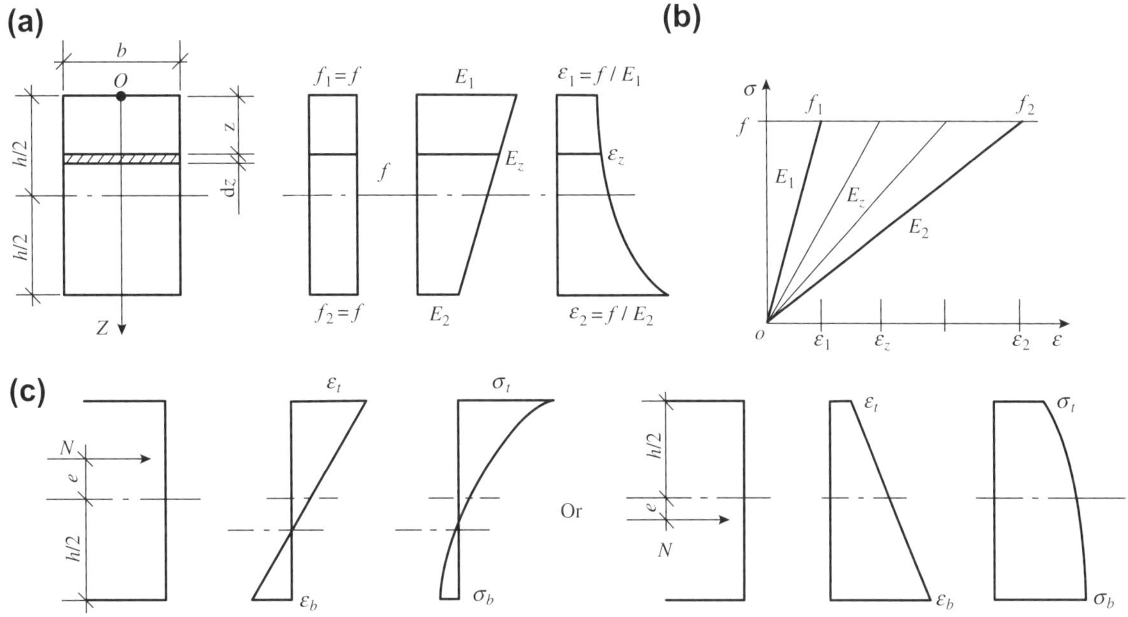

FIGURE 11-7 Analysis of a section composed of material with continuously variable behavior: (a) section; (b) variation of material behavior; (c) distribution of strain and stress on the section.

and the ultimate strain there is

$$\varepsilon_z = \frac{f}{E_z} \tag{11.9b}$$

Then, the stress–strain relationship at any point on the section is

$$\sigma = \varepsilon \cdot E_z \leq f \tag{11.9c}$$

When an axial compression (N) is acted eccentrically (e) on the structural member, the stress distribution is certainly nonlinear on its section, assuming the strain distribution is linear on the section. If the strains at the top and the bottom of the section (ε_t and ε_b) are taken as the two basic unknowns, the equilibrium equations are established and simplified as follows:

$$\left.\begin{aligned} &\sum X=0, \quad N=\frac{bh}{6}\left[\varepsilon_t\left(2E_1+E_2\right)+\varepsilon_b\left(E_1+2E_2\right)\right] \\ &\sum M=0, \quad M=\frac{bh^2}{12}\left(\varepsilon_t E_1-\varepsilon_b E_2\right) \end{aligned}\right\} \tag{11.10a}$$

The equations are solved and the strain ratio is obtained:

$$\frac{\varepsilon_t}{\varepsilon_b} = \frac{6E_2+\left(E_1+2E_2\right)(12e/h)}{6E_1-\left(2E_1+E_2\right)(12e/h)} \tag{11.10b}$$

Therefore, the stresses at the top and the bottom of the section are, respectively,

$$\left.\begin{aligned} \sigma_t=\varepsilon_t E_1=\frac{6E_2+\left(E_1+2E_2\right)(12e/h)}{E_1^2+4E_1E_2+E_2^2}\cdot\frac{NE_1}{bh} \\ \sigma_b=\varepsilon_b E_2=\frac{6E_1-\left(2E_1+E_2\right)(12e/h)}{E_1^2+4E_1E_2+E_2^2}\cdot\frac{NE_2}{bh} \end{aligned}\right\} \tag{11.11a}$$

and the stress at any point on the section is

$$\sigma = \left[\varepsilon_t - \left(\varepsilon_t-\varepsilon_b\right)\frac{z}{h}\right]\cdot E_z \tag{11.11b}$$

As the section of the structural member is composed of inhomogeneous material with continuously variable behavior, not only the top and the bottom sides of the section are certain failure control points but also every point on the section is possibly a failure control point, if the strain value at the point reaches its own ultimate value first. The position of the failure control point within the section is determined by the tangential point between the strain line distributed on the section caused by load and the ultimate strain curve of the material on the section (Fig. 11-8). As the strain values and slopes of both should be

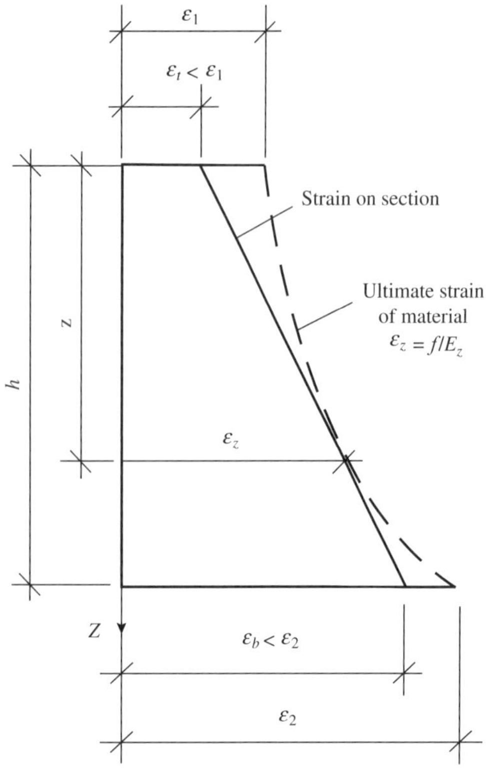

FIGURE 11-8 Determination of the failure control point on a section.

equal at the tangential point, two equations are established:

$$\left.\begin{aligned} \varepsilon_z &= \frac{f}{E_1 - (E_1 - E_2)(z/h)} = \frac{z}{h}\varepsilon_b + \left(1 - \frac{z}{h}\right)\varepsilon_t \\ \frac{d\varepsilon_z}{dz} &= \frac{(E_1 - E_2)f}{[E_1 - (E_1 - E_2)(z/h)]^2 h} = \frac{\varepsilon_b - \varepsilon_t}{h} \end{aligned}\right\} \quad (11.12)$$

Then, the position of the failure control point is solved:

$$\frac{z}{h} = \frac{E_1}{2(E_1 - E_2)} - \frac{(\varepsilon_t / \varepsilon_b)}{2(1 - (\varepsilon_t / \varepsilon_b))} \quad (11.13)$$

When Eqn (11.13) is substituted into Eqn (11.12), the ultimate strain value at the failure point is obtained.

The mechanical behavior of the section in which the material behavior is continuously variable (Fig. 11-7) can be analyzed using these basic formulas, and the strain, stress, curvature, failure control point, and ultimate strength (axial compression–bending moment) of the section are given. If the relationship between the moduli of elasticity at the top and the bottom sides of the section is $E_1 = 2E_2$, various conditions of the mechanical characteristics of the section obtained from the calculations are shown in Fig. 11-9. The ultimate axial compression–eccentricity relationship and the ultimate envelope of axial compression–bending moment of the inhomogeneous section are shown separately in Fig. 11-10.

According to the calculated results of the structural member of an inhomogeneous section composed of linear elastic material with continuously

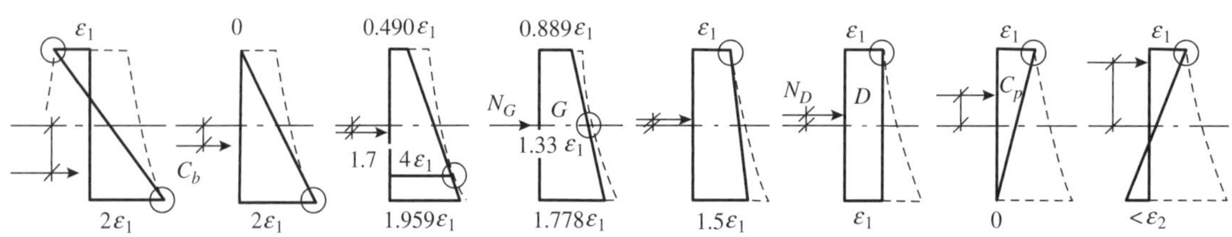

	(a)	(b)	(c)	(d)	(e)	(f)	(g)	(h)
e/h	−0.667	−0.125	−0.0476	0	0.0227	0.0556	0.200	>0.2
N/fbh	0.250	0.667	0.857	0.963	0.917	0.750	0.417	

FIGURE 11-9 Various conditions of mechanical character of a section with continuously variable behavior. Solid line, strain on section; dashed line, ultimate strain of material; circle, failure control point.

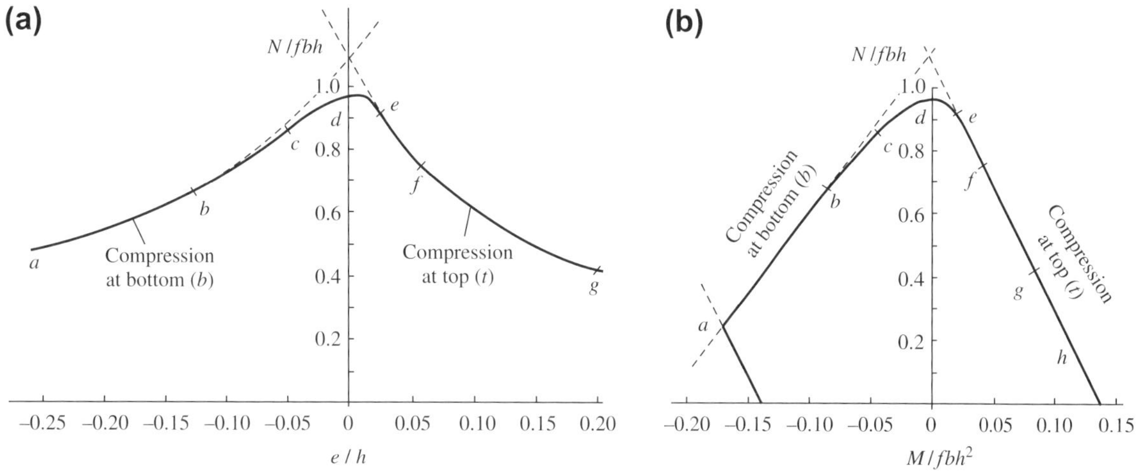

FIGURE 11-10 Ultimate strength of a section with continuously variable behavior: (a) relationship between the ultimate axial compression and eccentricity; (b) ultimate envelope of the axial compression–bending moment.

variable behavior, there are several important characteristics of the mechanical behavior:

1. When the strain distribution on the section agrees with the linear strain hypothesis, the stress distribution on the section should be continuous but nonlinear (Fig. 11-7(c)).
2. When the load is acted at the geometric center of the section (Fig. 11-9(d)), the strain on the section is nonuniform (i.e., the strain is smaller where the modulus of elasticity is greater and the strain is greater where the modulus of elasticity is smaller) and the curvature of the section is $1/\varrho \neq 0$.
3. The curve of the ultimate axial compression–eccentricity relationship and the ultimate envelope of axial compression–bending moment are asymmetrical about the ordinate axis (N) and the peaks deviate to the right side of the ordinate (i.e., the side of greater value of modulus of elasticity on the section).
4. The eccentricity of the optimum center (P) is between $0.0227h > e_P > 0$, and the optimum ultimate strength of the section is $N_P > N_G > N_D$; N_G and N_D are the ultimate strengths of the section when the loads are acted at the geometric ($e_G = 0$) and the deformation (e_D) centers, respectively.
5. When the eccentricity of the load varies from $e = -\infty$ to $+\infty$, the failure mode and control point of the structural member transit successively as shown below:

Eccentricity e/h	Failure mode and control point
$-\infty$ to -0.667	Tension at the top
-0.667 to -0.125	Compression at the bottom
-0.125 to $+0.0227$	Compression at the interior of the section (moves gradually from the bottom to the top)
$+0.0227$ to $+\infty$	Compression at the top

Correspondingly, the ultimate envelope of axial compression–bending moment of the section of the structural member is also composed of four segments of a straight line or a curve, which correspond to four failure modes.

The structural members of other kinds of inhomogeneous sections composed of various linear elastic or ideal elastoplastic materials can also be calculated and analyzed in accordance with the analytical method introduced above. It is difficult to find an analytical solution for a structural member of an inhomogeneous section composed of nonlinear materials. However, the mechanical behavior and its variable process during loading (or heating) of this kind of structural member can be obtained when the numerical method is used and the many calculations are conducted on a computer (see Chapter 12).

CONCLUSIONS

In engineering practice, different levels of damage are caused in the interior of a reinforced concrete structure as a result of occasional accidents (e.g., fire or earthquake) or bad durability of the materials. Nonuniform and asymmetrical distribution of the material behavior occurs in the cross section of some structural members because of the construction or bad manufacturing. Furthermore, phase differences in the stress history may occur on the cross section of some other structural members (e.g., strengthening a structural member) because of the construction or loading path. All of these can generally be called the structural member of an inhomogeneous material field section, and the mechanical behavior is more complicated than that of the structural member of homogeneous material.

Several important points of a mechanical character (i.e., deformation center, absolute strength center, optimum center, and core of the section) can be found on the section of a structural member of a general inhomogeneous material field. They do not coincide with the geometric center on the section and their positions vary constantly with the level of nonuniformity of the material field and the stress–strain relationship of the material. Knowing these characteristics well is helpful in understanding and determining the mechanical behavior of a structural member of inhomogeneous material.

The analytical method can be used to completely analyze the structural member of an inhomogeneous section of linear elastic material with simple variance. Two kinds of structural members of a rectangular section of inhomogeneous material are given in this chapter, and the variation process and transition regularity of the important characteristics of the sections are presented, e.g., stress distribution, curvature (deformation), failure mode and its control point, and the ultimate value of the axial compression–bending moment, when the eccentricity of an axial compression varies from $e = -\infty$ to $+\infty$. Both examples illustrate and demonstrate qualitatively the main phenomena and characteristics of the mechanical behavior of reinforced concrete structural members at elevated temperatures, as revealed by experiments (see Chapters 8 and 9).

REFERENCES

[11-1] Y. Guo, Non-uniform section reinforced concrete member analysis and distributed computing architecture research. Doctoral Thesis, Tsinghua University, Beijing (1999).

[11-2] Y. Guo, Z. Guo, Cross section analysis of reinforced concrete member of inhomogeneous material, Symposium of the Fifth National Conference on Basic Theory and Engineering Application of Reinforced Concrete Structures Tianjin University Press, Tianjin (1998) 202–206.

CHAPTER

12

Finite Element Analysis of the Loading History for Structures

12.1 BASIC ASSUMPTIONS AND CONSTITUTIVE RELATIONS OF MATERIALS

12.1.1 Basic Assumptions

The mechanical behavior of a single structural member and a statically indeterminate structure of reinforced concrete under the action of high temperature vary in a complicated fashion and deteriorate severely, as demonstrated clearly in the relevant tests (Chapters 8–10). When the structural members at elevated temperatures are the same geometric shape but are different in size, the distribution of the temperature field and the material behavior on their sections are not similar even under the same temperature–time curve, because of the thermal inertia of concrete. Therefore, the thermal behavior of a prototype structure in practical engineering is not similar to the test result of a model at reduced scale, and, obviously, it cannot be calculated using empirical formula regressed from the model.

There are two ways to understand and determine accurately the behavior of a structure at elevated temperatures (under fire), i.e., full-scale testing of a 1:1 model at elevated temperatures and theoretical analysis following the whole loading–heating history. However, it is difficult to test a structure integrally; only part of it can be tested because of the limited capacity of the experimental device, the test incurs high costs and is time consuming. Based on the thermal–mechanical constitutive relationships of the materials, which are demonstrated by many experiments, and a program for nonlinear finite element analysis, the mechanical history of the structure can be obtained after numerous calculations on a computer.

Analysis of the thermal behavior of a reinforced concrete structure has the following characteristics and difficulties:

1. A nonuniformly distributed temperature field on the section causes different levels of deterioration of the material at each point on it, and the behavior of the material varies continuously along the section.
2. The thermal–mechanical constitutive relationship of concrete is a complicated function consisting of four coupled factors: stress, strain, temperature, and time, and varies with the different temperature–stress paths.
3. Nonuniform thermal deformation on the section causes variations in tensile and compressive stresses and also causes loading or unloading conditions under the action of load.
4. The value of the creep of the material at elevated temperatures is considerable even within a short time period (hours) and varies constantly with stress, temperature, and time.
5. Deformation of a structural member at elevated temperatures is considerable; the influence of the geometric nonlinearity of the structure should be taken into account if necessary.

The computer program for nonlinear analysis of a reinforced concrete structure at room temperature can be compiled easily.[12-1,12-2] Only the simple constitutive (stress–strain) relationships of the materials are needed, without considering the complicated factors listed above, which are related to high temperature. Some existing computer programs for analyzing structures at elevated temperatures[3-9,5-3,10-1,12-3] do not take these complicated factors into account, although the stress–strain relationship varying with temperature of the material is used. Therefore, they cannot accurately reflect the practical behavior of the structure at elevated temperatures. Some other computer programs on the market (ADINAT,[12-4] ARC,[12-5] BERSAFE,[12-6] and ANSYS[12-7]) also have a function to analyze the thermal stress of a structure, but the influence of temperature on deformation and stress states of the structure are considered as an additional factor. In addition, the thermal constitutive relationship of the material is simplified too much and is quite different from the practical behavior of concrete.

The computer program NARCSLT (nonlinear analysis of reinforced concrete structures under loading and elevated temperatures)[1-12] introduced in this chapter represents the mechanical behavior of concrete and reinforcement materials at elevated temperatures (see Chapters 1–4) and the characteristics of structural members at elevated temperatures. Therefore, it can be used for complete analysis of the mechanical history of a structure under the actions of loading and high temperature.

The basic assumptions used in the program are:

1. A two-dimensional temperature field on the cross section of a structural member does not vary along its longitudinal axis, and is not related to the stress and the strain states of the material or cracking in the concrete.

The temperature field of a structural member depends only on the exchange of heat with the surrounding environment and is not related to the stress and strain of the material. When a crack appears in structural concrete and heat gets into it, the temperature distribution nearby is disturbed and the temperature along its longitudinal axis is different. The number of cracks in the structural member is limited and the depth of the cracks is small, therefore effect on the region is not considerable. Therefore, when the crack width is relatively small and the concrete cover has yet not spalled off, the temperature field on the section is still assumed to be unchanged by the crack. The method for determining the temperature field can be found in Chapters 6 and 7.

2. Planar deformation of a section

A member of a structure of a bar system is slender and the size of its section is far smaller than its length. Even though considerable and nonuniform thermal deformation is caused in the structural member under the action of high temperature, the section of the deformed member remains planar before the concrete cracks, because of the constraint effects from outside and inside the structural member. After a crack appears in the structural member, especially when it is approaching failure, the longitudinal reinforcement yields and the depth of the compression zone on the section is reduced; obviously, the cracked section does not conform to the condition of planar deformation. Tests on structural members at room temperature show that the deformation of a section conforms basically with the hypothesis of planar deformation, if the average strain of a piece of length (e.g., $> 0.4h$[12-8]) is considered.

3. No slip between the reinforcement and the concrete

Good adhesion between the reinforcement and the concrete is the basis for their working together. The strains of the reinforcement and concrete at an adjacent position in a structural member are equal and no relative slip occurs between them before the concrete cracks. However, after the crack is formed in the structural member, the strains in both materials near the crack are different and a relative slip appears between them. The bond stress–slip relationship between reinforcement and concrete at room temperature has been investigated fully,[0-2] but not at elevated temperatures due to the lack of experimental data. Therefore, it is assumed that no relative slip occurs between the reinforcement and adjacent concrete, and some constructions are designed to prevent relative slip between them, e.g., both ends of all reinforcements are anchored reliably within the support or joint zone, and the total deformation values are guaranteed to be equal. In addition, the influence of relative slip between reinforcement and concrete is considered properly when the calculation parameters are determined.

4. Tension of concrete utilized fully

The tension of concrete in a general structure at room temperature can be neglected during the calculations; the calculations are simplified and the error caused is small. The thermal deformation of concrete at elevated temperatures is considerable, and a large tension zone may appear on the section of a structural member and vary constantly under the actions of temperature and load. If the tension of the concrete is totally neglected, it can significantly change the stress distribution on the section and cause unreasonable results from the calculations. Therefore, only when the summation of the stress increments in the concrete is higher than the tensile strength of the concrete during the step by step analysis, is the concrete considered to be cracked and out of work.

12.1.2 Thermal–Mechanical Constitutive Model of Concrete

1. Basic equation

The coupling constitutive relationship of concrete under common actions of temperature and stress is complicated and the corresponding mathematic model suggested in Chapter 3 (Eqns (3.11)–(3.16)) is based on a great deal of results from experimental investigations. The total strain of concrete (ε_c) is composed of four components: the strain caused by the action of stress (ε_σ), freely expanding strain (ε_{th}), transient thermal strain (ε_{tr}), and short time creep (ε_{cr}). Therefore,

$$\varepsilon_c = \varepsilon_\sigma - \varepsilon_{th} + \varepsilon_{tr} + \varepsilon_{cr} \tag{12.1}$$

is established and after differentiation

$$d\varepsilon_c = d\varepsilon_\sigma - d\varepsilon_{th} + d\varepsilon_{tr} + d\varepsilon_{cr} \tag{12.2}$$

is obtained. The signs of compressive and tensile stresses, and the corresponding contraction and elongation strains, of concrete are positive and negative, respectively.

The functions and differentiations of the strain components are:

$$\sigma_c = f_\sigma(\varepsilon_\sigma, T), \quad d\sigma_c = \frac{\partial f_\sigma}{\partial \varepsilon_\sigma} d\varepsilon_\sigma + \frac{\partial f_\sigma}{\partial T} dT \tag{a}$$

$$d\varepsilon_\sigma = \frac{1}{\partial f_\sigma / \partial \varepsilon_\sigma}\left[d\sigma_c - \frac{\partial f_\sigma}{\partial T} dT\right] \tag{a'}$$

$$\varepsilon_{th} = f_{th}(T), \quad d\varepsilon_{th} = \frac{\partial f_{th}}{\partial T} dT \tag{b}$$

$$\varepsilon_{tr} = f_{tr}(\sigma_c, T), \quad d\varepsilon_{tr} = \frac{\partial f_{tr}}{\partial \sigma_c} d\sigma_c + \frac{\partial f_{tr}}{\partial T} dT \tag{c}$$

$$\varepsilon_{cr} = f_{cr}(\sigma, T, t), \quad d\varepsilon_{cr} = \frac{\partial f_{cr}}{\partial \sigma_c} d\sigma_c + \frac{\partial f_{cr}}{\partial T} dT + \frac{\partial f_{cr}}{\partial t} dt \tag{d}$$

Equations (a′), (b), (c), and (d) are substituted into Eqn (12.2) and

$$\begin{aligned} d\varepsilon_c = & \left(\frac{1}{\partial f_\sigma/\partial \varepsilon_\sigma} + \frac{\partial f_{tr}}{\partial \sigma_c} + \frac{\partial f_{cr}}{\partial \sigma_c}\right) d\sigma_c \\ & - \left(\frac{1}{\partial f_\sigma/\partial \varepsilon_\sigma}\frac{\partial f_\sigma}{\partial T} + \frac{\partial f_{th}}{\partial T} - \frac{\partial f_{tr}}{\partial T} - \frac{\partial f_{cr}}{\partial T}\right) dT \\ & + \frac{\partial f_{cr}}{\partial t} dt \end{aligned} \tag{e}$$

is obtained after it is rearranged. The formula for the stress increment is then established:

$$d\sigma_c = E_\sigma d\varepsilon_c + E_T\, dT + E_t\, dt \tag{12.3}$$

where

$$\left.\begin{aligned} E_\sigma &= \frac{\Delta f_\sigma}{\Delta \varepsilon_\sigma} H_\sigma \\ E_T &= -\left(\frac{\partial f_\sigma}{\partial T} + \frac{\partial f_\sigma}{\partial \varepsilon_\sigma}\frac{\partial f_{th}}{\partial T} - \frac{\partial f_\sigma}{\partial \varepsilon_\sigma}\frac{\partial f_{tr}}{\partial T} - \frac{\partial f_\sigma}{\partial \varepsilon_\sigma}\frac{\partial f_{cr}}{\partial T}\right) H_\sigma \\ E_t &= -\frac{\partial f_\sigma}{\partial \varepsilon_\sigma}\frac{\partial f_{cr}}{\partial t} H_\sigma \end{aligned}\right\} \tag{12.4}$$

and

$$H_\sigma = 1 \Big/ \left(1 + \frac{\partial f_\sigma}{\partial \varepsilon_\sigma}\frac{\partial f_{tr}}{\partial \sigma_c} + \frac{\partial f_\sigma}{\partial \varepsilon_\sigma}\frac{\partial f_{cr}}{\partial \sigma_c}\right) \tag{12.5}$$

2. Mathematical models of strain components

Equations (2.7) and (2.8) are taken as the formulas for the compressive stress–strain curve of concrete at elevated temperatures (i.e., f_σ):

$$\left.\begin{aligned} & x = \frac{\varepsilon_\sigma}{\varepsilon_p^T} \le 1, \quad y = \frac{\sigma_c}{f_c^T} = 2.2x - 1.4x^2 + 0.2x^3 \\ & x \ge 1, \quad y = \frac{x}{0.8(x-1)^2 + x} \end{aligned}\right\} \tag{2.8}$$

where f_c^T and ε_p^T are the compressive strength of concrete at elevated temperatures (Eqn (2.3)) and the corresponding peak strain (Eqn (2.5)), respectively. The tensile stress–strain (σ_t–ε_t) curve for concrete at elevated temperatures has not been reported in experimental investigations so far and a simplified model is used to meet the needs of calculation:

$$\left.\begin{array}{ll} \varepsilon_t \le \varepsilon_{p,t}^T, & \sigma_t = E_0\varepsilon_t \\ \varepsilon_{p,t}^T < \varepsilon_t < \varepsilon_{u,t}^T, & \sigma_t = f_t^T \\ \varepsilon > \varepsilon_{u,t}^T, & \sigma_t = 0 \end{array}\right\} \quad (12.6)$$

where f_c^T is the tensile strength of concrete at elevated temperatures (Eqn (1.4)), $\varepsilon_{p,t}^T = f_t^T / E_0$ is the tensile peak strain of concrete at elevated temperatures, E_0 is the modulus of elasticity of concrete at room temperature, $\varepsilon_{u,t}^T$ is the ultimate tensile strain of concrete at elevated temperature, and the value at room temperature is taken.

The mathematical models of other strain components, including freely expanding strain ε_{th}, transient thermal strain ε_{tr}, and short time creep ε_{cr}, can use the corresponding formulas introduced in Chapters 2 and 3, respectively:

$$\varepsilon_{th} = 28\left(\frac{T}{1000}\right)^2 \times 10^{-3} < 12 \times 10^{-3} \quad (2.1)$$

$$\varepsilon_{tr} = \frac{\sigma_c}{f_c}\left[72\left(\frac{T}{1000}\right)^2 - \left(\frac{T}{1000}\right)\right] \times 10^{-3} \quad (3.10)$$

$$\varepsilon_{cr} = \frac{\sigma_c}{f_c^T}\sqrt{\frac{t}{t_0}}\left(e^{6T/1000} - 1\right) \times 60 \times 10^{-6} \quad (2.24)$$

The symbols used in these formulas are the same as before.

12.1.3 Thermal–Mechanical Constitutive Model of Reinforcement

The total strain of reinforcement (ε_s) under the common actions of temperature and stress is composed of three factors: strain caused by the action of stress ($\varepsilon_{s,\sigma}$), freely expanding strain ($\varepsilon_{s,th}$), and short time creep ($\varepsilon_{s,cr}$):

$$\varepsilon_s = \varepsilon_{s,\sigma} + \varepsilon_{s,th} + \varepsilon_{s,cr} \quad (12.7)$$

The functions of the strain components are: $\sigma_s = f_{s,\sigma}(\varepsilon_{s,\sigma}, T)$, $\varepsilon_{s,th} = f_{s,th}(T)$, $\varepsilon_{s,cr} = f_{s,cr}(\sigma_s, T, t)$. The incremental thermal–mechanical constitutive relationship for reinforcement is also established according to the method used to derive the constitutive relationship for concrete:

$$d\sigma_s = E_{s,\sigma}\, d\varepsilon_s + E_{s,T}\, dT + E_{s,t}\, dt \quad (12.8)$$

where

$$\left.\begin{array}{l} E_{s,\sigma} = \dfrac{\partial f_{s,\sigma}}{\partial \varepsilon_{s,\sigma}} H_{s,\sigma} \\ E_{s,T} = \left(\dfrac{\partial f_{s,\sigma}}{\partial T} - \dfrac{\partial f_{s,\sigma}}{\partial \varepsilon_{s,\sigma}}\dfrac{\partial f_{s,th}}{\partial T} - \dfrac{\partial f_{s,\sigma}}{\partial \varepsilon_{s,\sigma}}\dfrac{\partial f_{s,cr}}{\partial T}\right) H_{s,\sigma} \\ E_{s,t} = -\dfrac{\partial f_{s,\sigma}}{\partial \varepsilon_{s,\sigma}}\dfrac{\partial f_{s,cr}}{\partial t} H_{s,\sigma} \end{array}\right\} \quad (12.9)$$

and

$$H_{s,\sigma} = 1 / \left(1 + \frac{\partial f_{s,\sigma}}{\partial \varepsilon_{s,\sigma}}\frac{\partial f_{s,cr}}{\partial \sigma_s}\right) \quad (12.10)$$

Among the mathematical models of the strain components of reinforcement that are needed for structural analysis are the formulas used for the tensile (compressive) stress–strain curve ($f_{s,\sigma}$), (Eqns (4.4)–(4.8)) and the function used for freely expanding strain ($f_{s,th}$) (Eqn (4.11)), and the function used for short time creep is introduced according to Dorn's theory[12-9]:

$$f_{s,cr}(\sigma_s, T, t) = \frac{\varepsilon_{t0}}{\ln 2}\cosh^{-1}\left(2^{Z\theta/\varepsilon_{t0}}\right) \quad (12.11)$$

where Z is the Zener–Hollomon parameter (1/hour), ε_{t0} is the creep parameter, θ is the time of temperature compensation (hours), represented as:

$$\theta = \int_0^t \exp - \frac{\Delta H}{Rt}\, dt \quad (12.12)$$

where R is the Mohr gas constant (J/(kg mol °C)), T is the temperature (°C), and ΔH is the activation energy of creep (J/(kg mol)). The value of each parameter can be found in the ACI guide.[12-9]

12.2 INCREMENTAL FORMAT OF THE CONSTITUTIVE RELATION OF A SECTION

12.2.1 Incremental Finite Element Formats of Stress, Strain, Temperature, and Time

When the nonlinear finite element method is used to analyze a structure and its members, each member is discretized into a certain number of beam elements along its longitudinal axis (X), and each beam element is also discretized into many small prism elements along its section. As the values of stress, strain, temperature, and their increments at every point within any prism element are not equal to one another, the center of the summation of the increments does not coincide, generally, with the geometric center of the element. When the size of the element is small enough, it is considered that the average values are uniformly distributed on the section of the element.

Analyses of a structure and its members are based on the analyses of their cross sections. Any section can be divided into rectangular elements, the sides parallel with the two perpendicular axes (Y and Z) that pass through the geometric center of the section. If the reinforcement is included in an element, the geometric center of the reinforcement should coincide with that of the element. According to the basic assumption of planar deformation, the deformation (displacement) of any section has three degrees of freedom, so the strain at the geometric center of the section (ε_0) and the angular rotations against the Y and Z axes (curvatures ϕ_y and ϕ_z) are taken as the basic unknowns (Fig. 12-1). Compressive strain on the section takes a positive sign and the curvature ϕ_y (ϕ_z) of the section takes a positive sign, if the angular rotation causes compressive strain, when $y \geq 0$ ($z \geq 0$), and tensile strain, when $y < 0$ ($z < 0$), on the section. Conversely, the strain and curvature take a negative sign.

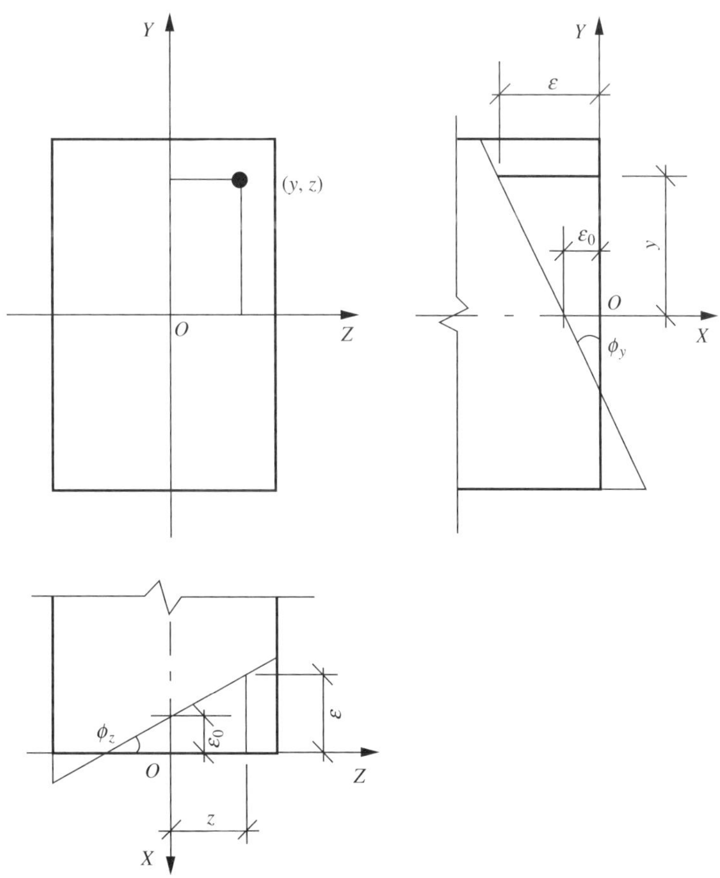

FIGURE 12-1 Basic variables of deformation of a section.

The strain at any point (y,z) on the section is

$$\varepsilon = \varepsilon_0 + y\varphi_y + z\varphi_z = [N(y,z)]\{u\} \quad (12.13)$$

where $[N(y,z)] = [\,1 \quad y \quad z\,]$ is the matrix of the shape function related to (y,z), $\{u\} = [\,\varepsilon_0 \quad \varphi_y \quad \varphi_z\,]^T$ is the vector of the basic variables.

The vectors of the strain and increments of the section after discretizing can be represented as:

$$\{\varepsilon\} = [N]\{u\} \quad (12.14)$$

and

$$\{\Delta\varepsilon\} = [N]\{\Delta u\} \quad (12.15)$$

where the vector of the strain is $\varepsilon = \{\varepsilon_1, \varepsilon_2, \varepsilon_i, \cdots, \varepsilon_n\}$, the vector of the strain increment is $\{\Delta\varepsilon\} = \{\Delta\varepsilon_1, \Delta\varepsilon_2, \cdots, \Delta\varepsilon_i, \cdots, \Delta\varepsilon_n\}$, and the matrix of the shape function is

$$[N] = \begin{bmatrix} 1 & 1 & \cdots & 1 & \cdots & 1 \\ y_1 & y_2 & \cdots & y_i & \cdots & y_n \\ z_1 & z_2 & \cdots & z_i & \cdots & z_n \end{bmatrix}^T$$

The number of rectangular elements divided from the section is n, and the coordinate, strain, and strain increment of the ith element are (y_i,z_i), ε_i, and $\Delta\varepsilon_i$, respectively.

Assuming no slip occurs between the reinforcement and the adjacent concrete, the strain at any point on the section is also the total strain of the concrete or reinforcement at that point, and their increments should also be equal:

$$\varepsilon = \varepsilon_c = \varepsilon_s \quad (12.16)$$

$$\Delta\varepsilon = \Delta\varepsilon_c = \Delta\varepsilon_s \quad (12.17)$$

The vectors of the stress increments of concrete and reinforcement on the section are obtained from Eqns (12.3), (12.8), and (12.17):

$$\{\Delta\sigma_c\} = [E_\sigma]\{\Delta\varepsilon)\} + [E_T]\{\Delta T\} + [E_t]\{\Delta t\} \quad (12.18)$$

$$\Delta\sigma_s = [E_{s,\sigma}]\,\Delta\varepsilon + [E_{s,T}]\,\Delta T + [E_{s,t}]\,\Delta t \quad (12.19)$$

where the incremental vectors of the stress of concrete and reinforcement, the temperature, and the time are, respectively:

$$\{\Delta\sigma_c\} = [\,\Delta\sigma_{c1} \quad \Delta\sigma_{c2} \quad \cdots \quad \Delta\sigma_{ci} \quad \cdots \quad \Delta\sigma_{cn}\,]^T$$

$$\{\Delta\sigma_s\} = [\,\Delta\sigma_{s1} \quad \Delta\sigma_{s2} \quad \cdots \quad \Delta\sigma_{si} \quad \cdots \quad \Delta\sigma_{sn}\,]^T$$

$$\{\Delta T\} = [\,\Delta T_1 \quad \Delta T_2 \quad \cdots \quad \Delta T_i \quad \cdots \quad \Delta T_n\,]^T$$

$$\{\Delta t\} = [\,\Delta t_1 \quad \Delta t_2 \quad \cdots \quad \Delta t_i \quad \cdots \quad \Delta t_n\,]^T$$

Obviously, the time increment (Δt_i) for every element should be equal, but the vector $\{\Delta t\}$ is still used to unify the format. The diagonal matrixes of the tangential moduli are:

$$[E_\sigma] = \begin{bmatrix} E_{\sigma 11} & & & & 0 \\ & \ddots & & & \\ & & E_{\sigma ii} & & \\ & & & \ddots & \\ 0 & & & & E_{\sigma nn} \end{bmatrix},$$

$$[E_T] = \begin{bmatrix} E_{T11} & & & & 0 \\ & \ddots & & & \\ & & E_{Tii} & & \\ & & & \ddots & \\ 0 & & & & E_{Tnn} \end{bmatrix}$$

$$[E_t] = \begin{bmatrix} E_{t11} & & & & 0 \\ & \ddots & & & \\ & & E_{tii} & & \\ & & & \ddots & \\ 0 & & & & E_{tnn} \end{bmatrix},$$

$$[E_{s,\sigma}] = \begin{bmatrix} E_{s,\sigma 11} & & & & 0 \\ & \ddots & & & \\ & & E_{s,\sigma ii} & & \\ & & & \ddots & \\ 0 & & & & E_{s,\sigma nn} \end{bmatrix}$$

$$[E_{s,T}] = \begin{bmatrix} E_{s,T11} & & & & 0 \\ & \ddots & & & \\ & & E_{s,Tii} & & \\ & & & \ddots & \\ 0 & & & & E_{s,Tnn} \end{bmatrix},$$

$$[E_{s,t}] = \begin{bmatrix} E_{s,t11} & & & & 0 \\ & \ddots & & & \\ & & E_{s,tii} & & \\ & & & \ddots & \\ 0 & & & & E_{s,tnn} \end{bmatrix}$$

where the diagonal matrixes of the areas of concrete and reinforcement for each element are:

$$[A_c] = \begin{bmatrix} A_{c11} & & & & 0 \\ & \ddots & & & \\ & & A_{cii} & & \\ & & & \ddots & \\ 0 & & & & A_{cnn} \end{bmatrix}$$

$$\text{and } [A_s] = \begin{bmatrix} A_{s11} & & & & 0 \\ & \ddots & & & \\ & & A_{sii} & & \\ & & & \ddots & \\ 0 & & & & A_{snn} \end{bmatrix}$$

If there is no reinforcement in the element (j), $A_{sjj} = 0$ is taken.

Equations (12.18) and (12.19) are substituted into Eqn (12.22), and

$$\{\Delta F\} = [K]\{\Delta u\} \tag{12.23}$$

is obtained after it is rearranged. This is the incremental format of the finite element between the basic unknowns ($\{\Delta u\} = [\Delta \varepsilon_0 \ \Delta \varphi_y \ \Delta \varphi_z]^T$) and the axial force, bending moment, temperature, and time of the section of the reinforced concrete structural member under common actions of temperature and load. This formula can be used for a structural member experiencing any temperature–load path. The incremental vector of the total load includes the incremental load vectors of loading, temperature, and time:

$$\{\Delta F\} = \{\Delta F_p\} + \{\Delta F_T\} + \{\Delta F_t\} \tag{12.24}$$

where

$$\{\Delta F_T\} = -[N]^T [E_{T,cs}]\{\Delta T\} \tag{12.25}$$

$$\{\Delta F_t\} = -[N]^T [E_{t,cs}]\{\Delta t\} \tag{12.26}$$

$$[E_{T,cs}] = [E_T][A_c] + [E_{s,T}][A_s] \tag{12.27}$$

$$[E_{t,cs}] = [E_t][A_c] + [E_{s,t}][A_s] \tag{12.28}$$

The summations of the stresses of all elements on the section are the internal forces of the section, which include axial force and bending moments in two directions:

$$\left.\begin{aligned} N &= \int_{A_c} \sigma_c \mathrm{d}A_c + \textstyle\sum_i \sigma_{si} A_{si} \\ M_y &= \int_{A_c} \sigma_c y \mathrm{d}A_c + \textstyle\sum_i \sigma_{si} y_{si} A_{si} \\ M_z &= \int_{A_c} \sigma_c z \mathrm{d}A_c + \textstyle\sum_i \sigma_{si} z_{si} A_{si} \end{aligned}\right\} \tag{12.20}$$

The increments of the internal forces can be written in the form:

$$\Delta F_p = \int_{A_c} [N(y,z)]^T \Delta \sigma_c \mathrm{d}A_c + \textstyle\sum_i [N(y_s, z_s)]^T \Delta \sigma_{si} A_{si} \tag{12.21}$$

where $\{\Delta F_p = [\Delta N \ \Delta M_y \ \Delta M_z]^T$ is the incremental vector of the internal force of the section or load, and $[N(y_s, z_s)] = [\ 1 \quad y_{si} \quad z_{si}\ ]$.

As the structure is discretized, Eqn (12.21) is changed into

$$\{\Delta F_p\} = [N]^T ([A_c]\{\Delta \sigma_c\} + [A_s]\{\Delta \sigma_s\}) \tag{12.22}$$

The total matrix of stiffness is

$$[K] = [N]^T [E_{\sigma,cs}] [N] \tag{12.29}$$

where

$$[E_{\sigma,cs}] = [E_\sigma] [A_c] + [E_{s\sigma}] [A_s] \tag{12.30}$$

12.2.2 Finding a Solution for the Incremental Finite Element Format

1. Characteristics of the incremental finite element format

A nonlinear problem, after discretizing into the finite elements, can result in a series of nonlinear algebra equations:

$$\Psi(\mathbf{a}) = \mathbf{K}(\mathbf{a}) \cdot \mathbf{a} - \mathbf{R} = 0 \tag{12.31}$$

where $a = [\ a_1 \quad a_2 \quad \dots \quad a_n\ ]^T$,

$\Psi = [\ \psi_1 \quad \psi_2 \quad \cdots \quad \psi_n\ ]^T$,

$\mathbf{R} = [\ R_1 \quad R_2 \quad \dots \quad R_n\ ]^T$, and $K(a)$ is a matrix of order $n \times n$.

Various numerical methods can be used to find the solution of Eqn (12.31), which is approached gradually by a series of solutions of linear equations. When the problem is related to the loading history, the incremental method is generally used for calculation.

For comparison with Eqn (12.23), let $\mathbf{R} = \lambda\bar{\mathbf{R}}$, and Eqn (12.31) is modified to be of incremental format:

$$\Psi(\mathbf{a}, \lambda) = \mathbf{K}(\mathbf{a})\,\mathbf{a} - \lambda\bar{\mathbf{R}} = 0 \tag{12.32}$$

where λ is a parameter describing the variation of the load.

Assuming $\mathbf{a}_m$ and $\mathbf{a}_m + \Delta\mathbf{a}_m$ are the solutions corresponding to the parameters λ_m and $\lambda_m + \Delta\lambda_m$, respectively, in Eqn (12.32), then

$$\Psi(\mathbf{a}_m, \lambda_m) = \Psi(\mathbf{a}_m + \Delta\,\mathbf{a}_m, \lambda_m + \Delta\,\lambda_m) = 0 \tag{12.33}$$

is obtained. When it is expanded into the Taylor series and the higher terms are neglected:

$$K_T(\mathbf{a}_m, \lambda_m)\,\Delta\,\mathbf{a}_m = \Delta\,\lambda_m\bar{\mathbf{R}} \tag{12.34}$$

is obtained, where

$$K_T(\mathbf{a}_m, \lambda_m) = \frac{\partial \Psi}{\partial \mathbf{a}}\Big|_{a = a_m}$$

There are no limits for **R** and **R** is generally independent of **a**; this assumption is also the basis for the solution of Eqn (12.31). If **R** and **R** are implicit in *a*, then Eqn (12.34) is similar to Eqn (12.23). Therefore, the nonlinear algebraic equations corresponding to Eqn (12.23) can be written in the form of Eqn (12.31), but **R** is implicit in **a** and cannot be written as

$$\mathbf{R} = \mathbf{K}^0(\mathbf{a})\,\mathbf{a} + \mathbf{R}^0 \tag{12.35}$$

where $R^0 = [\ R_1^0 \quad R_2^0 \quad \cdots \quad R_n^0\ ]$ is a vector that is not implicit in **a** and $\mathbf{K}^0(\mathbf{a})$ is a matrix of order $n \times n$.

Therefore, a new method for finding the solution of the incremental finite element format above has to be given.

2. Coupling treatment

It is seen from Eqn (12.24) that the incremental vector of the total load $\{\Delta F\}$ is implicit in the vector $\{u\}$; in practice, the incremental load vectors of temperature and time ($\{\Delta F_T\}$ and $\{\Delta F_t\}$) are implicit in $\{u\}$. The determination of the two vectors is dependent on the corresponding matrixes $[E_{T,cs}]$ and $[E_{t,cs}]$ (Eqns (12.25) and (12.26)), which are dependent on the vector $\{u\}$. On the other hand, a given increment step of $\{u\}$ is dependent on $\{\Delta F_T\}$ and $\{\Delta F_t\}$. Therefore, the relationship between them is coupled. If a direct iteration method is used to find the solution, not only is the convergence speed slow (or even divergence may occur) but also $\{\Delta F\}$ has to be calculated repeatedly in each iteration and additional work is needed.

In order to decouple, the Adams explicit method[12-10] is used first to estimate the incremental load vectors of the temperature and the time at the beginning of every increment step (e.g., the *m*th step):

$$\{\Delta F_T\}_m = \{\Delta T\}_m \sum_{i=0}^{r} \beta_{ri} [N]^T [E_{T,cs}]_{m-r} \tag{12.36}$$

$$\{\Delta F_t\}_m = \{\Delta t\}_m \sum_{i=0}^{r} \beta_{ri} [N]^T [E_{t,cs}]_{m-r} \tag{12.37}$$

where $\{\Delta F_T\}_m$, $\{\Delta F_t\}_m$, $\{\Delta T\}_m$, and $\{\Delta t\}_m$ are the corresponding incremental vectors at the *m*th increment step, and $[E_{T,cs}]_{m-r}$, and $[E_{t,cs}]_{m-r}$ are the corresponding matrixes at the $(m-r)$th increment step:

$$\beta_{ri}=(-1)^{i+1}\sum_{j=i}^{r}\begin{bmatrix} j\\ i\end{bmatrix} a_j,\quad a_j=(-1)^j\int_0^1\begin{bmatrix}-x\\ j\end{bmatrix}dx$$

where r is the number of increment steps utilized before the mth step.

When only the datum for the previous increment step is utilized, i.e., $r = 1$, Eqns (12.36) and (12.37) are changed into:

$$\{\Delta F_T\}_m=\frac{-3[N]^T[E_{T,cs}]_m+[N]^T[E_{T,cs}]_{m-1}}{2}\{\Delta T\}_m \quad (12.38)$$

$$\{\Delta F_t\}_m=\frac{-3[N]^T[E_{t,cs}]_m+[N]^T[E_{t,cs}]_{m-1}}{2}\{\Delta t\}_m \quad (12.39)$$

When the first increment step is calculated, i.e., $m = 1$, and no previous datum can be utilized, the incremental load vectors of temperature and time are calculated using the Euler method:

$$\{\Delta F_T\}_1=-[N]^T[E_{T,cs}]_1\{\Delta T\}_1 \quad (12.40)$$

$$\{\Delta F_t\}=-[N]^T[E_{t,cs}]_1\{\Delta t\}_1 \quad (12.41)$$

With these estimated values, the vector $\{\Delta F\}$ within each increment step is independent of the unknown vector $\{u\}$, and Eqn (12.23) can be solved using the general method.

3. Unbalanced force

A system of one degree of freedom (Fig. 12-2) is taken now as an example. The solution $\{u\}^1$ found from Eqn (12.23) is obviously not accurate for the incremental total load $\{\Delta F\}$. There is a difference between $\{\Delta F\}$ and the incremental vector of load (point B in the figure) corresponding to $\{u\}^1$, and the difference represents an unbalanced force:

$$\{\Delta R\}^1=\{\Delta F\}-\{F_\sigma\}+\{F\}_m \quad (12.42)$$

Differing from the condition at room temperature, the unbalanced force here is not the integration of the total stress, but is part of the internal force, which is integrated in the stress caused by the strain only. $\{F\}_m$ is the total accumulated value of $\{\Delta F_p\}$, $\{\Delta F_T\}$, and $\{\Delta F_t\}$ of the previous increment steps before the current increment step:

$$\{F\}_m=\sum_{i=1}^{m}(\{\Delta F_p\}+\{\Delta F_T\}+\{\Delta F_t\}) \quad (12.43)$$

Determination of the term $\{F_\sigma\}$ in Eqn (12.42) is difficult, so $\{\Delta R\}^1$ is calculated using an approximate method. When Eqn (12.43) is substituted into Eqn (12.42):

$$\{\Delta R\}^1=\{\Delta F_p\}+(\{\Delta F_T\}-\{\Delta\overline{F_T}\})+(\{\Delta F_t\}-\{\Delta\overline{F_t}\})+\{F_p\}-\{F_c\} \quad (12.44)$$

is obtained after rearranging, where $\{\Delta\overline{F_T}\}$ and $\{\Delta\overline{F_t}\}$ are the vectors of the internal forces obtained from integration of the stresses, and they are, respectively, the temperature and the time experienced from point A to point B in Fig. 12-2 and correspond to the stress caused by the strain.

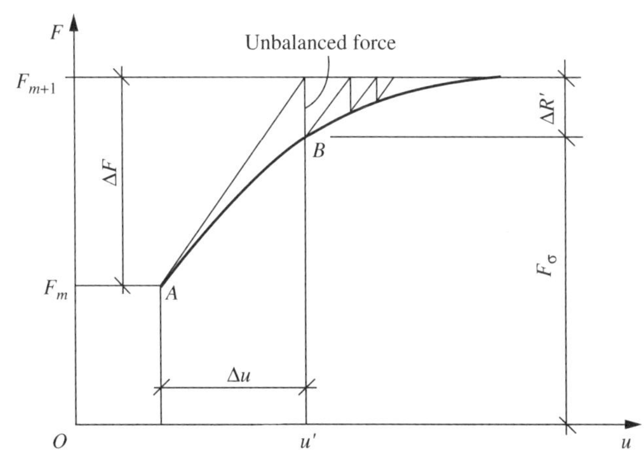

FIGURE 12-2 Incremental equation with one degree of freedom solved by the Euler method at the mth increment step.

$\{F_c\}$ is the total internal force on the section:

$$\{F_c\} = [N]^T([A_c]\{\sigma_c\} + [A_s]\{\sigma_s\})$$

When $\{\Delta F_T\} = \{\Delta \overline{F_T}\}$ and $\{\Delta F_t\} = \{\Delta \overline{F_t}\}$:

$$\{\Delta R\}^1 = \{\Delta F_p\} + \{F_p\} - \{F_c\} \quad (12.45)$$

is established. Then $\{\Delta R\}^1$ is obtained as $\{\Delta F_p\}$ and $\{F_p\}$ in the formula are known. It can be seen from Fig. 12-2 that $\{\Delta F_T\} - \{\Delta \overline{F_T}\}$, $\{\Delta F_t\} - \{\Delta \overline{F_t}\}$, and the unbalanced force tends toward zero simultaneously as the number of iterations increases.

4. Iteration within an increment step

Every step of the approximate linear treatment will cause some error, but the sensitivity of Eqn (12.23) is different from the others. The equilibrium equation of internal force on the section should be implicit in Eqn (12.23) and it can be satisfied only if $\{\Delta F\} = 0$. Generally, the usual method to find the solution does not follow this condition. It is assumed for Eqn (12.23) that the internal forces on the section are in equilibrium at the beginning of every increment step, but the total error accumulates constantly and the final solution obtained is far from the real solution (Fig. 12-3) as the number of steps for finding the solution increases. Generally, an improved method is to iterate within each increment step, using the Newton–Raphson (N–R) or modified Newton–Raphson (mN–R) method. When the latter is adopted:

$$\left.\begin{aligned} \{\Delta u\}^{k+1}_{m+1} &= ([K]^{-1})^0_{m+1}\begin{pmatrix} \{\Delta F\}_{pm+1} \\ +\{F_p\}_{m+1} \\ -\{F_c\}^k_{m+1} \end{pmatrix} \\ \{\overline{\Delta u}\}^{k+1}_{m+1} &= \sum_{i=1}^{k+1}\{\Delta u\}^i_{m+1} \end{aligned}\right\} \quad (12.46)$$

are used for the $(k+1)$th iteration within the $(m+1)$th increment step, where $([K]^{-1})^0_{m+1}$ is the inverse matrix of the total stiffness matrix of strain at the beginning of the $(m+1)$th increment step, $\{F_p\}_{m+1}$ is the load vector of the force on the section at the beginning of the $(m+1)$th increment step, $\{\Delta F_p\}_{m+1}$ is the incremental load vector of the force on the section at the beginning of the $(m+1)$th increment step, $\{F_c\}^k_{m+1}$ is the vector of the total internal force on the section after the kth iteration within the $(m+1)$th increment step,

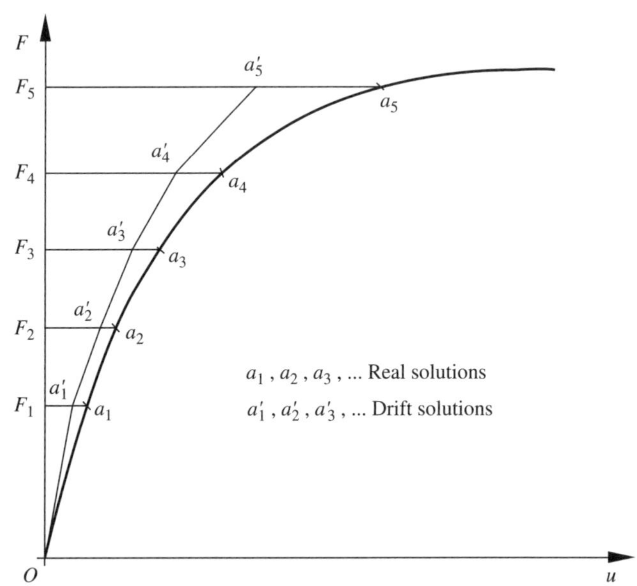

FIGURE 12-3 The drift of the solution of an incremental equation with the one degree of freedom system using the Euler method.

$\{\Delta u\}_{m+1}^{k+1}$ is the incremental vector of the basic variables obtained from the unbalanced force that occurs after the kth iteration within the $(m+1)$th increment step, and $\{\Delta \overline{u}\}_{m+1}^{k+1}$ is the kth correction value of the basic variables within the $(m+1)$th increment step.

5. Accelerating convergence within an increment step

When the mN–R method is used to find the solution of a series of nonlinear equations, the advantage is that the repeated formation and finding the inverse of the tangential stiffness matrix are avoided during every iteration, but the disadvantage is that the convergence speed of iterations is slow and the convergence is even difficult to reach, especially when the ultimate state of the structure is approaching. Therefore, the modified Aitken method of acceleration is used to increase the convergence speed.

The Aitken method of accelerating convergence is conducted once every other iteration. If the kth iteration is accelerated, the acceleration factor of the $(k+2)$th iteration is obtained from the unbalanced difference between the $(k+1)$th and $(k+2)$th iterations. The scheme of iteration accelerated within the $(m+1)$th increment step for a system with one degree of freedom is shown in Fig. 12-4 when the Aitken method is used. If the initial tangential stiffness is K_{m+1} and the local secant stiffness is K_s, the acceleration factor is taken as K_{m+1}/K_s. The accelerated $\Delta \overline{u}_{m+1}^{k+2}$ is obtained by multiplying the value of the $(k+2)$th iteration Δu_{m+1}^{k+2} by the acceleration factor.

The Aitken method of accelerating convergence within the $(m+1)$th increment step for a system with multiple degrees of freedom can be presented as:

$$\{\Delta \overline{u}\}_{m+1}^{k} = [\overline{a}^{k}]\{\Delta u\}_{m+1}^{k} \tag{12.47}$$

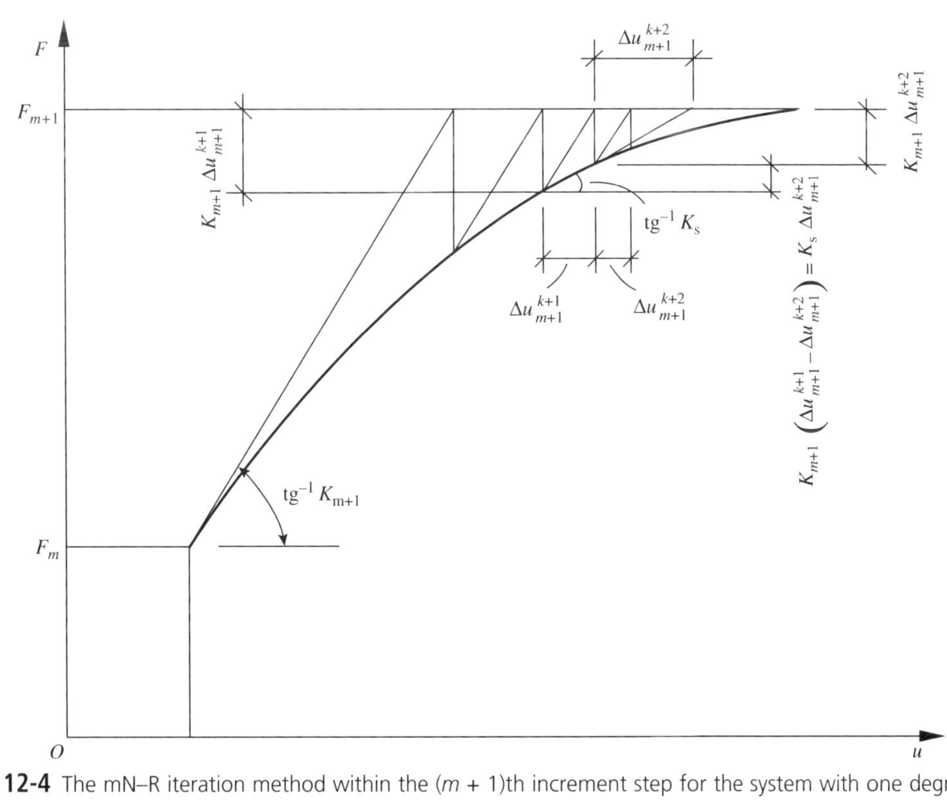

FIGURE 12-4 The mN–R iteration method within the $(m+1)$th increment step for the system with one degree of freedom when the Aitken method of acceleration is used.

where $\{\Delta \bar{u}\}^k_{m+1}$ is the incremental vector of the basic variables accelerated after the kth iteration within the $(m+1)$th increment step and $[\bar{a}^k]$ is the diagonal matrix. The formula for each element is:

$$\bar{a}^k_{ii} \begin{cases} 1, & k = 0,2,4,\ldots \\ \dfrac{\Delta u^{k-1}_{i,m+1}}{\Delta u^{k-1}_{i,m+1} - \Delta u^{k}_{i,m+1}}, & k = 1,3,5,\ldots \end{cases} \tag{12.48}$$

where $\bar{a}^k_{ii}$ is the acceleration factor corresponding to $\Delta u^k_{i,m+1}$, and $\Delta u^{k+1}_{i,m+1}$ and $\Delta u^k_{i,m+1}$ are the increments of the ith basic variable at the $(k-1)$ th and kth iterations, respectively, within the $(m+1)$th increment step.

The denominator term in Eqn (12.48) includes two elements with a minus between them; the value may be very small for a particular i and this causes $\bar{a}^k_{ii}$ to be large. To avoid this condition, the diagonal matrix $[\bar{a}^k]$ in Eqn (12.47) is replaced by a scalar, and then Eqns (12.47) and (12.48) are changed, respectively, into:

$$\{\Delta \bar{u}\}^k_{m+1} = \omega^k \{\Delta u\}^k_{m+1} \tag{12.49}$$

$$\omega^k = \begin{cases} 1, & k = 0,2,\ldots \\ \dfrac{\left(\{\Delta u\}^{k-1}_{m+1} - \{\Delta u\}^{k}_{m+1}\right)^T \{\Delta u\}^{k-1}_{m+1}}{\left(\{\Delta u\}^{k-1}_{m+1} - \{\Delta u\}^{k}_{m+1}\right)^T \left(\{\Delta u\}^{k-1}_{m+1} - \{\Delta u\}^{k}_{m+1}\right)}, & k = 1,3,\ldots \end{cases} \tag{12.50}$$

where ω^k is the acceleration factor at the kth iteration within the $(m+1)$th increment step.

12.2.3 Determination of Stress and Strain, and Their Increments When a Temperature Increment Occurs

1. Transformation of the stress–strain curves of concrete at elevated temperatures

When concrete experiences a path of loading at constant temperature, the strain increment caused by a stress increment can be calculated easily using Eqn (2.8). However, when concrete experiences a path of heating under constant load, not only does the temperature increment cause freely expanding strain (ε_{th}, Eqn (2.1)) and transient thermal strain (ε_{tr}, Eqn (3.10)) but also the value of the strain is influenced by the reduction in the concrete strength. It is difficult to clearly divide these strains from the experimental data of material behavior, but they can be determined by the transformation of the stress–strain curves of concrete at elevated temperatures.[1-12]

The compressive stress–strain curves of concrete at different temperatures (Fig. 2-5) have been measured from tests. The compressive strength of concrete reduces and the corresponding peak strain increases as the temperature increases. When the strain value of each stress–strain curve is divided separately by the corresponding peak strain, a family of curves with the same value ($\varepsilon/\varepsilon_p^T = 1$) of peak strain is obtained (Fig. 12-5(a)). This shows that the stress reduces

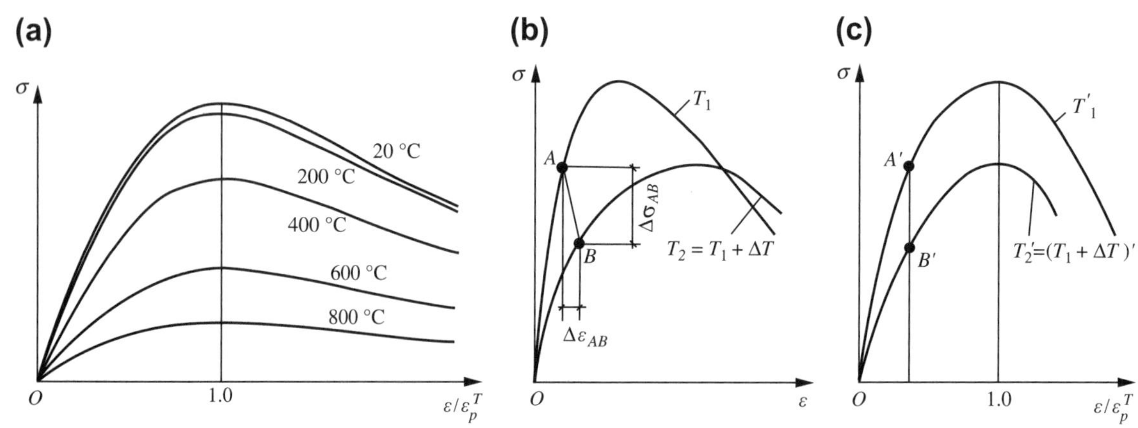

FIGURE 12-5 Transformation of the stress–strain curves of concrete at different temperatures: (a) relative peak strain of every curve is identical; (b) original curves; (c) transformed curves.

gradually when the temperature increases from T_1 to T_2, and the value of the relative strain remains constant, but the absolute value of the strain increases with that of the peak strain at higher temperature.

There are different stress–strain curves (Fig. 12-5(b)) for concrete at temperatures T_1 and T_2 (= $T_1 + \Delta T$), and they are transformed into curves T'_1 and T'_2, respectively, in Fig. 12-5(c). Point A on original curve T_1 is transformed into point A' on curve T'_1; point B', at which the value of the relative strain is equal to that of point A', can be found on transformed curve T'_2, and then point B is obtained on the original curve T_2 (Fig. 12-5(b)).

When the peak strain is $\varepsilon_p^{T_1}$ and the expression of stress is $\sigma = (\varepsilon, T_1)$ for the concrete at temperature T_1, the stress and strain of point A are, respectively, $\sigma_A = \sigma(\varepsilon_A, T_1)$ and ε_A. When there is a temperature increment ΔT and the temperature reaches $T_1 + \Delta T$, the increments to point B are:

$$\Delta \varepsilon_{AB} = \varepsilon_B - \varepsilon_A = \frac{\varepsilon_A}{\varepsilon_p^{T_1}} \cdot \frac{\partial \varepsilon_p}{\partial T}\Big|_{T=T_1} \Delta T \tag{12.51}$$

and

$$\Delta \sigma_{AB} = \left(\frac{\varepsilon_A}{\varepsilon_p^{T_1}} \frac{\partial \sigma_c}{\partial \varepsilon}\Big|_{(\varepsilon_A, T_1)} \frac{\partial \varepsilon_p}{\partial T}\Big|_{T=T_1} + \frac{\partial \sigma_c}{\partial T}\Big|_{(\varepsilon_A, T_1)} \right) \Delta T \tag{12.52}$$

where $\varepsilon_p(T)$ and $\sigma_c(\varepsilon, T)$ are known functions (see Eqns (2.5) and (2.8)).

2. Determination of the position of the stress–strain point when a temperature increment occurs

When the solutions of the finite element formulas for force, strain, temperature, and time increments are required, the load–-temperature path has to be divided into a certain number of increment steps. When the temperature is increased from T_i to $T_i + \Delta T$ within a certain increment step, a point on the locus of the stress–strain at the condition of temperature T_i has to be transformed into that at the condition of temperature $T_i + \Delta T$; the position of the corresponding yield point, e.g., point A, is also transformed into point B' (Fig. 12-6).

Assuming the unloading stiffness of the stress–strain curve at different temperatures is equal to the tangential stiffness at its origin, the transformation of any stress–strain point from the condition of temperature T_i into $T_i + \Delta T$ can be

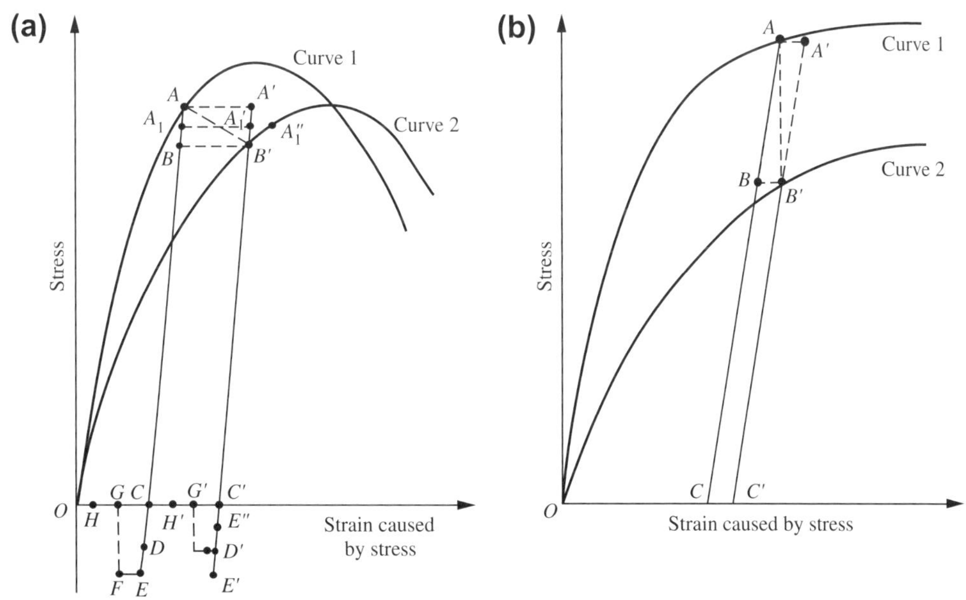

FIGURE 12-6 Locus transformation of the stress–strain point: (a) concrete; (b) reinforcement.

calculated using the following formulas, which are obtained from Eqns (12.51) and (12.52):

$$\varepsilon_c' = \varepsilon_A\left(1+\frac{1}{\varepsilon_p(T_i)}\frac{\partial\varepsilon_p}{\partial T}\bigg|_{T=T_i}\Delta T\right)+\frac{E_{c0}(\varepsilon_A-\varepsilon_c)+\begin{bmatrix}(\varepsilon_A/\varepsilon_p(T_i))(\partial\sigma/\partial\varepsilon)\big|_{(\varepsilon_A,T_i)}\\ \times(\partial\varepsilon_p/\partial T)\big|_{T=T_i}+(\partial\sigma/\partial T)\big|_{(\varepsilon_A,T_i)}\end{bmatrix}\Delta T}{E_{c0}+(\partial E_c/\partial T)\big|_{T=T_i}\Delta T} \tag{12.53}$$

$$\varepsilon_s' = \varepsilon_A + \frac{E_{s0}(\varepsilon_A-\varepsilon_s)+(\partial\sigma_s/\partial T)\,|_{(\varepsilon_A,T_i)}\,\Delta T}{E_{s0}+(\partial\sigma_s/\partial T)\,|_{T=T_i}\,\Delta T} \tag{12.54}$$

used, respectively, for concrete and reinforcement ($\partial\varepsilon_{sp}/\partial T=0$, Fig. 4-11) where ε_c and ε_s are values of the strains corresponding to the stresses at any point on the stress–strain loci of concrete and reinforcement at temperature T_i, ε'_c and ε'_s are the values of the strains corresponding to the stresses of concrete and reinforcement after transformation into $T_i + \Delta T$, E_{c0} and E_{s0} are the unloading stiffness of concrete and reinforcement at temperature T_i, σ and σ_s are stress functions of concrete and reinforcement at different temperatures, and ε_p and ε_{sp} are functions of strains at peak stresses of concrete and reinforcement.

It is seen from Fig. 12-6(a) (for concrete) that the segments *AB* and *DE* on curve 1 are transformed, correspondingly, into segments *A′B′* and *D′E′* on curve 2, and the absolute value of the stress on segment *A′B′* is greater than the absolute value of the yield limit (*B′*), which is impossible. If this happens, the point that is higher than the yield limit has to go back to the yield limit. For example, if point A_1 at temperature T_i is transformed into A'_1 at temperature $T_i + \Delta T$, point T'_1 has to be lowered to *B′* and, correspondingly, the strain caused by stress is reduced by

$$\Delta\varepsilon'_{A_1} = \frac{1}{E_{c0}+(\partial E_c/\partial T)\,|_{T=T_i}}\left(\sigma_{A'_1}-\sigma_{B'}\right) \tag{12.55}$$

where $\sigma_{A'_1}$ is stress at point A'_1, and $\sigma_{B'}$ is the yield limit corresponding to point A'_1 at temperature $T_i + \Delta T$.

After $\Delta\varepsilon'_{A_1}$ is obtained, let point *B′* move to point A''_1 along the stress–strain curve at temperature $T_i + \Delta T$ and the stress increment $(\sigma_{A'_1}-\sigma_{B'})$ is added; so, point A''_1 is the desired position at temperature $T_i + \Delta T$, corresponding to point A_1 of the stress–strain curve at temperature T_i.

3. Increment and iteration paths

It is assumed that the current increment step is *m* and the temperature is $\{T_i\}$, and the temperature is $\{T_{i-1}\}$ for the previous increment step. At the beginning of the increment step *m*, the stress–strain vector at temperature $\{T_{i-1}\}$ obtained from the last iteration of the previous increment step is transformed into that at temperature $\{T_i\}$ using the method described above, and the yield limit corresponding to each strain point at temperature $\{T_i\}$ is determined. Within each increment step, the path of the first iteration is called the increment path and the others are called iteration paths.

(1) Stress–strain point at the yield state

The position of the stress–strain point within the *m*th increment step for the *k*th element is shown in Fig. 12-7. Point *A* is the position at the beginning of the *m*th increment step and point B is the position of the stress–strain point after the *n*th ($n \geq 1$) iteration within the *m*th increment step. There are two situations when the next iteration is conducted: one of them is plastic loading i.e., $\sigma_k\Delta\varepsilon_k > 0$, and the other one is plastic unloading, i.e., $\sigma_k\Delta\varepsilon_k \leq 0$, where σ_k is the stress at point *A* (or *B*) and $\Delta\varepsilon_k$ is the stress–strain increment of concrete or reinforcement.

When $\sigma_k\Delta\varepsilon_k > 0$, both increment and iteration paths move along the stress–strain curve under the temperature condition T_i^k, which is the temperature of the *k*th element at temperature $\{T_i\}$. Then, the values of the strain caused by stress at point *A* at the beginning of the increment step and after the current iteration are put in storage. When $\sigma_k\Delta\varepsilon_k \leq 0$, the increment path is different from the iteration path. While the former is unloading along the *AC* direction (Fig. 12-7), the latter is unloading along *BA* rather than the *BC′* direction, and is unloading further along the

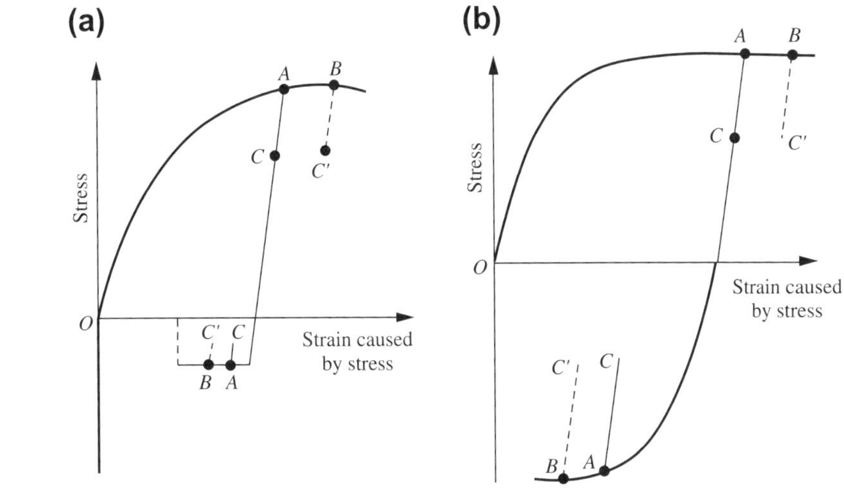

FIGURE 12-7 The position of the stress–strain point within the *m*th increment step for the *k*th element: (a) concrete; (b) reinforcement.

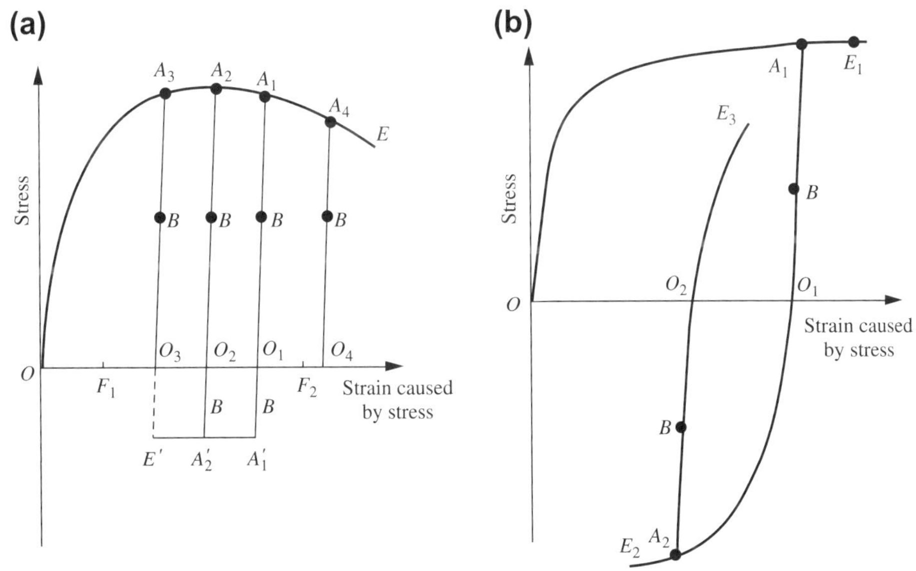

FIGURE 12-8 The position of the stress–strain point within the *m*th increment step for the *k*th element under an elastic state: (a) concrete; (b) reinforcement.

AC direction after point *A* is reached. Then, the values of the strain, caused by stress at point *A* at the beginning of the increment step and after the iteration, are recorded.

(2) Stress–strain point at the elastic state
When the stress–strain point is of elastic state, the directions of the increment and iteration paths are the same.

The position of the stress–strain point within the *m*th increment step for the *k*th element of elastic state is shown in Fig. 12-8. There are three loci of possible paths for concrete. When point *B* is of uncracked state of concrete and no plastic strain is caused by tensile stress, the increment and iteration paths move on the locus of $EA_1O_1A'_1E'O_3F_1$; when point *B* is of uncracked state but plastic strain is caused by tensile stress, the increment and iteration paths move on the locus of $EA_2O_2A'_2E'O_3F_1$; when point B is of cracked state of concrete, the increment and iteration paths move on the locus of $EA_3O_3F_1$ or $EA_4O_4F_2$. There are two loci of possible paths for reinforcement. When the stress of reinforcement at point B is greater than zero, the increment and iteration paths move on the locus of $E_1A_1O_1E_2$, and when the stress at point B is smaller than zero, the increment and iteration paths move on the locus of $E_2A_2O_2E_3$.

12.3 FINITE ELEMENT ANALYSIS OF A STRUCTURE

12.3.1 Equivalent Stiffness Matrix of a Beam Element

The finite element format of the constitutive relation of a section of structural member has been established above, and the equivalent stiffness matrix of each beam element has to be determined in advance, before the finite element analysis of the structure is begun.

The x axis of the local coordinates of a beam element is taken to coincide with the axis of its geometric center, and the displacement components of the element and its nodes are shown in Fig. 12-9. Polynomial expressions are used for displacement mode of the beam element:

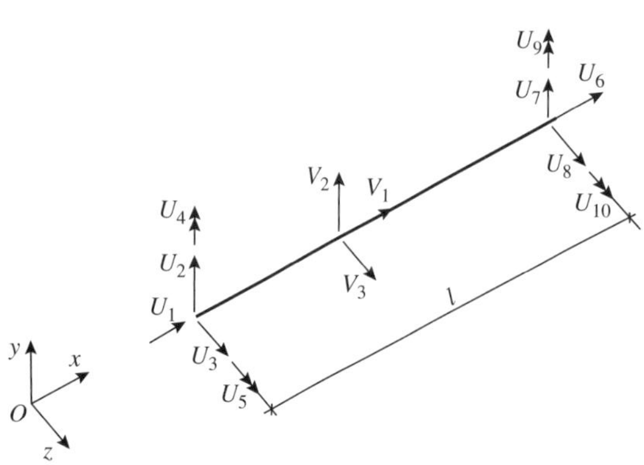

FIGURE 12-9 Displacement components of a beam element and its nodes.

$$\left.\begin{aligned} V_1 &= \alpha_1 + \alpha_2 x \\ V_2 &= \beta_1 + \beta_2 x + \beta_3 x^2 + \beta_4 x^3 \\ V_3 &= \gamma_1 + \gamma_2 x + \gamma_3 x^2 + \gamma_4 x^3 \end{aligned}\right\} \tag{12.56}$$

where α_i, β_i, and γ_i (i = 1,2,3,4) are generalized coordinates.

The formula for the displacements of the element represented by the displacements of its nodes is obtained from Eqn (12.56):

$$\{V\} = [N^b]\{U\}^e \tag{12.57}$$

where $\{V\} = [V_1\ V_2\ V_3]^T$ is the displacement vector of the beam element, $\{U\}^e = [U_1\ U_2\ U_3\ U_4\ U_5\ U_6\ U_7\ U_8\ U_9\ U_{10}]$ is the displacement vector of the nodes of the beam element, and $[N^b]$ is the shape function of the beam element:

$$\left[N^b\right] = \left[N_i^b\ \ N_j^b\right] = \begin{bmatrix} N_1^b & 0 & 0 & 0 & 0 & N_6^b & 0 & 0 & 0 & 0 \\ 0 & N_2^b & 0 & 0 & N_5^b & 0 & N_7^b & 0 & 0 & N_{10}^b \\ 0 & 0 & N_3^b & N_4^b & 0 & 0 & 0 & N_8^b & N_9^b & 0 \end{bmatrix} \tag{12.58}$$

where

$$N_1^b = 1 - \frac{x}{l},\quad N_2^b = N_3^b = 1 - \frac{3}{l^2}x^2 + \frac{2}{l^3}x^3$$

$$N_4^b = -N_5^b = -x + \frac{2}{l}x^2 - \frac{1}{l^2}x^3,\quad N_6^b = \frac{x}{l}$$

$$N_7^b = N_8^b = \frac{3}{l^2}x^2 - \frac{2}{l^3}x^3,\quad N_9^b = -N_{10}^b = -\frac{1}{l}x^2 + \frac{1}{l^2}x^3$$

Strain vector $\{u\}$ of the beam element is taken as the basic variable and its formula is

$$\{u\} = [L]\{V\} \tag{12.59}$$

where

$$[L] = \begin{bmatrix} -\dfrac{d}{dx} & 0 & 0 \\ 0 & -\dfrac{d^2}{dx^2} & 0 \\ 0 & 0 & -\dfrac{d^2}{dx^2} \end{bmatrix}$$

Equation (12.57) is substituted into Eqn (12.59) and

$$\{u\} = [B]\{U\}^e \tag{12.60}$$

is obtained, where

$$[B] = [L]\left[N^b\right] = \begin{bmatrix} a_i & 0 & 0 & 0 & 0 & a_j & 0 & 0 & 0 & 0 \\ 0 & b_i & 0 & 0 & e_i & 0 & b_j & 0 & 0 & e_j \\ 0 & 0 & c_i & d_i & 0 & 0 & 0 & c_j & d_j & 0 \end{bmatrix}$$

and

$$a_i = -a_j = \frac{1}{l}, \quad b_i = c_i = -b_j = -c_j = \frac{6}{l^2} - \frac{12}{l^3}x$$

$$d_i = -e_i = -\frac{4}{l} + \frac{6}{l^2}x, \quad d_j = -e_j = \frac{2}{l} - \frac{6}{l^2}x$$

If the incremental vector of load $\{\Delta P\}^e$, including that of the loads from outside, temperature, and time, is acted only on the nodes of the beam element,

$$\int_l \delta\{u\}^T\{\Delta F\}\,dx = \delta\{U\}^{eT}\{\Delta P\}^e \qquad (12.61)$$

is established according to the principle of virtual displacement, where $\{\Delta F\}$ is the incremental vector of the internal force on the section of the beam element and can be seen in Eqn (12.23). The total stiffness matrix of the strain in Eqn (12.23) becomes the stiffness matrix of the beam element here:

$$[D] = \begin{bmatrix} D_x & D_{xy} & D_{xz} \\ D_{xy} & D_y & D_{yz} \\ D_{xz} & D_{yz} & D_z \end{bmatrix}_V \qquad (12.62)$$

where D_x is the axial stiffness of the beam element, D_y, D_z are the flexural stiffness of the beam element with respect to the y and z axes, respectively, D_{xy}, D_{xz} are the coupling stiffness of the beam element between the axial and flexural stiffness with respect to the y and z axes, respectively, and D_{yz} is the coupling stiffness of the beam element between the flexural stiffness with respect to the y and z axes.

Substituting Eqns (12.23) and (12.62) into Eqn (12.61) gives:

$$[K]^e\{\Delta U\}^e = \{\Delta P\}^e \qquad (12.63)$$

where $\{\Delta U\}^e$ is the incremental vector of the displacement of nodes of the beam element and $[K]^e$ is the equivalent stiffness matrix of the beam element, and its expression is

$$[K]^e = \int_l [B]^T \begin{bmatrix} D_x & D_{xy} & D_{xz} \\ D_{xy} & D_y & D_{yz} \\ D_{xz} & D_{yz} & D_z \end{bmatrix} [B]\,dx \qquad (12.64)$$

As it is assumed that the stiffness of the section of the beam element is invariant along its length, Eqn (12.64) is integrated and becomes:

$$[K]^e = \begin{bmatrix} \frac{D_x}{l} & 0 & 0 & -\frac{D_{xz}}{l} & \frac{D_{xy}}{l} & -\frac{D_x}{l} & 0 & 0 & -\frac{D_{xz}}{l} & \frac{D_{xy}}{l} \\ & \frac{12D_y}{l^3} & \frac{12D_{yz}}{l^3} & -\frac{6D_{yz}}{l^2} & \frac{6D_y}{l^2} & 0 & -\frac{12D_y}{l^3} & -\frac{12D_{yz}}{l^3} & \frac{6D_{yz}}{l^2} & -\frac{6D_y}{l^2} \\ & & \frac{12D_z}{l^3} & -\frac{6D_z}{l^2} & \frac{6D_{yz}}{l^2} & 0 & -\frac{12D_{yz}}{l^3} & -\frac{12D_z}{l^3} & \frac{6D_z}{l^2} & -\frac{6D_{yz}}{l^2} \\ & & & \frac{4D_z}{l} & -\frac{4D_{yz}}{l} & -\frac{D_{xz}}{l} & \frac{6D_{yz}}{l^2} & \frac{6D_z}{l^2} & -\frac{2D_z}{l} & \frac{2D_{yz}}{l} \\ & & & & \frac{4D_y}{l} & -\frac{D_{xy}}{l} & -\frac{6D_y}{l^2} & \frac{6D_{yz}}{l^2} & \frac{2D_{yz}}{l} & -\frac{2D_y}{l} \\ & & & & & \frac{D_x}{l} & 0 & 0 & \frac{D_{xz}}{l} & -\frac{D_{xy}}{l} \\ & & & & & & \frac{12D_y}{l^3} & \frac{12D_{yz}}{l^3} & \frac{6D_{yz}}{l^2} & \frac{6D_y}{l^2} \\ & \textit{Symmetry} & & & & & & \frac{12D_x}{l^3} & \frac{6D_z}{l^2} & -\frac{6D_{yz}}{l^2} \\ & & & & & & & & \frac{4D_z}{l} & -\frac{4D_{yz}}{l} \\ & & & & & & & & & \frac{4D_y}{l} \end{bmatrix} \qquad (12.65)$$

This matrix is similar to the stiffness matrix of the element of a linear elastic structure at room temperature, but there are differences between them, i.e., the coupling stiffness between axial and flexural deformations is not zero generally, and every factor in this matrix is related to the histories of load, temperature, time, and deformation.

After the plastic hinge is formed at the end of a structural member, the angular rotations on both sides of the hinge are no longer equal. Therefore, the freedom of rotation at the plastic hinge has to be released, and the equivalent stiffness matrixes of the related beam elements have to be modified correspondingly.[1-12]

12.3.2 Finding a Solution for a Node Displacement Vector

Similar to the analysis of the usual finite element method, the local coordinates of every beam element are transformed first into integral coordinates of the structure, the equivalent stiffness matrixes for every beam element are collected for the total stiffness matrix of the structure $[K]$, and the incremental vectors of the load at the nodes of every beam element are added to that at the nodes of the structure $\{\Delta P\}$, which includes the vector of the unbalanced force after the last iteration within the previous increment step. Therefore, the nonlinear equation for the direct stiffness of the structure at the kth iteration within the ith increment step is

$$[K]_i^k\{\Delta U\}_i^k=\{\Delta P\}_i^k \qquad (12.66)$$

After $\{\Delta U\}_i^k$ is solved from the equation, the incremental vector $\{\Delta U\}_i^{ek}$ of the displacement at the nodes of each beam element is obtained and substituted into Eqn (12.63) to obtain the incremental vector $\{\Delta P\}_i^{ek}$ of the load at the nodes of all the beam elements. Then, the stiffness matrix and incremental vector of the displacement at the nodes of each beam element under $\{\Delta P\}_i^{ek}$ are calculated and added to the new total stiffness matrix $[K]_i^{k+1}$ and the incremental vector of the load at the nodes $\{\Delta P\}_i^{k+1}$ of the structure. Now, the vector of unbalanced force can be calculated:

$$\{r\}_i=\{\Delta P\}_i^{k+1}-[K]_i^{k+1}\{\Delta U\}_i^k \qquad (12.67)$$

If the norm of $\{r\}_i$ is smaller than the given convergence tolerance,

$$\{U\}_i=\{U\}_{i-1}+\{\Delta U\}_i^k \qquad (12.68)$$

is obtained and the calculation of the next increment step can be done, otherwise the iteration is continued until the norm of $\{r\}_i$ is smaller than the convergence tolerance.

12.3.3 Calculation Procedure

The NARCSLT computer program is compiled for nonlinear finite element analysis of the mechanical history of a reinforced concrete structure under common actions of temperature and load, and the calculation procedure is as follows:

1. The temperature and load are divided into a certain number of increment steps according to the time step, and the increments of temperature, load, and time within each increment step are not zero simultaneously. The increment vector of load $\{\Delta P\}$ at the nodes of the structure is not zero in any increment step, so Eqn (12.66) is a series of nonlinear equations at the kth iteration within the ith increment step, and the incremental vector of displacement at the nodes of the structure is not zero in any increment step.
2. Each member of the structure is divided into a certain number of beam elements along its length, and each beam element is divided again into a certain number of prism elements along its section. The areas of the cross sections of the concrete and reinforcement of each prism element are calculated.
3. The initial conditions of the structure are given.
4. The temperature field on the cross section of each beam element is calculated using the HTARC program (see Chapter 6). However,

this is not needed if the temperature increment within the current increment step is zero.

5. The vector of the stress–strain point on the section of every beam element obtained from the previous increment step is transformed into that under the temperature condition within the current increment step, and the vector of the yield limit of the section of every beam element under the current condition is found. This is not needed if the temperature increment within the current increment step is zero.
6. The tangential moduli, related to the strain, temperature, and time, of concrete and reinforcement on the section of a prism element of each beam element are calculated using Eqns (12.4) and (12.9).
7. The stiffness matrixes of the load, temperature, and time, and the total stiffness matrix of the section of a beam element are formed using Eqns (12.27)–(12.30).
8. The incremental load vectors of the temperature and time at the nodes of a beam element are calculated using Eqns (12.38)–(12.42).
9. The calculated result of Eqn (12.24) is added to the vector of the unbalanced force (the coordinates of which have been transformed) at the nodes of a beam element after the last iteration within the previous increment step, and the incremental load vector at the nodes of the beam element is obtained.
10. The equivalent stiffness matrix of every beam element is calculated using Eqn (12.65).
11. The local coordinates of every beam element are transformed into the integral coordinate of the structure for the equivalent stiffness matrix and the incremental vector of the load at the nodes.
12. The incremental vector of the load at the nodes of the structure is collected.
13. The total stiffness matrix of the structure is collected and the boundary condition of its displacement is introduced.
14. The vector of the unbalanced force is calculated using Eqn (12.67).
15. If the norm of the vector of the unbalanced force is smaller than the convergence tolerance given, the vector of the displacement at the nodes of the structure within the current increment step is calculated using Eqn (12.68), and the current condition of deformation of the structure is recorded. These are taken as the initial conditions for calculating the next increment step, which is started again from step 4 above and conducted until failure of the structure or the end of the last increment step given. If the norm of the vector of the unbalanced force is greater than the convergence tolerance given, the iteration is continued within the current increment step.
16. The incremental vector of the displacement at the nodes of the structure is calculated using Eqn (12.66).
17. The incremental vector of the displacement at the nodes of every beam element is determined.
18. The strain vectors caused by stresses of concrete and reinforcement on the section of every beam element are calculated.
19. The vector of the yield limit of the section of every beam element under the current temperature conditions is calculated and then return to step 6 above.

12.4 COMPARISON BETWEEN THE THEORETICAL CALCULATIONS AND THE EXPERIMENTAL DATA

1. A simply supported beam

The size, reinforcement construction, position of the load, and heating–time curve of a simply supported reinforced concrete beam can be seen in Chapter 8. The curve of the load–maximum deflection at the mid-span of the beam measured during the testing path of loading at constant temperature and the curve of the temperature–maximum deflection at the mid-span of the beam measured during the testing path of heating under constant load are plotted as the dashed lines in Fig. 12-10. The corresponding theoretical curves

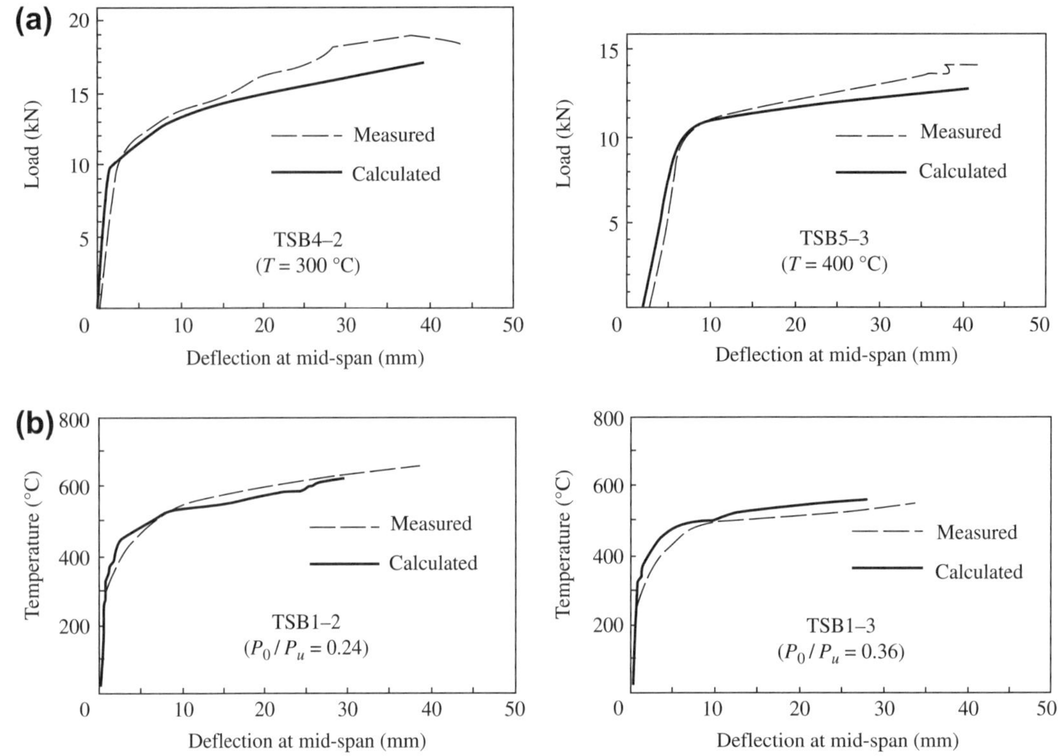

FIGURE 12-10 Comparison of the maximum deflection at the mid-span of a simply supported beam (TSB): (a) path of loading under constant temperature; (b) path of heating under constant load.

calculated by the NARCSLT program are plotted as the solid lines, and they agree well with the curves measured from the tests.

2. Continuous beam

The reinforced concrete continuous beams of two spans are tested under the path of heating under constant load, and the size, material, and construction of the specimens can be seen in Chapter 10; the testing conditions of the specimens, such as the position of the load, the level of constant load, and the number of heated spans, are different from one another. The measured curves for the maximum deflection at the mid-span and the reaction at the end support of the heated span, which vary with temperature, of some specimens are shown separately as dashed lines in Fig. 12-11(a) and (b). The corresponding curves calculated by the NARCSLT program are plotted as solid lines. It is found after comparison that the variation regularity of the theoretical curves is consistent with that of the curves measured under various test conditions, and the corresponding values on both curves are similar. However, the reaction at the end support of the continuous beam is sensitive during calculation, so the error is slightly greater.

3. Frame

The reinforced concrete frames of a single bay and single story are tested under the path of heating under constant load, and the size and construction of the specimens can be seen in Chapter 10. The ratio between the linear stiffness of its beam and column or the level of constant load of the specimens is different. The variation curves for relative deflection at the mid-span of the beam and the bending moment and shear force on the bottom section of the column measured during the heating process are shown as dashed lines in Fig. 12-12. The corresponding curves calculated

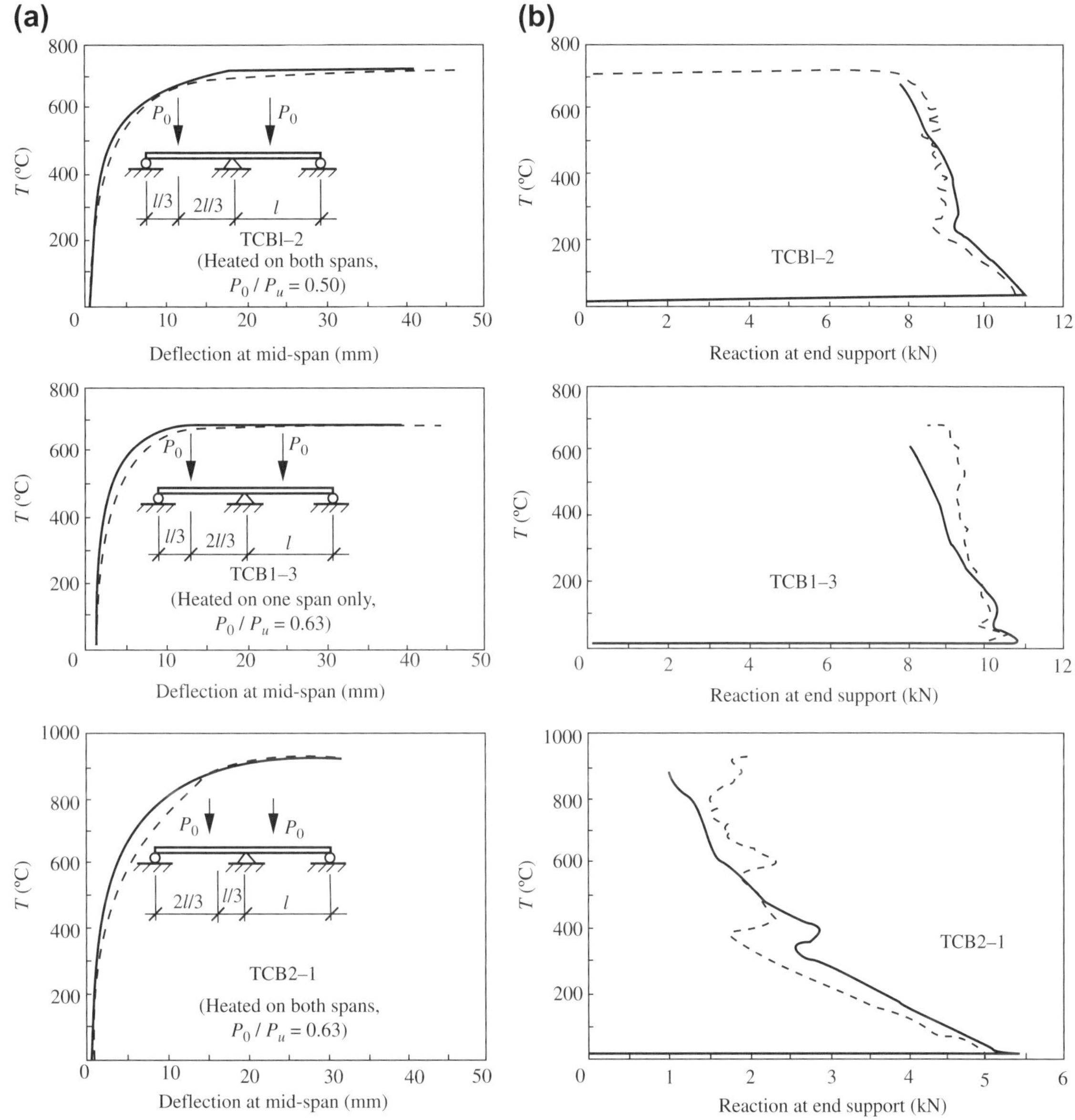

FIGURE 12-11 Comparison between the theoretical and measured values of continuous beam specimens (TCB): (a) maximum deflection at the mid-span; (b) reaction at the end support. Solid line, calculated; dashed line, measured.

by the NARCSLT program are plotted as solid lines. The theoretical and measured curves experience several reverses, and their variation regularities and the corresponding values are similar.

Based on the comparisons between the main experimental results and the theoretical calculations for various simple and statically indeterminate structural members experiencing different temperature–load paths and variable test conditions, it is demonstrated that the NARCSLT program is reasonable and feasible and can be used for the analysis of internal forces and deformation of a reinforced concrete structure under any temperature–load path.

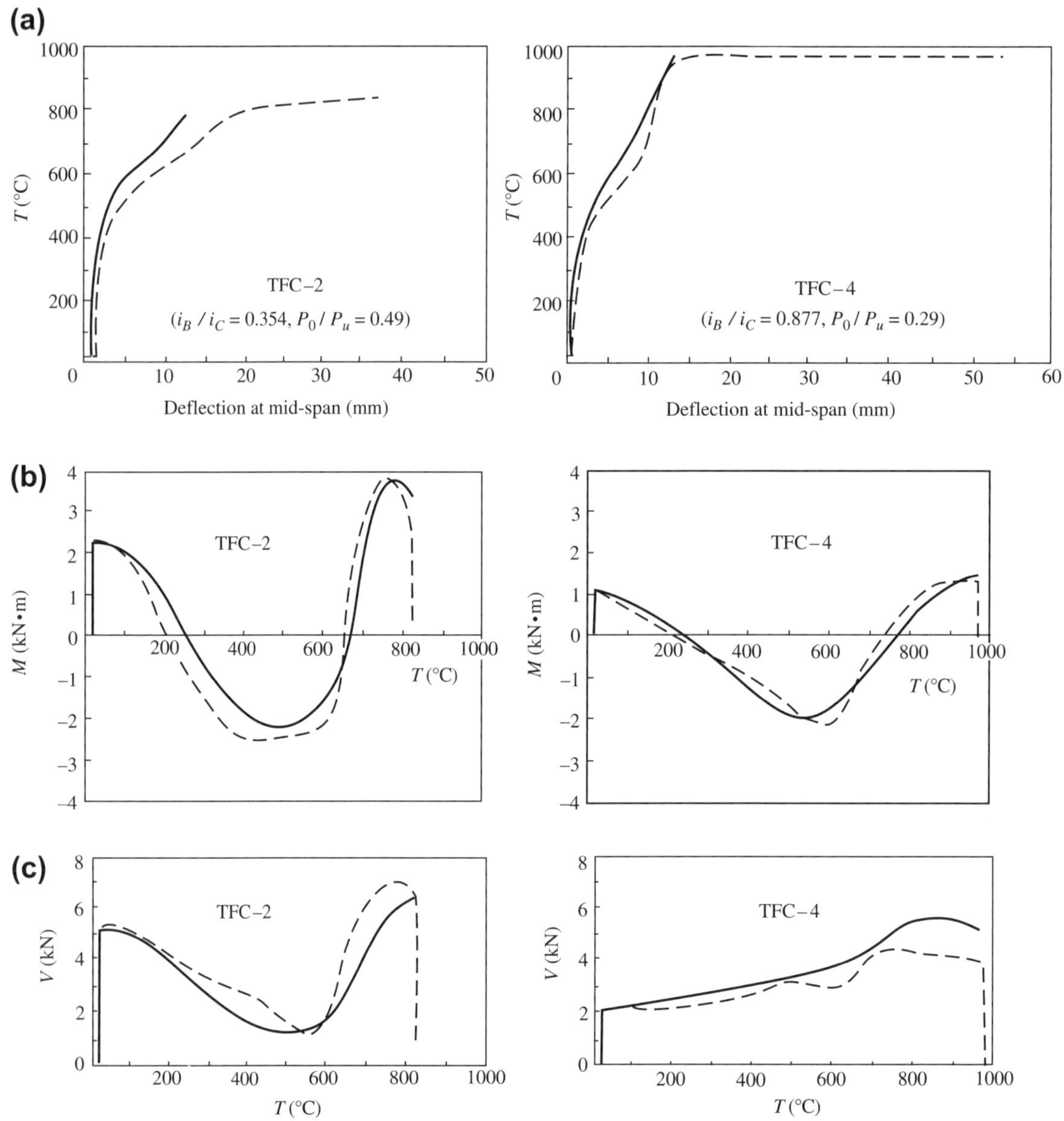

FIGURE 12-12 Comparison between the theoretical and measured values of frame specimens (TFC): (a) relative deflection at the mid-span of a beam; (b) bending moment on the bottom section of a column; (c) shear force on the bottom section of a column. Solid line, calculated; dashed line, measured.

CONCLUSIONS

The behavior responses of a reinforced concrete structural member, especially a statically indeterminate structure, under common actions of temperature and load are quite variable and complicated. Determining the strength at elevated temperatures or the fire endurance by testing a full-scale model of every structure in practical engineering is impossible, and also not necessary. Therefore, a reasonable and accurate theoretical system has to be established based on experimental investigations of the behavior of materials and structural members at elevated temperatures. The solution can be found after analysis and calculation. The nonlinear finite element analysis program (NARCSLT) introduced in this chapter can be used to calculate the mechanical history of a reinforced concrete structure under different temperature–load paths, and the variation regularity of the structural behavior can be given.

The main procedure of the NARCSLT program is:

- Every member of the structure is discretized into beam elements along its length, and every beam element is also discretized into small prism elements along its cross section.
- A predetermined temperature–load path is divided into the increment steps of temperature, load, and time.
- The HTARC program (see Chapter 6) is introduced to analyze the temperature field on the cross section.
- The coupling temperature–stress constitutive model of concrete and the temperature–stress constitutive model of reinforcement are used, which are demonstrated by the experimental investigations.
- Incremental finite element formats of stress, strain, temperature, and time are established for the analysis of the section.
- The transformation rule of the position of the stress–strain point is provided when a temperature increment occurs.
- The coupled series of equations are decoupled.
- Some measures are used to accelerate the calculation process of iteration.
- The equivalent stiffness matrix of the beam element and the total stiffness matrix of the integral structure are derived.
- The basic equation of a finite element is solved and the basic unknowns (displacements at nodes) are obtained, then the indices of various types of behavior of the structure and its members are calculated.
- After the program is compiled and debugged, it is used to analyze the test specimens of several simply supported beams and statically indeterminate continuous beams and frames. The theoretical values (curves) obtained agree well with the experimental results.

REFERENCES

[12-1] C. Wang, Z. Teng (Eds.), Theory of Reinforced Concrete Structure, Construction Industry Press of China, Beijing (1985).

[12-2] B. Zhu, Z. Dong, Nonlinear Analysis of Reinforced Concrete, Tongji University Press, Shanghai (1985).

[12-3] B. Ellingwood, J.R. Shaver, Reliability of reinforced concrete beams subjected to fire, Journal of Structural Engineering 103 (ST9) (1977) 1047–1059.

[12-4] K.J. Bathe, ADINAT — Finite-Element Program of Dynamic Nonlinear Analysis with Automatic Increment of Temperature, Translated by Computer Centre of Chinese Science Academy, Department of Engineering Mechanics, Tsinghua University, and Department of Shipbuilding, Shanghai Jiaotong University (1980).

[12-5] O. Buyukozturk, J.J. Connor, ARC — A computer program for three dimensional finite element analysis of reinforced, prestressed, and refractory concrete structures, MIT, Department of Civil Engineering, Research Report R78-33 (1978).

[12-6] T.K. Hellon, S.S. Protheroe, The BERSAFE Finite Element System, Computer Aided Design 6 (1974) 15–24.

[12-7] J. Chen, G. Cai (Eds.), Computer Aided Engineering Analysis — SYS User Guidebook, Railway Press of China, Beijing (2001).

[12-8] Research Group for Strength of Eccentric Compression Member. Experimental investigation on normal cross-section strength of reinforced concrete eccentric compression member. Proceedings of Reinforced Concrete Structures (Volume II), Edited by Chinese Academy of Building Science. Construction Industry Press of China, Beijing (1981) 19–61.

[12-9] ACI Committee 216, Guide for Determining the Fire Endurance of Concrete Elements, American Concrete Institute, Detroit (1981) 216 R-81.

[12-10] Q. Li, N. Wang, D. Yi, Numerical Analysis, Huazhong Technology Institute Press, Wuhan (1982).

CHAPTER

13

Practical Calculation Methods for the Ultimate Strength of Members and Structures at Elevated Temperature

13.1 BASIC ASSUMPTIONS AND DETERMINATION OF EQUIVALENT SECTION

The mechanical history of a reinforced concrete structure and its member at elevated temperatures can be obtained accurately using the nonlinear finite element method. Although it is possible from a theoretical point of view, the calculation process is complicated. The temperature–time curve and variability on the space during a building fire in real life is undetermined, the thermal and mechanical behavior of the structural materials is variable and scattered, and the thermal–mechanical constitutive relations are not yet satisfied, therefore the accuracy of the results of theoretical analysis is still difficult to guarantee in a practical sense. On the other hand, ultimate strength is the most important mechanical behavior of the structure and its member during or after a fire accident (at elevated temperatures), and it is also the most interesting problem for the structural engineers dealing with the accident. As the ultimate strength of the structure is related directly to its safety, it is necessary to develop an approximate method of engineering accuracy and a clear concept; and the ultimate strength of a structural member at elevated temperatures must be easy to use.

According to existing experimental investigations and theoretical analyses, the failure pattern and the strain and stress distributions on the section at the ultimate state of the reinforced concrete structural member during high temperature or after cooling are similar to that at room temperature. Therefore, the calculation principle and method for a structural member at room temperature can also be used for that at elevated temperatures, but the strength and deformation indices of the reinforcement and concrete have to be modified correspondingly because of deterioration in the materials, which depends upon the temperature distribution in the section.

The basic assumptions used to calculate the ultimate strength of the structural member during or after high temperature are as follows:

1. The temperature field on the section is known. The temperature field on the section of a structural member can be obtained from theoretical analysis (see Chapter 6), according to the ISO standard temperature–time curve, or the actual heating process and time, or fire endurance (h) needed. It can also be determined directly using various charts given in the relevant design code, manual, or professional book,[0-6,0-7] or Chapter 7 in this book.

   When the temperature field on a section is calculated, generally the action of the reinforcement area is not considered and the influence of the stress and crack conditions is also neglected. The temperature of the reinforcement on the section is taken as that of concrete at the same position. When the residual strength of the structural member after

Experiment and Calculation of Reinforced Concrete at Elevated Temperatures

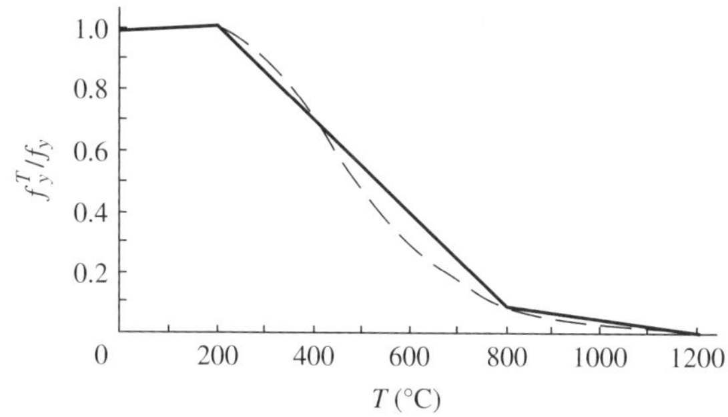

FIGURE 13-1 Calculated strength of reinforcement at elevated temperatures.

cooling is calculated, the maximum temperature experienced and its distribution on the section should also be determined.

2. The strain on the section is distributed linearly, i.e., the condition of planar deformation is satisfied.
3. No slip occurs between the reinforcement and the adjacent concrete.
4. The tensile strength of concrete at elevated temperatures is neglected ($f_t^T = 0$).

The calculated values for the strength of reinforcement and concrete under and after high temperature are simplified according to the corresponding experimental results. The tensile and compressive strengths of reinforcement at elevated temperatures are equal, and the value taken is shown in Fig. 13-1; the formula is:

$$\left.\begin{array}{l} T \le 200\ ^\circ\text{C}:\ f_y^T = f_y \\ 200\ ^\circ\text{C} < T \le 800\ ^\circ\text{C}:\ f_y^T = \left[1 - 0.9\left(\dfrac{T-200}{600}\right) f_y\right] \\ 800\ ^\circ\text{C} < T \le 1200\ ^\circ\text{C}:\ f_y^T = \left(\dfrac{1200-T}{4000}\right) f_y \\ T \ge 1200\ ^\circ\text{C}:\ f_y^T = 0 \end{array}\right\} \tag{13.1}$$

When the reinforcement is cooled down to normal temperature after being heated to a high temperature, the strength value is taken as the same value as if it did not experience a high temperature, i.e., the strength at room temperature.

The compressive strength of concrete at elevated temperatures (f_c^T) varies with temperature and the relationship between them is a complicated curve (see Fig. 2-7). It may be replaced, approximately, by a trapezoid or step shape[13-1,13-2,9-12] (Fig. 13-2), according to different temperature values and mechanical conditions of the structural member when it is calculated.

When the temperature on the section of a reinforced concrete structural member is distributed nonuniformly, different compressive strengths (f_c^T) are caused correspondingly on the section, so the calculation of the ultimate strength of the member is complicated. When the section is transformed into an equivalent homogeneous section of concrete, the method and formulas in the existing design code[0-1] can also be used to calculate various structural members at elevated temperatures. This is convenient for the engineer to understand and use.

A structural member of a rectangular section and a one-dimensional temperature field (e.g., a floor slab or a wall panel), is presented as an example to illustrate the method for determining the equivalent section. First, the positions of several relevant isothermal lines are calculated, e.g., b_2, b_3, b_5, and b_8 (Fig. 13-3(a)), according to the predetermined temperature distribution on the section. Based on the equivalent principle of ultimate strength of the section, i.e., the resultant value and its acting position of compressive stress of concrete should be equivalent, the actual width of every temperature zone on the section is reduced and multiplied by the ratio between the respective strength and the strength at room temperature (f_c^T/f_c; Fig. 13-2), so the equivalent trapezoid, single or double T-flange section is obtained (Fig. 13-3(b)). The ultimate strength of the structural member with the equivalent section can then be calculated as if it is a member of homogeneous concrete (of strength f_c), and is not different from that at room temperature.

All the reinforcements (e.g., A_{s1} and A_{s2}) on the equivalent sections are kept at the original position, and the tensile or compressive strength of each reinforcement depends on the temperature at its own position (see Fig. 13-1).

When the temperature field on the section of a structural member is two-dimensional, its equivalent section can also be determined by the same

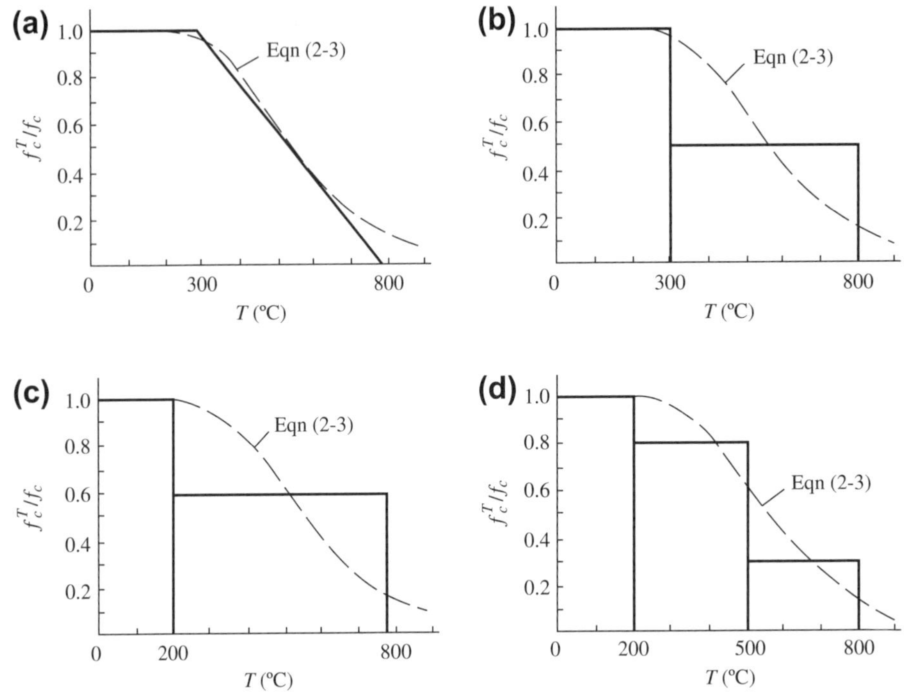

FIGURE 13-2 Calculated compressive strength of concrete at elevated temperatures: (a) trapezoid; (b) two steps[13-1,13-2]; (c) two steps[9-12]; (d) three steps.[9-12]

principle and method as shown in Fig. 13-4 and Fig. 13-6 later in this chapter.

When the basic assumptions and the method to determine the equivalent section described above are used, some other assumptions are included implicitly; e.g., the action of thermal stress on the section is neglected, the influence of different temperature–stress paths of concrete on its strength is not considered, the ultimate compressive stress block of concrete on the section is taken to be approximately rectangular, and its strength is f_c.

13.2 CENTRAL COMPRESSIVE MEMBERS WITH PERIPHERY EXPOSED TO HIGH TEMPERATURE

When a central compressive column of a rectangular section is subjected to the action of high temperature (fire) on its periphery, the temperature field and the stress distribution on its section are symmetrical about the two perpendicular axes, and the strain is distributed uniformly on the section. Selecting the calculated strength of concrete at elevated temperatures (f_c^T, see Fig. 13-2), and determining the positions of the isothermal lines needed (see Chapter 7), the shape of these lines is rounded to a rectangle, e.g., $b_3 \times h_3$, $b_8 \times h_8$, etc., in Fig. 13-4; the formula is then established for the ultimate strength of the column.

If the calculated strength of concrete at elevated temperatures is taken as that of a trapezoid (Fig. 13-2(a)), the ultimate stress state on the section of the column is shown in Fig. 13-4(b), and the bearing capacity of concrete is the volume of the stress block of a truncated pyramid. (The calculated strength of concrete varies linearly when the temperature increases from 300 °C to 800 °C (Fig. 13-2(a)), but the isothermal lines of temperature ranging from 300 °C to 800 °C are distributed at different intervals on the section (e.g., Fig. 7-12). Therefore, the ultimate stress block on the section should be a truncated pyramid with a slightly

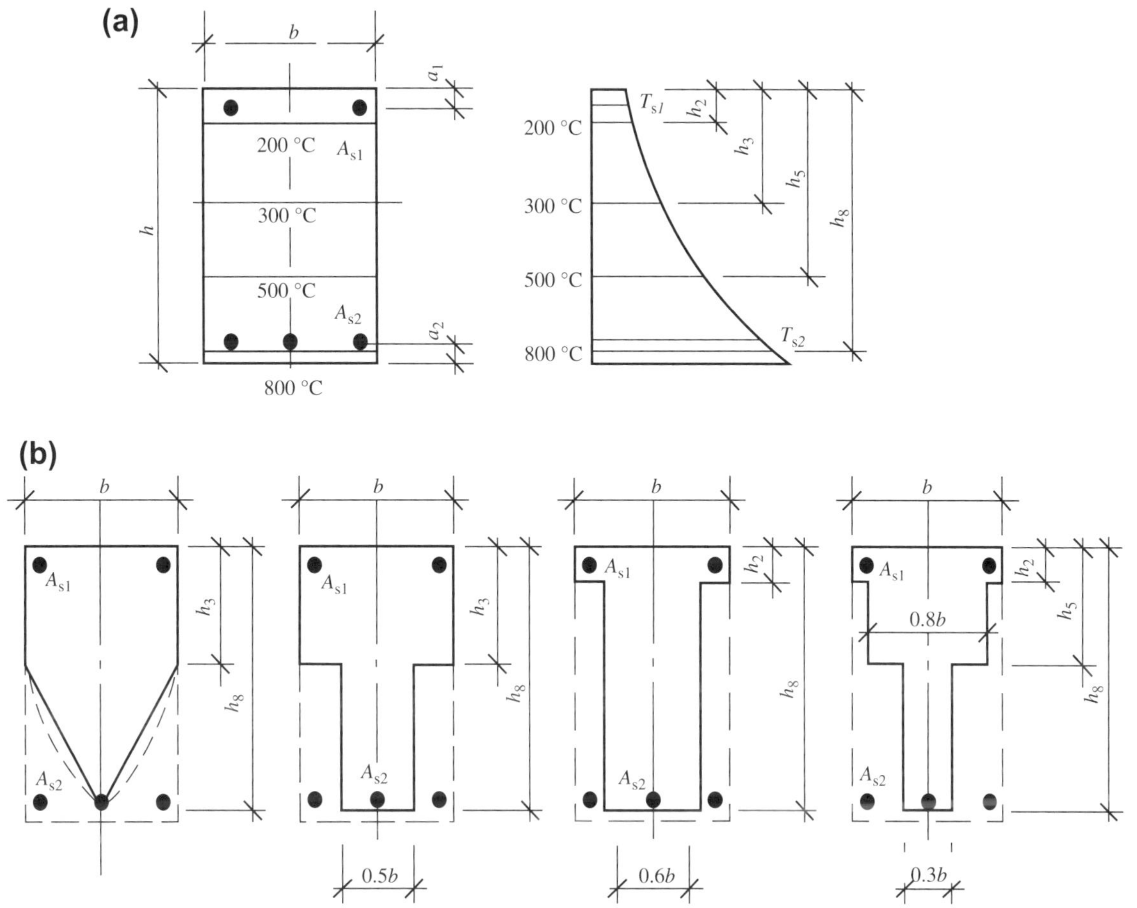

FIGURE 13-3 Determination of an equivalent section: (a) temperature distribution on a section; (b) equivalent sections corresponding to Fig. 13-2.

convex surface (dashed line in Fig. 13-4(b)), if it is calculated accurately according to the value of the distributed temperature. However, when the stress block is taken approximately as a truncated pyramid of planar surface (solid line in Fig. 13-4(b)), the error caused is limited.) The yield strength (f_{yi}^T) of each reinforcement (area of A_{si}) at elevated temperatures is determined separately according to the temperature at its position on the section. So the total ultimate strength of the column is:

$$N_{u1}^T = \left[\frac{1}{3}(b_3h_3 + b_8h_8) + \frac{1}{6}(b_3h_8 + b_8h_3)\right] f_c + \sum_i f_{yi}^T A_{si} \quad (13.2)$$

If the calculated strength of concrete at elevated temperatures is taken in two steps (Fig. 13-2(b)), the ultimate stress state on the section of the column is shown in Fig. 13-4(c) and the ultimate strength of the column is:

$$N_{u2}^T = \frac{1}{2}(b_3h_3 + b_8h_8) f_c + \sum_i f_{yi}^T A_{si} \quad (13.3)$$

The difference between the ultimate strengths of the column, which correspond to the two types of calculated strengths of concrete at elevated temperatures, is:

$$N_{u2}^T - N_{u1}^T = \frac{1}{6}(b_8 - b_3)(h_8 - h_3) f_c > 0 \quad (13.4)$$

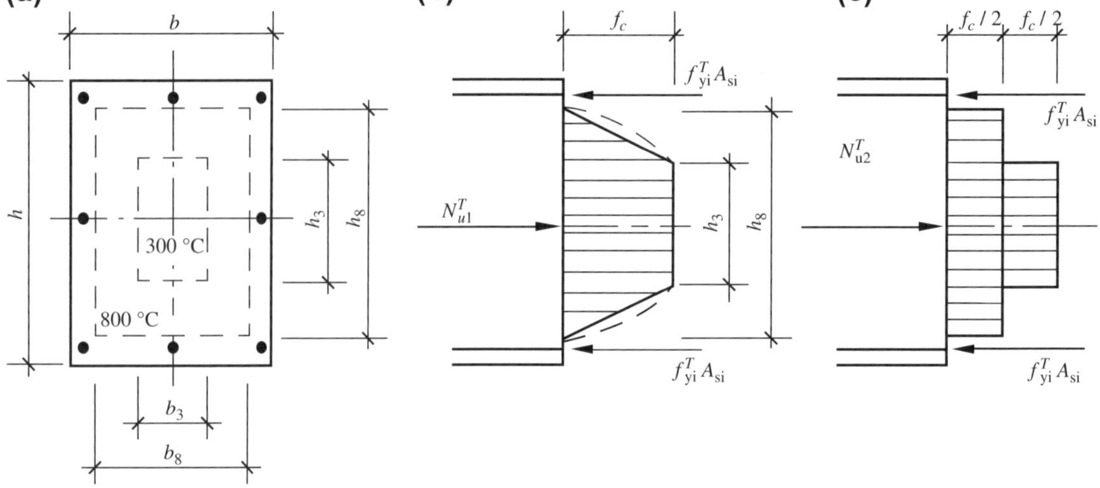

FIGURE 13-4 Central compressive column with the periphery exposed to high temperature: (a) isothermal lines on section; (b) trapezoid stress block; (c) step stress block.

Therefore, the ultimate strength of the column at elevated temperatures calculated by the latter condition is slightly higher, but the relative difference between them is small in engineering practice.

13.3 FLEXURAL MEMBERS

A slab with one surface and a beam of rectangular section with three surfaces exposed to high temperature are introduced as examples to derive the basic formulas of ultimate bending moment (M_u^T), according to the basic assumptions and the method for determining the equivalent section. The same method and procedure can also be used for beams in other situations, such as sections with different shapes, different temperature conditions, and different calculated strengths of concrete at elevated temperatures.

13.3.1 Slabs with One Surface Exposed to High Temperature

In a reinforced concrete slab with one surface exposed to high temperature, the isothermal lines of temperatures 300 °C and 800 °C (h_3 and h_8) at time t are known, the trapezoid calculated strength of concrete at elevated temperatures (Fig. 13-2(a)) is selected, and the temperatures of the upper and the lower reinforcements (A_{s1} and A_{s2}) on the section are T_{s1} and T_{s2}, respectively; the corresponding calculated strengths f_{y1}^T and f_{y2}^T are determined from Fig. 13-1. Then the equivalent section and the ultimate stress state on the section of the slab are obtained as shown in Fig. 13-5. The basic equations of the ultimate bending moment of the section can be established for the two situations separately, i.e., tension zone (+M) and compression zone (−M) exposed to high temperature.

1. Tension zone exposed to high temperature

As the temperature of the reinforcement on the tension zone is high (Fig. 13-5(b)), the strength (f_{y2}^T) is low and the depth of the compression zone on the section is small. In general, when $x \le h_3$ or $f_{y2}^T A_{s2} \le f_c b h_3 + f_{y1}^T A_{s1}$, the equilibrium equations are established as:

$$\left.\begin{aligned} &\sum N=0:\ \ f_c bx + f_{y1}^T A_{s1} = f_{y2}^T A_{s2} \\ &\sum M=0:\ \ M_u^T = f_c bx\left(h_0 - \frac{x}{2}\right) + f_{y1}^T A_{s1}\left(h_0 - a_1\right) \end{aligned}\right\} \tag{13.5}$$

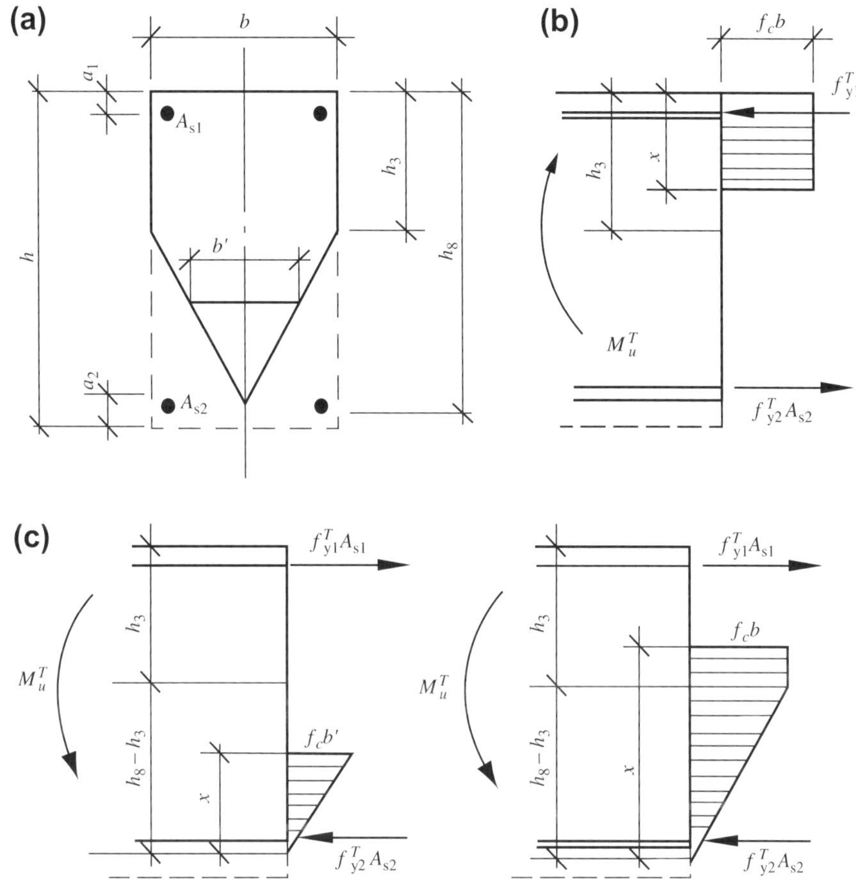

FIGURE 13-5 Calculation diagram for the ultimate bending moment of a slab with one surface exposed to high temperature (trapezoid calculated strength of concrete): (a) equivalent section; (b) tension zone exposed to high temperature; (c) compression zone exposed to high temperature.

If the result calculated from the equations is $x \le 2a_1$, the action of compressive reinforcement (A_{s1}) can be neglected and the equations are simplified to:

$$\left.\begin{aligned} &f_c b x = f^T_{y2} A_{s2} \\ &M^T_u = f_c b x \left(h_0 - \frac{x}{2}\right) \end{aligned}\right\} \tag{13.6}$$

2. Compression zone exposed to high temperature
When the depth of the compression zone (Fig. 13-5(c)) is $x \le h_8 - h_3$ or $f^T_{y1}A_{s1} \le \frac{1}{2} f_c b\,(h_8 - h_3) + f^T_{y2}A_{s2}$, the equivalent width at neutral axis of the section is $b' = [x/(h_8 - h_3)]\,b$, and the basic equations are:

$$\left.\begin{aligned} &\sum N = 0:\ f_c \frac{bx^2}{2\,(h_8 - h_3)} + f^T_{y2}A_{s2} = f^T_{y1}A_{s1} \\ &\sum M = 0:\ M^T_u = f_c \frac{bx^2}{2\,(h_8 - h_3)}\left(h_8 - \frac{2}{3}x - a_1\right) \\ &\qquad + f^T_{y2}A_{s2}\,(h - a_1 - a_2) \end{aligned}\right\} \tag{13.7}$$

When the depth of compression zone is $x > h_8 - h_3$ or $f^T_{y1}A_{s1} > \frac{1}{2} f_c b\,(h_8 - h_3) + f^T_{y2}A_{s2}$, the basic equations become:

$$\left.\begin{aligned} &f_c\left[\frac{1}{2}b\,(h_8 - h_3) + b\,(x - h_8 + h_3)\right] + f^T_{y2}A_{s2} = f^T_{y1}A_{s1} \\ &M^T_u = f_c\left[\begin{aligned} &\frac{1}{6}b\,(h_8 - h_3)\,(h_8 + 2h_3 - 3a_1) \\ &+ \frac{1}{2}b\,(x - h_8 + h_3)\,(h_8 + h_3 - x - 2a_1) \end{aligned}\right] \\ &\quad + f^T_{y2}A_{s2}\,(h - a_1 - a_2) \end{aligned}\right\} \tag{13.8}$$

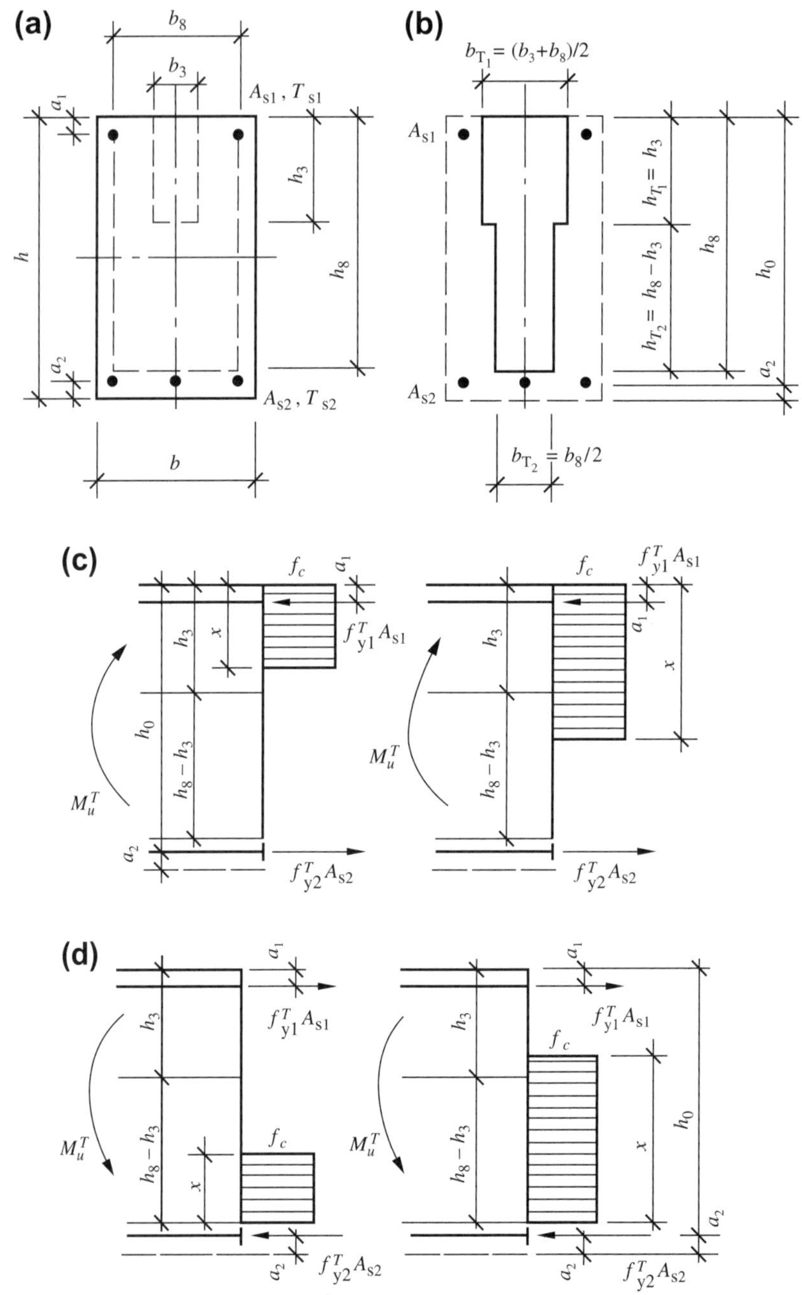

FIGURE 13-6 Calculation diagram for the ultimate bending moment of a beam with three surfaces exposed to high temperature (two-step distribution of concrete strength): (a) isothermal lines on section; (b) equivalent section; (c) tension zone exposed to high temperature; (d) compression zone exposed to high temperature.

13.3.2 Beams with Three Surfaces Exposed to High Temperature

A reinforced concrete beam is of rectangular section ($b \times h$) and three surfaces are exposed to high temperature. When the two-step calculated strength of concrete at elevated temperatures (Fig. 13-2(b)) is selected, the formula is

$$\begin{aligned} &T \le 300\ °\text{C}:\ f_c^T = f_c \\ &300\ °\text{C} < T \le 800\ °\text{C}:\ f_c^T = 0.5 f_c \\ &T > 800\ °\text{C}:\ f_c^T = 0 \end{aligned} \qquad (13.9)$$

When the isothermal curved lines for temperatures of 300 °C and 800 °C on its section at time t (see Chapter 7) are rounded approximately to rectangles, the lengths of their sides are b_3, h_3 and b_8, h_8 (Fig. 13-6(a)), respectively. According to the principle described above, an equivalent T-shaped section (Fig. 13-6(b)) is obtained, and the width and thickness of the upper flange are:

$$b_{T_1} = b_3 + \frac{1}{2}(b_8 - b_3) = \frac{1}{2}(b_3 + b_8)$$

and $h_{T_1} = h_3$, respectively, and the width and depth of the web are, respectively, $b_{T_2} = (1/2)\,b_8$ and $h_{T_2} = h_8 - h_3$.

The areas of the reinforcements (A_{s1} and A_{s2}) and their positions on the section remain invariable and the calculated strengths (f_{y1}^T and f_{y2}^T) depend on the temperature (T_{s1} and T_{s2}, respectively). The ultimate bending moment of the section of the beam is calculated separately for two conditions, i.e., tension zone ($+M$) and compression zone ($-M$) exposed to high temperature.

1. Tension zone exposed to high temperature

The calculation diagram of ultimate stress state on the section is shown in Fig. 13-6(c) and the basic equations are as follows. When the depth of the compression zone is $x \le h_3$ or $f_{y2}^T A_{s2} \le f_c b_{T_1} h_3 + f_{y1}^T A_{s1}$:

$$\left.\begin{aligned} &f_c b_{T_1} x + f_{y1}^T A_{s1} = f_{y2}^T A_{s2} \\ &M_u^T = f_c b_{T_1} x \left(h_0 - \frac{x}{2}\right) + f_{y1}^T A_{s1}(h_0 - a_1) \end{aligned}\right\} \qquad (13.10)$$

When the depth of the compression zone is $x > h_3$ or $f_{y2}^T A_{s2} > f_c b_{T_1} h_3 + f_{y1}^T A_{s1}$, the basic equations are:

$$\left.\begin{aligned} &f_c\left[b_{T_1} h_3 + b_{T_2}(x - h_3)\right] + f_{y1}^T A_{s1} = f_{y2}^T A_{s2} \\ &M_u^T = f_c b_{T_1} h_3 \left(h_0 - \frac{1}{2} h_3\right) \\ &\quad + f_c b_{T_2}(x - h_3)\left(h_0 - \frac{h_3 + x}{2}\right) \\ &\quad + f_{y1}^T A_{s1}(h_0 - a_1) \end{aligned}\right\} \qquad (13.11)$$

In general, the condition of $x > h_3$ is rare, because the temperature of the reinforcement on the tension zone of the beam is high and the tensile strength (f_{y2}^T) is low. If $x < 2a_1$ is obtained after calculation, the action of compressive reinforcement can be neglected (i.e., $A_{s1} = 0$) and the ultimate bending moment is calculated again.

2. Compression zone exposed to high temperature

The calculation diagram of the ultimate stress state on the section is shown in Fig. 13-6(d) and the basic equations are as follows. When the depth of the compression zone is $x \le h_8 - h_3$ or $f_{y1}^T A_{s1} \le f_c b_{T_2}(h_8 - h_3) + f_{y2}^T A_{s2}$:

$$\left.\begin{aligned} &f_{y1}^T A_{s1} = f_c b_{T_2} x + f_{y2}^T A_{s2} \\ &M_u^T = f_c b_{T_2} x \left(h_8 - \frac{x}{2} - a_1\right) + f_{y2}^T A_{s2}(h_0 - a_1) \end{aligned}\right\} \qquad (13.12)$$

When the depth of the compression zone is $x > h_8 - h_3$ or $f_{y1}^T A_{s1} > f_c b_{T_2}(h_8 - h_3) + f_{y2}^T A_{s2}$, the basic equations are:

$$\left.\begin{aligned} &f_{y1}^T A_{s1} = f_c\left[b_{T_2}(h_8 - h_3) + b_{T_1}(x - h_8 + h_3)\right] + f_{y2}^T A_{s2} \\ &M_u^T = f_c b_{T_2}(h_8 - h_3)\left(\frac{h_8 + h_3}{2} - a_1\right) \\ &\quad + f_c b_{T_1}(x - h_8 + h_3) \\ &\quad \left(\frac{h_8 + h_3 - x}{2} - a_1\right) + f_{y2}^T A_{s2}(h_0 - a_1) \end{aligned}\right\} \qquad (13.13)$$

If other kinds of approximate calculated strength of concrete at elevated temperatures (e.g., Fig. 13-2) are used, the basic equations of the ultimate bending moment of the beam can be established using the same method. The calculated results will be slightly different but similar to the experimental results.

13.4 ECCENTRIC COMPRESSIVE MEMBERS

13.4.1 Determination of the Ultimate Envelope of Axial Compression–Bending Moment

When an eccentric compressive member of a reinforced concrete structure is under the action of asymmetrical temperature along the direction of depth on its section, e.g., one or three surfaces are exposed to fire, the ultimate envelope of axial compression–bending moment ($N_u^T - M_u^T$, e.g., Fig. 9-13) is also asymmetrical about the ordinate. There are five asymmetrical characteristic points on the envelope: the peak point (P) is located on the right side of the ordinate and represents the optimum strength (N_p^T) of the member, which corresponds to the optimum center ($e_p > 0$); two points of intersection in the positive and negative directions of the abscissa (M and M') represent, respectively, the values of the ultimate bending moment of the member ($N = 0$) with the tension zone and the compression zone exposed to high temperature; the others are two boundary points (B and B'), which distinguish between the failure patterns of larger and smaller eccentricity compressions, with eccentricities $e_B > 0$ and $e_B' < 0$, ($e_B \neq | e_{B'} |$), which correspond to the axial compression acting on the sides of the low and the high temperature zones, respectively, on the section (see Fig. 13-7).

During the loading of the structural member, the tensile reinforcement located on the high temperature zone yields first when the eccentricity of axial compression is $e \geq e_B$, and the tensile reinforcement located on the lower temperature zone yields first when the eccentricity of axial compression is $e \leq e_B'$ (algebraic value). Afterward, the concrete and reinforcement on the compression zone reach successively their own compressive strength, hence the failure pattern of larger eccentricity compression or flexure is caused. The stress distribution on the section at the ultimate state of these structural members is determined, and the stress diagram on the compression zone of concrete can be taken approximately as a rectangle. Then, two equilibrium equations are established and solved, and the ultimate strength (N_u^T, M_u^T) of the member is obtained.

When the eccentricity of axial compression is $e_B > e > e_B'$, the concrete on the near side of axial compression, whether on the low or the high temperature zone on the section, reaches its

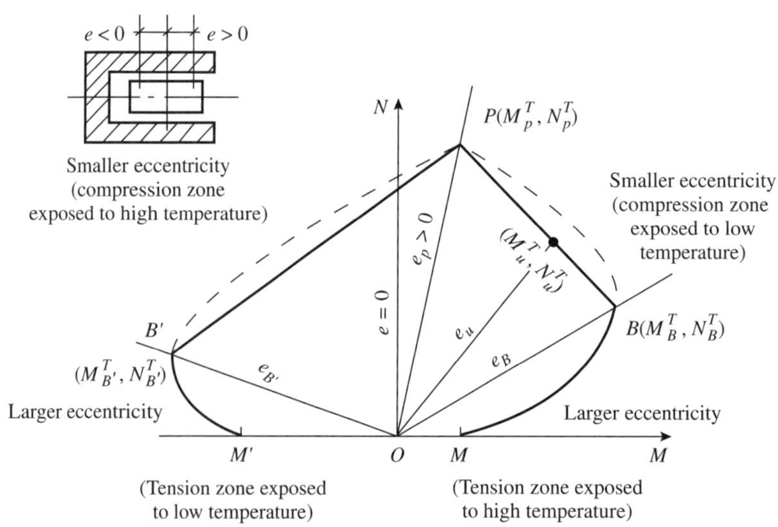

FIGURE 13-7 Ultimate envelope of the axial compression–bending moment of a structural member at elevated temperatures.

compressive strength first, hence the failure pattern of smaller eccentricity compression is caused. In the meantime, the concrete and reinforcement on the further side of axial compression cannot reach their own strength whether compressive or tensile. Therefore, the stress distribution on the section of these structural members at the ultimate state varies with the eccentricity (e) of axial compression and is transformed from the condition of tension on the high temperature zone and compression on the low temperature zone to the condition of compression on the whole section, and again to the condition of tension on the low temperature zone and compression on the high temperature zone. In addition, the stress values of concrete and reinforcement on most parts of the section are unknown, so it is difficult to establish the equilibrium equations accurately and to find the ultimate strength. The same problem exists for the structural member of smaller eccentricity compression at room temperature. In the existing design code for concrete structures,[0-1] the rectangular stress block (strength of concrete f_c) of one part of or even the whole section is assumed and used for approximate calculation, which seems imperfect from the theoretical point of view.

There is a general regularity for the strengths of materials and structural members in structural engineering, which is demonstrated by many experimental investigations and theoretical analyses: the ultimate envelope of various strengths should be convex, e.g., the envelopes of concrete strength under bi- and tridimensional stress states[0-3] and the ultimate envelopes of composite strength of various structural members, which include the envelopes of axial force–bending moment (N–M), axial force–shear force (I–V), shear force–torque (V–T), bending moment–torque (M–T), and bending moment–shear force–torque (M–V–T).[0-2] The ultimate envelope of axial compression–bending moment of a reinforced concrete structural member at elevated temperatures (see Fig. 9-13) is not exceptional, so the envelope segments $\widehat{B'P}$ and $\widehat{PB}$ within the range of the failure pattern of smaller eccentricity compression is also convex and the curvatures are small. If they are replaced by the straight lines $\overline{B'P}$ and $\overline{PB}$, respectively, the result of the calculation should be safe with limited error, the calculation method is simplified, and the assumption of excessive approximation for ultimate stress block on the section is avoided.

When the optimum strength (M_p^T, N_p^T) of the structural member and the ultimate strengths $((M_B^T, N_B^T)$ and $(M_{B'}^T, N_{B'}^T))$ at both boundary points (B and B'), respectively, which distinguish between the failure patterns of larger and smaller eccentricity compressions, are known, the formula of the straight envelope line $\overline{BP}$, corresponding to the failure pattern of smaller eccentricity compression, is:

$$\left(M_B^T - M_p^T\right) N_u^T + \left(N_p^T - N_B^T\right) M_u^T = M_B^T N_p^T - M_p^T N_B^T \tag{13.14}$$

where M_u^T, N_u^T are the coordinate values of any point on the straight line BP, i.e., the ultimate strength of the structural member.

When $M_{B'}^T$ is taken as a negative value and replaces M_B^T in Eqn (13.14), the formula of straight envelope line $\overline{B'P}$ is obtained.

If the eccentricity of a structural member at ultimate state e_u is known, the ultimate strength can be calculated by the formulas below, which are derived from Eqn (13.14):

$$\left.\begin{aligned} N_u^T &= \frac{M_B^T N_p^T - M_p^T N_B^T}{\left(M_B^T - M_p^T\right) + \left(N_p^T - N_B^T\right) e_u} \\ M_u^T &= N_u^T e_u \end{aligned}\right\} \tag{13.15}$$

According to the analyses above, the method to determine the ultimate envelope of axial compression–bending moment of the structural member at elevated temperatures or to calculate the ultimate strength is as follows:

1. The flexural member ($N = 0$) is divided into two conditions, i.e., tension zone ($+M$) and compression zone ($-M$) exposed to high temperature, and both ultimate strengths can be calculated separately using Eqns (13.5)–(13.13).

2. Two boundary points on both sides of the ordinate (corresponding eccentricities e_B and e_B'), distinguishing between the failure patterns of larger and smaller eccentricity compression, are determined and the corresponding ultimate strengths are calculated.
3. When the eccentricity of a compressive structural member at ultimate state (see Eqn (13.22)) is $e_u \geq e_B$ or $e_u \leq e_B'$, the equilibrium equations (Eqns (13.17)–(13.20)), which fit with the failure pattern of larger eccentricity compression, are used to find the ultimate strength.
4. The optimum strength of a section of a structural member is calculated using Eqn (13.21).
5. The formula for a straight envelope line (Eqn (13.14)), corresponding to the failure pattern of smaller eccentricity compression, is established and the ultimate strength of the section is calculated using Eqn (13.15).

13.4.2 Calculation of the Ultimate Strength of a Section

A structural member of rectangular section with three surfaces exposed to high temperature is taken as an example; the methods and formulas for calculations of the boundary points distinguishing two failure patterns, the ultimate strength of the section with larger eccentricity, and its optimum strength are given.

When the cross section of a structural member ($b \times h$) is given, the temperature field is determined first according to the condition of fire or the fire endurance needed. Then the equivalent section is obtained (Fig. 13-8(a)) after the calculated strength of concrete at elevated temperatures (Fig. 13-2) is selected. The failure characteristic of the structural member on the boundary between larger and smaller eccentricity compressions is that when the reinforcement in the tension zone reaches its yield strength (f_y^T) and the concrete on the outer side in the compression zone of the section simultaneously reaches its ultimate value of strain ($\varepsilon_{c,u}^T$).

The boundary depth of the compression zone (x_u) of the section can be obtained directly based on the basic assumption of strain distributing linearly on the section. If the nonuniformly distributed stress on the compression zone of the section at the ultimate state is simplified to an equivalent rectangular stress block, the corresponding boundary depth of the compression zone is changed to $x_B = 0.8x_u$.[0-1,0-2] Since the effective depths of the section and the values of ultimate strain of the materials are different for both conditions, i.e., tension zone ($+M$, $e_B > 0$) and compression zone ($-M$, $e_B' < 0$) exposed to high temperature, the boundary depths of the compression zone of both conditions are calculated separately:

tension zone exposed to high temperature ($e_B > 0$):

$$x_B = \frac{0.8\varepsilon_{c,u}^T}{\varepsilon_{c,u}^T + \varepsilon_{y2}^T} h_0 \tag{13.16a}$$

compression zone exposed to high temperature ($e_B' < 0$):

$$x_{B'} = \frac{0.8\varepsilon_{c,u}^T}{\varepsilon_{c,u}^T + \varepsilon_{y1}^T} (h_8 - a_1) \tag{13.16b}$$

where ε_{y2}^T, ε_{y1}^T are the yielding strains of tensile reinforcements on the high and the low temperature zones, respectively, and they can be calculated using Eqn (4.8), and $\varepsilon_{c,u}^T$ is the ultimate compressive strain of concrete on the outer side of the compression zone of equivalent section. The value may be taken as $\varepsilon_{c,u} = 3.3 \times 10^{-3}$ at room temperature,[0-1] and the value of $\varepsilon_{c,u}^T / \varepsilon_{c,u}$ at elevated temperatures can be calculated using Eqn (2.5); $\varepsilon_{c,u}^T \leq 6.6 \times 10^{-3}$ is suggested.[9-6]

When a reinforced concrete structural member at elevated temperatures fails with the pattern of larger eccentricity compression, the ultimate stress distribution on its section can be simplified as shown in Fig. 13-9, the equilibrium equations can be established, and then the ultimate strength

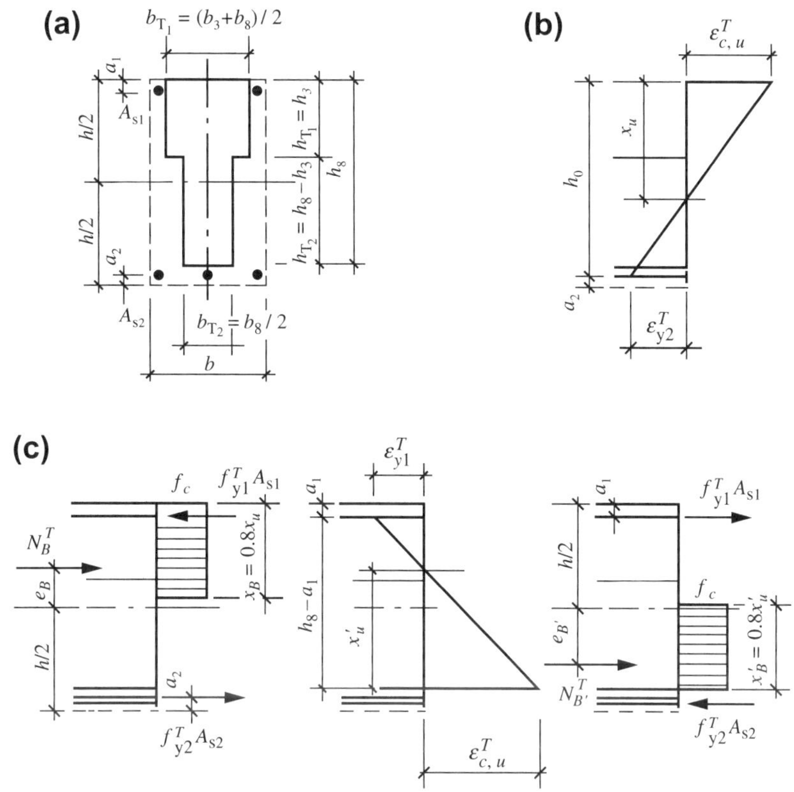

FIGURE 13-8 Strain and stress states on a section of a structural member at elevated temperatures and at boundary points between two failure patterns: (a) equivalent section; (b) positive boundary $e_B > 0$ (tension zone exposed to high temperature); (c) negative boundary $e_{B'} < 0$ (compression zone exposed to high temperature).

can be found separately for the conditions of the tension and compression zones exposed to high temperature.

1. Tension zone exposed to high temperature ($e_u \geq e_B$ or $x \leq x_B$)

When $x \leq h_{T_1} = h_3$ or $N_u^T \leq f_c b_{T_1} h_{T_1} + f_{y1}^T A_{s1} - f_{y2}^T A_{s2}$, the equilibrium equations are:

$$\left.\begin{aligned} N_u^T &= f_c b_{T_1} x + f_{y1}^T A_{s1} - f_{y2}^T A_{s2} \\ N^T\left(e_u + \frac{h}{2} - a_2\right) &= f_c b_{T_1} x\left(h_0 - \frac{x}{2}\right) + f_{y1}^T A_{s1}(h_0 - a_1) \end{aligned}\right\} \quad (13.17)$$

where e_u is the eccentricity of the structural member at the ultimate state (Eqn (13.22)).

When $x > h_{T_1} = h_3$ or $N_u^T > f_c b_{T_1} h_{T_1} + f_{y1}^T A_{s1} - f_{y2}^T A_{s2}$, the equilibrium equations change to:

$$\left.\begin{aligned} N_u^T &= f_c\left[b_{T_1} h_{T_1} + b_{T_2}(x - h_{T_1})\right] + f_{y1}^T A_{s1} - f_{y2}^T A_{s2} \\ N_u^T\left(e_u + \frac{h}{2} - a_2\right) &= f_c\left[\begin{aligned} & b_{T_1} h_{T_1}\left(h_0 - \frac{h_{T_1}}{2}\right) \\ & + b_{T_2}(x - h_{T_1})\left(h_0 - \frac{x + h_{T_1}}{2}\right) \end{aligned}\right] + f_{y1}^T A_{s1}(h_0 - a_1) \end{aligned}\right\} \quad (13.18)$$

and the ultimate bending moment of the section is $M_u^T = N_u^T e_u$.

2. Compression zone exposed to high temperature ($e_u \leq e_{B'}$ or $x \leq x_{B'}$)

When $x \leq h_{T_2} = h_8 - h_3$ or $N_u^T \leq f_c b_{T_2} h_{T_2} + f_{y2}^T A_{s2} - f_{y1}^T A_{s1}$, the equilibrium equations are:

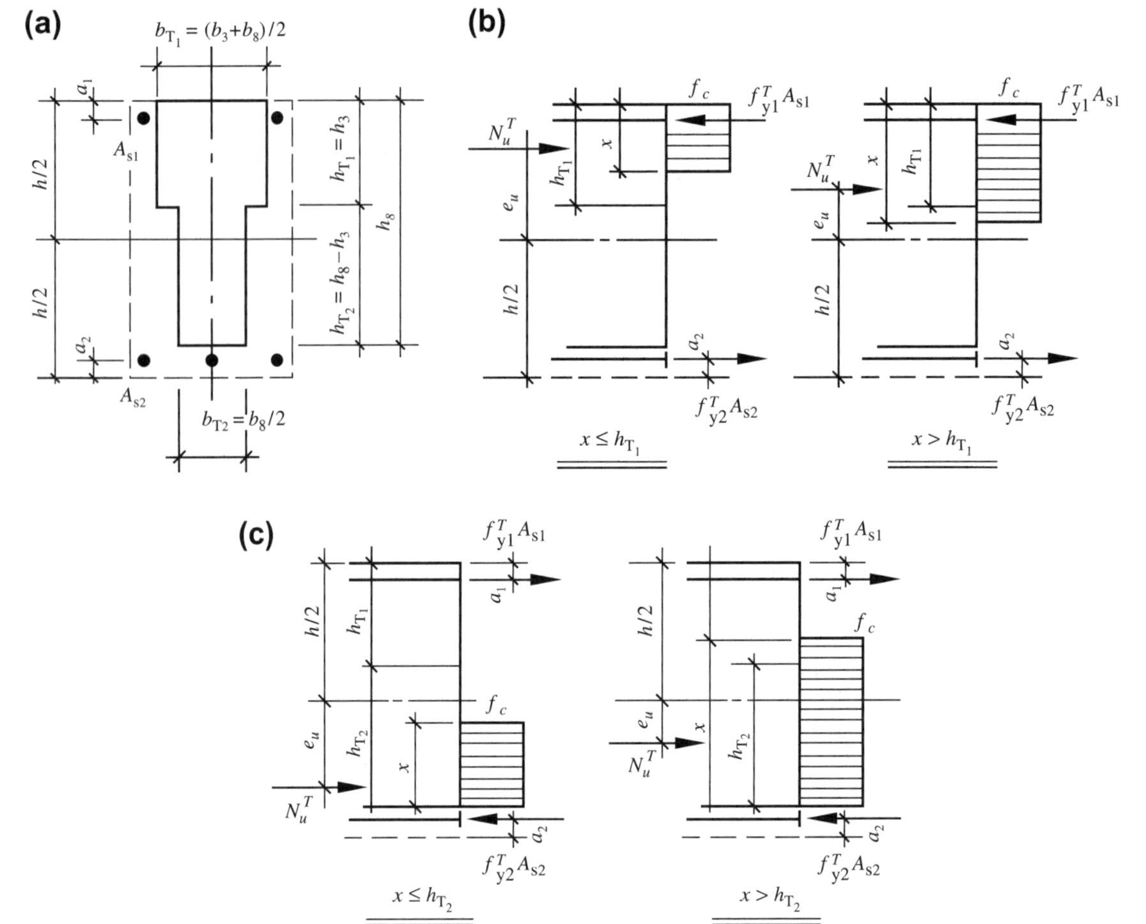

FIGURE 13-9 Calculation diagram of the ultimate strength of a section of a structural member with large eccentricity compression and at elevated temperatures: (a) equivalent section; (b) tension zone exposed to high temperature; (c) compression zone exposed to high temperature.

$$\left.\begin{aligned} N_u^T &= f_c b_{T_2} x + f_{y2}^T A_{s2} - f_{y1}^T A_{s1} \\ N_u^T\left(e_u + \frac{h}{2} - a_2\right) &= f_c b_{T_2} x\left(h_{T_1} + h_{T_2} - \frac{x}{2} - a_1\right) \\ &\quad + f_{y2}^T A_{s2}(h_0 - a_1) \end{aligned}\right\} \tag{13.19}$$

When $x > h_{T_2} = h_8 - h_3$ or $N_u > f_c b_{T_2} h_{T_2} + f_{y2}^T A_{s2} - f_{y1}^T A_{s1}$, the equilibrium equations change to:

$$\left.\begin{aligned} N_u^T &= f_c[b_{T_2} h_{T_2} + b_{T_1}(x - h_{T_2})] + f_{y2}^T A_{s2} - f_{y1}^T A_{s1} \\ N_u^T\left(e_u + \frac{h}{2} - a_1\right) &= f_c\left[b_{T_2} h_{T_2}\left(\frac{2h_{T_1}+h_{T_2}}{2} - a_1\right)\right. \\ &\quad \left.+ b_{T_1}(x - h_{T_2})\left(\frac{2h_{T_1} + h_{T_1} - x}{2} - a'\right)\right] \\ &\quad + f_{y2}^T A_{s2}(h_0 - a_1) \end{aligned}\right\} \tag{13.20}$$

If the depth of the compression zone on the section of the eccentric compressive member at elevated temperatures calculated from the equations above is greater than its boundary depth, i.e., $x > x_B$ or $x > x_B'$, the failure pattern of the member is of smaller eccentricity compression. Therefore, the ultimate strength of the member can be calculated from Eqns (13.14) and (13.15) in Section 13.4.1. In these equations the ultimate strengths at boundary points distinguishing two failure patterns are needed and can be calculated as follows: N_B^T and M_B^T (tension zone exposed to high temperature, $e_B > 0$) are calculated by taking $x = x_B$, which is substituted into Eqn (13.17) or Eqn (13.18); N_B^T and M_B^T (compression zone exposed to high temperature, $e_B < 0$) are calculated by taking $x = x_B'$, which is substituted into Eqn (13.19) or Eqn (13.20).

The optimum strength of the section corresponding to the peak point (P) on the envelope and its acting position can be calculated based on the equivalent homogenous section (Fig. 13-9(a)). According to Section 11.2, the optimum center (P) of a section of homogenous material coincides with the geometry center (G) and deformation center (D) of the same section. Therefore, the calculation formulas are obtained:

$$N_p^T = f_c\,(b_{T_1}h_{T_1} + b_{T_2}h_{T_2}) + f_{y1}^T A_{s1} + f_{y2}^T A_{s2}$$
$$e_p = \frac{h}{2} - \frac{1}{N_p^T}\frac{1}{2}\left[\begin{array}{l} f_c b_{T_1} h_{T_1}^2 + f_c b_{T_2} h_{T_2}^2 \left(h_{T_1} + \frac{h_{T_2}}{2}\right) \\ + f_{y1}^T A_{s1} a_1 + f_{y2}^T A_{s2}\,(h - a_2) \end{array}\right]$$
$$M_p^T = N_p^T e_p \tag{13.21}$$

where e_p is the optimum eccentricity.

The ultimate eccentricity (e_u) of the structural member, appearing in the formulas above, is the sum of the initial eccentricity ($e_0 = M/N$) of the load and its additional eccentricity (i.e., lateral deflection Δ_u^T) at the ultimate state:

$$e_u = e_0 + \Delta_u^T \tag{13.22}$$

If the enlarged coefficient of eccentricity of the structural member at elevated temperatures is taken as η, then:

$$\eta = \frac{e_u}{e_0} = 1 + \frac{\Delta_u^T}{e_0}$$

The additional eccentricity (Δ_u^T, see Fig. 9-10) of the structural member at the ultimate state under the common actions of temperature and load is given from the experiments on an eccentric compressive structural member at elevated temperatures. Then, the general regularity of the enlarged coefficient of eccentricity of the structural member at elevated temperatures is confirmed as below: the value of η increases with temperature on the section, the value of η decreases as the initial eccentricity increases, the value of η of positive eccentricity ($e_0 > 0$) is slightly greater than that of the negative eccentricity ($e_0 < 0$) if the absolute values of both eccentricities are equal, and a certain amount of ultimate eccentricity ($\Delta_u^T \neq 0$) is still caused due to the action of temperature, even if the initial eccentricity of the specimen $e_0 \to 0$.

The enlarged coefficient of eccentricity of a concrete structural member at room temperature shows similar regularity, i.e., it increases as the depth of the cross section decreases and it decreases as the eccentricity increases. The relevant formulas listed in the existing design code for a reinforced concrete structure,[0-1] which are normally used for structures at room temperature, are suggested after trying, comparing, and analyzing to calculate the enlarged coefficient of eccentricity of the structural member at elevated temperatures. These formulas are:

$$\eta = 1 + \frac{1}{1400\,(e_0/h_0)}\left(\frac{l_0}{h}\right)^2 \zeta_1\zeta_2 \tag{13.23}$$

$$\zeta_1 = \frac{0.5 f_c A}{N} \le 1.0$$

$$\zeta_2 = 1.15 - 0.01\frac{l_0}{h} \le 1.0$$

where l_0 is the calculated length of the structural member obtained by following the relevant provisions in the design code, and h and h_0 are the total and effective depths of the cross section, which are determined from the equivalent section.

The minimum value of additional eccentricity should be taken as 20 mm or $h/30$ for a compressive structural member with central load or smaller eccentricity.

According to the calculations described above, specimens heated to 800 °C[9-3] are calculated and checked, and the theoretical envelope is drawn in Fig. 13-10 and compared with the measured data. It is seen that the variation regularities of both are similar, the values of both are approaching, and the result obtained tends toward safety.

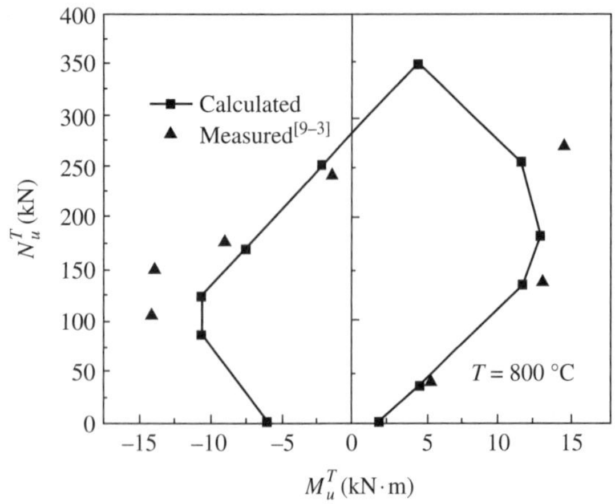

FIGURE 13-10 Demonstration of the calculation method.[13-2]

13.5 STATICALLY INDETERMINATE STRUCTURES

13.5.1 General Principle of Plastic Limit Analysis

According to experimental investigations of a statically indeterminate reinforced concrete structure, e.g., continuous beam and frame (see Chapter 10), at elevated temperatures under both paths of loading at constant temperature and heating under constant load, the specimens fail because several plastic hinges are formed successively at some critical cross sections and the segments nearby, and a movable mechanism of one degree of freedom then results. The failure mechanism may be a local one (e.g., in one span only of a continuous beam or in a beam of the frame) or an integral one (e.g., in both spans of a continuous beam or a combined beam–column mechanism of the frame, see Fig. 10-17). Therefore, the plastic limit analysis method (or method of equilibrium at ultimate state) used for the statically indeterminate structure at room temperature can also be used for analyzing the ultimate strength of the statically indeterminate structure at elevated temperatures, and the calculation can be conducted easily.

The basic assumptions used for the plastic limit analysis method[13-3–13-5] of a statically indeterminate structure at room temperature are:

1. The bending moment–curvature ($M - 1/\varrho$) relationship of a cross section of a structural member is ideal elastoplastic; this means that the curvature increases greatly with unvaried value of bending moment and is not influenced by the axial and shear forces on the section, after the ultimate bending moment (M_u) is reached.
2. The loads acted on the structure increase proportionally, i.e., the relative ratios between the loads are unvaried, and they increase monotonically and unloading never appears.
3. The deformation of the structure is small before the formation of its failure mechanism and the internal forces of the structure are not influenced by its deformation.

The basic method of plastic limit analysis is as follows: one or several possible failure mechanisms of one degree of freedom and corresponding positions of the plastic hinges are determined or selected in advance for the statically indeterminate structure; the principle of virtual work or the ultimate equilibrium equation is used to calculate the ultimate strength (P_u) of the structure when the ultimate bending moments (M_u) at the plastic hinges are known. Although the nonlinear deformation history of the structure need not be considered during the calculation and the actions of other factors (e.g., initial stress on the section, initial internal forces of the structure, settlement of the support) are not accounted for, the accuracy of the ultimate strength calculated for the structure is not affected.

The mechanical requirements, which have to be satisfied for the ultimate state or a real failure mechanism of a statically indeterminate structure, are as follows:

1. Equilibrium condition: all the forces from outside and the internal forces on the cross section acting on the structure itself or any part of it are in equilibrium.
2. Ultimate condition: the value of the bending moment (and other internal forces) on every

cross section has to be equal or smaller than the available ultimate value ($M \le M_u$).
3. Mechanism condition: plastic hinges of sufficient number are formed in the structure itself or in some part of it and a movable mechanism of one degree of freedom results.

The load satisfying these conditions is the ultimate load of the structure.

The load value calculated from the equilibrium condition for one of the possible failure mechanisms selected is called a possible ultimate load, which satisfies conditions 1 and 3. When the value of the internal force calculated from the equilibrium condition for the load value selected is equal to or smaller than the ultimate value of that on the corresponding section, the load is called a statically admissible load, and conditions 1 and 2 are satisfied.

It has been proved theoretically[133] that the possible ultimate load is the upper bound of the ultimate load and the minimum value is the value of the ultimate load (upper bound theorem), while the statically admissible load is the lower bound of the ultimate load and the maximum value is the value of the ultimate load (lower bound theorem). Therefore, the value of the possible ultimate load is certainly greater than that of the statically admissible load, and they provide the range of the approximate and accurate solutions of the ultimate load. When both of them are equal and all the conditions above are satisfied, the accurate value of the ultimate load is obtained.

13.5.2 Characteristics of Plastic Hinges at Elevated Temperatures

The plastic limit analysis of a statically indeterminate reinforced concrete structure under common actions of temperature and load are the same as that of the structure at room temperature, if the influence of the heating–loading path is not taken into account. However, the thermal–mechanical behavior of the structure, especially the mechanical condition of the plastic hinge at elevated temperatures, is complicated, and this should be considered carefully during the plastic limit analysis.

When a simply supported reinforced concrete beam with three surfaces exposed to high temperature is acted by a positive bending moment (i.e., tension zone on its section exposed to high temperature), the bending moment–curvature ($M - 1/\varrho$) relation under the path of loading at constant temperature is shown in Fig. 8-7, and the temperature-curvature ($T - 1/\varrho$) relation under the path of heating under constant load is shown in Fig. 8-10. The curves vary gently and the curvature increases considerably after the tensile reinforcement yields and before the ultimate state is reached. This shows that a considerable angular rotation appears near the critical section, so a plastic hinge is formed, which is called a tensile hinge at high temperature. When the beam is acted by a negative bending moment (i.e., tension zone on its section exposed to lower temperature), the bending moment–curvature ($M - 1/\varrho$) relation under the path of loading at constant temperature is shown in Fig. 8-9; similar deformation regularity is seen and the plastic hinge formed is called a tensile hinge at low temperature.

The tensile hinges at the high and the low temperatures are formed because the first of the tensile reinforcements on the section yields; the tensile hinges are of unidirectional rotation, so they can rotate only along the same direction as the bending moment. The value of the ultimate bending moment there remains unvarying. However, the ultimate values of the bending moments at the tensile hinges at high temperature (i.e., tension zone exposed to high temperature) and at low temperature (i.e., compression zone exposed to high temperature) differ greatly (see Fig. 8-8), and should not be confused with each other.

With regard to the influence of the heating–loading path on the behavior of the plastic hinge, the ultimate bending moment–temperature relationship (Fig. 8-11) of the beam specimen shows that the curve corresponding to the path of heating under constant load is always above the curve corresponding to the path of loading at constant temperature, and the maximum difference in the ultimate strength between them appears in the temperature range of 300–525 °C. If the temperature is $T < 300$ °C or $T > 525$ °C, the heating–loading path makes no obvious difference to the ultimate strength of the beam. When the fire resistance of a structure is analyzed, the temperature of the fire is generally far greater than 525 °C, so the influence of the heating–loading path may not be considered for the plastic limit analysis. The ultimate bending moment of the beam section can be determined simply according to the path of loading at constant temperature, and the calculation is then simplified and the result obtained tends toward safety.

When the structural member is acted on by an axial compression and bending moment (N, M) together or by an eccentric compression (N, $e_0 = M/N$), the axial force (or bending moment $M = Ne$) deflection (curvature) of its section under the path of loading at constant temperature is shown in Fig. 9-9.

The compressive specimens of the failure pattern of larger eccentricity (Fig. 9-13), including the specimens with tension zones exposed to high temperature ($e_0 > 0$) and to low temperature (or compression zone exposed to high temperature, $e_0 < 0$), fail also because the first tensile reinforcement yields, and considerable plastic deformation (curvature) is caused before its failure; the ultimate bending moment varies slightly. Therefore, both conditions of these specimens may also be considered approximately as the tensile hinges at high and the low temperature, respectively.

As far as the compressive specimens of the failure pattern of smaller eccentricity are concerned, their failure is controlled by the compressive concrete on the side of low temperature ($e_0 > 0$) or high temperature ($e_0 < 0$), separately. Once the ultimate strength (N_u^T) of the specimen is reached, the value of the ultimate strength cannot be maintained and reduces gradually (although the ultimate bending moment $M_u^T = N_u^T e_u$ on its section reduces slightly), as its deformation (curvature) develops continuously. The reducing amplitude of its ultimate strength depends on the temperature reached and the eccentricity of the load of the specimen. The higher the temperature and the larger the eccentricity of the load, the more gently the deformation increases and the less the ultimate strength reduces.

When reinforced concrete frames are tested at elevated temperatures, it is found that a considerable plastic angular rotation can also result within a longer longitudinal area of the column because its stiffness (and cross section) is small and the compressive strain of concrete at high temperature is high, so a compressive hinge at high temperature is formed (see Section 10.3.1 and Fig. 10-17).

If the compressive hinge at high temperature is the last one (or last batch) among all the plastic hinges appearing during the process of a movable mechanism of the structural forms, the ultimate strength ($M_u^T = N_u^T \cdot e_u$) at the hinge can still be used for the plastic limit analysis. However, if the compressive hinge at high temperature has appeared in advance, the ultimate strength at the hinge is reduced as the plastic angular rotation occurs there during the process of other plastic hinges forming, successively. When the movable mechanism of the structure is reached eventually, the actual strength at the hinge (section) is smaller than its ultimate strength ($< N_u^T$) and the ultimate bending moment there is also smaller slightly ($M_u^T = N_u^T \cdot e_u$). This should be considered carefully during the plastic limit analysis.

The ultimate strengths and deformations of the eccentrically compressive structural members under different heating–loading paths are compared and it is demonstrated that the ultimate envelope of the axial compression–bending moment of the member under the path of heating under constant load is outside that under the path of loading at constant temperature (Fig. 9-15), i.e., the ultimate strength of the former is greater than that of the latter; and the deformation at the ultimate state of the specimen under the path of heating under constant load is also greater than that under the path of loading at constant temperature (Fig. 9-14(b)). Therefore, when the ultimate strength of the column section under the path of loading at constant temperature is used simply for the plastic limit analysis of the structure, the calculation is then simplified and the result obtained tends toward safety.

13.5.3 Method and Procedure of Plastic Limit Analysis

The upper bound method of plastic limit analysis is generally used to calculate and determine the ultimate strength of a statically indeterminate structure at elevated temperatures. The continuous beam and simple frame are used as examples to explain the calculation procedure of the method.

1. Calculation of the temperature field

According to the temperature–time curve (e.g., the ISO standard curve) of the peripheral medium of the structure and the fire endurance (h) desired, the temperature field on the section of every structural member is calculated (see Chapters 6 and 7).

2. Selection of possible failure mechanisms

One or several kinds of movable mechanisms of one degree of freedom, which are possible geometrically, are selected and the positions of the plastic hinges are determined correspondingly for the statically indeterminate structure, based on the degree of indeterminacy, and the mechanical characteristics. The mechanisms composed may be the whole structure or part of the structure, and the number of plastic hinges is at least one more than the degree of indeterminacy.

Each span of a continuous multi-span beam may compose one possible failure mechanism. When the plastic hinges are formed successively at the positions of the maximum bending moment in the mid-span and at both end supports of one span of the continuous beam (Fig. 13-11(a)), one movable mechanism of one degree of freedom is composed. A simple frame of one bay and one story is a statically indeterminate structure of three degrees of indeterminacy; several kinds of failure mechanism (Fig. 13-11(b) and (c)) are possible. When only vertical loads act symmetrically on the frame, the local beam-type failure mechanism of three plastic hinges may be composed, and two of the plastic hinges are formed at the ends of the beam or at the upper ends of the columns, depending on the relative strengths of the beam and column sections. In addition, a special failure mechanism of combined beam–column type with five plastic hinges may be composed (Fig. 10-17) and cause integral failure of the frame. When a horizontal load also acts on the frame, a side-displacement failure mechanism or a combined beam–column mechanism[13-6] of four plastic hinges may be composed.

3. Calculation of the ultimate strength (bending moment) of a critical section of a structural member

The ultimate bending moment (M_u^T) of a structural member at elevated temperatures is calculated approximately using the equivalent section method (see earlier in this chapter), based on the size, reinforcement, strengths of materials, and the temperature field of its section.

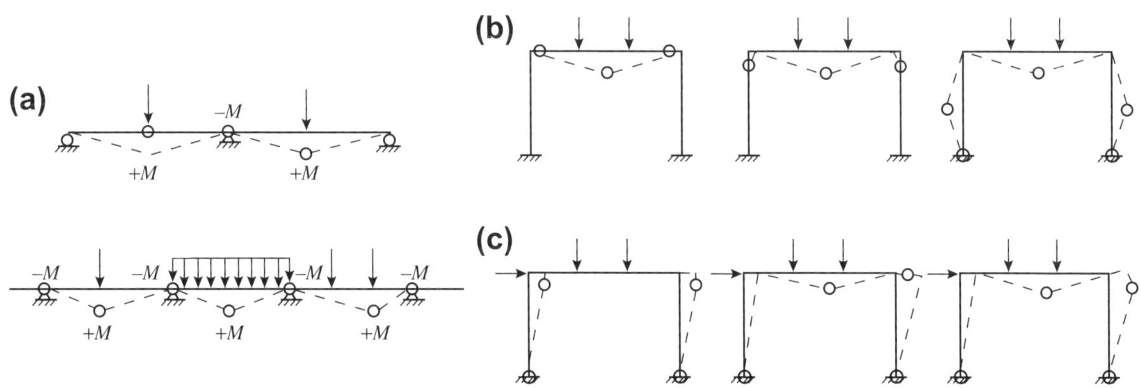

FIGURE 13-11 Possible failure mechanisms of a continuous beam and simple frame: (a) continuous beam; (b) frame (vertical load only); (c) frame (vertical and horizontal loads).

4. Calculation and determination of the ultimate load of a structure at elevated temperatures

The equilibrium equation at the limit state or the equation of virtual work is established for every possible failure mechanism of the structure, and the corresponding value of the limit load is calculated. The minimum among all the limit loads corresponding to all possible failure mechanisms is chosen as the ultimate load (upper bound solution) of the statically indeterminate structure at elevated temperatures.

When one span of a continuous multi-span beam is considered and two concentrated loads are acting symmetrically on the span, the movable mechanism is composed (Fig. 13-12(a)) after three plastic hinges are formed successively at the sections of both supports (A and B) and the maximum bending moment (C) in the span. The value of the ultimate load (P_u) of this span can be calculated after the ultimate bending moments at the corresponding sections (M_A, M_B, and M_C) are determined.

(1) Method of equilibrium at the ultimate state

Suppose the beam is simply supported at both ends; the bending moment at the cross section (C), where the load is acted, is

$$M_0 = P_u \beta l$$

When the ultimate bending moments at the sections of both supports and the mid-span of the beam are known, they satisfy the equilibrium condition at the ultimate state (Fig. 13-12(b)):

$$(1-\beta)M_A + \beta M_B + M_C = M_0$$

Solving these two equations, the value of the ultimate load of this span is obtained:

$$P_u = \frac{1}{l}\left(\frac{1-\beta}{\beta}M_A + M_B + \frac{1}{\beta}M_C\right) \tag{13.24}$$

(2) Method of virtual work principle

Suppose a virtual displacement ($\delta = 1$) occurs under point C where the load is applied. The virtual angular rotations of the plastic hinges at both supports and the mid-span of the beam are, respectively (Fig. 13-12(a)):

$$\theta_A = \frac{1}{\beta l},\ \theta_B = \frac{1}{(1-\beta)l}$$

and

$$\theta_C = \frac{1}{\beta l} + \frac{1}{(1-\beta)l} = \frac{1}{\beta(1-\beta)l}$$

According to the principle of virtual work, the virtual work done by the external loads should be equal to that done by the internal forces, so:

$$P_u\left(1 + \frac{\beta}{1-\beta}\right) = M_A\theta_A + M_B\theta_B + M_C\theta_C \tag{13.25}$$

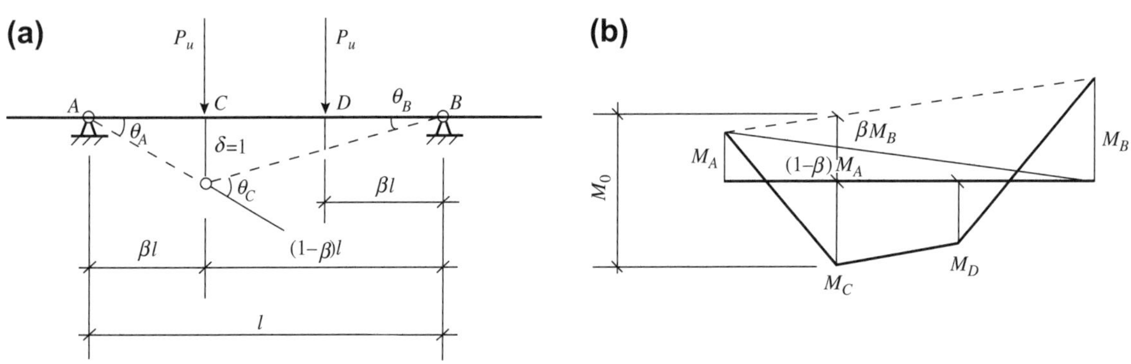

FIGURE 13-12 Calculation model of the ultimate load of a continuous beam: (a) failure mechanism and virtual displacement; (b) bending moment diagram at the ultimate state.

is established. After the virtual angular rotations are substituted into Eqn (13.25), the value of ultimate load (P_u) obtained is the same as Eqn (13.24).

If the continuous beam is multi-span and each span has a different span length, temperature, or load, the value of the ultimate load of every span at elevated temperatures can be calculated individually based on the principles and methods introduced above. The minimum of the ultimate loads corresponding to all the spans is exactly the ultimate load needed for the continuous beam. When a frame of different shape and load is analyzed, the value of the ultimate load can be calculated individually for every one of the possible failure mechanisms, e.g., as shown in Fig. 13-11, and the minimum among all the possible ultimate loads is the actual ultimate load of the frame.

There is another situation. When the load acting on a statically indeterminate structure is known in advance, the fire endurance of the structure needs to be calculated. Therefore, the fire endurance of the structure is assumed to be several periods (e.g., 1*h*, 2*h*, 3*h*, ...), the corresponding temperature field on the section and the ultimate bending moment of each structural member are calculated separately for each period, and the various possible failure mechanisms are selected; then the value of the ultimate load of the structure at elevated temperatures is calculated and determined for every period assumed. Comparing the loads calculated with the load known, the fire endurance (*h*) of the structure is then obtained.

13.5.4 Calculation Examples and Demonstration

A fire accident occurring in a practical building is very complicated, the initial data are difficult to collect completely and accurately, and the analysis and comparison are also not convenient. Therefore, the experimental results for the continuous beam and frame introduced above are used as examples to demonstrate the calculations.

1. Continuous beam of two spans

Five specimens of a continuous beam with two spans are tested under the path of heating under constant load. The factors investigated include the position of the load, initial loading level, and the number of the span heated (see Table 10-1). The details of the specimens, initial data and results of the experiments, and their calculation parameters can be found in Chapter 10. The relevant data for specimen TCB2-1 heated on both spans are: the load is located $2l/3$ from the side support, the initial loading level $P_0/P_u = 0.63$, and the ultimate temperature measured from the test is 937 °C (see Table 10-4).

The specimen is put into the experimental furnace and then heated; the elevating temperature–time curve in the furnace is shown in Fig. 13-13(a) and the elevating speed within 200 min is far lower than that of the ISO standard curve. Specimen TCB2-1 fails when the ultimate temperature 937 °C is reached after 162 min of heating time. Obviously, the temperature field on the section of the specimen cannot be obtained from the charts in Chapter 7 but should be calculated instead using the HTARC program (see Chapter 6). The result obtained is presented in Fig. 13-13(b).

When the calculated strength of concrete at elevated temperatures is taken in two steps (Fig. 13-2(b)), the equivalent section of the specimen is obtained (Fig. 13-13(c)). After the temperature of tensile reinforcement is determined from the temperature field on the section of the specimen (Fig. 13-13(b)), the yield strength at elevated temperatures is calculated based on Fig. 13-1. Using the relevant formulas introduced in Section 13.3.2, the ultimate bending moments at elevated temperatures calculated are $M_C^T = 0.555$kNm for the section in the mid-span (tension zone exposed to high temperature) and $M_B^T = 6.649$kNm for the section near the support (compression zone exposed to high temperature).

The span length and the load acting on the continuous beam with two spans are symmetrical, so the value of the ultimate load for one span only needs to be calculated. When the tensile hinges at high and the low temperature are formed,

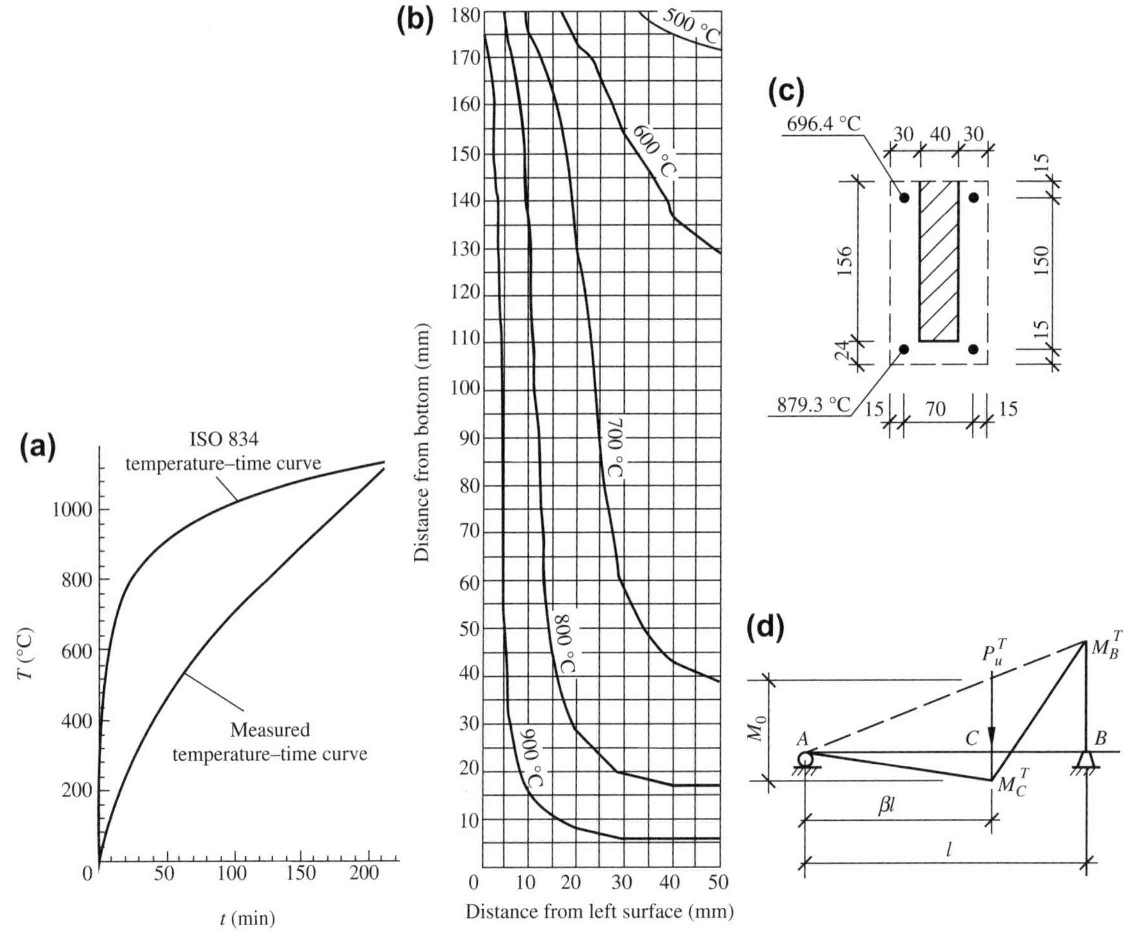

FIGURE 13-13 Calculation of the ultimate load of a continuous beam specimen (TCB2-1): (a) temperature–time curve; (b) temperature field on a section at the ultimate state; (c) equivalent section; (d) failure mechanism.

respectively, on the sections at the maximum bending moment (point C) in the span and at the middle support (point B), as the end support of the specimen is of hinge ($M_A = 0$), so the unique possible movable mechanism of one degree is composed (Fig. 13-13(d)) and the specimen fails. According to the equilibrium principle at the ultimate state, the equation

$$M_0 = P_u^T \beta (1-\beta) l = M_C^T + \beta M_B^T$$

is established and the value of the ultimate load of the continuous beam at elevated temperatures is obtained:

$$P_u^T = \frac{1}{l}\left[\frac{1}{\beta(1-\beta)}M_C^T + \frac{1}{(1-\beta)}M_B^T\right] \tag{13.26}$$

The values of the ultimate loads of all the specimens in a continuous beam are calculated and listed in Table 13-1, and the ratios between the calculated and measured values are within 0.75–1.05. Considering the complicacy and deviation of the mechanical behavior of a statically indeterminate structure of reinforced concrete at elevated temperatures, it is confirmed that the results calculated by the practical methods of the equivalent section and plastic limit analysis are of acceptable accuracy and tend toward safety. The calculated

TABLE 13-1 **Ultimate Loads of Continuous Beam Specimens**

Number of specimens	Temperature in furnace (°C)	Heating time (min)	Temperature of tensile reinforcement in mid-span (°C)	Ultimate bending moment at mid-span section M_C^T (kN m)	Value of ultimate load (kN)		
					Calculated P_u^T	Measured P_u^S	Ratio P_u^T/P_u^S
TCB1-1	950.0	165	892	0.533	10.31	10.0	1.03
TCB1-2	743.2	108	679	2.151	16.38	20.0	0.82
TCB1-3	697.9	96	628	2.749	18.62	20.0	0.93
TCB2-1	937.0	162	879	0.555	18.70	25.0	0.75
TCB2-2	490.9	54	404	5.413	36.92	35.0	1.05
TCB2-3	823.0	131	766	1.109	20.78	24.7	0.84

The ultimate bending moment at the support section varies slightly with the temperature in the furnace and is taken as a constant: $M_B^T = 6.649$ kN m.

error is caused by various factors: the actual temperature field on the section varies along the length of the specimen and has a low value near the support; the specimen is calculated based on the path of loading at constant temperature but it is actually tested under the path of heating under constant load; errors exist in the values taken for the thermal parameters of concrete and for the temperature of reinforcement; measuring errors also exist in various experimental data, including the size of the section, the position of the reinforcement, the position and value of the load, and the temperature in the furnace.

2. Frame with a single bay and a single story

Four specimens of a single bay and single-story frame are tested under the path of heating under constant load. The factors that influence the investigation are the initial loading level and ratio between the linear stiffness of the beam and the column of the frame (see Table 10-2). The initial data, experimental parameters, and main results for the specimens can be found in Section 10.3 and the elevating temperature–time curve in the furnace is the same as shown in Fig. 13-13(a). The temperature fields on the sections of the beam and column (which are different sizes) of the frame specimens at the maximum experimental temperature are calculated separately using the HTARC program (see Chapter 6). When the calculated two-step strength of concrete at elevated temperatures (Fig. 13-2(b)) is also assumed, the corresponding equivalent sections are obtained, and then the ultimate values of the positive (tension zone exposed to high temperature) and negative (compression zone exposed to high temperature) bending moments of the beam and column can be calculated, respectively (Table 13-2).

The failure mechanism of the frame specimens obtained from the tests (see Fig. 10-17) are used for the calculation, i.e., they are of beam type for specimens TFC-2 and TFC-3 and of combined beam–column type for specimens TFC-4 and TFC-5. When the ultimate load of the frame is calculated according to the principle of virtual work, the two diagrams of virtual displacement shown in Fig. 13-14 are used for both failure mechanisms.

The equation of virtual work is

$$2P_u^T \times \frac{2}{3} = 2M_D^T\frac{2}{l} + M_E^T\frac{4}{l}$$

for the failure mechanism of beam type, and the formula of its ultimate load is obtained:

$$P_u^T = \frac{3}{l}\left(M_D^T + M_E^T\right) \tag{13.27}$$

The equation of virtual work is:

$$2P_u^T \times \frac{2}{3} = 2M_A^T\frac{2}{l} + 2M_B^T\frac{4}{l} + M_E^T\frac{4}{l}$$

TABLE 13-2 Ultimate Loads of Frame Specimens

Number of specimens	Temperature in furnace (°C)	Heating time (min)	Ultimate bending moment of section (kN m)				Value of ultimate load (kN)		
			M_A^T	M_A^T	M_D^T	M_E^T	Calculated P_u^T	Measured P_u^S	Ratio P_u^T/P_u^S
TFC-2	820.0	130	—	—	4.352	0.945	10.59	13.70	0.77
TFC-3	979.4	232	—	—	3.130	0.295	6.85	17.30	0.40
TFC-4	964.1	228	3.808	0.367	—	0.314	8.83	9.80	0.90
TFC-5	952.2	184	3.948	0.459	—	0.392	9.56	9.30	1.03

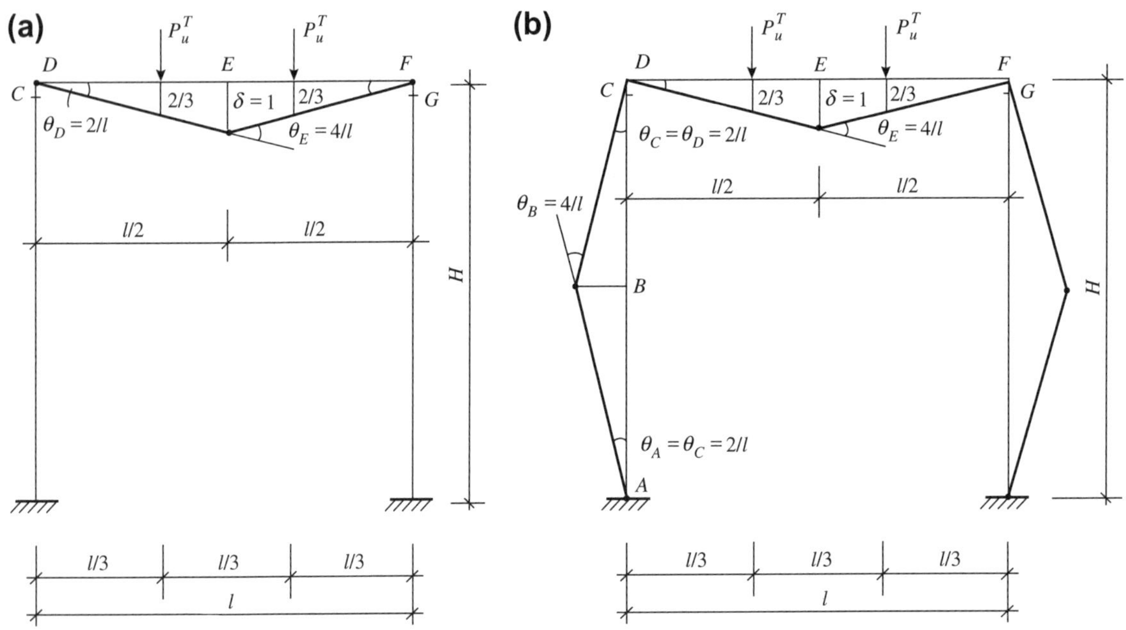

FIGURE 13-14 Failure mechanisms and virtual displacements of frame specimens: (a) TFC-2 and TFC-3; (b) TFC-4 and TFC-5.

for the failure mechanism of combined beam–column type, and the formula of its ultimate load is:

$$P_u^T = \frac{3}{l}\left(M_A^T + 2M_B^T + M_E^T\right) \tag{13.28}$$

The symbols for ultimate bending moments in these formulas correspond to various sections of the frame shown in Fig. 13-14. However, the position and rotating direction of the plastic hinges should be considered. The tensile hinges at the high and the low (or compression zone exposed to high temperature) temperature have to be distinguished and different values of ultimate bending moments should also be given.

The values of the ultimate loads calculated for all the frame specimens are listed in Table 13-2. The values calculated by the practical method described above agree well with the experimental results, except for a few specimens. One of the reasons for calculation error is that axial compression is present on the beam of the frame because the longitudinal thermal deformation is restrained by both columns; the same applies to the calculation for the continuous beam.

CONCLUSIONS

Eccentrically compressive (central and flexural) members of reinforced concrete are the most popular basic structural members used in structural engineering. The history of their mechanical behavior under common actions of elevated temperatures and load (internal force) is complex and the calculations are tedious. A simple and practical approximation method is necessary and possible for calculating the most important index of mechanical behavior, i.e., the ultimate strength at elevated temperatures.

According to the method of equivalent sections suggested in this chapter, the method of calculating the strength of concrete at elevated temperatures (f_C^T/f_C) is simplified, and the cross section of nonuniform strength of concrete, which is caused by the nonuniformly distributed temperature, is then transformed into an equivalent section of homogeneous concrete. The relevant methods and formulas introduced in the existing design code for reinforced concrete structures can then be used to calculate the ultimate strength of a structural member at elevated temperatures.

The calculated strength of concrete at elevated temperatures can be simplified into a trapezoid shape or into two steps (Fig. 13-2(a)–(c)); the results from the calculations based on it show acceptable accuracy. If a structural member is kept at elevated temperatures for a longer time and the area at higher temperature on the section increases, using three steps (Fig. 13-2(d)) to calculate the strength of concrete at elevated temperatures is suggested.[9-12]

The formulas for the ultimate strength of a structural member at elevated temperatures given in this chapter can be used only for a centrally compressive column of rectangular section with the periphery exposed to high temperature, a slab and beam with one or three surfaces exposed to high temperature, and an eccentrically compressive column with three surfaces exposed to high temperature. If a structural member is under different thermal conditions, e.g., a nonrectangular section, a nonstandard temperature–time curve, other conditions of the periphery exposed to high temperature, and different calculated strength of concrete at elevated temperatures, the same principle described above can be used to determine the equivalent section and ultimate strength. When the fire endurance (or the maximum temperature or duration of a fire) and the residual strength after cooling of a structure are considered, the same principle and method introduced above can also be used to determine the equivalent section, and the calculations are then completed.

The value of the ultimate load of a statically indeterminate structure of reinforced concrete can be calculated using the method of upper bound solution of plastic limit analysis. Various types of possible movable mechanisms of one degree of freedom and the positions of the corresponding plastic hinges are selected in advance. The equation is established based on the equilibrium condition at the ultimate state or the principle of virtual work and is then resolved individually for every mechanism. The minimum among all the solutions is the actual ultimate load of the structure. The experiments demonstrate that the practical calculation method introduced can satisfy the requirement of accuracy for engineering applications.

REFERENCES

[13-1] J. Yang, X. Shi, Z. Guo, Simplified calculation of ultimate load bearing capacity of reinforced concrete beams under high temperature, Industrial Construction 32 (3) (2002) 26–28.

[13-2] J. Yang, X. Shi, Z. Guo, Simplified calculation of ultimate load of eccentrically compressed reinforced concrete members under high temperature, Building Structure 32 (8) (2002) 23–25.

[13-3] Y. Long, S. Bao (Eds.), Textbook for Structural Mechanics, Second ed., Higher Education Press, Beijing (2001).

[13-4] P.G. Hodge, Plastic Analysis of Structure, Translated by Jiang Y., Xiong Z., Science Press, Beijing (1965).

[13-5] B. Xu, X. Liu (Eds.), Plastic Limit Analysis of Structure, Construction Industry Press of China, Beijing (1985).

[13-6] B.G. Neal, The Plastic Methods of Structural Analysis, Chapman & Hall, London (1956).

CHAPTER 14

Fire Resistance Analysis and Damage Grade Evaluation of a Structure

14.1 PROBLEMS OF FIRE RESISTANCE ANALYSIS IN STRUCTURAL ENGINEERING

Fire accidents in a building generally cause enormous losses of property and material resources, and even regrettable loss of human life. In order to reduce the occurrence of fire accidents and the associated losses, a fire prevention design should be incorporated in a building following the relevant design code.[0-11,14-1,14-2] In addition, fire resistance analysis should be performed of the structure of a building, if necessary.

Currently, fire resistance analysis is not conducted in general structural engineering during the design process. One main reason for this is that there is no mandatory requirement in the design code. In addition, the relevant technical code, the bases of design and checking calculations, and reliable and practical methods of analysis are not yet available (in China).

There are three types of situations with different characteristics that require fire resistance analysis.

1. Fire resistance design for the proposed building

The fire endurance grade (divided into four grades) of a building is determined to fit the requirement of the code of fire prevention[0-11,14-1] during its design according to the importance of the building, the possibility of fire, the safety of the building during fire, and the difficulty of extinguishing the fire. The fire endurance (*h*) required for every type of structural member in the building can be decided accordingly. Then, a standard fire endurance test (ISO temperature–time curve) of the main structural member is conducted to measure the available fire endurance. If the test does not satisfy the requirement of the code, the design should be modified or the construction improved, until its fire endurance as required by the code is satisfied.

However, the test for fire endurance of a structural member is expensive, and it is not possible to test all the structural members under various conditions. Therefore, it is more economical and reasonable to use an accurate and reliable analysis method to calculate the ultimate strengths at elevated temperatures or the fire endurance of the structural members and the structure, and to check whether the fire prevention code requirements are satisfied.

2. Safety evaluation of an existing building under a fire accident

After the circumstances of the existing building are collected in detail, the fire load can be calculated based on the conditions of the rooms and combustibles within them, and the temperature–time curve of the potential fire can be determined.[0-12–0-14] The fire endurance of the structural members and the structure are obtained individually from the model test or theoretical analysis and are compared with that assigned in the code. The safety regions and damage regions of different levels of fire resistance in the building are analyzed and divided; corresponding measures, such as heat insulation, fire obstruction, or partial strengthening,

Experiment and Calculation of Reinforced Concrete at Elevated Temperatures

are undertaken for the structure and its members where the fire endurance is not satisfactory. The safety and fire resistance of the structure is then ensured.

3. Analysis of the damage level of the structure of a building after a fire accident

Once a fire accident occurs in a building, the range and level of the damage in the structure should be investigated and analyzed to determine the safety of the building and the possibility of using it again, or to be used as the main basis of strengthening measures. Therefore, after a fire, a structure should be observed and surveyed carefully, and some tests should be conducted in situ, if necessary, in order to fully evaluate various important phenomena and data. These include: the reason, time, and position of the fire, how the fire spread and was extinguished, the damaged regions in the building, the type and quantity of materials burned, the possible maximum temperature reached and the duration of the fire, the exterior damage levels of the structure and its members (e.g., cracking, loosening of concrete, separation of the surface layer, damage at the edges and corners of the section, falling of the concrete cover, baring and bending of the reinforcement, defects in the member), and even explosive spalling off and falling of inner concrete, and local holes and collapse. The number and position of the structural members damaged and the numerical indices of the damage level (e.g., width of crack, deflection of the member, spalling thickness of concrete, area of hole) are measured and recorded. The residual strength of the concrete after the fire is estimated by empirical observation and analysis or obtained from a nondestructive test. A cylindrical sample of concrete may be taken by drilling if necessary, and then tested in the laboratory. Even a few structural members can be selected and tested under loading in situ to examine the mechanical behavior after the fire.

All the results surveyed in situ are included in the report of the damage in the structure after the fire. Then, some significantly damaged and typical structural members are selected to conduct the fire resistance analysis, and the residual strengths are calculated to check the safety to decide whether the structural members can still be used, or if they should be used as the basis for preparing strengthening measures.

The behavior of fire resistance (or endurance) or damage after a fire in a structure and its members need to be analyzed for all the situations listed above. The results of the investigations provided in this book can now be used in the fire resistance analysis of the structure. The main contents and general method and procedure of fire resistance analysis are as follows:

1. Determination of the temperature–time (T–t) curve and the duration of the fire

The ISO standard curve (Eqn (5.1)) can be used for a new building during its design, and the temperature–time curve can be calculated from the actual fire load for an existing building when the initial data are available. For the structure after a fire, the possible maximum temperature reached and the duration of the fire are determined based on the data surveyed and analyzed.

2. Analysis of the temperature field of a structural member

Generally, the temperature field of a structural member is assumed to be unvaried along its longitudinal axis, so only a two-dimensional temperature field on a section needs to be analyzed. Both the accurate theoretical solution and the calculation charts or data provided in various references (see Part 2) can be used.

3. Calculation of the ultimate strength at elevated temperatures or fire endurance of a structural member and the structure

The indices of mechanical behavior and constitutive relations of structural materials at elevated temperatures or after a fire (see Part 1) are determined based on their type and properties. Then, the ultimate strength or fire endurance of a single member and the whole structure can be calculated using the accurate theoretical solution (see Chapter 12) or the practical method (see Chapter 13).

4. Evaluation of the fire damage grade of a structure

Comparing the calculated results or the data surveyed in situ, which are introduced above, with the evaluation table of fire damage grades of a structure (see Table 14-1), the damage grades of every structural member and the whole structure are evaluated individually. Then, the building is divided into a certain number of regions according to the fire damage grades, and they are classified into different categories.

5. Conclusion

The safety of a structure is evaluated integrally, and the possibility of using it again or the necessity of strengthening can be decided accordingly. Strengthening measures for the structure can be suggested, if necessary.

14.2 EVALUATION OF THE FIRE DAMAGE GRADE OF A STRUCTURAL MEMBER

14.2.1 Typical Examples of a Building Fire

When a fire accident occurs in a building on the ground, it is easily seen and the alarm can be given early; it is also convenient for the firefighters to evacuate the building, extinguish the fire, and limit damage to the property, as the periphery of the building is opened. According to the statistical data for China,[0-12] 80% of building fires are put out within 1 h and 95% are put out within 2 h. The shorter the duration of the fire, the smaller the economical loss and the structural damage. The main bearing structure of reinforced

TABLE 14-1 Fire Damage Grades and Their Evaluation Criteria in a Concrete Structure and Its Members

	Grade	I	II	III	IV
Damage	**Level**	**Slight**	**Medium**	**Serious**	**Extreme**
Duration of fire (h)		<2	2–4	4–10	>10
Concrete	Color on surface	Blackened by smoke	Slight white	White-gray	Light red
	Crack	Fine, a few	Fine, many	Obvious, more	Wide, more
	Surface layer, edge and corner	Still complete	A few layers lost	Broken, slightly	Broken significantly
	Depth damaged (mm)	<25	25–50	60–120	>150
Reinforcement	Concrete cover	Complete	Loose	Fallen off slightly	Fallen off considerably
	Deformation	Few	Small	Great	Much greater
	T_{max} experienced (°C)	<400	400–800	800–1000	>1000
Structure and its member	Appearance damaged	Not obvious	Obvious	Considerable	Serious
	Ultimate strength	Reduced slightly	Reduced obviously	Lost considerably	A few remain
Repairing and strengthening measures		Treat the surface	Remove loose surface layer, clean up surface of reinforcement, cast concrete for strengthening	Clean up damaged concrete, add reinforcement or enlarge cross section, cast concrete for strengthening	Significant measures or rebuild

concrete, e.g., beam, column, and frame, will not be damaged significantly. Some examples of fires that occurred in high-rise buildings abroad show that even though the duration of a fire is 8–10 h, the main frame structure of reinforced concrete remains undamaged, although the combustibles and fittings in the rooms are completely burnt out.

As far as underground reinforced concrete structural engineering is concerned, fire accidents occur easily and are difficult to put out, because of the special environment. For example, the underground space is possibly moist, and the electrical equipment, components, and circuits easily become damp and the insulation has deteriorated; the loads of artificial lighting and other electrical equipment are considerable and the burning times are longer. The potential danger of a fire accident caused by electrical equipment is enormous when the equipment is aged. Only a few people stay at night in many underground spaces (e.g., market and storehouse), so normally the fire has already developed vigorously before it is discovered. When the internal area (and volume) of the underground space is high but the entrance and exit are limited, these narrow areas are usually filled with spouts of smoke at high temperature and even poisonous gas during a fire, which makes it difficult for the firefighters to enter the site and take direct measures to extinguish the fire. Closing up all the entrances and exits, or isolating the fire region, is the only measure that can be used for most underground fire accidents. The fire will then automatically be put out when the combustibles or oxygen in the interior of the underground space are burnt out. Therefore, the duration of a fire in underground engineering is longer and considerable loss and serious damage of the structure result. These need special consideration and several typical examples are reported below (from examples and investigation reports of fire accidents by civil air defense and other underground engineering (brief report), compiled by the Design Code for Fire Prevention Group, Civil Air Defense Engineering, 1988.3).

An underground trade center was ring shaped, the peripheral length was 435.6 m, and the top was an arched structure. It was an underground building of three floors, and its total area was 9000 m^2 with 12 entrances and exits. The first floor was a market; the width and height were (4–6) m and 3.6 m, respectively, and the combustible goods in the shop were about 60 kg/m^2. The second floor was a passageway 3 m wide with vendors' stands. The third floor was an entertainment area 9 m wide, and it was 15 m underground. It was rebuilt from civil air defense works in 1986, and no fire prevention measures, which are required according to the design code for fire prevention,[14-2] were set up during the design. The fire accident occurred at midnight on 15 September, 1988, started by a spark from an electrical wire in a shop on the first floor. After the fire truck reached the building, the firefighters were obstructed by heavy smoke from the exit and could not enter the site of the fire. They blocked off the fire region and controlled the fire gradually. The fire burnt out eventually after 17 h. The survey after the fire showed that the burnt area of the commercial hall on the first floor was about 270 m^2, the concrete cover fell down, and the reinforcement was exposed on the top part of the arched structure within a length of 180 m, and the maximum depth of concrete spalling off was 70–80 mm.

A cave warehouse built in 1976 had an arch of reinforced concrete on the top. The length and net span were 290 m and 10 m, respectively, and the top of the arch was 6.5 m in height. The main storage goods were products of cotton, fur, leather, and timber, and the combustibles were 400–500 kg/m^2. There were no fire prevention measures included in the design of the building, except a water pond that was built outside the cave. On 23 September, 1984, somebody committed arson, i.e., a burning cigarette end caused a fire in the cave 180 m from the gate. When the smoke from the gate was found about 1.5 h later, the firefighters arrived, but several firefighters died or were injured because of heavy smoke and high temperatures. Then, the gate of the cave was closed until the fire burnt out after 41 days. The survey after the fire reported that about 40 m^2 of the concrete top slab, which was 200 mm thick, had collapsed and large areas of the walls on both sides had cracked significantly.

A fire accident occurred in an underground railway station under construction because a spark from electrical welding caused timber mold and electrical cable to burn. The fire was put out 16 hours later. The examination after the fire showed that the concrete within an area of 484 m^2 spalled off and the maximum thickness of the spalled off concrete reached 170 mm; the reinforcements within an area of 272 m^2 were exposed and deflected downward.

A fire lasted for 10 h in an underground garage between two high-rise apartment buildings. The concrete cover of the top slab (300 mm thick) within a considerable area fell down and the reinforcement was exposed and deflected downward. The maximum depth of concrete burnt reached 120–150 mm, but the beam and column were only damaged slightly.

There are similar examples of fire accidents throughout the world. A large underground shopping center (Akasaray)[14-3] in Istanbul, Turkey, with an area of 72 m × 144 m, had four entrance–exits. The structure was a slab–column cast in situ system. In 1975, a fire accident was caused at midnight by a spark from a short circuit of some electrical equipment. Although the firefighters came to the site in time, they could not reach the fire region, so the fire spread extensively. Then all the entrance–exits were closed after 14 h and water was poured in through holes on the top slab, but this was not effective. The fire gradually reduced after 36 h and burnt out after 63 h. The examination after the fire showed that the top slab was damaged significantly, a considerable area of concrete spalled off and the maximum depth reached was 160 mm, and the reinforcement was exposed and deflected downward, but the column and its capital were only damaged slightly.

The fire resistance behavior of steel and timber structures in buildings is poor. Generally, the fire endurance of a bared steel structure is not longer than 1 h. Even if heat insulation treatment or fire prevention (obstruction) is used in the steel structure, the fire endurance is normally still less than 4 h. The twin towers of the World Trade Center in New York were the tallest buildings in the world. When they were struck separately by two airplanes on 11 September, 2001, only local damage occurred on the upper parts. However, after a large quantity of fuel stored in one of the airplanes caused a fire in one of the buildings, the strength of the steel structure reduced sharply, and both towers collapsed successively within 1 h and were in total ruins.

It is seen from the examples of serious fire accidents introduced above that the fire resistance behavior of reinforced concrete structure is much better than that of steel and timber structures, i.e., longer fire endurance, less loss of ultimate strength, and the buildings can be repaired and strengthened after a fire because of the thermal inertia of concrete, the integrity of the structure, and the larger size of the section of structural member. Even so, reinforced concrete structures can suffer serious damage and local collapse, if the duration of the fire is long.

14.2.2 Evaluation Criteria for Fire Damage Grade

It is practical to investigate and test the damage level of a structure that has experienced a fire accident. Comprehensive analysis and classification of the damage conditions of a structure after a fire is necessary to establish the general regularity of the damage level of the structure, which varies with the duration of the fire.

The experimental data on the fire resistance of various structural members and structures (at elevated temperatures) are used to research the behavior response and damage level for different durations and maximum temperatures of the fire. Many quantitative results, e.g., percentage reduction of the ultimate strength, depth burnt and loss of concrete, and crack and deformation conditions, are then obtained. In addition, the temperature field on the section of a structural member and its variation with the duration of fire, obtained from experimental measurements or theoretical analysis (see Part 2), can be used to determine or estimate the strength loss, cracking, depth of concrete spalled off (fallen down), loss of strength of reinforcement, and reduction in the effective cross section.

Up to now there is no legal code in China that can be used to divide and determine the fire damage grades of various structural members and the structure. Referring to the relevant literature in China and elsewhere and the comprehensive results of surveying, testing, and theoretical research introduced above, the fire damage levels of a concrete structure and its member could be divided into four grades: slight, medium, serious, and extreme damage. The evaluation criteria for each damage grade is given in Table 14-1, based on the duration of the fire, the appearance of the concrete and the reinforcement damaged, and the overall conditions of the structure and its members. Repair or strengthening measures are also suggested in principle in the table for damaged structures of different grades.

Some items in the evaluation criteria of fire damage grades of reinforced concrete structures can be presented quantitatively, e.g., duration of fire, depth of damaged concrete, and maximum temperature the reinforcement experienced. These are the main factors to determine the damage level of the structure. However, other factors are difficult to find quantitatively, so they are qualitative. The damage condition of a reinforced concrete structure after a fire is complicated, e.g., the damage levels of the structural members at various parts of the structure are different, the damage levels of a horizontal beam and slab are different from that of a vertical column and wall, and their influences on the ultimate strength and safety of the whole structure are also different. Therefore the fire resistance behavior of a structure during a fire accident or after the building has cooled has to be analyzed completely if an accurate fire damage level needs to be known.

14.3 PRACTICAL EXAMPLE OF FIRE RESISTANCE ANALYSIS AND DAMAGE GRADE EVALUATION

A city, an important political, economic, and military center in Central Plains, China, is located at the hub of railway communications and the railway station is built in the central part of the city. An underground multi-story building is to be built under the public square by the station to save land. It will be used as a garage and business district in peacetime to satisfy the requirements of communication and shopping in the station area, and it could be used to protect people and disperse the population during wartime. Therefore, the building is of great importance. To ensure the safety of the structure, fire resistance analysis and evaluation of the fire damage grade for the building has been requested, after the preliminary design of the structure, including the civil air defense design, is completed.

14.3.1 The Structural Briefing

The building will be built in stages according to the national plan, and the total land area will be 107 m × 238 m after it is built. The major part of the building is two floors underground, and the remainder is three floors. The first and second floors will be used as a business center and garage, respectively, and the third floor is reserved for an underground railway station. During the architectural design, the fire prevention districts, emergency exits, and dispersing passages are determined and the firefighting equipment, including a sprinkler system, is set up according to the requirements of the fire prevention code.[14-2] Therefore, the functions of fire prevention and firefighting in this building are much improved compared with similar buildings constructed previously.

The main structural system of the building is slab–column (without beams) of reinforced concrete cast in situ, and 18 districts in the plan are divided by construction joints. The configuration of a standard district (Fig. 14-1) is:

- the cross section of a square column is 600 mm × 600 mm
- the top area of the column capital is 2.6 m × 2.6 m
- the distances between every two adjacent columns are 6 m in one direction and 7 m or 6 m in another direction
- the net heights of the first and second floors are 4.2 m and 4.0 m, respectively, and

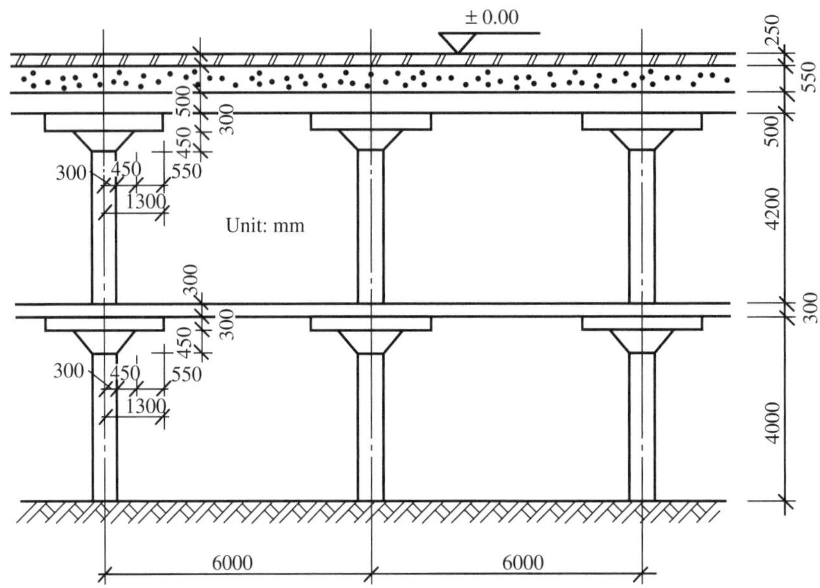

FIGURE 14-1 Size of the slab–column structure in a standard area of an underground building.

- the thicknesses of the top slabs of the first and the second floors are 500 mm and 300 mm, respectively. There are concrete pavements 250 mm thick and a covering of earth 550 mm thick on the top of the building.

The loads acting on the top slabs of the first and second floors in peacetime are 30 kN/m^2 and 12 kN/m^2, respectively. The structural materials used are grade C30 concrete and grade II reinforcement. When the fire resistance analysis was conducted, the design of the main structure had not been completed and accurate data, e.g., the design values of the internal forces, the reinforcement ratios, and constructions of the beam and column, were not finally determined.

Considering the characteristics of the underground building and the lessons from previous fire accidents, once a fire accident occurs, the duration is possibly quite long. When the fire resistance of the structure is analyzed and the possible damage grade is evaluated, the duration of fire is classified into ten ranks (i.e., 0.5, 1, 1.5, 2, 3, 4, 6, 8, 10, and 12 h), and the damage levels of the structure for different durations of fire are determined.

14.3.2 Analysis of the Temperature Field on a Section of the Structural Member

The basic assumptions used for the analysis of temperature fields of the slabs and column are as follows:

- The temperature–time relation of the fire is taken as that of the ISO standard curve (see Eqn (5.1)).
- Only the two-dimensional temperature field on the section of the column is calculated and is unvaried along its longitudinal axis; only the one-dimensional temperature field along the thickness of the slab is calculated and is not related to the position on the slab plan.
- The structural members are considered to be composed of homogeneous concrete. Neither the action of internal reinforcement nor the influences of variations in the internal forces, deformation, and cracking are taken into account.

Based on these assumptions, the basic differential equation of heat conduction is established (see Eqn (5.4) and Table 5-2), and the combined numerical method for resolution (see Chapter 6) is used. The difference method is used to discretize the time (temperature) field, the finite element method is used to discretize the space field, the cross section is divided into rectangular elements connected by the nodes, the temperature values at the nodes are taken as the unknowns, and the calculation and accumulation are conducted for every time (temperature) step successively. Therefore, the HTARC program is compiled and all the calculations are completed by the computer.

The values of the thermal parameters of concrete, especially those at higher temperatures, are needed during the calculations, but accurate values measured from tests are not available in China. Therefore, the values suggested in the European code[2-6] for concrete with siliceous aggregate are referenced and used as follows: Eqn (5.2a) for the coefficient of heat conduction λ_c, Eqn (5.3) for the mass heat capacity C_c, the mass density $\varrho_c = 2400$ kg/m^3, and Eqn (7.1) for the coefficient of heat transfer β_T.

The temperature distribution along the thicknesses of the top slabs of the first and the second floors of the building under different values of fire endurance (h) are shown separately in Fig. 14-2. It is found after comparing both figures that the temperature distributions within a depth of 200 mm from the surface exposed to fire are less different for slabs of different thicknesses (≥300 mm).

The temperature distribution along lines AA (symmetrical axis) and BB on the section of the square column varies as shown in Fig. 14-3. The periphery of the column section is exposed to fire, therefore the temperature gradient is obviously smaller than that of the top slab. The line BB is about the position of the longitudinal reinforcements of the column, and the temperature values of most parts on it are similar, but the temperature near the corner of the section is higher.

According to the practical calculation method for the ultimate strength of the reinforced concrete structural members at elevated temperatures (see Chapter 13), the positions of certain isothermal lines (e.g., temperatures of 300 °C, 500 °C, and 800 °C) on its section need to be determined. Based on the calculated results introduced above (Fig. 14-2), the curves of the depths, at which certain temperature values are reached and which vary with the duration of the fire, can be drawn for the top slab with the bottom surface exposed to fire (Fig. 14-4). These will be used for further calculations.

As for the square column with the periphery exposed to fire, the isothermal lines for certain temperatures (e.g., 300 °C, 800 °C) on the section can be drawn for various durations of fire (Fig. 14-5). Only one-fourth or one-eighth of its section is shown in the figure, as the section is symmetrical bidirectionally. It is clear from the figure that as the duration of fire increases, the temperature on its surface increases continuously; the isothermal line at the same temperature (e.g., 800 °C) moves gradually inside the section, and its enveloped area reduces gradually, so the effective strength of the section decreases continuously.

14.3.3 Analysis of Ultimate Strength of Slabs at Elevated Temperatures

The slab–column system is a kind of two-way structure, so the reinforcements of the slab are set up in two perpendicular directions, but the ultimate strength in both directions can be calculated separately as with the one-way slab. The slab–column system of this underground building is multi-floor, multi-span, and cast in situ; only one surface of the top slabs for the first and the second floor is exposed to fire when a fire accident occurs only in the corresponding floor. Nevertheless, positive and negative bending moments appear together in both directions, and the tension and compression zones exposed to fire (elevated temperatures) appear in both directions and should be calculated.

According to the results of experimental investigations and theoretical analysis, the ultimate strength of the shear resistance of a slab at room temperature is generally higher than that of the bending resistance, because the reinforcement ratio in the slab is low. This is also true for the slab at elevated temperatures, so it is not necessary to check the ultimate shear strength of the slab of this building.

The ultimate bending moment in each direction of the slab (M_u^T) can be calculated

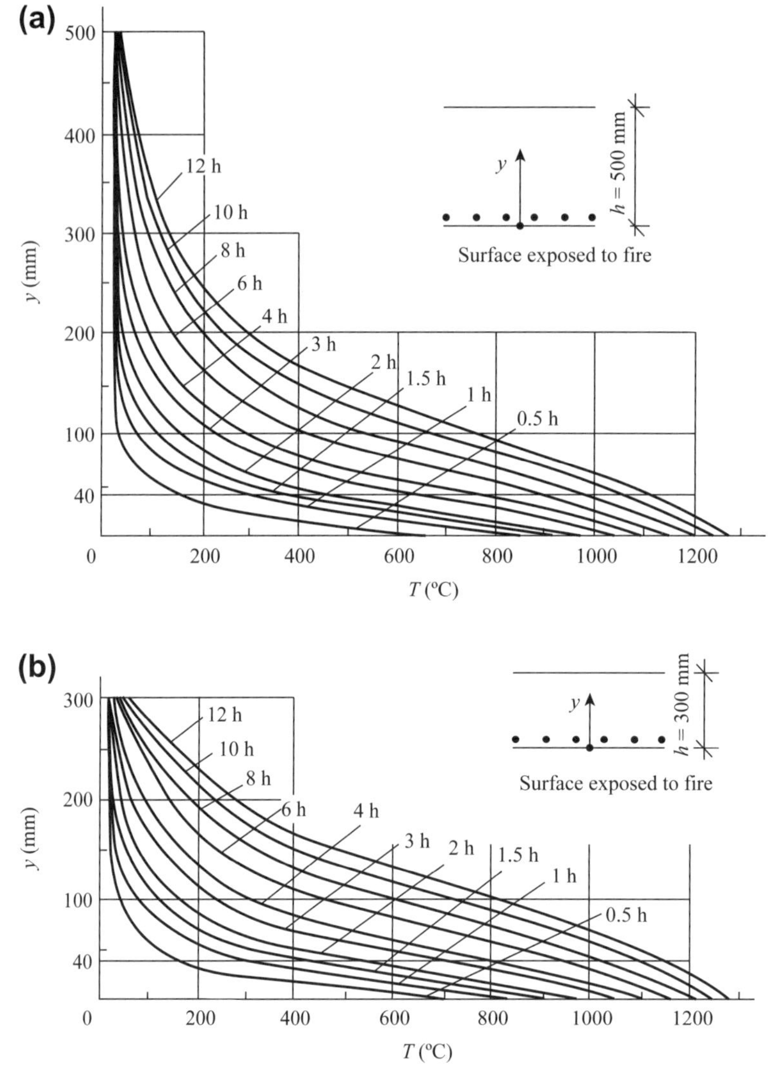

FIGURE 14-2 Temperature distributions on sections of slabs: (a) top slab of the first floor; (b) top slab of the second floor.

approximately using the method of equivalent section (see Chapter 13). The basic assumptions are presented in Section 13.1, the calculated (both tensile and compressive) strength of the reinforcement at elevated temperatures is shown in Fig. 13-1, the calculated compressive strength of concrete at elevated temperatures is simplified into trapezoid distribution (see Fig. 13-2(a)), and the equivalent section obtained is the same as shown in Fig. 13-5(a). The stress diagrams at the ultimate state of the cross sections with tension and compression zones exposed to high temperature are the same as shown in Fig. 13-5(b) and (c), respectively, and the corresponding formulas for calculating the ultimate bending moment are Eqns (13.5)–(13.8).

The data used for the slab of the building are:

- thickness of slab (h): 300 and 500 mm
- position of reinforcement (a): 40 and 50 mm

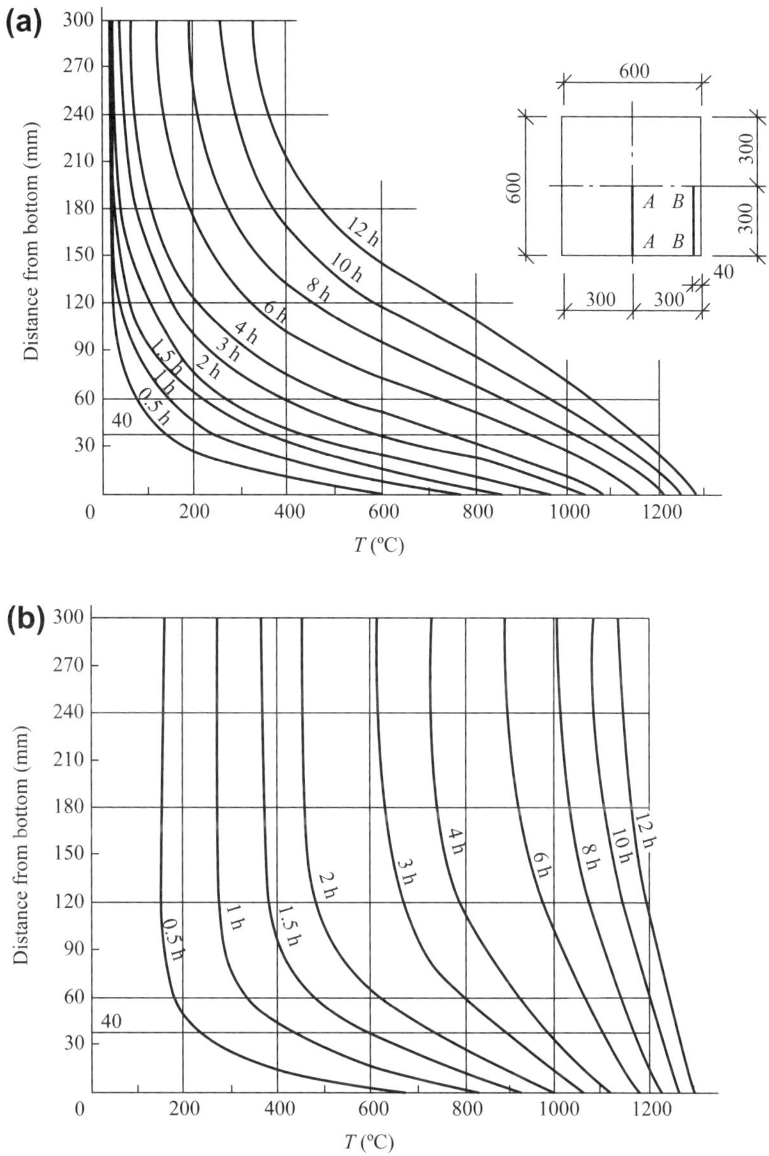

FIGURE 14-3 Temperature distribution on a section of a column: (a) along line *AA*; (b) along line *BB*.

- reinforcement ratio (ϱ): 0.3 and 0.5%
- strengths of materials: reinforcement, $f_y = 310$ N/m^2; concrete, $f_c = 15$ N/m^2

The ultimate values of the positive and negative bending moments of the slab for different durations of fire are calculated using the method of equivalent section, and the ratios between them and the corresponding values of the slab at room temperature (M_u^T / M_u) are shown in Fig. 14-6. Some conclusions can be drawn as follows.

When the slab (h = 300 mm, 500 mm) is acted by a positive bending moment and the thickness of the concrete cover is normally 25 mm (or a = 40 mm), the ultimate strength remains basically unvaried if the duration of fire is less than 1 h.

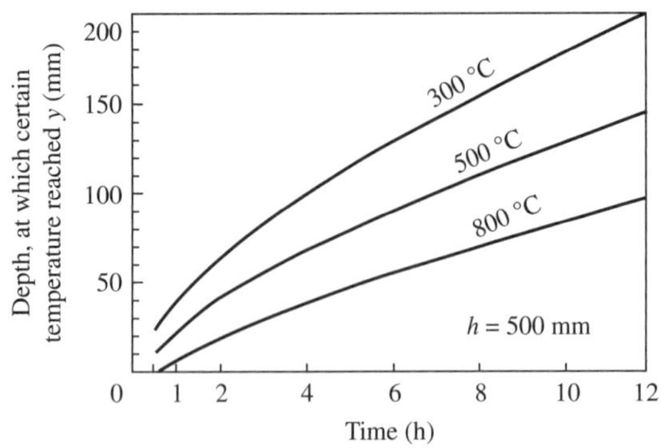

FIGURE 14-4 Relationship between the depth of the slab at which certain temperature values are reached and the duration of the fire.

However, the ultimate strength reduces sharply for a fire of longer duration, and will be only 17% and less than 10%, respectively, of that at room temperature, if the duration of the fire is 4 and 6 h.

If the thickness of the concrete cover of the slab is increased by 10 mm (or $a = 50$ mm), the ultimate strength is increased by 18–20% compared with that of a normal cover, when the duration of the fire is 2–4 h. However, there is no obvious difference between them when the duration of the fire is longer than 6 h.

When the slab is acted by a negative bending moment, the ultimate value reduces gradually as the fire lasts, and still retains 50–72% of that at room temperature, when the duration of the fire reaches 12 h. The variation of the ultimate value of the bending moment depends mainly on the thickness of the slab, and a thicker slab is more advantageous. Comparatively, the reinforcement ratio and the thickness of the concrete cover (or a) have less influence.

14.3.4 Analysis of the Ultimate Strength of a Column at Elevated Temperatures

The structural system of the building is regular and uniform and the vertical load carried is considerable. Therefore, an arbitrary column, the periphery of which is exposed to fire when a fire accident occurs, is considered approximately as a central compressive column, and the relative ultimate strength during the fire (N_u^T/N) can be calculated and checked.

The temperature field on the section of the column for every given duration of fire endurance (t) can be obtained from the available analysis results. Then, the isothermal lines for temperatures of 300 °C and 800 °C on the section (e.g., Fig. 14-5) are found and rounded to square ones, and the ultimate strength of the column at elevated temperatures (N_u^T) can be calculated using Eqn (13.2) or Eqn (13.3).

The column of this building is of square section, and the stress distribution on the section at the ultimate state is the same as shown in Fig. 13-4(b), if the calculated compressive strength of concrete at elevated temperatures is taken approximately as a trapezoid (Fig. 13-2(a)). When the volume of the stress block is approximated to that of a truncated pyramid, the ultimate strength of the column at elevated temperatures can be calculated by the formula:

$$N_u^T = \frac{1}{3} f_c \left(A_3 + A_8 + \sqrt{A_3 A_8} \right) \tag{14.1}$$

The ratio between it and that at room temperature ($N_u = f_c A$) is:

$$\frac{N_u^T}{N_u} = \frac{1}{3} \left(\frac{A_3}{A} + \frac{A_8}{A} + \sqrt{\frac{A_3 A_8}{A^2}} \right) \tag{14.2}$$

where A is the total area of the section of the column, A_3 is the area of the section at which the temperature ≤300 °C, and A_8 is the area of the section at which the temperature ≤800 °C and includes A_3.

When these formulas are used for the calculations, the isothermal lines on the section need not be rounded to square ones.

The main longitudinal reinforcements of the column are located inside its section, but near its periphery, and are exposed to high temperature during a fire accident (Fig. 14-3(b)). Therefore, they can only bear limited axial force and are neglected in these formulas.

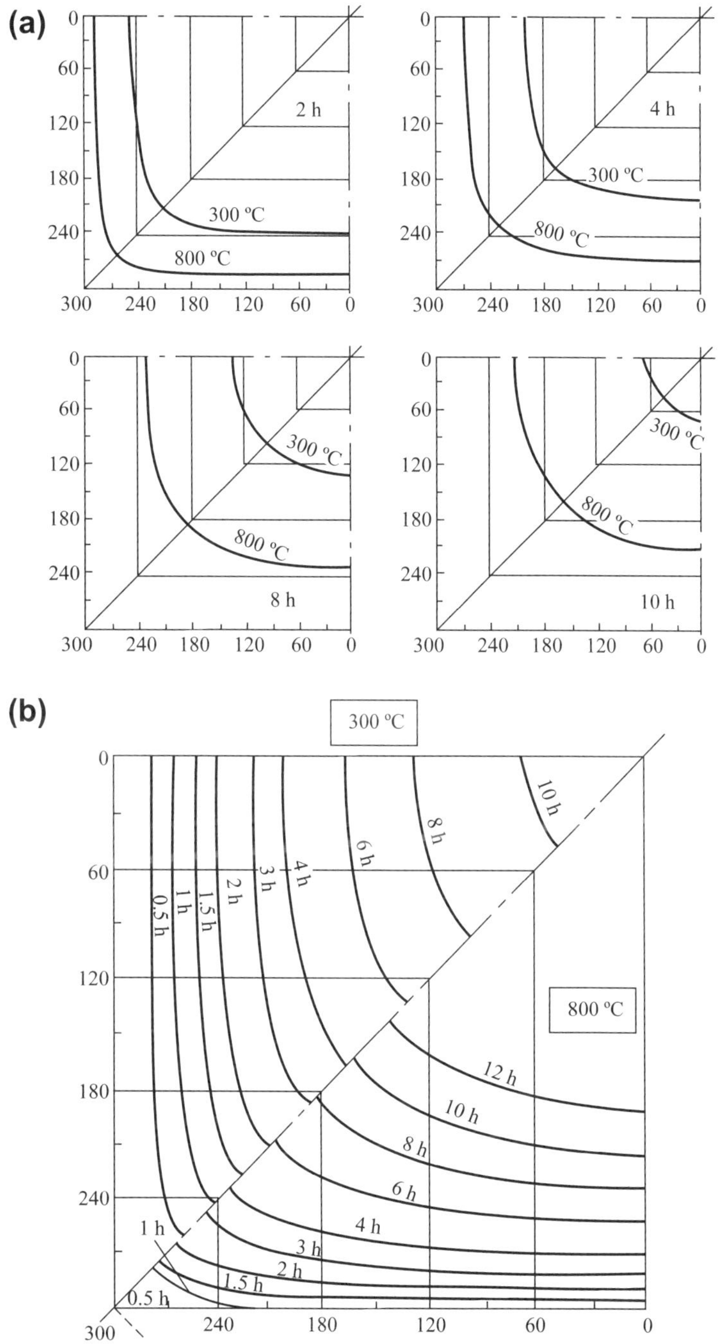

FIGURE 14-5 Isothermal lines on a section of a column with periphery exposed to fire: (a) temperature field on a section at different times during a fire; (b) time needed to reach a certain temperature.

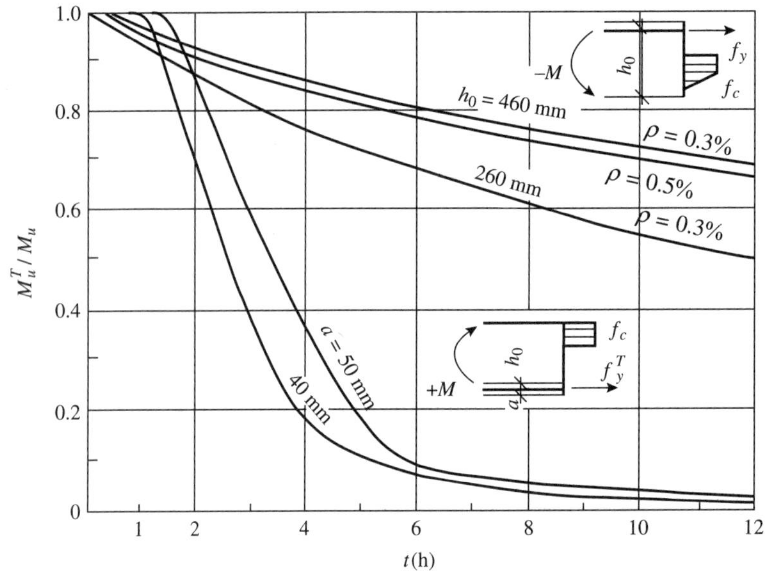

FIGURE 14-6 Relationship between the ultimate bending moment and duration of fire in the slab.

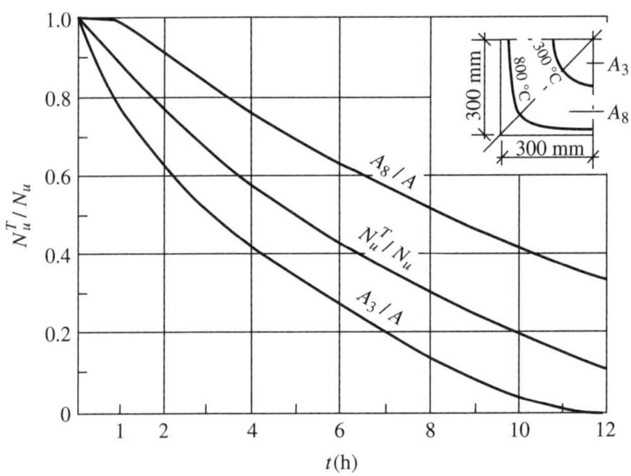

FIGURE 14-7 Relationship between the ultimate axial force of a column and the duration of the fire.

The relative values of the ultimate axial forces (N_u^T / N_u) of the column for different durations of fire calculated from these formulas are plotted in Fig. 14-7. It is seen that the ultimate strength of the column reduces continuously as the fire lasts; about 75%, 45%, and 11% of the ultimate strength of the column at room temperature remains after 2, 6, and 12 h, respectively.

14.3.5 Fire Damage Evaluation of a Structure

The evaluation criteria of fire damage grades (Table 14-2) for a reinforced concrete structure and its members include several important quantitative indices. To complete the fire resistance analysis for this building, the values of the relevant indices, corresponding to different

TABLE 14-2 Analysis of Fire Damage Grade of Slab–Column Structure

Duration of fire (h)		0.5	1.0	1.5	2.0	3.0	4.0	6.0	8.0	10.0	12.0
Top slab of the second floor											
$h = 300$ mm	T_s (°C)	150	290	400	500	640	800	930	1000	1070	1120
$a = 40$ mm	$+M_u^T / M_u$	1.000	1.000	0.849	0.674	0.386	0.172	0.072	0.048	0.031	0.019
$\varrho = 0.3\%$	P_u^T / P_u	0.985	0.965	0.873	0.771	0.602	0.472	0.382	0.336	0.297	0.264
	d (mm)	—	10	15	25	35	45	65	85	100	115
	Grade	I	I	I	II	II	III	III	III	IV	IV
Top slab of the first floor											
$h = 500$ mm	T_s (°C)	150	290	400	500	640	800	930	1000	1070	1120
$a = 40$ mm	$+M_u^T / M_u$	1.000	1.000	0.849	0.674	0.386	0.172	0.072	0.048	0.031	0.019
$\varrho = 0.3\%$	P_u^T / P_u	0.995	0.985	0.897	0.802	0.638	0.517	0.443	0.410	0.383	0.358
	d (mm)	—	10	15	25	35	45	65	85	100	115
	Grade	I	I	I	II	II	III	III	III	IV	IV
$h = 500$ mm	T_s (°C)	110	215	300	400	535	670	840	940	1020	1070
$a = 50$ mm	$+M_u^T / M_u$	1.000	1.000	0.996	0.851	0.584	0.359	0.092	0.064	0.045	0.030
$\varrho = 0.5\%$	P_u^T / P_u	0.984	0.971	0.958	0.876	0.727	0.601	0.442	0.406	0.376	0.350
	d (mm)	—	10	15	25	35	45	65	85	100	115
	Grade	I	I	I	II	II	III	III	III	IV	IV
Column 600 mm × 600 mm	N_u^T / N_u	0.932	0.868	0.816	0.755	0.667	0.580	0.447	0.325	0.206	0.113
	d (mm)	—	5	10	15	30	45	65	85	105	125
	Grade	I	I	I	II	II	III	III	III	IV	IV

durations of fire, can be obtained from the results calculated using the following methods:

- The maximum temperature of the main reinforcement (T_s °C) can be found in the diagrams of the temperature distribution on the sections of the structural members (Figs. 14-2 and 14-3) based on the position (or value of a) of the reinforcement.
- The depth of damaged concrete (d, mm) is taken as the average distance from the surface exposed to fire to the position on the section at which the temperature reaches 700 °C.
- The ratios between the ultimate strength of the structural members at elevated temperatures and at room temperature are taken as M_u^T / M_u for a statically determinate beam and slab and N_u^T / N_u for the column, and the ratio of the ultimate load P_u^T / P_u should be taken for a statically indeterminate structure.

The slab–column system of this building is a statically indeterminate structure. The ultimate state is reached only when the yield lines are formed successively in the span (positive bending moment) and at the support (negative bending moment) regions of the slab and a failure mechanism is then composed. If the reinforcement ratios of the sections in the span and at the supports are equal, the relative value of the ultimate load (P_u^T / P_u) of the slab is taken approximately as the average value of the relative ultimate bending moments M_u^T / M_u in the span and at the support.

The quantitative criteria for the ultimate strength for different damage grades of the structure and its members at elevated temperatures or after a fire are not given in Table 14-1. The damage grades for the various structural members of this building are evaluated, referring to the relative ultimate load as follows:

Damage grade	I	II	III	IV
P_u^T / P_u	>0.8	0.8–0.5	0.5–0.3	<0.3

Based on the method and evaluation criteria presented, the top slabs and columns in the first and the second floors of this building are calculated separately, and the fire damage grades are determined. The main results of the calculations and evaluations are listed in Table 14-2.

It is seen from the data listed in the table that:

- The differences in the thickness (300 mm or 500 mm) and reinforcement ratio (0.3% or 0.5%) of the slab have less influence on the fire damage level.
- When the thickness of the concrete cover of the reinforcement in the slab is increased (i.e., from $a = 40$ mm to $a = 50$ mm), the damage level can be reduced if the duration of the fire is less than 6 h, but is not improved obviously when the duration of the fire is longer than 6 h.
- The fire damage level of the column is about the same as that of the slab.

According to the results of the fire resistance analysis and damage evaluation of the structural members introduced above, the fire damage grade of the reinforced concrete slab–column cast in situ structure of this building can be evaluated integrally as follows:

Duration of fire (h)	<2	2–4	4–10	>10–12
Damage grade	I	II	III	IV

Several additional explanations are presented below for the fire resistance analysis and damage evaluation of this building.

- One of the main bases for the analysis and evaluation above is the ISO standard temperature–time curve of fire (see Eqn (5.1)); correspondingly the temperature elevates monotonically with the duration of fire and reaches about 1300 °C at $t = 12$ h. However, when a fire accident occurs in a building, generally the temperature value is not increased monotonically within a long duration of fire, but will be a process of increasing–steady–decreasing, and, eventually, going out (see Fig. 5-2), because the supply of combustibles and oxygen in the room are limited.
- Some favorable factors, such as fire prevention measures, heat insulation, and fire control (sprinkler), may be used in the architectural design of the building, but the effects are not

considered in the fire resistance analysis of the structure.

- Due to the lack of materials and detailed data in the structural analysis, the fire resistance analysis of the building depends on the ultimate strength of the structural members at elevated temperatures.

CONCLUSIONS

Fire resistance analyses of a structure and its members need to be conducted for various situations in structural engineering, such as fire resistance design of a proposed building, safety evaluation of an existing building under fire, and evaluation of the damage level in a structure after a fire. The fire (high temperature) resistance behavior or fire endurance can then be obtained.

Fire (high temperature) resistance analyses of a structure and its members include the determination of temperature–time curve and duration of fire, analysis of the temperature field, calculation of the mechanical behavior (mainly the ultimate strength or fire endurance) at elevated temperatures, and evaluation of the fire damage grades. The necessary methods and data for analyzing the fire resistance of a reinforced concrete structure and its members are presented in this book.

When a reinforced concrete structure is subjected to a fire accident, various damage phenomena appear successively and its damage level depends mainly on the duration of the fire and the maximum temperature experienced. It is suggested that the levels of fire damage in a reinforced concrete structure and its members should be divided into four grades, based on the survey of the damage condition of the building after the fire and the results of relevant experimental and theoretical research. The evaluation criteria for grading fire damage are suggested and explained by an example of engineering practice.

REFERENCES

[14-1] National Standard of People's Republic of China Design Code for Fire Prevention of Civil High-Rise Building GB 50045–95, Plan Press of China, Beijing (1995).

[14-2] National Standard of People's Republic of China Design Code for Fire Prevention of Civil Air Defense Engineering GB 50098–98, Plan Press of China, Beijing (1998).

[14-3] E. Arioglu, K. Anadol, A. Candogen, An Underground Shopping Centre Fire and After Fire Repair Project, American Concrete Institute, Chicago (1983). Special Publication SP 80-11.

INDEX

Page references followed by “f” indicate figure, by “b” indicate box, and by “t” indicate table.